BIOFUELS - STATUS AND PERSPECTIVE

Edited by **Krzysztof Biernat**

Biofuels - Status and Perspective
http://dx.doi.org/10.5772/58662
Edited by Krzysztof Biernat

Contributors

Emad Shalaby, Koji Yamane, Kiyoshi Kawasaki, Eda Ondul, Nadir Dizge, Bulent Keskinler, Nedim Albayrak, Omotola Oluwafunmilayo Babajide, Leslie Petrik, Farouk Ameer, Dr. Mushtaq Ahmad, Hammad Ahmad Jan, Muhammad Zafar, Shazia Sultana, Kifayat Ullah, Muhammad Aqeel Ashraf, Nitin Raut, Talal Al Balushi, Reena Raut, Suren Panwar, Shuguang Deng, Kurt A. Rosentrater, Lílian Lefol Nani Guarieiro, Bo Hu, Lincoln Oliveira, Jose Castanheiro, Kingsley Iroba, James A. Dyer, Ronald Hatfield, Jorge Sepúlveda, Armando Tibigin Quitain, Mohd Azlan Mohd Ishak, Siti Nur Ain Mohd Hassan, Siti Nurlia Ali, Mohd Fauzi Abdullah, Ali H. Jawad, Wan Izhan Nawawi Wan Ismail, Khudzir Ismail, Asnida Yanti Ani, Nur Nasulhah Kasim, Anna I. Maiorova, Aleksandr Vasil'Ev, Oganes Chelebyan, Didem Özçimen, Thiago Nogueira, Denise Cordeiro, Rodrigo Munoz, Adalgiza Fornaro, Antonio Miguel, Maria Andrade, José Eduardo De Oliveira, Gisela Montero, Margarita Stoytcheva, Marcos Alberto Coronado Ortega, Conrado Garcia, Jesús Cerezo, Lydia Toscano, Ana MarÃa Vázquez, José A. León, Araceli Amaya- Chávez, Krzysztof Biernat

Published by InTech

Janeza Trdine 9, 51000 Rijeka, Croatia

Publishing Process Manager Iva Simcic

Technical Editor InTech DTP team

Cover InTech Design team

First published October, 2015

Printed in Croatia

Additional hard copies can be obtained from orders@intechopen.com

Biofuels - Status and Perspective, Edited by Krzysztof Biernat

p. cm.

ISBN 978-953-51-2177-0

For EU product safety concerns:
IN TECH d.o.o., Prolaz Marije Krucifikse Kozulić 3, 51000 Rijeka, Croatia,
info@intechopen.com or visit our website at intechopen.com.

Meet the editor

Krzysztof Biernat Ph.D. (Mech.Eng.), is a professor of the Automotive Industry Institute (PIMOT), acting as Coordinator of Polish Technology Platform for Biofuels. He is member of Coordinating Committee of Polish Technology Platform for Bioeconomy and a member of Scientific Board. Prof. Biernat is also a Deputy Director of Institute of Ecology and Bioethics at state University Cardinal Stefan Wyszynski in Warsaw. He represented Poland in European Technology Platform and in Renewable Heating and Cooling European Technology Platform. He is also a member in American Council on Renewable Energy. Prof. Biernat is a an expert in Organization States of America and Operational Programmes: Infrastructure and Environment, Innovation Economy and Human Capital. He specializes in chemical thermodynamics of environmental processes as well as obtaining technologies, quality evaluation and use of exploitative liquids, including biofuels, and biorafinery systems. He received many national and foreign distinctions, decoration and orders for scientific and pro-innovative activities. He is an author of above 200 publications in area of properties and exploitative conditionings of fuels, biofuels and other liquids as well as environmental protection. He has supervised a few dozen of master's and engineering thesis's in this area. K. Biernat has expert licence in the field of fuel and energy economy. Prof. Biernat is a member of many national and foreign scientific societies including American Chemical Society and American Association for the Advancement of Science.

Contents

Preface

Fuel for internal combustion engines and stationary equipment, including biofuels as alternative fuels, should meet several requirements, directly or indirectly related to the provision of adequate performance, fuel economy and environmental protection. Due to the tightening requirements for the greenhouse gas emissions reduction in the world, there are conducted scientific research leading to effective production of biofuels as substitutes for fossil fuels. Initially it was assumed that the basic raw material for biofuels production will be wide range of biomass. According to the European definition, biomass means the biodegradable fraction of products, waste and residues of biological origin from agriculture (including vegetal and animal substances), forestry and related industries including fisheries and aquaculture, as well as the biodegradable fraction of industrial and municipal waste. Thus, assuming biomass as a primary source of raw materials for the biofuels production, in this definition there there are included two basic paths of raw materials and the corresponding conversion technologies, namely the processes of BtL (*biomass to liquid*), alternatively BtG (biomass to gas) and WtL (waste to liquid), alternatively WtG (waste to gas). There are also introduced *bioliquids* as liquid biofuels used for energy purposes other than for transport, including the production of electricity, heat and cold. Therefore, the processes leading to the bioliquids production, from so defined biomass as a raw material are located in the BtE (*biomass to energy*) and WtE (*waste to energy*).

Taking into account past experience and the development of biofuel production technology, specifically alternative fuels derived from biological materials and civilization waste (xtL processes) seeks primarily to reduce CO_2 emissions, or its optimal balance in the production processes and fuel combustion. The introduction of the "xtL processes" was dictated by confirmed of biomass as a raw material deficiency, which could prevent the obtaining from biofuels in the quantities for meeting the indicator goals, both in terms of percentile share in the total volume of transport fuels, as well as in significant reduction in carbon dioxide emissions. The problem of biomass shortage as a raw material is very crucial especially in Europe, where the fulfillment of the goals requires the import of biomass and certain biofuels (ethanol, biodiesel and vegetable oils). Therefore, it is anticipated that raw materials perspective for the production of biofuels will include: all waste substances (biodegradable in the beginning), including waste biomass (lignocellulosic feedstock), and - in the long term - waste carbon dioxide, and even steam. Due to the environmental conditions the energy crops as raw material should be considered carefully. The modern state of the art, describes as safe crops (cultures): Jatropha, Camelina, halophytes and algae, which may be a raw material for biofuels. The need to search for new raw materials required for the production of biofuels (and bioliquids), is particularly important in industrialized countries. In these countries, the need for the introduction of biofuels has been legally recognized and simultaneously there has been the development of waste biomass processing technologies.

Considerations for raw materials, its properties, environmental impact and development prospects in the production and use of biofuels, have been included in this monograph, consisting of 25 chapters. The thematic scope is very comprehensive and mutually engaging, so it was difficult to clear dividing the content into monothematic sections, what hence may occur slight repetition of some of the problems in separate sections.

The section on Biofuels First Generation - Coventional, included chapters 1 to 5. This section starts panoramic chapter, showing the capacity to produce biodiesel from non-food sources of biomass. The second chapter presents the possibilities of biocatalytic technology production of biodiesel. Two more chapters include a discussion of biodiesel production technology in Sub-Saharan African Countries and Pakistan. Section ends with a chapter 5, showing the problems of the ethanol production from corn in the United States with an evaluation of the using by-products from this process possibility.

The second section comprises a Second Generation Biofuels - Advancement and contains three chapters (6 to 8). Chapters 6 and 8 show the complementary points of overwiew in the field of processes for the production of second generation biofuels from algae (microalgae) as raw material. This section ends with chapter 8, showing the sub and supercritical fluid technologies for the production of transportation biofuels.

Chapters 9 and 10 are the Alternative Fuels section. As it was already mentioned, these fuels because of increasing amounts of industrial wastes, mainly plastics, textiles, etc. can be an important source of energy. This section contains discussion of the scope of the study of polystyrene in biodiesel solubility, and presents the overview of obtaining alternative fuels in coliquefaction processes with the biomass and coal, which was implemented in Malaysia.

The next section provides an overview of the problems in the field of biofuels Quality and Exploitation and consists of chapters 11 to 16. This section represents chapters that discuss: the impact of biofuels on burning of the particle emissions, thermoanalytical methods for assessing the quality of biodiesel, the quality of biodiesel produced from sunflower oil, the kinetics of the transesterification process used cooking oils using different alcohols and the use of photoacoustic spectroscopy for the quantitative assessment of the type of biomass. Section is closed by the chapter 16, showing the range of possibilities for the study of the use of certain types of biofuels in aviation.

The penultimate section of monograph is devoted to the problems in the field of assesment on the Environmental Impact and Policy and contains three chapters (17, 18 and 19). This section describes considerations in the field of environmental impact assesment of feedstock for biofuel production, especially rape cultivation and animal husbandry, in the Prairie provinces in Canada, as well as evaluating the long-term of biofuels impact on the state of the atmosphere in Brazil. This section is closed by chapter 9, which is the review of conditions in the field of biodiesel production in Mexico.

The last section of the monograph dedicated to the Biorefinery Processes, contains chapters from 20 to 25. Introductory chapter to this section is chapter 20, devoted to reviewed biorefinery technology, also in terms of their impact on sustainable development. The other chapters discuss sub-processes, ranging in biorefinery processes, including processes for initial use and preparation of raw materials, i.e. processes HTL - hydrothermal upgrading or liquefaction of biomass with the use of water under sub and supercritical conditions, processes of use waste glycerol to produce hydrogen (Chapter 22) and oxygen fuel additives (Chapter 23), processes for removal and use the phosphorus from animal manure (raw materials for the preparation of biogas and derivatives). This section and the entire monograph closes chapter 25, which discusses sustainable methods of obtaining biomass for energy purposes.

Presented monograph is the result of research and analytical work of many authors. The creation of this monograph require a lot of work, both on organization and coordination stage. Therefore, on behalf of myself and all the other authors, I express many thanks to Ms. Iva Simcic from InTech Publishing. Her engagement led to the involvement of this, I hope, interesting and needed monograph.

In closing, I would like to thank my wife, Danuta, for understanding, patience and supportive words of encouragement that helped me to carry out the all work related with the edition of this monograph.

Prof, Krzysztof Biernat
Institute of Ecology and Bioethics of CSWU
Automotive Industry Institute
Warsaw, Poland

Section 1

Biofuels First Generation-Conventional

Chapter 1

A Review of Selected Non-Edible Biomass Sources as Feedstock for Biodiesel Production

Emad A. Shalaby

Additional information is available at the end of the chapter

http://dx.doi.org/10.5772/59178

1. Introduction

Biofuels are being given serious consideration as potential sources of energy in the future. Due to recent petroleum crisis and unavailability of petroleum diesel, the need for petroleum diesel is increasing day by day hence there is a need to find out an appropriate solution. The use of renewable resources for energy production is a strategic focus of governmental institutions to restructure the national energy economies, and many efforts are undertaken to increase the share of renewable energies within the nation; biodiesel is presently the most widely accepted alternative fuel for diesel engines due to its technical, environmental and strategic advantages. Biodiesel is known as a carbon neutral fuel because the carbon present in the exhaust is originally fixed in the atmosphere. Moreover, different vegetable oils can be used for biodiesel production like soybean oil, jatropha oil, rapeseed oil, palm seed oil, sunflower oil, corn oil, peanut oil and cottonseed oil in addition to the possibility for use of waste cooking oil, and algal species (micro and macroalgae). Compared to fossil-based diesel fuels, biodiesel possesses many advantages such as cleaner engine emissions, biodegradable, renewable and superior lubricating property. However, until now biodiesel is not yet commercialized all over the world. And this may be due to the high cost of the raw material. Biodiesel Free on Board (FOB) costs between 0.65 and 1 U.S. dollars (USD/L). The biodiesel production cost is nearly 1.5–2 times more than petroleum-based diesel depending upon feedstock oils. It is reported that nearly 70%–95% of the total cost of biodiesel production arises from the cost of raw material; i.e., vegetable oil or animal fats. The cost of biodiesel can be reduced if we use available, non-coasted, unused biomass such as non-edible oils as frying oils, acid oils and agriculture waste oils instead of edible oils.

2. Non-edible biomass sources

2.1. Date palm seeds

The date (*Phoenix dactylifera*) has always played an important role in the economy and social life of the people of arid and semiarid regions of the world. *Phoenix dactylifera* (date or date palm), a flowering plant species in the palm family Arecaceae, is one of the member of the genus *Phoenix*, widely cultivated for its edible fruit. Dates have been a staple diet in the Middle East for thousands of years. Date seeds, considered as waste product, are either discarded or used as fodder for domestic farm animals. Egypt is considered the first rank in the production of date as shown in Table 1. The total world production of date fruits was about 6.64 million tons in 2007 [1], with Egypt, Iran, Kingdom of Saudi Arabia, UAE and Pakistan being the main producing countries. Habib and Ibrahim [2] found that, the relative percentage of date seeds to the date fruits ranges 6.10%–11.47%. The same author also estimated the annual world production of date seeds at about 697,000 tons in 2007. Date seeds are commonly considered as a waste product (Figure 1) that is either discarded or used in animal feed. However, date seeds have been shown to possess extractable high value added components [3]. Date seeds, which are low cost agriculture by-products, can be used for production of activated carbon.

Figure 1. Date palm seeds pictures.

In an attempt to make use of this costless date seeds for biodiesel production, Shanab et al [4] found that the oil content (extracted with petroleum ether) in the four date seed cultivars ranged between 3.31% and 4.48% as recorded in Table 2.

Ramly showed the highest oil content (4.48%) followed in descending order by Amhat (3.88%) then Sewy (3.34%) and Haiany (3.31%).

It is evident from these results and previous investigations that, date seeds do not provide high yield of oil [2], but the oil may serve as a potential source of some important phytochemicals [3].

Transesterification of the extracted date seed oils by Methanol/KOH mixture led to the production of fatty acid methyl esters (biodiesel).

2.1.1. Physico-chemical properties of produced biodiesel from date palm seeds

Iodine value (IV) is a measure of unsaturation degree. The obtained results revealed that date seed cultivars showed different iodine values (mg I_2/g), which ranged between 90.4 mg/g in Sewy, 88.5 in Amhat, 86.3 in Haiany and 75.3 mg/g in Ramly (Table 3).

These variable contents may be due to high content of saturated fatty acids in their oils, which was confirmed by the GC analysis of the four date seed cultivars.

Shalaby and El-Gendy [5] reported that waste cooking oil and its methyl esters have low iodine values (~ 62, 60 mg I_2/100g oil). Also, JUS [6] illustrated that, methyl esters used as biodiesel fuel must have an iodine value less than 120 mg I_2/100g of the oil sample (JUS EN 14214). This means that the iodine values of the produced biodiesel from date seed oils in this study (75.3–90.4 mg I_2/g) as well as from *Phoenix canariensis* seed oil (76.66 mg I_2/g) were due to the high contents of saturated fatty acids leading to low or even no oxidation tendency in these oils [7].

The low Acid values (AV) recorded in the four cultivars of date seed oils ranged 0.5–0.92 mgKOH/g, and these results indicated that these oils can be stored for longer periods without deterioration, which seems in conformity with the results reported by Ojeh [8].

Acid value of biodiesel was shown to be higher than that of standard petrodiesel, but it meets the standard limits of EN 14215 and D-6751, indicating that the free fatty acids content will not cause operational problems such as corrosion and pump plugging caused by corrosion and deposit formation.

Variable saponification values (SV) in the studied four date seed cultivars (185.4–202.3 mgKOH/g) of the *P. canariensis* seed oil (191.28) indicated the presence of low molecular weight triacylglycerols, which are in conformity with those reported by Eskin *et al.*, [9] and Oomah *et al.*, [10] who declared that their saponification values are similar to values of canola oil and raspberry seed oil.

More or less similar higher heating values (HHV, Table 4) of the produced fatty acid methyl esters (39.84 kg/g–40.50 kg/g) indicated that they have nearly similar stability values due to their low polyunsaturated fatty acids contents in their oils, which were clearly observed in the GC analysis of fatty acid methyl esters (Table 5).

Infrared spectrum (IR) of the produced biodiesel showed the presence of ester group (-C-O-) at bands 1027, 1165 and 1745, as well as the absence of the hydroxyl peak, which can be correlated to the transesterification process of date seed oil (Figure 2).

These results were concomitant with those reported by Shalaby and El-Gendy (2012) on waste cooking oil methyl ester.

Fatty acid composition of the esterified products using gas chromatography (GC) revealed the presence of eight saturated fatty acids ($C_{8:0}$–$C_{18:0}$) in the three date seed cultivars: Amhat, Ramly and Haiany, while cultivar Sewy has only seven saturated fatty acids as recorded in Table 5.

The highest percentage of saturated fatty acids was shown in Amhat cultivar (79.517%) followed in descending order by those of Haiany (68.135%), Sewy (45.764%) and Ramly (20.986%).

The fatty acids caprylic ($C_{8:0}$) and capric ($C_{10:0}$) constituted the largest percentage of saturated fatty acids in the studied date seeds. Ramly cultivar was the only date seed containing Arachidic fatty acid ($C_{20:0}$) of 0.441%.

The recorded unsaturated fatty acids were oleic ($C_{18:1}$) and Linoleic ($C_{18:2}$), which were present in variable relative percentages in only three of the four date seed cultivars (24.64% in Ramly to 0.573% in Sewy), while linoleic acid was absent in Amhat cultivar (Oleic represented by 1.88%).

Ramly recorded the most pronounced content of unsaturated fatty acids (20.99% Oleic + 3.65% Linoleic), and the least percentage of saturated fatty acids (20.986%).

Our results go parallel with those published by Nehdi *et al.* [11] on date palm seeds and seed oils.

Yong and Salimon [12] reported that liquid oils with high oleic fatty acid content (as Ramly oil) normally have good flavor and frying stability.

Oleic acid is important in nervous cell construction and can be changed in the organism into a set of compounds close to prostaglandin, which have an important role at the vessel level and for blood coagulation [13]. So, Ramly can be of importance in this concern.

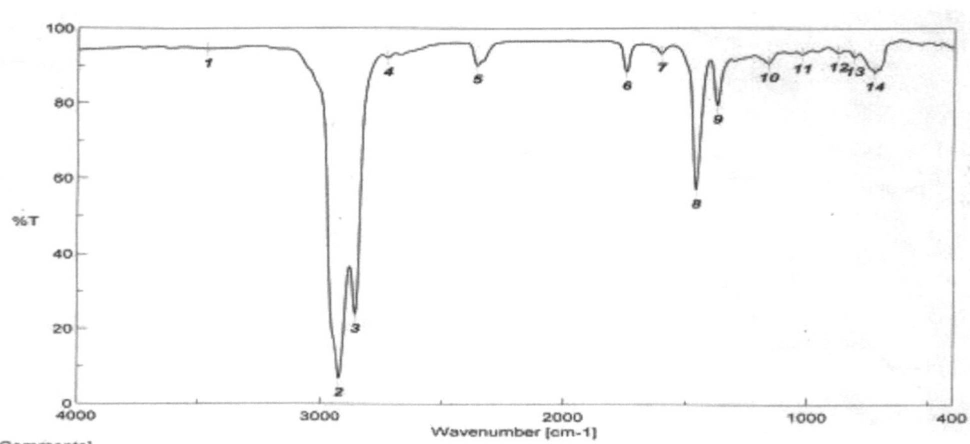

Figure 2. Infrared spectrum of Fatty acid methyl ester (FAME) of date palm seed.

Country	Production (tons)	% World
Egypt	1,352,950	17.2
Saudi Arabia	1,078,300	13.7
Iran	1,023,130	13.0
United Arab of Emirrate	775,000	9.8
Pakistan	759,200	9.6
Algeria	710,000	9.0
Iraq	566,829	7.2
Sudan	431,000	5.4
Oman	276,400	3.5
Libya	161,000	2.0
Arab countries	5,351,479	68.11
Asia	4,804,126	61.1
Africa	3,011,205	38.3
America	26,003	0.3
Europe	16,121	0.2
World	7,857,455	100

Table 1. Total date production by different countries and continents [1].

Species	Oil (%)	Biodiesel %	Glycerol %
Amhat	3.88	2.05	1.26
Ramly	4.48	2.43	1.43
Siwy	3.34	1.90	1.33
Haiany	3.31	1.84	1.34

Table 2. Oil contents, Biodiesel and glycerol (%) in different date seed cultivars [4].

Species	Acid value (mg KOH/g)	Iodine value (mg I_2/g)	Saponification value (mg KOH/g)
Amhat	0.92	88.5	185.4
Ramly	0.50	75.3	199.0
Siwy	0.67	90.4	192.5
Haiany	0.72	86.3	202.3

Table 3. Some chemical properties of methyl esters of date seed [4].

Species	HHV (Kj/g)	CN
Amhat	40.50	60.52
Ramly	40.14	55.42
Siwy	40.18	57.03
Haiany	39.84	54.61

Table 4. Higher heating value (HHV) and cetane number (CN) of fatty acid methyl esters of date seed cultivars [4].

Fatty acids	RT	Relative % of Fatty acids in different Date seeds			
		Amhat	Ramly	Sewy	Haiany
C8:0	7.3	50.77	-	-	47.35
C10: 0	11.16	28.37	-	57.06	26.88
C11:0	13.36	5.44	1.07	-	8.9
C12:0	15.51	2.77	1.55	9.69	4.82
C13:0	17.51	5.93	22.9	14.9	5.9
C14:0	19.61	2.62	4.52	10.38	4.72
C15:0	21.45	0.97	4.61	3.08	2.7
C16:0	23.27	0.79	7.63	2.48	0.39
C17:0	24.9	-	1.50	1.10	-
C18:0	26.63	-	0.85	-	0.37
C18:1	27.16	2.31	45.9	1.04	1.22
C18:2	28.06	-	7.98	0.19	0.48
C18:3	29.28	-	-	-	-
C20:0	29.76	-	0.96	-	-
SFA		97.69	46.12	98.77	98.3
USFA		2.31	53.88	1.23	1.7
SFA/USFA ratio		42.3/1	0.85/1	80.3/1	57.8/1

*R_t: Retention time; SFA: Saturated fatty acids; USFA: Unsaturated fatty acids

Table 5. Fatty acids content in different commercial date seeds cultivars [4].

2.2. Non-edible plants (e.g., Jatropha)

Jatropha is a genus of flowering plants in the spurge family, Euphorbiaceae. Jatropha is a drought-resistant perennial, growing well in marginal/poor soil and considered as one of the best candidates for future biodiesel production. Jatropha the wonder plant produces seeds (Figure 3) with an oil content of around 27%–40%. The oil can be combusted as fuel without being refined. It burns with clear smoke-free flame, tested successfully as fuel for simple diesel engine. Oil contains also insecticide. It can also be used as a bio-pesticide and for medicinal purposes. It is found to be growing in many parts of the country, is rugged in nature, can survive with minimum inputs, and is easy to propagate, and these growth conditions give this plant the big chance for selection as biofuel sources.

Figure 3. Pictures of Jatropha tree and seeds.

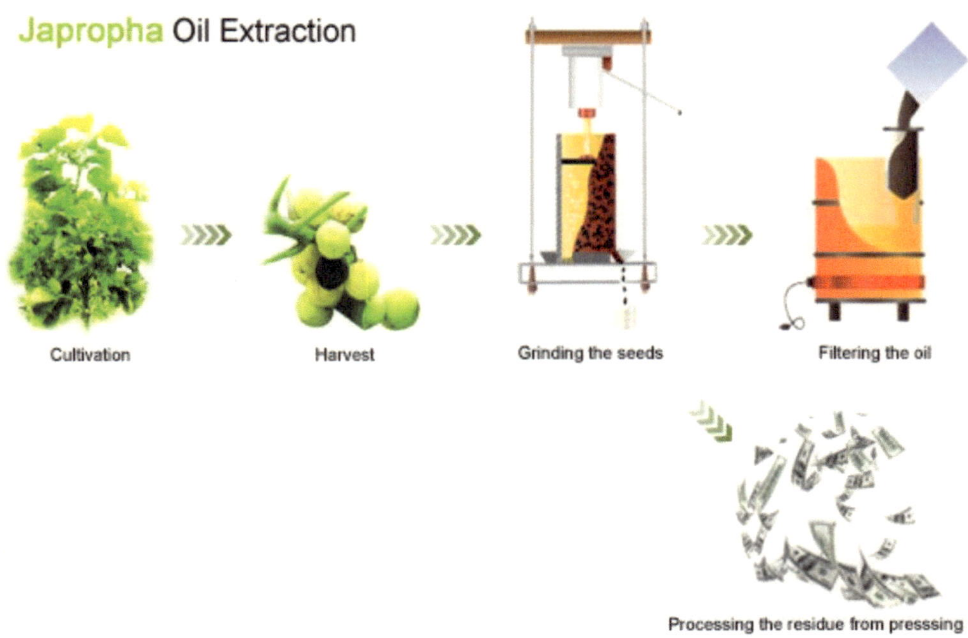

Figure 4. Jatropha oil extraction method. (http://www.aumkiipure.com/oil-extraction.html)

The suitability of a particular oil type from Jatropha species depends on its chemical and physical properties and the plant origin. The fatty acids composition is very important for

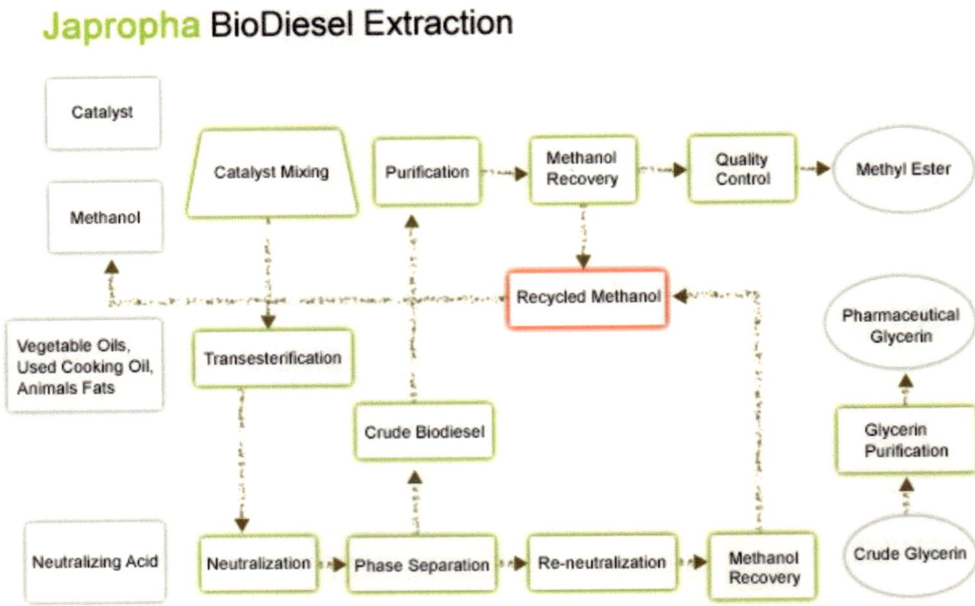

Figure 5. Jatropha biodiesel extraction methods (http://www.aumkiipure.com/oil-extraction.html).

physical and chemical properties of produced biodiesel. The fatty acid composition of jatropha oil has been reported in Table 6. The Gas Chromatographic analysis of the fatty acid compositions of the oils showed oleic acid to be significantly higher ($p < 0.05$) than linoleic, palmitic, stearic and myristic acids.

Also, medically, it is used for different diseases due to its antioxidant, anticancer and antibacterial activities. These activities may be due to the presence of various active ingredients including phenolic compounds and flavonoid groups.

Fatty acids	Formula	Structure	Wt%
Myristic	$C_{12}H_{28}O_2$	14:0	0.5-1.4
Palmitic	$C_{16}H_{32}O_2$	16:0	12-7.0
Stearic	$C_{18}H_{36}O_2$	18:0	5.0-9.7
Oleic	$C_{18}H_{34}O_2$	18:1	37-63
linoleic	$C_{18}H_{32}O_2$	18:2	19-41

Table 6. Fatty acid profile of jatropha oil

Oils extracted (Figure 4) from jatropha seeds as one of non-edible crops are potential feedstock's for biodiesel production. From all non-edible oil plants, jatropha, karanja, mahua and castor oils are the most often used in biodiesel production (Figure 5). In many countries, edible oils are not produced in enough amounts to meet the nutrition requirements for human use and must be imported. Hence, the price of biodiesel produced from edible oils is much higher than that of petrodiesel. Therefore, non-edible oils from plants like jatropha, karanja, neem, mahua and other plants (Figure 6) are the only possibility for biodiesel production.

Figure 6. Selected non-edible fruits bearing plants.

2.3. Agricultural residues as biomass sources

From the different economy sectors, there is different industrial, agricultural, forestry, municipal residues. Of these residues (Figure 7), agricultural residues produced in large quantities every year. Wheat, Rice, sugar cane, maize, soybeans and groundnuts are just a few examples of crops that generate considerable amounts of residues. These biomass residues are an important source of energy both for domestic as well as industrial purposes. Juninger et al. [14] mentioned that, if only all process-based agricultural residues alone would be used, they could contribute between 25% and 40% of the total primary commercial energy production in various Southeast Asian countries

Until now, bagasse and rice husk are considered the two major biomass resources being used for power generation (approx. 15–17 and 5.0 Mio t/year, respectively). Moreover, forest residues and wood wastes represent a large potential resource for energy production; the primary advantage of using forest residues for power generation is that an existing collection of infrastructure is already set up to harvest wood in many areas.

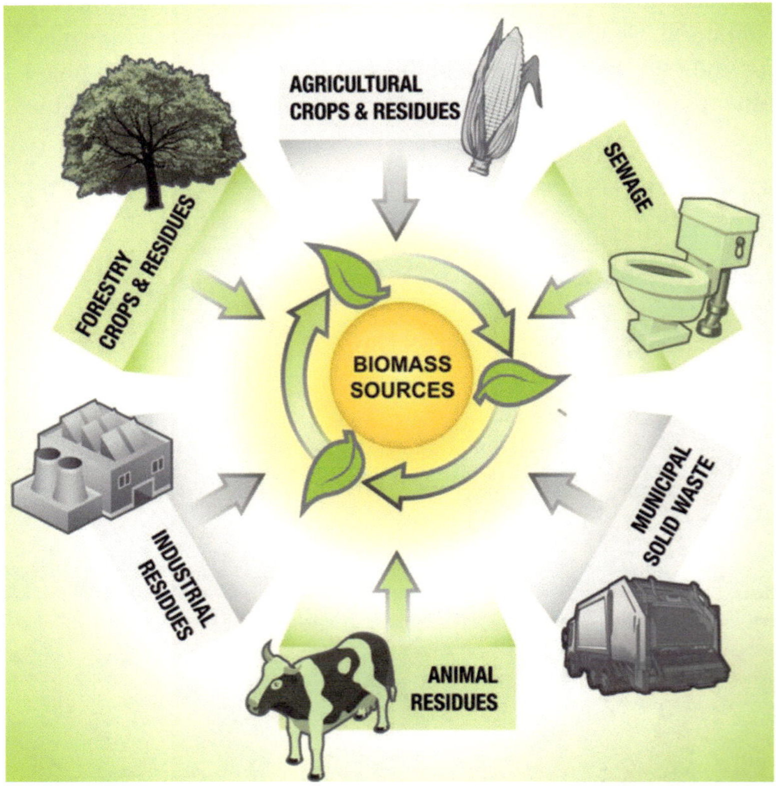

Figure 7. The major expected biomass sources for biofuel production (http://i2.wp.com/www.bioenergyconsult.com/wpontent/uploads/2012/04/biomass_resources1.jpg).

2.4. Waste cooking oils

Recycling of waste cooking oils is increasingly being carried out to produce biofuel (Figures 8 and 9). Opportunities for businesses and consumers to recycle used cooking oil ("yellow grease") have increased due to the price of waste cooking oils (WCO), which is 2–3 times cheaper than virgin vegetable oils. A significant advantage is that, biofuels derived from waste cooking oil typically burn clean, have low carbon content and do not produce carbon monoxide. This helps communities to reduce their carbon footprints (there was on average of a decrease of 14% for CO2, 17.1% for CO and 22.5% for smoke density when using biodiesel from recycling of waste cooking oils.). The recycling of cooking oil also provides a form of revenue for restaurants, which are sometimes compensated by cooking oil recyclers for their used deep fryer oil. However, the optimum conditions for biodiesel production (methanol/oils ratio and concentration of catalyst) are inconsistent. They strongly depend on the properties of WCO.

The processes of such oils and fats pose a significant challenge because of their disposal problems and possible contamination of the water and other resources.

http://dx.doi.org/10.5772/59178

Using Gc analysis of biodiesel produced from waste cooking oil, 60% of the fatty acids were found to be monounsaturated (C18:1). Polyunsaturated fatty acids were found to be approximately 26% (C18:2, C18:3). However, 8% fatty acids were saturated. (Palmitic acid and stearic acid were the major saturated fatty acids) as reported by Chhetri et al. [15].

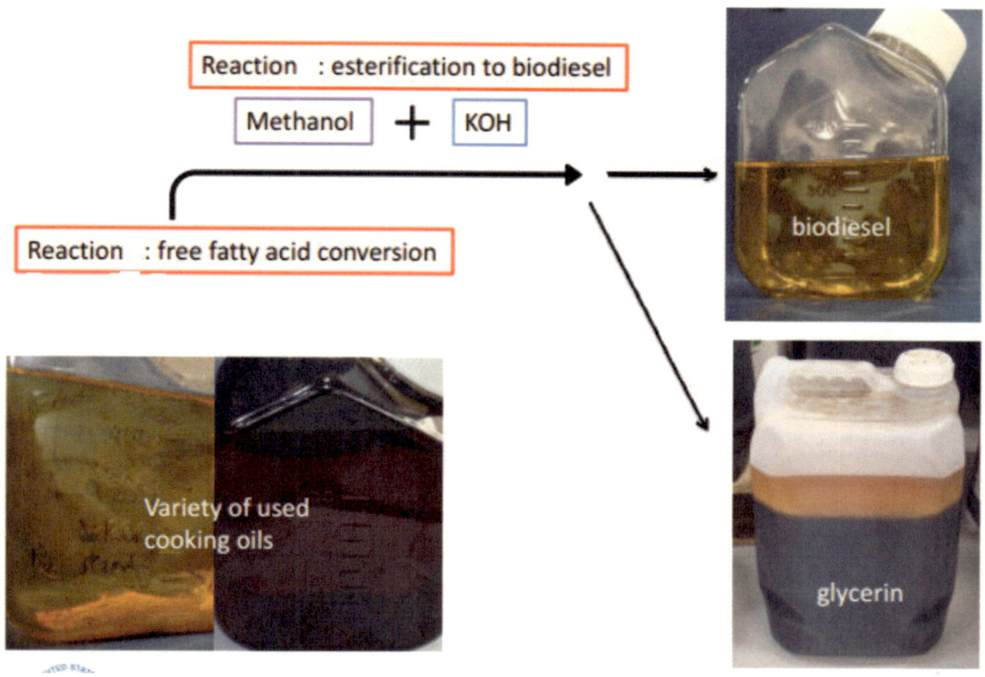

Figure 8. Biodiesel production from waste cooking oils.

The chemical properties of waste cooking oil and produced biodiesel are very important factors for using the biodiesel in industrial sector. From these properties; Acid value (AV) refers to the content of free acids in the sample, which have an influence on fuel aging. It is measured in terms of the quantity of KOH required to neutralise sample. Predojevic [7]reported that, the base catalyzed reaction is reported to be very sensitive to the content of free fatty acids, which should not exceed a certain limit recommended to avoid deactivation of the catalyst, formation of soaps and emulsion. Sharma *et al.* [16] found that, AV of the feedstock for alkaline transesterification has to be reduced to less than 2 mg KOH/g (1%), while only few examples of transesterification with feedstock acid value of up to 4.0 mg KOH/g (2%) were found. The authors also reported that, the limit of free fatty acids is a bit relaxed and the value a little beyond 1% (i.e. 2 mg KOH/g) when using waste cooking oil as feedstock for biodiesel production. In addition, the acid value of WCO feed stocks was 5 mg KOH/g oil. From this result it could be concluded that the used WCO had values above the recommended (2 mg KOH/g). However, this value did not turn out to be limiting for the efficiency of the applied two-stage process, as it will be discussed along with the obtained product yield and purity later.

Furthermore, The AV of the produced Fatty acids methyl ester (FAME) was 0.48 mg KOH/g oil. The recorded acid value of the produced biodiesel is higher than that of standard petro-diesel, but it meets the standard limits of biodiesel EN14215 and biodiesel D-6751, indicating that the free fatty acid content will not cause operational problems, such as corrosion and pump plugging, caused by corrosion and deposit formation [5].

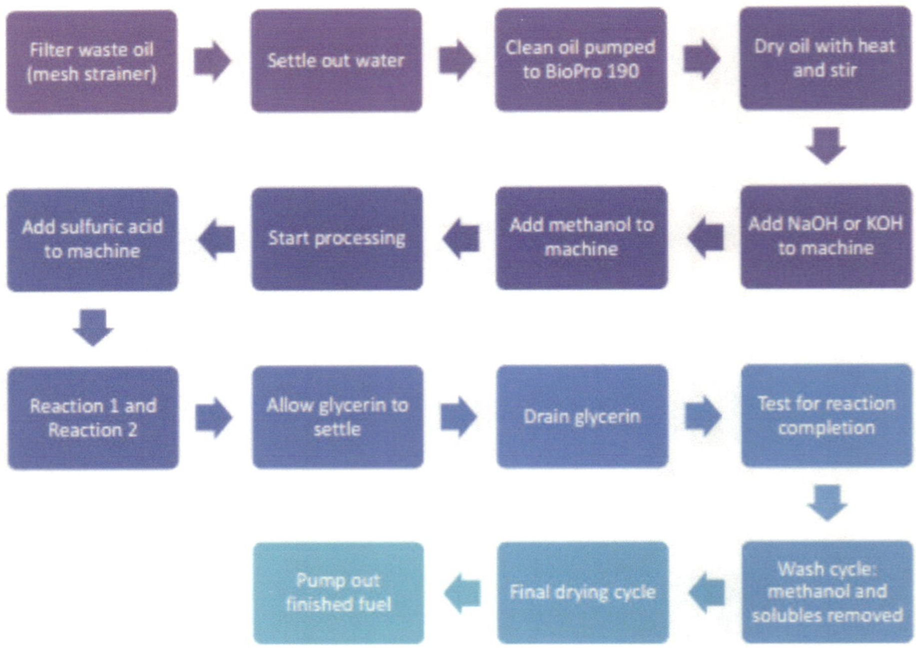

Figure 9. The processes for biodiesel and glycerin production from waste cooking oils

2.5. Algae (especially microalgae)

Microalgae are photosynthetic microorganisms that are able to rapidly generate biomass from solar energy, CO_2 and nutrients in bodies of water. This biomass consists of important primary metabolites such as sugars, oils and lipids, for which process path-ways exist for the production of high-value products including human and animal feed supplements, transport fuels, industrial chemicals and pharmaceuticals. Algae are capable of producing 30 times the amount of oils and lipids per unit area of land as compared to terrestrial oilseed crops (microalgae are capable of producing much higher amount oil per unit area of land, compared to many terrestrial oilseed crops, such as soybean, coconut and palm [17].

Many microalgae are exceedingly rich in oil, which can be converted to biodiesel using existing technology (Figure 10). More than 50% of their biomass as lipids, sometimes even up to 80%, and oil levels of 20%–50% are quite common [18]. Lipids production and biodiesel extraction from algae depend on algal species and extraction solvent system[19].

http://dx.doi.org/10.5772/59178

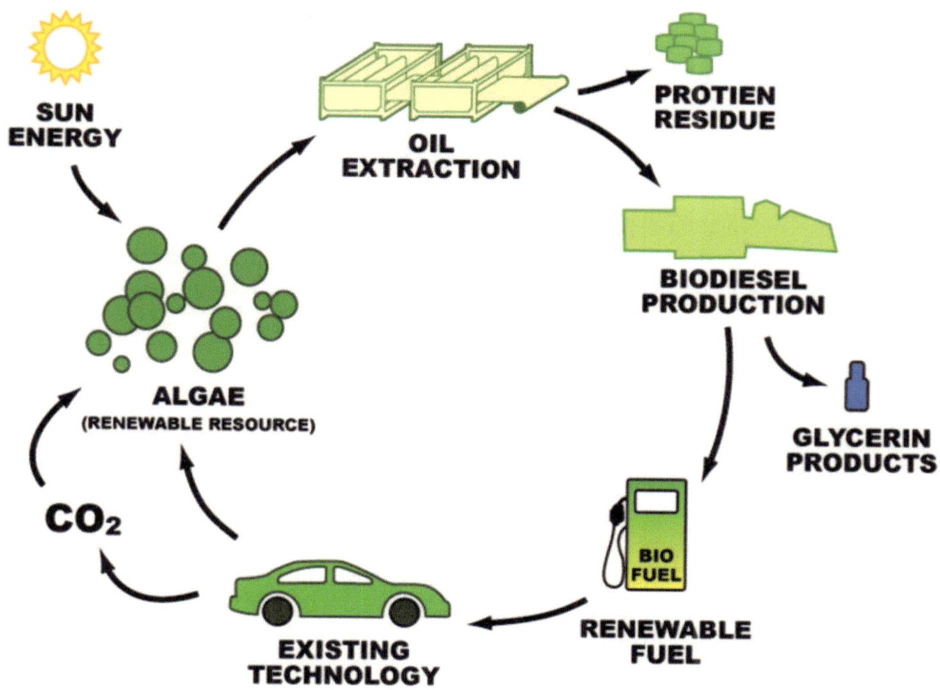

Figure 10. Process for conversion of algal oils to biodiesel. (http://refuelingthefuture.yolasite.com/third-generation-bio-fuels.php)

2.5.1. The benefits of Algae over previous generations

- Algae are one of the most promising long-term, sustainable sources of biomass and oils for fuel, food, feed and other co-products.
- Easy cultivation and growth.
- Easy to enhance the oil production using biotechnological tools (cultivation under biotic and a biotic stress).
- It is currently one of the most researched topics. This means algae fuel can become even more efficient than it already is. This also means that there will be new, more practical methods for growing and processing algae.
- Low cost of producing biodiesel from algae oils.
- Algae are not a food crop. Therefore, is does not compete with food consumption.
- Although an alga does not reduce carbon emissions, it does stop the introduction of new carbon dioxide into the atmosphere as it displaces the use of fossil fuels.
- Algae can be grown in sea water. Therefore, algae farms do not have to compete with agricultural farmland.

- Algae can be cultivated in different wastewater sources. This can be used as an alternative water treatment method. This method is cost effective. It requires a low amount of energy. It reduces sludge, and in the end, it cleans the water and produces algae for biofuel use.

Results in Table 7 show the lipid amounts extracted from eight algal species (5 macro and 3 microalgae) by the two extraction methods described in the experimental section. The red macroalga *Jania rubens* produced 2.8% lipid when extracted using hexane: ether (1:1, v/v) system whereas the recovery was doubled (4.4%) using the extraction system based on chloroform / methanol (2:1, v/v). No significant difference in the total lipid content was obtained from the red algae *Galaxaura* and *Gelidium* using both systems for extraction (2.4, 3.1 and 2.5, 3.0, respectively). The red seaweed *Asparagopsis taxiformis* and the green *Ulva* produced 1.2-fold increase in extracted lipid percentage and the brown macroalga *Colpomenia sinuosa*, produced a 1.52-fold increase in lipids when extracted by chloroform / methanol (2:1, v/v) system, as illustrated clearly in Tables 1 and 2. Moreover, the microalga *Dictyochloropsis splendida* showed 2.4 % of total lipid when extracted with hexane / ether (1:1, v/v) system, but on using chloroform/methanol (2:1, v/v) as extraction mixture, the percentage of total lipids increased 5.2 times to reach 12.5 % showing at the same time the highest biodiesel production (8.75 %) of the eight algal species used in this investigation. The cyanbacterium *Spirulina platensis* produced a 3-fold increase in lipid content using the chloroform/methanol-based method (Tables 8 and 9).

Algal species	**Algal pictures**	**Chloroform/ methanol (2:1, v/v)**	**Hexane/ether (1:1, v/v)**
Jania rubens		4.4±0.12	2.8±0.04
Galaxaura oblongata		2.5±0.09	2.4±0.01
Gelidium latifolium		3.0±0.0	3.1±0.02

http://dx.doi.org/10.5772/59178

Algal species	Algal pictures	Chloroform/ methanol (2:1, v/v)	Hexane/ether (1:1, v/v)
Asporagopsis taxiformis		4.1±0.08	3.4±0.05
Ulva lactuca		4.2±0.1	3.5±0.1
Colpomenia sinuosa		3.5±0.05	2.3±0.03
Dictyochloropsis splendida	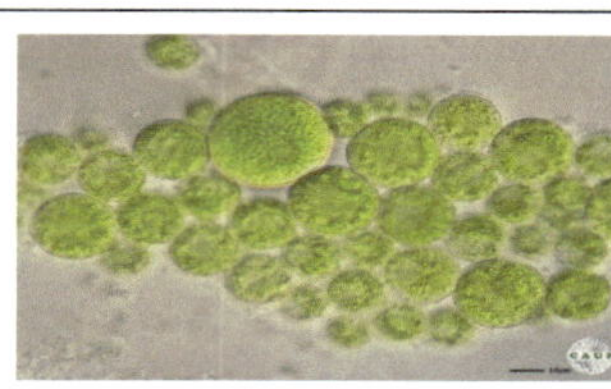	12.5±0.23	2.4±0.14
Spirulina platensis	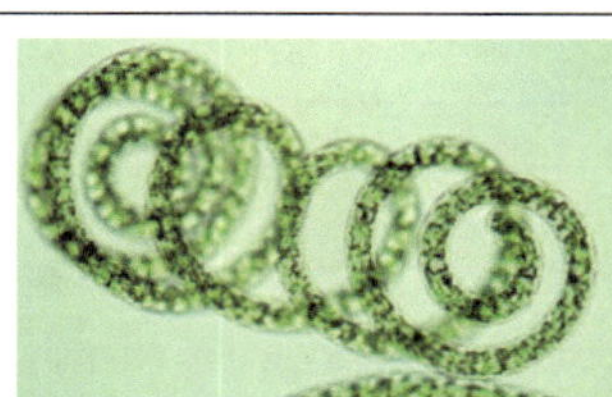	9.2±0.25	3.0±0.10
	LSD	**0.3261**	**0.3261**

Each value is presented as mean of triplet treatments, LSD: Least different significantly at P ≤ 0.05 according to Duncan's multiple range test

Table 7. Comparison between lipid percentages (%) produced by eight algal species using hexane/ether (1:1, v/v) and chloroform/methanol (2:1, v/v) extraction systems [21].

The obtained results illustrated in Table 8 revealed that the solvent mixture hexane/ether were not the most suitable system for lipid biodiesel extraction from algae because these solvents were unable to extract polar lipids. On the contrary chloroform/methanol system extracted greater percentage of lipid (non-polar and polar lipids) and gave place to higher biodiesel yields by transesterification. The lowest biodiesel production was observed in the red seaweed *J. rubens* (0.25%) followed in ascending order by *Gelidium latifolium* (1.3%), *Galaxaura oblongata* (2.06%) and A. taxiformis (3.64%). While the green macroalga Ulva lactuca and the brown seaweed Colpomenia sinuosa produced comparable biodiesel percentages (3.8 and 3.1%, respectively), the greatest yield of biodiesel was achieved by the green microalgae *D. splendida* (8.75%) followed in descending order by the cyanbacterium *S. platensis* (7.5%) as illustrated in Table 2. Using chloroform/methanol (2:1, v/v) solvent system, we are able to produce not only biodiesel in large percentage but also a sediment containing glycerin and pigments. The produced biodiesel have slightly alkaline pH values ranging 7.5–8.5 in all preparations. Our results concerning the green microalga *D. splendida* (produced 12.5% lipids) agreed with those obtained by Hossain and Salleh [20] who reported that the green filamentous alga Oedogonium sp produced higher lipid.

Biodiesel colour	**Sediment %**	**Biodiesel%**	**Lipid %**	**Algal sp.**
Light brown	± 0.05 → 4.2[a]	0.25±0.01	4.4±0.12	*Jania rubens*
Light green	0.08±0.0	2.06±0.02	2.5±0.09	*Galaxaura oblongata*
Yellow	1.6±0.01	1.3±0.0	3.0±0.0	*Gelidium latifolium*
Dark green	0.40±0.01	± 0.10 → 3.64[c]	4.1±0.08	*Asporagopsis taxiformis*
Light green	0.44±0.0	3.8±0.12	4.2±0.1	*Ulva lactuca*
Yellow	0.31±0.05	3.1±0.05	3.5±0.05	*Colpomenia sinuosa*
Colourless	3.75±0.08	8.75±0.24	12.5±0.23	*Dictyochloropsis splendida*
Light green	1.66±0.06	7.5±0.30	9.2±0.25	*Spirulina platensis*
	0.1786	**0.3314**	**0.3261**	**LSD**

Each value is presented as mean of triplet treatments, LSD: Least different significantly at $P \leq 0.05$ according to Duncan's multiple range test.

Table 8. Total lipid, biodiesel, sediments percentage and biodiesel colour of different algal species using the extraction solvent system Chloroform/methanol (2:1, v/v) (Afify et al., 2010).

3. Conclusion

From this chapter we conclude that non-edible biomass can be used as a suitable and cheap feedstock for producing biodiesel. Most of non-edible biomass has a considerable amount of low-chain fatty acids, which gives special features to biodiesel. The most important advantages of the current biomass are production of biodiesel with high cetane number, low viscosity. These factors can increase engine's output and decrease pollution compared with other biodiesel fuels. The present experimental results support that methyl ester of non-edible biomass can be successfully used as a biodiesel in the internal combustion engine.

Author details

Emad A. Shalaby

Address all correspondence to: dremad2009@yahoo.com

Biochemistry Department, Faculty of Agriculture, Cairo University, Giza, Egypt

References

[1] FAO (2007). Statistical Databases. http:// faostat.fao.org (accessed 15.08.09).

[2] Habib, H.M. and Ibrahim, W. (2008). Nutritional Quality Evaluation of Eighteen Date Pit Cultivars. International Food Science and Nutrition, 60:99-111.

[3] Habib, H.M. and Ibrahim, W.H. (2011). Effect of Date Seeds on Oxidative Damage and Antioxidant Status *in vivo*. J. Sci. Food Agric., 91:1674-1679.

[4] Shanab, S.M.M., Hanafy E.A., & Shalaby, E.A. (2014). Biodiesel Production and Antioxidant Activity of Different Egyptian Date Palm Seed Cultivars. Asian Journal of Biochemistry, 9: 119-130.

[5] Shalaby, E. & El-Gendy, N. (2012). Two Steps Alkaline Transesterification of Waste Cooking Oil and Quality Assessement of Produced Biodiesel. International J of Chemical and Biochemical Sci., 1: 30-35.

[6] JUS EN 14214:2004. Automotive Fuels. Fatty Acid Methyl Esters (FAME) for dengines-Requirements and Test Methods. Standardization Institute, Belgrade, Serbia.

[7] Predojevic, Z.J. (2008). The Production of Biodiesel from Waste Frying Oils: A Comparison of Different Purification Steps. Fuel, 87: 3522-3528.

[8] Ojeh, O. (1981). Effects of Relining on the Physical and Chemical Properties of Cashew Kernel Oil. J. Fats Oils Technol., 16: 513-517.

[9] Eskin, N.A.M.; McDonald, B.E.; Przybylski, R.; Malcolmson, L.J.; Scarth, R. and Mag, T. (1996). Canola Oil, In: Hui, Y.H. (ed.). Bailey's Industrial Oil and Fat Products, Vol. 2, 5th ed., John Wiley and Son. New York. pp. 1-95.

[10] Oomah, B.D., Ladet, S., Godfrey, D.V., Liang, J. & Girard, B. (2000). Characteristics of Raspberry (*Rubus idaeus* L.) Seed Oil. Food Chemistry, 69: 187-193.

[11] Nehdi, I., Omri, S., Khalil, M.I. & Al-Resayes, S.I. (2010). Characteristics and Chemical Composition of Date Palm (*Phoenix canariensis*) Seeds and Seed Oil. Industrial Crops and Products, 32: 360-365.

[12] Yong, Y. & Salimon, J. (2006). Characteristics of *Elateriospermum tapos* Seed Oil as a New Source of Oil Seed. Ind. Crops Prod., 24: 146-151.

[13] Veeresh Babu, S.V., Veeresh, B., Patil, A. & Warke, B. (2009). Lauric Acid on Myristic Acid Prevent Testosterone Induced Prostatic Hyperplasia in Rats. Eur. J. Pharmacol., 22: 193-199.

[14] Junninger, M, Faaij, A, Van den Broek, R., Koopmans, A, & Hulscher, W, Fuel Supply Strategies for Large-Scale Bio-Energy Projects in Developing Countries, Electricity Generation from Agricultural and Forest Residues in Northeastern Thailand.

[15] Chhetri, AB.; Watts,KC.; and Islam, MR. (2008). Waste Cooking Oil as an Alternate Feedstock for Biodiesel Production. Energies, 1: 3-18; DOI: 10.3390/en1010003.

[16] Sharma, Y. C., Singh, B. & Upadhyay, S. N. (2008). Advancements in Development and Characterization of Biodiesel: a Review. Fuel, 87: 2355-2373.

[17] Sheehan, J., Dunahay, T., Benemann, J., Roessler. P. (1998). A Look back at the US Department of Energy's Aquatic species Program – Biodiesel from Algae. National Renewable Energy Laboratory (NREL) Report: NREL/TP-580-24190. Golden, CO.

[18] Chisti Y. (2007). Biodiesel from Microalgae. Biotechnology Advances, 25: 294-306.

[19] Afify AMR, Shalaby EA, Shanab SMM. (2010). Enhancement of biodiesel production from different species of algae. Grasas Y. Aceites, 61 (4): 416-422.

[20] Hossain, A.B.M., Salleh, A. (2008). Biodiesel Fuel Production from Algae as Renewable Energy. Am. J. Biochem. and Biotechn., 4(3): 250-254.

Chapter 2

Biocatalytic Production of Biodiesel from Vegetable Oils

Eda Ondul, Nadir Dizge, Bulent Keskinler and
Nedim Albayrak

Additional information is available at the end of the chapter

http://dx.doi.org/10.5772/59949

1. Introduction

Alcoholysis of vegetable oils is an important reaction to produce fatty acid alkyl esters which are excellent substitutes for diesel fuel and valuable intermediates in oleochemistry [1, 2]. Fatty acid methyl esters, a mixture of mono-alkyl esters are also known as biodiesel, obtained from both vegetable oils such as sunflower oil, canola oil, soyabean oil, jatropha oil, palm oil, rapeseed oil, peanut oil, cotton seed oil and animal fats such as beef tallow, and lard. Biodiesel can also be produced from other sources such as waste cooking oil, algae, and greases [3]. Biodiesel production has attracted considerable attention in the past two decades because of biodegradable, renewable, non-toxic, and environmentally friendly and socially responsible fuel [4]. Biodiesel can be produced by several methods: direct use or blending, microemulsion, thermal cracking (pyrolysis), and transesterification including acid-catalyzed processes, base-catalyzed processes, lipase-catalyzed processes, non-ionic base-catalyzed processes, and heterogeneously catalyzed processes [5, 6]. Among these methods, alkali catalyzed process including an alkali catalyst (usally NaOH, KOH, or sodium methoxide) has been accepted industrially due to its high conversion of triglycerides to methyl esters in a short reaction time and high reaction rates. In spite of these advantages of chemical transesterification process, it also possesses some disadvantages such as the need to eliminate the catalyst and salt from the biodiesel phase, to remove saponification products, the difficulty of recycling glycerol, and their energy-intensive nature, leading to development of alternative processes [7-9]. Alcoholysis is also carried out under acidic conditions, but this process requires higher reaction temperatures. In order to overcome these drawbacks, recently, enzymatic transesterification has attracted much attention for biodiesel production since it produces high purity product and provides an easy separation from the by-product glycerol. The use of enzymes (lipases) as catalysts in biodiesel production overcomes the problems inherent to alkali catalysts. It is reported that the enzymatic reactions are insensitive to free fatty acid (FFA) and water content in the raw material [10]. So far, many attempts have been made to develop enzymatic process by using either extracellular or intracellular lipase as a biocatalyst [1, 11]. Lipases (EC 3.1.1.3), also defined as triacylglycerol acylhydrolases, catalyze the hydrolysis of ester bonds in long chain triacylglycerols (TAGs) to produce free fatty acids (FFAs) and glycerol. In general, the active site of lipases is formed by serine, aspartic (or glutamic) acid and histidine amino acid groups. Interfacial activation, which is unique to the class of lipases for its use in transesterification of fats and oils, takes place in presence of a substrate and lipase active site structure. Lipases are used in a wide range of fields due to their ability in utilizing all mono, di, and triglycerides as well as the FFA, low product inhibition, high activity and yield in non-aqueous media, low reaction time, temperature and alcohol resistance, but the high cost of enzyme remains a barrier for its industrial applications [10]. In order to decrease the cost of the process, the enzyme can be immobilized on a suitable carrier and reused many times. So far, many techniques and different carriers have been employed for immobilization of lipases to produce biodiesel. They have been successfully immobilized on porous kaolinite particle, biomass support particles, macroporous resin, gel-entrapped, celite, silica, and Eupergit C250L [12-16]. Several oils have been catalyzed with lipase enzymes until now. Lipase catalyzed production of biodiesel from soybean oil, sunflower oil, palm oil, kernel oil, coconut oil, rice bran oil, mixture of vegetable oils, grease and tallow oil, microbial oil, and waste oil containing vegetable oils have been reported in the past decades [17-25]. In this chapter, focus will be given toward enzymatic biodiesel production from various vegetable oils. *Thermomyces lanuginosus* lipase and *Candida antarctica* lipase A were immobilized on cotton cloth which is a low cost carrier. Transesterification of sunflower, canola, and waste cooking oil with methanol and ethanol was carried out by continuous operation system. The essential aim of this study was to investigate the production of biodiesel from vegetable oils by enzymatic transesterification with immobilized lipases on fibrous matrix by polyethyleneimine in a packed bed bioreactor at industrial scale.

2. Enzymatic transesterification

In transesterification reaction, one ester is converted into another ester. This conversion occurs as transfer of an acyl group. The acyl group transfer can take place between one ester and another ester (interesterification), an ester and an acid (acidolysis), or an ester and an alcohol (alcoholysis). In broad terms, the transesterification reaction between TAGs and alcohol to produce biodiesel is a sequence of three consecutive and reversible reactions, by which diacylglycerol (DAG) and monoacylglycerol (MAG) are formed as intermediates. Enzymatic synthesis of biodiesel has been usually performed at moderate temperature between 20 and 60 °C. When transesterification process is completed, the by-product glycerol (lower phase) is simply separated from the biofuel (upper phase) and neither neutralization nor deodorization of the product is necessary. However, an overdose of alcohol provides higher yield of biodiesel [26]. Biocatalysis has been considered a trend for sustainable synthesis technology due to biologic origin of the catalyst, selectivity and the possibility of reusing agro-industrial residues for biocatalyst production, which classifies the method as a green process [27]. Enzymatic catalysis has been applied for biodiesel which starts its industrial scale operation in China [28].

http://dx.doi.org/10.5772/59949

However, some factors such as substrate type, solvent type, alcohol type, water content of reaction medium, the reaction temperature, immobilization type and the lipase concentration influence the conversion of enzymatic transesterification reaction. In the literature, different lipases have been used upto now for biodiesel synthesis but it is hard to make any generalizations about the optimal reaction conditions. This is because, lipases obtained from different sources tend to respond differently to changes in the reaction medium [29-39]. Costs of chemical biodiesel production have still been lower than those of the enzymatic processes, however, if the pollution of natural environment is also taken into consideration, these costs are comparable. In the enzyme-catalyzed biodiesel production, the high enzyme cost significantly impacts the process profitability. The cost of commercial products for industrial use of enzymes is approximately 1, 000 $/kg which is significantly higher than that of the alkali catalyst (0.62 $/kg). Biodiesel fuel is expensive in comparison with petroleum-based fuel as 60-80% of the cost is associated with the feedstock oil [40]. Production of cheaper, robust lipase preparations and development of systems providing the long-term, iterative use of these biocatalysts can give rise to the replacement of chemical processes with enzymatic ones [28]. Currently, the high cost of biodiesel is the biggest obstacle to commercialization. The main reason is highly purified straight vegetable oil (SVO) used as a feedstock and this problem can be overcome by using used/waste vegetable oils that is much cheaper than SVO. Another obstacle in biodiesel production is the high food prices for oil. Both problems can be solved by using waste/used oil thereby gaining cost advantage. In addition, evaluation of the waste oil in terms of biodiesel can help to solve the problem of waste oil disposal. However, high free fatty acid (FFA) content of feedstock is the main problem encountered when using alkali catalyst. On the other hand, enzymatic transesterification does not have this limitation and hence can be used with waste/used oil. Moreover, almost all FFAs present in the waste/used oil can be converted to biodiesel in high yield using this approach [41].

2.1. Alcohol

Various types of acyl acceptors, alcohols, primary short-chain alcohols like methanol, ethanol, propanol, and butanol, as well as secondary alcohols like isopropanol and 2-butanol, straight and branched-chain, esters can be employed in transesterification using lipases as catalysts [42]. The prerequisites for selecting the alcohol for industrial-scale biodiesel production are that it must be cheap and in plentiful supply. Due to their price and availability, methanol and ethanol have been the most used alcohols for industrial biodiesel production. Currently, only methanol and ethanol, meet these two requirements. Ethanol is renewable and less toxic than methanol but methanol is preferred in biodiesel production because it is less expensive and more readily available in most countries than ethanol [30, 42]. However, these two alcohols are the stronger denaturing agents than longer aliphatic alcohols and inactivate enzymes. Besides, the rate of lipase-catalyzed transesterification reaction usually increases with the length of hydrocarbon chain of alcohol [30]. Meanwhile, the short alcohol chain causes lipase deactivation. It is believed that this is because the essential water layer around them which is essential for the optimum conformation of the enzyme is stripped off [43]. Most of the refined plant oils can be converted into fatty acid methyl esters to meet the specifications of biodiesel standard by stepwise alcohol addition to prevent an irreversible lipase inactivation [44]. Shimada et al. (1999) reported that the lipase from *Candida antarctica* (Novozym 435) in a solvent-free system was deactivated irreversibly when the methanol concentration exceeded its solubility level. They found that stepwise addition of methanol prevented lipase deactivation. A three-step addition process converted 98.4% of the oil to its corresponding methyl esters in 48 h and the immobilized lipase was set to be reused for 50 batches [1]. Watanabe et al. (2000) investigated the influence of methanol using a two-step strategy on biodiesel synthesis. It has been shown that one-third of the alcohol adding at the beginning of the reaction caused slow conversion into biodiesel (10 h reaction time). After that, the rest of the alcohol was added in a single step and biodiesel conversion increased by the presence of the biodiesel since its solubility increased [21]. In another study [45] two lipases from *Pseudomonas fluorescens* and *Pseudomonas cepacia* (now *Burkholderia cepacia*) were used and they provided 58% and 37% conversion in the presence of 1:8 oil/methanol molar ratio in a solvent-free system, respectively. However, they have been shown to be completely inactive for another six lipases tested under these conditions. It is clear that the excess alcohol above and beyond the stoichiometric ratio increases the reaction rate, but too much alcohol may also deactivate the enzyme [30]. There are also some arguments against using excess alcohol in industrial-scale processes, such as higher energy consumption, larger equipment requirements, and the need to treat the unreacted alcohol. To prevent the alcohol deactivating the enzyme, many researchers have used organic solvents in the reaction medium to increase the solubility of the alcohol and reduce its concentration [12, 16, 18, 46].

2.2. Water content

The effect of water content is essential for enzymatic reactions due to formation of hydrogen bonds which are fundamental in the interactions for maintaining the conformation of the enzymes. Water has strong influence on the catalytic activity and stability of the lipase. Therefore, the transesterification yields depend on the size of interfacial area which can be increased by the addition of certain amounts of water as well as the availability of an oil-water interface. However, lipases increase the hydrolysis reaction in aqueous medium and excess water causes the decrease of the transesterification yield by promoting the hydrolysis reaction [42]. The ideal water content in the reaction medium varies greatly depending on the enzyme and the reaction medium, and so must be studied on a case-by-case basis. Water content in reaction mixture can be determined by either water activity or as weight percentage of feedstock oil. Water activity is the ratio of vapor pressure of a given system [38]. Optimum water content for the transesterification reaction is very important. The optimum water content in the reaction depends upon the lipase type and feedstock, immobilization technique and solvent type [47]. For example, Kaieda et al. 2001 found that the water concentrations that resulted in the best conversions were 8-20% for *Candida rugosa* lipase, 4-20% for *Pseudomonas fluorescens* lipase, and 1-2% for *Pseudomonas cepacia* lipase [35]. Deng et al. (2005) studied several immobilized commercially available lipases and reported that the conversion obtained from the transesterification reaction with all the other lipases (*Thermomyces lanuginosus, Rhizomucor miehei, Pseudomonas cepacia,* and *Pseudomonas fluorescens*) with the exception of *Candida antarctica* was higher when anhydrous ethanol was replaced with hydrous ethanol (4% water) [48]. It is also very important to take into consideration the amount of the water present in the reagents and even in the enzyme in order to design appropriate reaction medium. Studies of lipase reutilization at different water concentrations have to be carried out since water can influence enzyme stability, making it crucially important for designing an economically feasible process [48]. Some authors have suggested that adding water into the enzymatic reaction medium can protect lipases against deactivation in the presence of short-chain alcohols [13, 19].

2.3. Organic solvent use

The use of organic solvents in enzymatic biodiesel synthesis improves mutual solubility of hydrophobic compounds (e.g. TAG and biodiesel), triglycerides and hydrophilic compounds (e.g. alcohols and glycerol). Organic solvents also protect enzymes for denaturation resulted high concentrations of alcohols [42]. Solvents also serve to reduce the viscosity of the reaction medium, enabling a higher diffusion rate to be achieved and reducing mass transfer problems. Therefore, a suitable solvent must be found, which both enhances the catalytic activity of the enzyme and keeps it stable. Thus, the presence of a solvent renders a high yield and reduces the enzyme inhibition by alcohol [47]. The most suitable non-polar hydrophobic organic solvents such as n-heptane, petroleum ether, isooctane, n-hexane and cyclohexane were used for enzymatic biodiesel synthesis and immobilized lipases showed high degree of efficiency in the presence of non-polar solvents. But when using hydrophobic solvents, glycerol is insoluble and remains in the reactor and it is adsorbed to the immobilized lipase. The polar hydrophilic organic solvents are much less useful in enzyme-catalyzed biodiesel production as they strongly interact with the essential water microlayer around the enzyme molecules influencing its native structure, thereby, leading to denaturation [42]. Recently, processes of transesterification, which is well known for its compatibility with lipases, have been also conducted in less conventional solvents, e.g. in supercritical gases like butane (C_4H_{10}) and carbon dioxide (CO_2). CO_2 is also regarded as a green solvent owing to its low toxicity, non-flammability, and its environmentally good-natured character [30].

2.4. Biocatalysis type

Recently, lipases have been studied for biodiesel production as whole-cell immobilized lipases. Each type of biocatalyst has its strengths and weaknesses when it comes to reducing the contribution of the biocatalyst in the final cost of the biodiesel. Recent studies have been focusing on improving catalysis performance and stability of the enzyme with the aim to reduce the lipase cost in the biodiesel conversion process. Different approaches have been developed for application mode of lipases. Solid state fermentation, whole-cell biocatalyst and immobilized lipase in different supports are the main studied modes. The application of solid state fermentation was created for reducing cost in lipase production and could be used as a catalyst in batch and continuous operation. The solid state fermentation of agricultural residues permits for cost-efficient production and low-price when compared to commercial enzymes. Since solid state fermentation avoids the extraction, purification, and immobilization steps in enzyme production with satisfactory catalytic results in transesterification reaction [27].

2.4.1. Free biocatalysis

Microbial lipases have gained wide industrial importance and they now share about 5% of the world enzyme market after proteases and carbohydrases. Lipases of microbial origin are more stable than plant and animal lipases and are available in bulk at lower cost compared to lipases of other origin. Yeasts lipases are easy to handle and grow compared to bacterial lipases. Among the yeast lipases, *Candida rugosa* has gained good commercial importance. The most commonly used biocatalyst for biodiesel production are the microbial lipases that are produced by a number of fungal, bacterial, and yeast species [40]. Free enzymes are far cheaper than immobilized lipases. They can be purchased in an aqueous solution composed of the enzyme solution plus nothing more than a stabilizer to prevent enzyme denaturation (e.g. glycerol or sorbitol) and a preservative to inhibit microbial growth (e.g. benzoate) [49].

2.4.2. Immobilized biocatalysis

Immobilization of lipases was carried out using entrapment, physical adsorption, ion exchange, and crosslinking. Carriers for lipase immobilization include polyurethane foam, silica, sepabeads, cellulosic nanofibers. Based on the criteria for selecting the immobilization technique and carrier dependings on the source of lipase, the type of reaction system (aqueous, organic solvent or two-phase system), and the bioreactor type (batch, stirred tank, membrane reactor, column and plug-flow) can be designed. The literature is replete with various lipase producing microorganisms, enzyme immobilization methods, and physical carriers. The challenge will be to select a carrier and immobilization technique that will allow maximum lipase activity, retention, and stability on the oil substrate. Among the immobilization method, adsorption technique is the simplest and most widely used technique for lipase immobilization. Adsorption method consists of bonding the lipase to the immobilization support surface through weak forces such as van der Waals or hydrophobic interactions. However, the main disadvantage of this technique is enzyme desorption from the support due to low bond strength between the enzyme and the support [40].

2.4.3. Whole-cell biocatalysis

In recent years, whole-cell immobilized lipases have been studied for biodiesel production. This method is cheaper as it does not require the enzyme purification and isolation steps from fermentation broth. The efficiency of the transesterification process could be increased by using microbial cells that produce intra-cellular lipase as whole-cell biocatalysts [40, 46]. Filamentous fungi have been identified as robust whole-cell biocatalysts for biodiesel production: among these *Rhizopus* and *Aspergillus* have been most widely used [42]. There are several recent works reporting the utilization of bacteria, yeast and fungi as whole-cell biocatalysts in biodiesel process [27].

3. Experimental

3.1. Material

Commercial powder lipase from *Thermomyces lanuginosus* (fungal lipase, lyophilized, specific activity of 1400 U/mg solid) and *Candida antarctica* lipase A (specific activity of 2500 U/mg solid) were purchased from Codexis Inc. (Pasadane, CA). Sunflower oil and canola oil were supplied by a local company; waste cooking oil (WCO) was used following filtration to remove particles

http://dx.doi.org/10.5772/59949

remaining in the oil after domestic use. Polyethyleneimine [PEI; $(C_2H_5N)_n$] as 50% (w/v) (average molecular weight 750.000) and glutaraldehyde (GA, acidic aqueous solution as 25% (w/v)) were obtained from Sigma (USA) and AppliChem (Darmstadt, Germany), respectively. All other chemicals were of analytical reagent grade and used without further purification.

3.2. Experimental procedure

3.2.1. Lipase immobilization

Figure 1 shows multi-layer immobilization of *T. lanuginosus* lipase on cotton cloth by aggregation with polyethyleneimine (PEI). 1 mL of PEI solution (pH=11), containing 2 mg of PEI, was added to each 0.1 g piece of cotton cloth. The PEI solution volume was at a sufficient level to completely wet the cloth. After adsorption of PEI, 50 mg of enzyme (5 mL of 10 mg/mL enzyme solution) was added. Upon the addition of enzyme to PEI-adsorbed cotton, a "milky" turbid solution was formed. The flasks were put into a shaker-incubator (Heidolf Unimax 1010, Germany) at 150 rpm at room temperature (25±1 °C) for 5 min. The white turbidity disappeared within 5 min and the coupling solution was completely clarified. The clarified coupling solution was slowly decanted and PEI-enzyme coated cottons were dipped in a cold GA solution (2.5% (w/v), pH= 3.5) for cross-linking at 5 min. The cross-linked cottons were washed with distilled water and potassium phosphate buffer (1 M, pH= 7). It is important to note that there was no washing step until the completion of GA cross-linking [25, 39]. Based on the degree of immobilization tests, about 80-90% of the enzyme was immobilized in this procedure. The actual enzyme loading was determined as 180 mg of *T. lanuginosus* lipase per 1 g of cotton cloth.

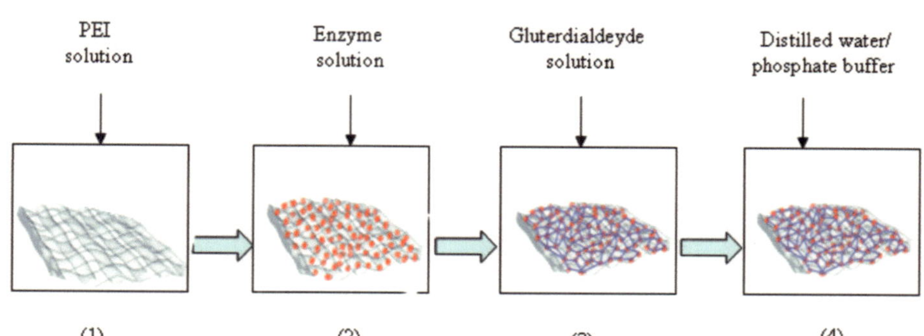

1. Adsorption : PEI solution + Cotton cloth
2. Immobilization : PEI adsorbed cloth + Enzyme
3. Cross-linking : Enzyme adsorbed cotton cloth + Glutardialdehyde
4. Wash : Enzyme adsorbed cotton cloth + Distilled water and phosphate buffer

Figure 1. The procedure for PEI-multilayer lipase immobilization on cotton cloth fibrils

3.2.2. Biodiesel production with immobilized lipase-catalyzed transesterification

Production of biodiesel by enzymatic catalyzed transesterification from various vegetable oils was studied in a packed bed reactor (Figure 2). A small piece of immobilized cotton cloth (1 g) was placed in the glass column reactor (1 cm diameter x 12 cm height) with a water jacket maintained at a constant temperature (30 °C). Substrate mixture (oil and alcohol) was continuously recirculated throughout the immobilized enzyme reactor with a peristaltic pump at a flow rate of 50 mL/min by adding of alcohol in three-steps. Immobilized cotton cloths were washed by *tert*-butanol before adding of alcohol to each reaction medium. Reaction was continued for 10 h. Samples were taken from the flask at appropriate time intervals and analyzed for fatty acid methyl esters (FAMEs) and glyceride contents by high performance liquid chromatography (HPLC) [50].

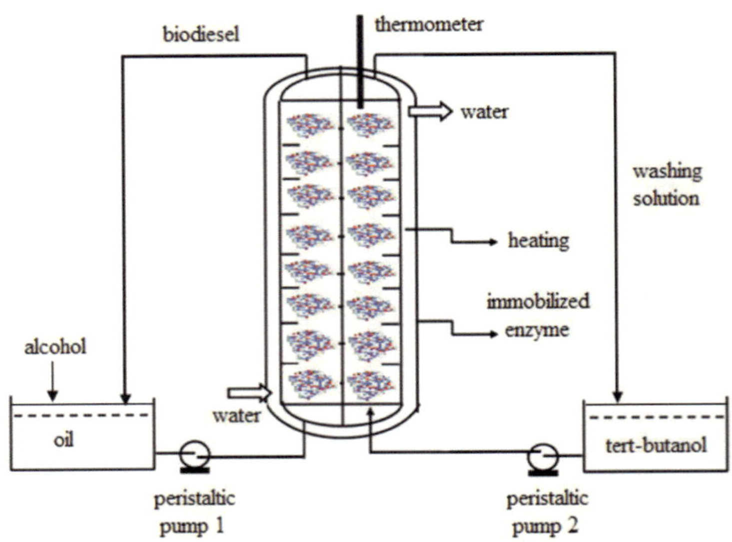

Figure 2. Schematic diagram of the enzymatic column reactor used in the transesterification reaction

4. Results and discussion

4.1. Lipase screening

Firstly, lipase screening was performed to find the lipase that has the best catalytic activity in the transesterification of sunflower oil. The most active lipase was then used in further transesterification studies. Two lipases, *Thermomyces lanuginosus* and *Candida antarctica* lipase A, were screened for their transesterification activity. The screening results for the tested lipases are presented in Figures 3 and 4. As can be seen from the figures, among the tested lipases, *T. lanuginosus* lipase showed the highest activity in the transesterification reaction of sunflower oil with methanol at 30 °C. *C. antarctica* lipase A showed very little conversion in the transesterification reaction with both ethanol (44%) and methanol (28%). After 10 h of reaction with *T. lanuginosus* lipase, the product contained 91.3% of methyl esters, 2% monoglycerides and diglycerides, and 6.7% triglycerides (Figure 3). However, 28% methyl esters, 5% monoglycerides and diglycerides and 67% triglycerides were obtained with *C. antarctica* lipase A using methanol (Figure 4).

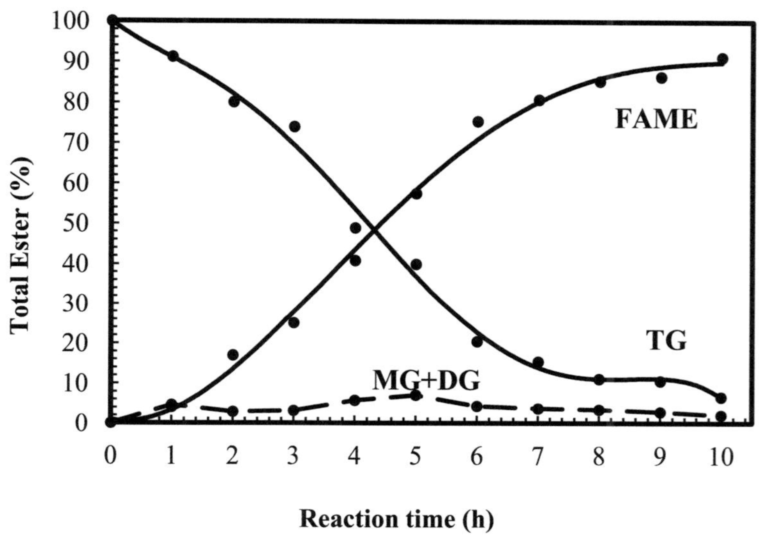

Figure 3. Immobilized *T. lanuginosus* lipase catalyzed transesterification of sunflower oil, 1:3 M ratio of methanol, 30 °C reaction temperature, 10 h reaction time (FAME: Fatty acid methyl ester; TG: triglyceride; DG: diglyceride; MG: monoglyceride)

4.2. Effect of reaction parameters on transesterification

4.2.1. Effect of alcohol type

It is well known that excessive short-chain alcohols such as methanol might inactivate lipase seriously. However, at least three molar equivalents of methanol are required to complete conversion to its corresponding methyl esters of the oil [26, 30]. Experiments were performed to determine the yield of methyl or ethyl esters by varying the alcohol type using *T. lanuginosus* lipase. The reaction was carried out to avoid enzyme inactivation by adding methanol stepwise. The molar ratio of sunflower oil to alcohol was kept constant in the concentration of 1:3 for both methanol and ethanol. Results are summarized in Figure 5. As was expected, all alcohol types resulted in an increase in the yield of esters. However, the formation of esters reached a maximum level using methanol. The biodiesel conversions were about 91.3% of methyl esters and 82.8% of ethyl esters for methanol and ethanol, respectively.

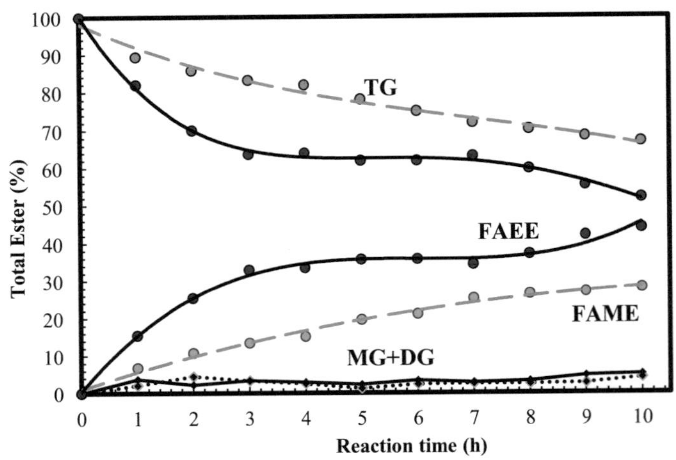

Figure 4. Immobilized *Candida antarctica* lipase A catalyzed transesterification of sunflower oil, 1:3 M ratio of methanol and ethanol, 30 °C reaction temperature, 10 h reaction time (FAME: Fatty acid methyl ester; FAEE: Fatty acid ethyl ester; TG: triglyceride; DG: diglyceride; MG: monoglyceride)

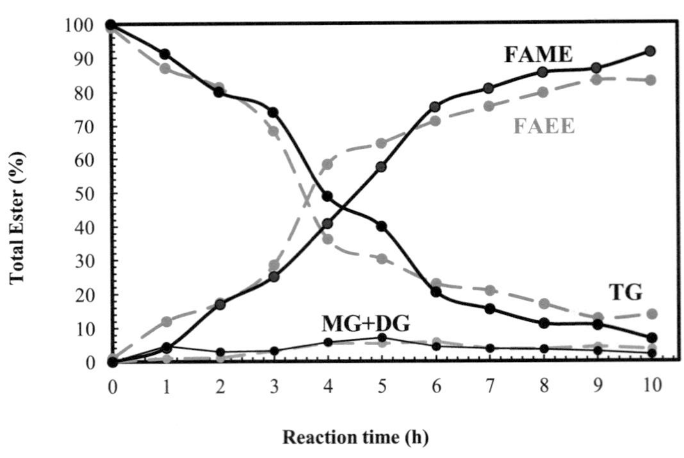

Figure 5. Effect of alcohol types on immobilized *T. lanuginosus* lipase catalyzed transesterification of sunflower oil, 1:3 M ratio of methanol and ethanol, 30 °C reaction temperature, 10 h reaction time (FAME: Fatty acid methyl ester; FAEE: Fatty acid ethyl ester; TG: triglyceride; DG: diglyceride; MG: monoglyceride)

4.2.2. Effect of water concentration

The effect of water content was examined in the range of 0-2 g and at constant molar ratio of oil to methanol with sunflower oil. The reactions were carried out according to the reaction setup described earlier. The results presented in Figure 6 indicated that water was not required to activate the *T. lanuginosus* lipase. A maximum ester yield (81%) could be achieved at without water reaction conditions. Water concentration in reaction mixture is a characteristic and one

http://dx.doi.org/10.5772/59949

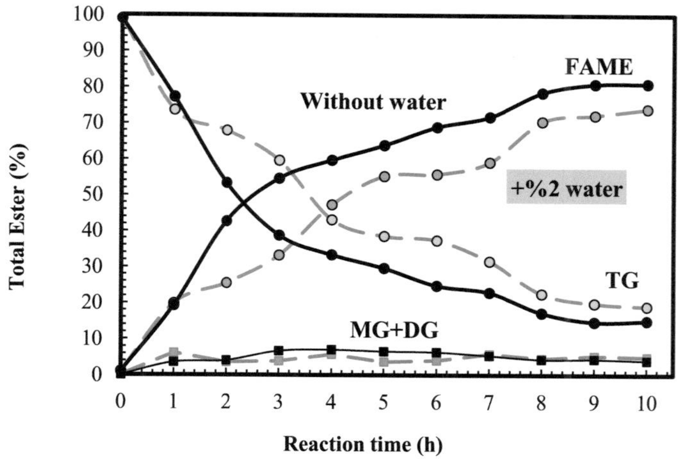

Figure 6. Effect of water content on immobilized *T. lanuginosus* lipase catalyzed transesterification of sunflower oil, 1:3 M ratio of methanol, 30 °C reaction temperature, 10 h reaction time (FAME: Fatty acid methyl ester; TG: triglyceride; DG: diglyceride; MG: monoglyceride)

of the most important factors affecting lipase-catalyzed transesterification reaction rate and yield of biodiesel synthesis [12, 30, 32, 33]. Fukuda et al. [15] reported that the presence of excess water in the reaction mixture reduces the transesterification reaction rate.

4.2.3. *Effect of reaction temperature*

Experiments were performed to determine the effect of temperature on catalytic activity of immobilized *T. lanuginosus* lipase in transesterification reaction. Temperatures in the range of 30-60 °C were examined with results shown in Figure 7. It was found that the enzyme lost its activity dramatically when temperature was increased above 40 °C. Optimal temperature observed for biodiesel production was 30 °C. Studies of biodiesel production performed with immobilized lipase under laboratory conditions have generally indicated use of temperatures between 30-40 °C [20, 26, 34-37]. On the other hand a number of studies have used temperatures between 40-50 °C [5, 9]. Biodiesel yields of enzymes have increased with increasing of temperature, but enzymes have denatureted and decreased in efficiency in most temperatures [26].

4.2.4. *Effect of oil type*

The results depicted in Figure 8 shows that sunflower oil provided the highest methyl ester yield (91.3%) in reactions with methanol, among sunflower, canola, and waste cooking oil. However, the initial reaction rate was higher for canola oil and waste cooking oil than sunflower oil. Free fatty acids formed soaps with alkali salts when alkali-catalyzed process was used to produce biodiesel from waste cooking oils. Use of waste cooking oil in the production of biodiesel with immobilized lipase to cotton cloth has been effective enough in providing substantial methyl ester yield. Since hydrophilic feature of carrier used in immobi-

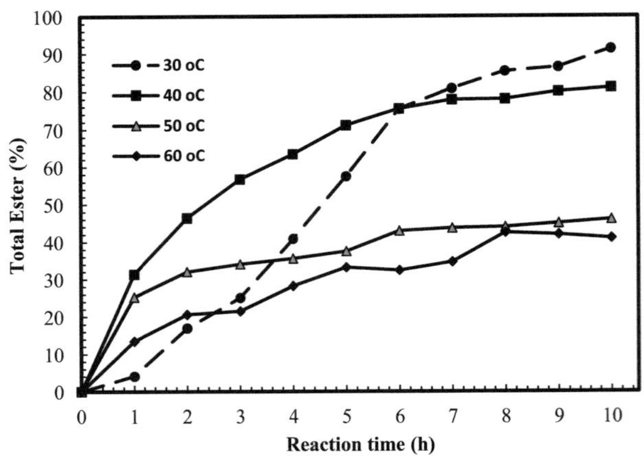

Figure 7. Effect of temperature on immobilized *T. lanuginosus* lipase catalyzed transesterification of sunflower oil, 1:3 M ratio of methanol, 10 h reaction time

lization process may adsorb the water on cotton cloth in the reaction medium. Fatty acid methyl ester yields from canola and waste cooking oil were 79.9% and 81%, respectively.

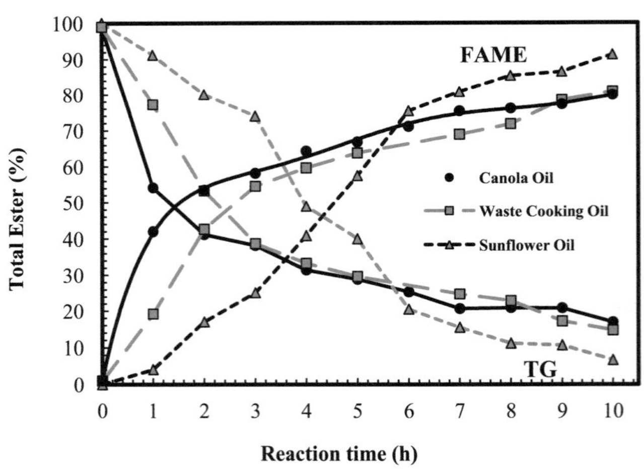

Figure 8. Effect of oil types on immobilized *T. lanuginosus* lipase catalyzed transesterification of sunflower oil, 1:3 M ratio of methanol, 30 °C reaction temperature, 10 h reaction time (FAME: Fatty acid methyl ester; TG: triglyceride)

4.2.5. Effect of washing with tert-butanol

The effect of washing with *tert*-butanol of immobilized cotton cloths during transesterification reaction of sunflower oil and 1:3 molar ratio of oil to methanol is presented in Figure 9. It was shown that immobilized lipase decreased its activity from 91.3% to 77.5% when washed with *tert*-butanol during 10 repeated reactions at 30 °C, each lasting 10 h. However, immobilized lipase decreased its activity dramatically from 91.3% to 61.9% when unwashed with *tert*-

butanol. Activity of lipase increased by washing of immobilized cotton cloths with *tert*-butanol of before methanol addition, since the washing process removed hydrophilic glycerol that occurred during reaction, which couldn't be limited to the diffusion of substrate to lipase molecule. The methanol migrates from the reaction mixture to glycerol layer, and the lipase is inactivated by higher concentration of methanol in the glycerol layer [35].

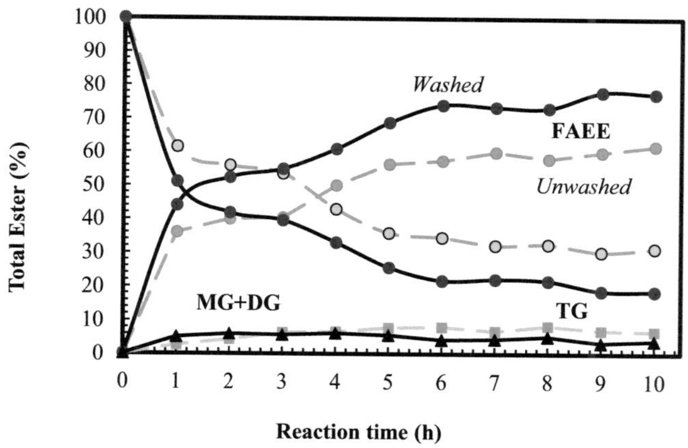

Figure 9. Effect of washing with *tert*-butanol on immobilized *T. lanuginosus* lipase catalyzed transesterification of sunflower oil, 1:3 M ratio of methanol, 30 °C reaction temperature, 10 h reaction time, 10th repeated use (FAME: Fatty acid methyl ester; TG: triglyceride; DG: diglyceride; MG: monoglyceride)

5. Conclusions

The method of enzyme immobilization involving polyethyleneimine (PEI)-enzyme aggregate formation was developed and glutaraldehyde was used as a cross-linking agent between free amine groups increasing the enzyme stability. In the present study, a high biodiesel yield was obtained with *T. lanuginosus* lipase immobilized on cotton cloths. To avoid strong methanol inhibition, it was fed 3 steps into reaction medium and a high conversion (91.3%) was achieved. It was also found that glycerol has a significant inhibitory effect on the transesterificaiton reaction. The immobilized enzyme was rinsed with *tert*-butanol to over come this drawback. Additionally, immobilized lipase could be used at least 10 times (100 h) without any activity limitation. Since the immobilization method is very simple and cheap, it could also be used for the immobilization of other enzymes.

Acknowledgements

The study was supported by the TUBITAK, The Scientific and Technological Research Council of Turkey (Project No: MAG-107M487).

Author details

Eda Ondul[1*], Nadir Dizge[2], Bulent Keskinler[3] and Nedim Albayrak[4]

*Address all correspondence to: edaondul@yyu.edu.tr

1 Yuzuncu Yil University, Department of Food Engineering, Van, Turkey

2 Mersin University, Department of Environmental Engineering, Mersin, Turkey

3 Gebze Technical University, Department of Environmental Engineering, Kocaeli, Turkey

4 Hitit University, Department of Food Engineering, Corum, Turkey

References

[1] Shimada Y, Watanable Y, Samukawa T, Sugihara A, Noda H, Fukuda H, Tominaga Y. Conversion of vegetable oil to biodiesel using immobilized Candida antarctica lipase. J Am Oil Chem Soc 1999;76: 789-93.

[2] Köse Ö, Tüter M, Aksoy HA. Immobilized Candida antarctica lipase-catalyzed alcoholysis of cotton seed oil in a solvent-free medium. Bioresour Technol 2002;83: 125-9.

[3] Pearl GG. Animal fat potential for bioenergy use. Bioenergy 2002. The Tenth Biennial Bioenergy Conference, 22-26 Sept 2002, Boise, ID.

[4] Lai C, Zullaikah S, Vali SR, Ju Y. Lipase-catalyzed production of biodiesel from rice bran oil. J Chem Tech Biotechnol 2005;80: 331-7.

[5] Vargas RM, Sercheli R, Schuchardt U. Transesterification of vegetable oils: a review. J Braz Chem Soc 1998;9(1): 199-10.

[6] Hanna MA, Ma F. Biodiesel production: A review. Bioresour Technol 1999;70: 1-15.

[7] Zhang Y, Dubé MA, McLean DD, Kates M. Biodiesel production from waste cooking oil. 1. Process design and technological assessment. Bioresour Technol 2003;89: 1-16.

[8] Hama S, Yamaji H, Fukumizu T, Tamalampudi S, Kondo A, Noda H, Fukuda H. Biodiesel-fuel production in a packed-bed reactor using lipase-producing Rhizopus oryzae cells immobilized within biomass support particles. Biochem Eng J 2007;34: 273-8.

[9] Chiu CW, Dasari MA, Sutterlin WR, Suppes GJ. Removal of residual catalyst from simulated biodiesel's crude glycerol for glycerol hydrogenolysis to propylene glycol. Ind Eng Chem Res 2006;45: 791-5.

[10] Aarthya M, Saravananb P, Gowthamana MK, Rosea C, Kaminia NR. Enzymatic transesterification for production of biodiesel using yeast lipases: An overview. Chem Eng Res Design 2014;92: 1591-601.

[11] Matsumoto T, Takahashi S, Ueda M, Tanaka A, Fukuda H, Kondo A. Preparation of high activity yeast whole cell biocatalysts by optimization of intracellular production of recombinant Rhizopus oryzae lipase. J Mol Catal-B: Enzym 2002;17: 143-9.

[12] Ivanov AE, Schneider MP. Methods for the immobilization of lipases and their use for ester synthesis. J Mol Catal-B: Enzym 1997;3: 303-9.

[13] Iso M, Chen B, Eguchi M, Kudo T, Shrestha S. Production of biodiesel fuel from triglycerides and alcohol using immobilized lipase. J Mol Catal-B: Enzym 2001;16: 53-8.

[14] Noureddini H, Gao X, Philkana RS. Immobilized Pseudomonas cepacia lipase for biodiesel fuel production from soybean oil. Bioresour Technol 2005;96: 769-7.

[15] Fukuda H, Hama S. Tamalampudi S, Fukumizu T, Miura K, Yamaji H, Kondo A. Lipase localization in Rhizopus oryzae cells immobilized within biomass support particles for use as whole-cell biocatalysts in biodiesel-fuel production. J Biosci Bioeng 2006;101(4): 328-3.

[16] Wang F, Gao Y, Tan TW, Nie KL. Immobilization of lipase on macroporous resin and its application in synthesis of biodiesel in low aqueous media. Chin J Biotechnol 2006; 22(1): 114-8.

[17] Mittelbach M. Lipase catalyzed alcoholysis of sunflower oil. J Am Oil Chem Soc 1990;61(3): 168-70.

[18] Okuba M, Takahashi M. Production of submicron-size monodisperse polymer particles having aldehyde groups by the seeded aldol condensation polymerization of glutaraldehyde (II). Colloid Polym Sci 1994;272: 422-6.

[19] Nelson LA, Foglia TA, Marmer WN. Lipase-catalyzed production of biodiesel. J Am Oil Chem Soc 1996;73: 1191-95.

[20] Kaieda M. Samukawa T, Matsumoto T, Ban K, Kondo A., Shimada Y., Noda H, Nomoto F, Ohtsuka K, Izumoto E, Fukuda H. Biodiesel fuel production from plant oil catalyzed by Rhizopus oryzae lipase in a water-containing system without an organic solvent. J Biosci Bioeng 1999; 88(6): 627-31.

[21] Watanable Y, Shimada Y, Sugihara A, Noda H, Fukuda H, Tominaga Y. Continuous production of biodiesel fuel from vegetable oil using immobilized Candida antarctica lipase. J Am Oil Chem Soc 2000;77(4): 355-9.

[22] Abigor RD, Uadia PO, Foglia TA, Haas, KC, Okpefa JE, Obibuzor JU. Lipase-catalyzed production of biodiesel fuel from some Nigerian lauric oils. Biochem Soc Trans 2000;28: 979-81.

[23] Kamini NR, Lefuji H. Lipase catalyzed methanolysis of vegetable oils in aqueous medium by *Cryptoccus* spp. S-2. Process Biochem 2001;37: 405-10.

[24] Lara-Pizarro AV, Park EY. Lipase-catalyzed production of biodiesel fuel from vegetable oils contained in waste activated bleaching earth. Process Biochem 2003;38: 1077-82.

[25] Albayrak N, Yang ST. Production of galacto-oligosaccharides from lactose by Aspergillus oryzae β-galactosidase immobilized on cotton cloth. Biotechnol Bioeng 2002;77(1): 8-19.

[26] Shimada Y, Watanable Y, Sugihara A, Tominaga Y. Enzymatic alcoholysis for biodiesel fuel production and application of the reaction to oil processing. J Mol Catal-B: Enzym 2002;17: 133-42.

[27] Teixeira CB, Junior VM, Macedo GA. Biocatalysis combined with physical technologies for development of a green biodiesel process. Renew Sustain Energ Rev 2014;33: 333-43.

[28] Tan T, Lu J, Nie K, Zhang H, Deng L, Wang F.. Progress on biodiesel production with enzymatic catalysis in China. Sheng Wu Gong Cheng Xue Bao 2010;26(7): 903-6. Review. Chinese.

[29] Bajaj A, Lohan P, Jha PN, Mehrotra R. Biodiesel production through lipase catalyzed transesterification: an overview. J Mol Catal B Enzym 2010; 62: 9-14.

[30] Antczak MS, Kubiak A, Antczak T, Bielecki S. Enzymatic biodiesel synthesis-key factors affecting efficiency of the process. Renew Energ 2009;34: 1185-94.

[31] Salis A, Pinna M, Monduzzi M, Solinas V. Biodiesel production from triolein and short chain alcohols through biocatalysis. J Biotechnol 2005;119: 291-9.

[32] Balcio VM, Paiva AL, Malcata FX. Bioreactors with immobilized lipases: State of the art. Enzyme Microb Tech 1996;18(5): 392-416.

[33] Wehtje E, Adlercreutz P. Lipases have similar water activity profiles in different reactions. Biotechnol Lett 1997;19: 537-40.

[34] Samukawa T, Kaieda M, Matsumoto T, Ban K, Kondo A, Shimada Y, Noda H, Fukuda H. Pretreatment of immobilized Candida antarctica lipase for biodiesel fuel production from pant oil. J Biosci Bioeng 2000;90(2): 180-3.

[35] Kaieda M, Samukawa T, Kondo A, Fukuda H. Effect of methanol and water contents on production of biodiesel fuel from plant oil catalyzed by various lipases in a solvent-free system. J Biosci Bioeng 2001;91: 12-5.

[36] Dossat V, Combes D, Marty A. Continuous enzymatic transesterification of high oleic sunflower oil in a packed bed reactor: influence of the glycerol production. Enzyme Microb Tech 1999;25: 194-200.

[37] Du W, Xu Y, Liu D, Zeng J. Comparative study on lipase-catalyzed transformation of soybean oil for biodiesel production with different acyl acceptors. J Mol Catal-B: Enzym 2004;30: 125-9.

[38] Abdulla R, Ravindra P. Immobilized Burkholderia cepacia lipase for biodiesel production from crude Jatropha curcas L. oil. Biomass Bioenergy 2013;56: 8-13.

[39] Ondul E, Dizge N, Albayrak N. Immobilization of Candida antarctica A and Thermomyces lanuginosus lipases on cotton terry cloth fibrils using polyethyleneimine. Colloids Surf B 2012;95: 109-14.

[40] Christopher PL, Kumar H, Zambare VP. Enzymatic biodiesel: Challenges and opportunities. Appl Energ 2014;119: 497-520.

[41] Al-Zuhair S, Almenhali A, Hamad I, Alshehhi M, Alsuwaidi N, Mohamed S. Enzymatic production of biodiesel from used/waste vegetable oils: Design of a pilot plant. Renew Energ 2011;36: 2605-14.

[42] Gog A, Roman M, Tos M, Paizs C, Irimie FD. Biodiesel production using enzymatic transesterification: Current state and perspectives. Renew Energ 2012;39: 10-6.

[43] Zeng J, Du W, Liu X, Liu D, Dai L. Study on the effect of cultivation parameters and pretreatment on Rhizopus oryzae cell-catalyzed transesterification of vegetable oils for biodiesel production. J Mol Catal B: Enzym 2006;43: 15-1.

[44] Hamaa S, Kondo A. Enzymatic biodiesel production: An overview of potential feedstocks and process development. Bioresour Technol 2013;135: 386-95.

[45] Salis A, Pinna M, Monduzzi M, Solinas V. Comparison among immobilised lipases on macroporous polypropylene toward biodiesel synthesis. J Mol Catal B: Enzym 2008;54: 19-26.

[46] Ranganathan SV, Narasimhan SL, Muthukumar K. An overview of enzymatic production of biodiesel. Bioresour Technol 2008;99: 3975-81.

[47] Guldhe A, Singh B, Mutanda T, Permaul K, Bux F. Advances in synthesis of biodiesel via enzyme catalysis: Noveland sustainable approaches. Renew Sustain Energ Rev 2015;41: 1447-64.

[48] Deng L, Xu X, Haraldsson GG, Tan T, Wang F. Enzymatic production of alkyl esters through alcoholysis: A critical evaluation of lipases and alcohols. J Am Oil Chem Soc 2005;82: 341-7.

[49] Nielsen PM, Brask J, Fjerbaek L. Enzymatic biodiesel production: Technical and economical considerations. Eur J Lipid Sci Technol 2008;110: 692-700.

[50] Turkan A, Kalay S. Monitoring lipase-catalyzed methanolysis of sun flower oil by reversed-phase high-performance liquid chromatography: Elucidation of mechanisms of lipases. J Chromatogr A 2006;1127: 34-44.

Chapter 3

Technologies for Biodiesel Production in Sub-Saharan African Countries

Omotola Babajide, Leslie Petrik and Farouk Ameer

Additional information is available at the end of the chapter

http://dx.doi.org/10.5772/59859

1. Introduction

The importance and relevance of generating liquid fuels from biomass and producing biodiesel from vegetable oils given the enormous environmental advantages the process offers cannot be overemphasized. The increasing popularity of biodiesel has generated a great demand for successful commercial production methods, which, in turn, calls for the development of technically and economically sound process technologies. The applicability of various technologies employed currently for biodiesel production under diverse catalytic and process systems is explored, with particular attention paid to sustainable protocols for sub-Saharan Africa. Africa holds abundant renewable energy sources with an estimated 1.1 million gigawatts of potential hydropower capacity, geothermal and solar energy as well as substantial biomass potential. Irrespective of this, however, in reality there appears to be a rather limited utilization of these renewable energy sources as modern renewable energy sources account for less than 2% of primary energy demand in sub-Saharan Africa as far back as in the 1990s [1]. Sub-Saharan Africa also has significant amounts of renewable energy that could still be exploited, as renewable energy technologies (RETs) have demonstrated their capacity to meet energy needs where the conventional energy supply options have proved unsuccessful. Modern biomass technologies in Africa need to be developed to their optimum potential as this specifically would offer sub-Saharan Africa self-reliance with respect to energy supplies at the national and local levels with economic, ecological or environmental, social and security benefits [2]. Energy consumption in Africa is largely dominated by combustible renewable resources and if Africa were to take its rightful place in the developing renewable energy market, the advantages would be abundant. A majority of African countries obtain their main household energy resource from biomass used for various activities such as cooking, drying and space heating. According to reports by Energy for Sustainable Development in Africa, primary energy consumption has been increasing steadily to about 24% [3]. While liquid fuels remain the largest source of energy, fossil fuels are expected to meet much of the global energy required for sustainable development. Predictable high world oil prices in the nearest future, though prices are currently very low, may lead many energy users to shift from liquid fuels when feasible will make a way for renewable energy sources [4]. Among the many possible resources, biodiesel has received the most attention as a promising substitute for conventional petrol-diesel fuel.

2. Biofuels

Biofuels are energy sources that are produced from biomass—the living matter of plants or organic waste. Biofuel comprises of liquid or gaseous fuels used basically for the transport sector, which are primarily produced from biomass. They include ethanol, methanol, biodiesel, hydrogen and methane, respectively. Liquid biofuels to fuel vehicles, engines while gaseous fuels cells for electricity generation. An estimated 54 billion litres of biofuels was recorded in 2007 as a result of increased world production and these gains meant biofuels accounted for 1.5% of global supply of liquid fuels, up 0.25 % from the previous year [5]. First generation biofuels include two main types:

- Bioethanol produced from plant sugars of biofuel crops, such as sugarcane and maize, which ferments to produce ethanol. The ethanol produced can be further blended with 5–10% gasoline fuel for use in normal cars; however, higher percentages of ethanol are needed for specially adapted cars [6].
- Biodiesel synthesized from a chemical reaction between triglycerides and an alcohol. The use of triglycerides without any modification directly in modified diesel engines cut biodiesel processing costs and eliminates excessive glycerol by-product from the process.

The transesterification of vegetable oils has been widely known and employed since the 19th century. In fact, the process currently in use for making biofuels from biomass still follows former protocols and the feedstock utilized for their preparation remain very similar. Peanut, hemp and corn oil and animal tallow conventionally used for biodiesel have been partially replaced by soybean; rapeseed, recycled oil while forest wastes, trees and sugarcane are used for bioethanol. According to Luque et al. [7], the history of biofuels is however more political and economic than technological. In the 18th century, Rudolph Diesel firstly demonstrated using peanut oil as diesel for his compression ignition engine at a world exhibition in Paris. These particular vegetable oils were employed as fuels in diesel engine before engine alterations were made, enabling the use of currently known diesel fuel. Biofuels were believed to would offer a solution to fuel issues in the transportation industry not only by Diesel but also by Henry Ford who was convinced that renewable resources were essential to the success of future automobile designs to be run with ethanol. During the World War II, countries engaged in the war utilized biomass fuels in their machines and despite their use during this time, biofuels have remained inconsequential due to the discovery of fossil fuels and a concrete petrol-based industry that made the world very dependent on petroleum. The potential of biofuels reappeared in public awareness, brought back by various bioenergy programs established in Brazil in 1975 and international organizations such as the International Energy Agency (IEA) Bioenergy, established in 1978 by the organization for Economic Co-operation and Development (OECD), which aimed to expand bioenergy research, development and implementation between countries that have related national programs. The United Nations (UN) International Biofuels Forum is constituted by a number of countries including Brazil, China, India, the United States, and European Union along with South Africa, which is the only country from the African continent. The United States, Brazil, France, Sweden and Germany are world leaders in biofuel development and consumption. The worldwide advancement of the biofuels industry, however, over the recent years has mostly been triggered by a global crisis produced by the increased use of fossil fuels, resulting in limitation of supply, previously high inflationary prices and negative environmental impact. However, towards 2008, interest in biofuels underwent a huge surge largely accredited to the compulsion towards mandatory fuel blends in Europe and other developed nations [8]. Sub-Saharan Africa's biofuel production potential was emphasized by several research studies, which drove the intense interest for African biofuel investments [9–11] and stimulated international investors to acquire large regions of land for feedstock plantations in sub-Saharan Africa [12–14]. The environmental and social sustainability of biofuels has been under extensive scientific scrutiny with initiatives emerging to develop guidelines for sustainable production according to the communiqué of the Roundtable on Sustainable Biofuels [15,16] with regards the market regulations from consumer countries and producer countries (e.g. South Africa and Mozambique) [17]. According to Mwakasonda, Southern Africa possesses largely untapped potential of biofuels in the world [18] with biodiesel development in sub-Saharan Africa still regarded to be in its initial stages. This has initiated deliberations and discussion among policy makers, development practitioners and other stakeholders to foster the development of biofuels in Africa. Bio-fuels are considered as a source of foreign exchange especially for oil-deprived countries via the development and use of locally produced renewable fuel, reduction of demand for imported petroleum, boosting of local agriculture production and additional markets and revenue to farmers, leading consequently to the increase of some rural people's purchasing powers and quality of life, beneficial environmental impact through the use of organic municipal solid waste materials to generate a higher value end-product, reduced level of carbon dioxide emitted by motor engines and subsequent preservation of the atmosphere [19]. It is worthy of mention that despite these benefits, biofuels/biodiesel production in sub-Saharan Africa has met with challenges and caused some unforeseen problems. For example, in several African countries, the biofuel policy priorities at national level revolved around energy security and rural development as well as attempts to take advantage of carbon finance where available. Despite policy and investors interest, several early biofuel ventures collapsed. According to research studies carried, there is evidence to suggest that lack of proper agronomic knowledge, lack of appropriate institutions to regulate nascent biofuel sector, investor caution, lack of market development and a growing understanding of the potential environmental and socio-economic impacts has left poor local communities poor or even poorer [20]. Studies have also shown that poverty outcomes are directly related to the loss of access of local communities to natural ecosystems implying that that there may be significant linkages between the environmental and socioeconomic performance of biofuel projects [21].

2.1. Biodiesel

Biodiesel presents the following advantages: improved fuel performance, increased lubricity, higher cetane number and flashpoint as compared to fossil-diesel, lower toxicity to living organisms, reduced exhaust emissions, and its versatility for use as fuel [22–25]. It is a local renewable source of energy that reduces importation of energy and affords improved security of energy supply. It is highly biodegradable [26, 27]. It also improves the quality of the environment with less harmful soot generated from the exhaust of vehicles. Biodiesel when mixed with fossil diesel creates a biodiesel blend suitable for diesel engines with minimal modifications as the superior lubricity of biodiesel increases functional engine efficiency; with a low viscosity and lower carbon monoxide emissions [28], biodiesel is user-friendly, non-toxic, and free of sulfur [29] and aromatics [30]. Possessing a high flash point eases storage and the presence of a higher amount of oxygen in biodiesel fuel guarantees the complete combustion of hydrocarbons. The production of biodiesel (methyl esters) from vegetable oils represents an alternative means of producing liquid fuels from biomass, and one which is growing rapidly in commercial importance and relevance due to the fluctuations in petroleum prices and the environmental advantages the process offers. Biodiesel can be produced from a variety of feedstock, vegetable oils, waste cooking oils and animal fats. These oils typically consist of C_{14}–C_{20} fatty acid triglycerides. In order to produce a fuel that is suitable for use in diesel engines, these triglycerides are usually converted into the respective mono alkyl esters by base-catalyzed transesterification with short chain alcohol, usually methanol.

3. Catalytic processes

Catalysts are mainly classified as homogeneous or heterogeneous groups. Homogeneous catalysts act in the same phase as the reaction mixture, whereas heterogeneous catalysts act in a different phase from the reaction mixture. Heterogeneous catalysts have the advantage of easy separation and reuse due to the advantage of being in a different phase from the reaction medium. Presently, the biodiesel industry is dominated by homogeneous catalytic applications due to the basic convention and less time required for the conversion of oils to their respective methyl esters. NaOH and KOH are mainly used because they are easily soluble in methanol, forming sodium and potassium methoxide respectively while enhancing transesterification reactions to completion. When the acid value (AV) of the oil is high, an acid catalyst such as hydrochloric acid or sulfuric acid is used to lower the AV and then the alkali catalyst is utilized for biodiesel synthesis.

3.1. Homogeneous/heterogeneous catalysis

Conventional biodiesel production is done via base-catalysed transesterification using homogeneous alkaline catalysts. This is the most commonly used technique as it is considered to be the most economical process [31]. The fact that homogeneous catalysts cannot be reused makes conventional heterogeneous catalysis preferable as it offers a series of advantages among which are easy separation of products, reusability of catalysts and reduced amount of waste water during biodiesel production. Heterogeneous catalysis is thus considered to be a green process, as the purification steps of products are much more simplified with high yields of methyl esters being achieved [32]. Economically, heterogeneous catalysis poses a better alternative with regards the reduction of the costs associated with the purification and separation of reaction products in biodiesel production. This development could lead to a cost-effective process that is environmentally friendly. For example, the high catalytic activity of calcium containing natural and biological materials (eggshells, limestone calcite, cuttlebone, dolomite and hydroxyapathite) for transesterification reactions have been reported [33]. Waste coal fly ash-based catalysts and fly ash-based zeolites have also proved to be successful catalysts and offer an alternative route for the utilization of South African fly ash in transesterification reactions for biodiesel production [34, 35].

3.2. Process systems

3.2.1. Ultrasound technology in biodiesel production

Chemical reactions involving ultrasound homogenization/technology has been studied and has developed as an expanding research area in the field of biodiesel production. Utilizing ultrasound technology for producing biodiesel employs ultrasonic field known to produce chemical and physical effects that arise from the collapse of cavitation bubbles. The mixing intensity is considered a very important parameter to enhance a progressive transesterification reaction due to the fact that oil and alcohol are immiscible. Thus low-frequency sonication can be used to create emulsions from immiscible liquids and this effect could be employed for biodiesel preparation. The use of ultrasound technology has thus proven to provide a latent alternative to conventional biodiesel production process via transesterification by initiating substantial mechanical energy required for mixing and activating the transesterification reaction for biodiesel synthesis [36]. One major problem, however, in the transesterification of oils using methanol is the fact that alcohol is a poor solvent for fatty materials. Thus the need for homogenization of the reaction mixture since this reaction can only occur in the interfacial region between the liquids (due to the fact that fats and alcohols are not totally miscible); hence, the application of vigorous mixing is required to increase the area of contact between the two immiscible phases. This method of homogenization has been found successful for both biodiesel production batch processes and continuous operation [37–39]. The use of ultrasound technology in biodiesel production presents an efficient, swift and economically functional process with reduced reaction time, static separation time as well as generally higher yields over the conventional process [38, 40]. The effects of use of mechanical stirring, ultrasound technology and hydrodynamic cavitation on soybean biodiesel yields were studied under the following parameters (catalyst, KOH; feedstock, soybean oil; solvent, methanol; alcohol/oil molar ratio of 6:1, temperature of 45°C) by Lifka et al. [41] and Ji et al. [38]. Results obtained showed that the application of ultrasound technology enabled faster reaction periods and higher yields in comparison to mechanical stirring. Conversely, Thanh et al. [42] report identical biodiesel yields during alkaline transesterification of vegetable oils via both mechanical stirring and ultrasound processes using KOH. Employing ultrasound processes and solid catalysts in biodiesel synthesis as compared to conventional batch processes has also been found to reduce reaction time drastically. Kumar et al. [43] also report a 98.53% biodiesel yield when a combination of an ultrasound process and a solid catalyst is employed as compared to a conventional batch process. Figure 1 shows a schematic of biodiesel production using ultrasound technology. Production of biodiesel under the ultrasonic processing possess the following advantages, reduction in processing time, less amount of alcohol, minimal catalyst, faster separation time and reduced reaction temperature as revealed in several studies [37–43].

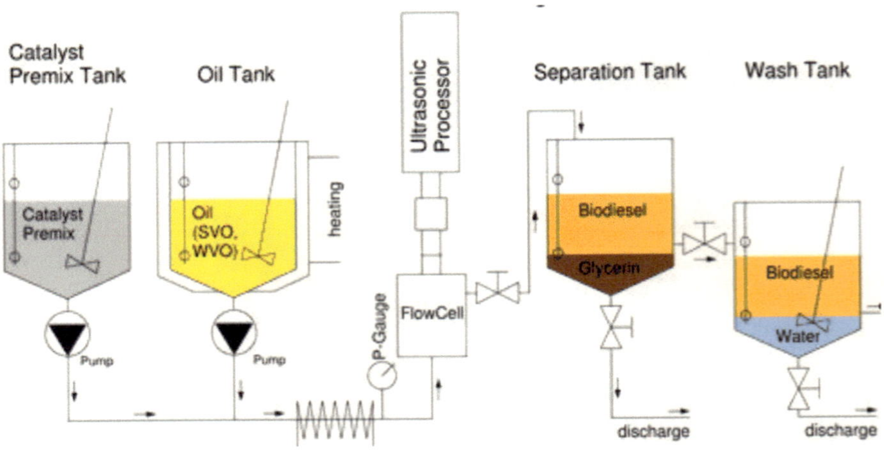

Figure 1. Schematic of biodiesel production using an ultrasound (Reproduced with permission from ©Hielscher Ultrasonics)

3.2.2. Application of jet mixing in biodiesel production

The application of jet mixing in commercial biodiesel production has been developed and applied to improve mixing and enhance mass transfer between vegetable oils/ animal fats with methanol/ethanol in stirred tank reactors in the presence of basic or acidic catalysts [44]. The challenges related to conventional biodiesel production processes include the limitation of reaction rate by mass transfer between triglycerides and alcohol due to factors of immiscibility; high conversion limitations in the absence of product removal since transesterification itself is a reversible reaction; and disadvantage of batch mode operations of small and medium-scale biodiesel commercial processes (against the advantages of continuous operation) necessary for the development of process intensification technologies. This technology has been applied successfully in South Africa for commercial biodiesel production and has recorded shorter reaction profile, low molar ratio of alcohol to oil as well as lower catalyst concentrations. The operating cost and energy consumptions required to purify biodiesel could be minimized while the recovery of excess alcohol and catalyst by-products during downstream processing made more efficient.

http://dx.doi.org/10.5772/59859

Figure 2. Stirred tank reactors using the jet mixing technology

3.2.3. *Jet reactor*

A variety of practical engineering applications have employed the use of impinging jets to enhance heat transfer due to the high local heat transfer coefficient it presents. These applications include quenching of metals and glass, cooling of turbine-blades, cooling and drying of paper as well as the cooling of electronic equipment. Extensive review and numerous studies that expound heat transfer enhancement via impinging jets can be found in literature [45-47]. Jet rectors are reactors based on the impinging jet technology. This system provides intense mixing under pressure with different nozzle sizes which were specially developed for steel cutting (with water) and further developed for the mining sector especially in the goldmine sector of South Africa. The application of jet reactors as shown in Figure 2 was further developed in a continuous process reactor mainly for the production of biodiesel in South Africa by Nieuwoudt [48]. The jet-loop system (schematic shown in Figure 3) has been optimized in different studies and upscaled in medium- and large-scale biodiesel commercial plants in southern Africa [48].

3.2.4. *Membrane technology*

Biodiesel production using membrane reactors is a new concept that is still being tested for optimal conditions. However, there are numerous prospects to finding paramount permutation between catalyst and membrane. Design and optimization studies will be required to improve the membrane reactor for commercial small-scale operations, especially in sub-Saharan Africa. Therefore, a process is required to simultaneously overcome the shortcom-

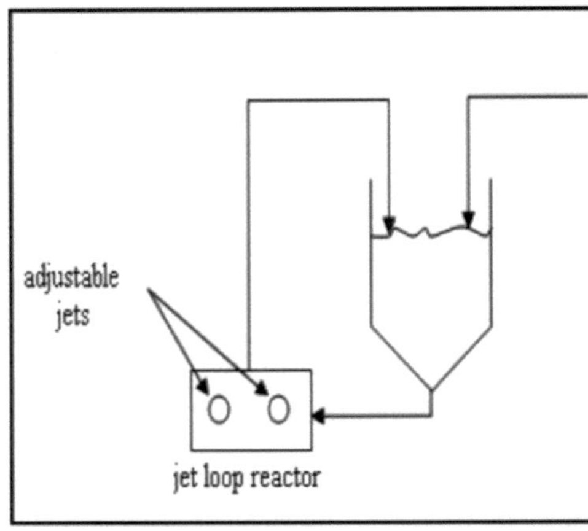

Figure 3. Schematic diagram of the jet reactor

ings of feedstock and the use of homogenous catalysts with the aid of membrane technology and heterogeneous catalysts. Different chemical reaction processes has been successfully applied using the membrane reactor technology, which suggests an apparent success in biodiesel production. The transesterification of lipids is a classic reversible chemical reaction that could also be combined with membrane reactor technology [49]. These membranes can be either organic in nature (i.e. polymeric) or inorganic, with inorganic membranes being better than the former in terms of their excellent thermal stability [50]. It has also been found out that different pore-sized membranes could retain canola oil using a reactor with a high purity of canola biodiesel obtainable. This method (as shown in Figure 4) has clear advantages over conventional means as it ends with a fatty acid methyl ester (FAME)-rich phase, a controlled contact of incompatible reactants, and an elimination of undesired side reactions. There is also an integration of reaction and separation into a single process, thereby reducing separation costs and recycle requirements, and an enhancement of thermodynamically limited or product inhibited reactions resulting in higher conversions. The FAME-rich phase still contains FAME, methanol, glycerol and water, causing a problem of downstream processing in terms of separation [50, 51]. This work by Dube et al. clearly demonstrates that a membrane reactor can be used to alleviate many of the difficulties highlighted and successfully carry out the transesterification of lipids to biodiesel. Preliminary laboratory testing on the application of membrane technology has also been conducted by a group of researchers in South Africa. The process developed was able to simultaneously overcome the shortcomings of waste cooking oil feedstock and the use of homogenous catalysts with the aid of membrane technology and heterogeneous catalysts. The membrane was remarkably able to block un-reacted feed and impurities from entering the permeate, which gave rise to a higher purity FAME product [52].

3.2.5. Reactive distillation technology

Reactive distillation can be applied successfully in biodiesel production as a potent process intensification technique since the reactions leading to the end-product are controlled by chemical equilibrium. This process has been found to be highly advantageous in esterification-

http://dx.doi.org/10.5772/59859

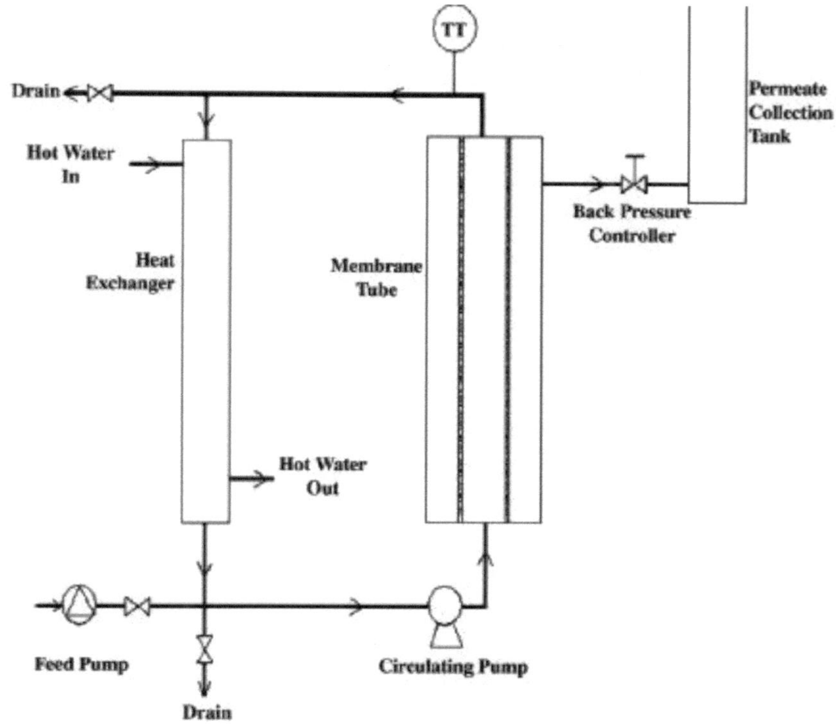

Figure 4. Schematic diagram of a separation membrane reactor [52]

type reactions with high free fatty acid feedstock [53]. The use of excess methanol becomes unnecessary with this method as this can shift the reaction equilibrium towards ester production by continuous removal of the water by-product [54]. An additional flash evaporator and a decanter are used to guarantee the high-purity biodiesel product from multiple feedstocks, which may include a biodiesel reactor, a decanter, a flash evaporator and a distillation column. Since methanol and water are much more volatile than the fatty ester and acid, these will separate easily at the top [55]. Researchers have considered various aspects of this technique, including optimization of reaction conditions, heat integration, use of thermally coupled distillation columns [56] as well as dual reactive distillation processes [56] with respect to biodiesel catalysis. A complex distillation column with a side rectifier has been shown capable of carrying out a reactive distillation process for the production of biodiesel as demonstrated in a novel integrated reactive separation process for FAME synthesis (Figure 5). This integrated biodiesel process is based on reactive separations powered by heterogeneous catalysts offering significant advantages such as minimal capital investment and operating costs, as well as limited catalyst-related waste streams and eliminating soap formation. This novel technology reported efficiently uses the raw materials (including low-cost feedstock, i.e. waste cooking oil) while considerably reducing the energy requirements for biodiesel production—85% lower compared to the baseline studies [57]. Using pure free fatty acid as feedstock, Kiss et al. [54] report an energy saving of up to 45% due to heat integration inclusion and a significant reduction in steam consumption in the raw material pre-heaters. This process presents the following advantages over conventional biodiesel production processes, shorter reaction time, high unit productivity, no additional alcohol requirement, lower capital and operational costs (due to the elimination of additional separation units as a single column is adequate), elimination of neutralization and separation steps of catalysts when solid acid catalysts are used. It also offers additional substantial advantages, such as higher reaction rate and selectivity, avoidance of azeotropes and reduced energy consumption as well as solvent usage [51]. The apparent advantage that this method has over other methods of processing biodiesel stimulates further study, especially for continuous high-volume production in sub-Saharan Africa.

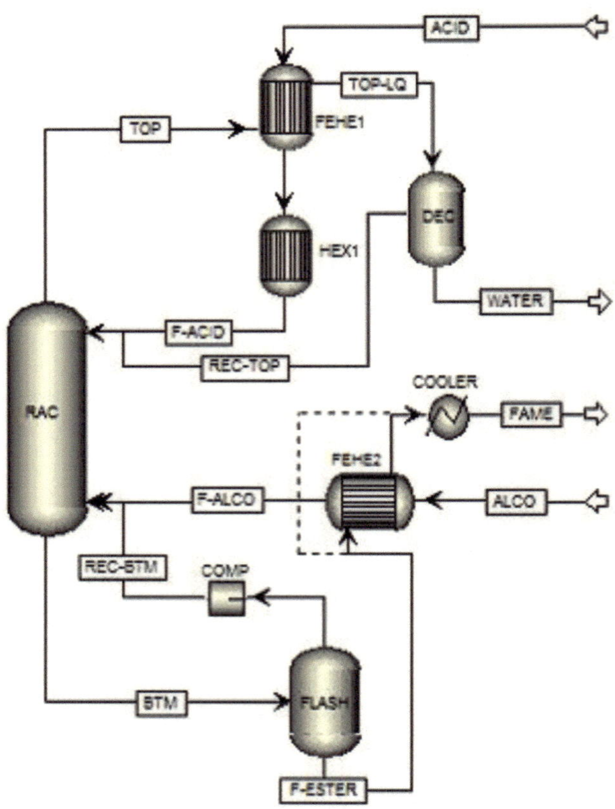

Figure 5. Schematic diagram of an integrated reactive separation process for FAME synthesis [57]

4. Feedstock: Potential and requirement

A major strategy to ensure the sustainability and success of the biodiesel industry is reliant upon the availability of adequate supplies of reasonably priced feedstock. Raw material being the main driver in determining the total production cost of biodiesel, there is therefore need to focus attention on the type, availability and the use of raw materials (which include vegetable or animal fat and oils). Another issue to contend with is the competition between food production and energy production as strong views are expressed in research and academic circles. In order for the production of biodiesel to be sustainable in sub-Saharan Africa, the choice of feedstock must correlate with the availability of such in specific environments. A brief biofuel potential in the SADC region of Africa is presented in Table 1. The various potential feedstock for biodiesel production that are currently available in southern Africa, which can be considered suitable, include but not limited to cashew nut, sesame seeds, castor oil, pumpkin, rapeseed, avocados, coconut, soybean, cotton seed, sunflower and maize. A highlight of a few of these feedstocks is done in the following sections.

Country	Population (million)	Land size (million Ha)	Forest area	Forestry (%)	First generation biofuels	Next generation biofuels	Potential for next generation biofuels
Angola	12.53	124. 67	59.104	47.41	●		●
Botswana	1.84	56.673	11.943	21.07	●		●
Comoros	0.73	0.217	0.005	2.30			
Congo D.R.	66.52	234.541	133.610	56.97	●		●
Madagascar	20.04	58.704	12.838	22.08	●		●
Malawi	13.93	11.848	3.402	36.16	●		●
Mauritius	1.27	0.204	0.037	18.23	●		●
Mozambique	21.29	80.159	19.262	24.57	●		●
Namibia	2.09	82.542	7.661	9.31	●		●
South Africa	43.79	121.991	9.203	7.58	●		●
Swaziland	1.13	1.736	0.541	31.45	●		●
Tanzania	40.21	94.509	35.257	39.90	●		●
Zambia	11.67	75.261	42.452	57.11	●		●
Zimbabwe	12.38	39.058	17.540	45.34	●		●
TOTAL =	251.55	860.479	352.863	41.01	●		●

Table 1. Status of biofuel potential in SADC region [58]

4.1. Canola

Canola oil is an efficient biodiesel feedstock with excellent cold-flow properties due to the low saturation of its triglyceride content. About 44% oil can be extracted from the canola seed when crushed as compared to only 18 % for soybeans; a relatively popular biodiesel feedstock and 30% for sunflower. A great advantage that the use of canola as a feedstock source offers is that nothing goes to waste. The oil cake, which accounts for about 60% of the by-product, can be used as a protein-rich animal feed and the leftover glycerol from the oil can be used in producing soaps, cosmetics and other personal care products. ELIDZ is embarking on a canola-based biodiesel project together with the Eastern Cape Development Corporation, AsigSA along with the Department of Agriculture [59]. The project is expected to generate a number of jobs in the rural areas. Canola has the advantage of being a nitrogen-fixing winter crop, which can be alternated with maize, thereby increasing the maize yields, which would have the added benefit of increasing food security. On the topic of food security, a larger proportion of maize, which is currently excluded from biofuel production, can be utilized in the food production process as animal feed at a lower cost (since the raw product has already created other income and requires less transportation) and could additionally result in lower meat production costs.

4.2. Sunflower and soybean

Other biodiesel feedstock sources include the sunflower and soybean and they are extensively cultivated in the South African Development Community (SADC) region of Africa. Their impact on employment is high as they are generally grown by both large and smallholder farmers for food crop or as industrial crops for small- and medium-scale enterprises in biodiesel [60]. This expanded use for the biodiesel industry creates surplus demand and encourages large productions. With respect to yields, sunflower offer greater yields but at a higher price than soybean as yields are highly influenced by seed selection, plant density as well as pest and weed control.

4.3. Jatropha

Jatropha is regarded as a bio-energy feedstock without adequate scientific knowledge on the shrub and this has resulted in disappointed farmers in countries like Mozambique and Zambia, where out-grower schemes have failed [61]. Southern Africa's climatic conditions favour the production of a wide range of bio-fuel feedstocks but a study on preferred feedstocks carried out by WWF in five SADC countries prioritized Jatropha, sweet sorghum and sugarcane [62]. Jatropha was the most preferred feedstock largely because of its portrayal as a "miracle crop" that grows on marginal soils with limited to no management. The shrub grows wild but can be cultivated for bio-diesel production. It is drought tolerant, suited to well-drained soils and survives on a wide range of terrains and soil types. Its seed oil and other vegetative parts are, on the other hand, poisonous [63]; however, high temperature treatment can reduce toxicity. Oil from the seed can be processed into bio-diesel, soap and candles. In Malawi, Zambia and Zimbabwe, Jatropha is grown as a live fence/hedge by smallholder farmers. It has been established under plantation conditions by private companies in countries such as Mozambique and Tanzania. However, it poses a number of challenges. Its commercial cultivation is yet to take off and very few large-scale commercial plantations have been harvested, processed and reported [64]. It is worthy of mention that Jatropha's agronomic requirements, seed yields and economic returns are largely unknown. The political position towards Jatropha, however, has been strengthened by a national government initiative to support bio-diesel production while the benefits from job creation as well as the use of the end product could come in very positively, particularly for farmers. In addition to the oil produced from Jatropha, the cake remaining after the seeds are processed is a good organic fertilizer after composting it and can be used to make paper, cosmetics, toothpaste, embalming fluid, and cough medicine, among other items as well. One concern however is that the seeds are highly flammable and therefore the process should not be located near to any sugar or paper producing operations.

4.4. Algae

Algae can be grown using waste materials such as sewage and without displacing land currently used for food production. The production of algae to harvest oil for biodiesel has not been conducted on a commercial scale, but the potential is promising. This "second generation" biodiesel feedstock has the potential to dramatically expand the resource base for the production of biodiesel in the future, hence contributing to solving complications of air pollution from CO_2 and alleviating crises of food competition through energy production. A comprehensive study by Thurmond [65] found that algae and the potential of microalgae as an alternative and sustainable energy source) may offer an immense resolution to meeting large-scale and sustainable feedstock supply in developed continents like North America, Europe and Asia, which may thereafter be applicable to Africa. Recent reports show that a genetically engineered bacterium developed by scientists in the United States can produce ethanol biofuel from coarse, wild growing switch grass rather than using vital food crops such as maize [66].

4.5. Waste oil

Recycled oil is a primary feedstock used for the production of biodiesel in most small- to medium-scale biodiesel plants in Southern Africa; however, its insufficient availability presents a big limitation to a large-scale biodiesel production process. Due to the numerous usages of recycled oil (e.g. yellow grease), its collection for use as various other potential feedstock (for the manufacture of soap, cleansing creams, inks, glues, solvents, paint thinner, rubber, lubricants and detergents) makes it highly competitive. The promising potential of using recycled oils as a livestock feed additive has also been identified as it makes livestock feed look fresh and lubricated while also reducing wear and tear on milling machinery. It is worthy of mention to note that apart from quantity constraint, the quality of the oil could have a knock on effect on the quality of the biodiesel produced [67]. A viable biodiesel industry in southern Africa cannot survive primarily on recycled oils due to their limited supply and availability. To meet the long-term feedstock need of the industry in southern Africa, a dedicated biodiesel oil-seed will need to be identified and developed. It is proposed that the best (monopolistic) feedstock should be one that empowers all players across the board from small to large, at the same time a breeding ground for sustainable biofuels industry.

5. Conclusion

In light of the discussions and review presented in this chapter, it is critical for indigenous biodiesel industry to continuously improve on aspects that will strengthen its prospects with respect to better market penetration and production technologies. Having highlighted specific issues related to newly improved biodiesel production processes that could be effectively utilized, It is important to note that there still exists a significant discourse as to the sustainability of a single feedstock to sufficiently supply or meet national biofuel /biodiesel targets however seemingly realistic that target may seem. Challenges brought about by the unavailability of suitable land for the cultivation of a single feedstock as well as competing demands from other important areas (like food and forestry) preclude such production targets. Biodiesel as the most promising substitute for conventional fossil diesel is dependent on the availability of an array of feedstock that offers the following advantages; wide geographic diversity, better resilience to natural disasters and other production shocks. Conclusively, intensive research on providing improved technologies to provide possible advancement to enhance sustainable biodiesel development in sub-Saharan Africa is expedient without jeopardizing security, public health or the environment.

Author details

Omotola Babajide[1,2*], Leslie Petrik[1] and Farouk Ameer[1]

*Address all correspondence to: obabajide@uwc.ac.za

1 Environmental and Nano Sciences Research Group, Chemistry Department, University of the Western Cape, Bellville, Cape Town, South Africa

2 School of Sciences, Adeyemi University of Education, Ondo, Nigeria

References

[1] Elmissiry, M.M. and Punungwe, J. 1998. The position of renewable energy in Zimbabwe. http://www.punungwe.s5.com/renergy.html (accessed on 28 August 2011).

[2] Amigun, B., Musango, J. and Stafford, W. Biofuels and Sustainability in Africa. Renewable and Sustainable Energy Reviews 2011;15:1360–1372.

[3] United States Economic Commission for Africa (UNECA) 2006: Energy for Sustainable Development, Atmosphere/Air Pollution, Climate Change and Industrial Development Report. African Regional Implementation Review for the 14th Session of the Commission on Sustainable Development (CSD-14). http://www.uneca.org/csd/CSD4 (accessed on 12 September 2014).

[4] GENI, 2003: International Energy Demand. Global Energy Network Institute. http://www.geni.org/globalenergy/library/energytrends/currentusage/index.shtml (accessed on 12 September 2013)

[5] Monfort, M.C. Global trends in seafood and sustainability 2008: market movements and trends, Seafood Choice Alliance Conference, Barcelona, Spain, 2008.

[6] Bourne J.K.2007. Green Dreams: Making fuel from crops could be good for the planet —after a breakthrough or two. National Geographic http://www.cbsd.org/cms/lib07/PA01916442/Centricity/Domain/1539/Article%20I%20-%20Green%20Dreams.pdf (accessed on 10 May 2014)

[7] Luque R., Herrero-Davila L, Campelo J.M, Clark J.H., Hidalgo J.M, Luna J.M. and Romero A.A.: Biofuels: A technological perspective. Energy Environ. Sci. 2008;1: 542–564.

[8] von Maltitz, G., Haywood, L., Mapako, M. And Brent, A. 2009 Analysis of opportunities for biofuel production in sub-Saharan Africa. CIFOR, Bogor, Indonesia.

[9] Smeets, E.M.W., Faaij, A.P.C., Lewandowski, I.M. and Turkenburg, W.C. A bottom-up assessment and review of global bio-energy potentials to 2050. Progress in Energy and Combustion Science 2007; 33(1): 56–106.

[10] [10] Batidzirai, B., Faaij, A. and Smeets, E. Biomass and bioenergy supply from Mozambique, Energy for Sustainable Development 2006; 10(1): 54–81.

[11] Johnson, F.X. and Matsika, E. Bio-energy trade and regional development: the case study of bio-ethanol in southern Africa. Energy for Sustainable Development 2006;10(1): March.

[12] Schutter O.D. 2009 Large-scale land acquisitions and leases: a set of core principles and measures to address the human rights challenge. www2.ohchr. org/english/issues/food/BriefingNotelandgrab. pdf (accessed on 23 May 2014).

[13] World Bank 2010. Raising global interest in farmlands. Can it yield sustainable and equitable benefits? World Bank, Washington, DC.

[14] Friis, C. and Reenberg, A. 2010. Land grab in Africa: emerging land system drivers in a teleconnected world. GLP Report No. 1. GLPIPO,Copenhagen, Denmark.

[15] Royal Society 2008. Sustainable biofuels: prospects and challenges. Policy document 01/08. Royal Society, London.

[16] Gallagher, E. 2008. The Gallagher review of the indirect effects of biofuels production. Renewable Fuels Agency, East Sussex, UK.

[17] von Maltitz, G. and Staffford, W. Assessing opportunities and constraints for biofuel development in sub-Saharan Africa. 2011; Working Paper 58, CIFOR, Bogor, Indonesia.

[18] Mwakasonda, S. 2007: Moving toward sustainable biofuel programs in Africa. The first high-level biofuel seminar in Africa, Addis Ababa, Ethiopia.

[19] Babajide, O. 2012: Optimisation of biodiesel production via different catalytic and process systems. Unpublished PhD Thesis, University of the Western Cape, South Africa. Available online at http//etd.uwc.ac.za

[20] ESPA (2014) Unravelling biofuel impacts on ecosystem services, human wellbeing and poverty alleviation in Sub-Saharan Africa. http://www.espa.ac.uk/projects/ne-l001373-1 (accessed on the 12th of October 2014).

[21] Willis, K.J., Bennett, K.D., Burrough, S.L., Macias-Fauria, M. and Tovar C. Determining the response of African biota to climate change: using the past to model the future. Philosophical Transactions of the Royal Society B: Biological Sciences 2013; 368(1625):201–204.

[22] Ma, F. and Hanna, M.A. Biodiesel production: a review. Bioresource Technology 1999;70:1–15.

[23] Zhang, Y., Dube, M., McLean, D.D. and Kates, M. Biodiesel production from waste cooking oil: 2. Economic assessment and sensitivity analysis. Bioresource Technology 2003; 90(3): 229–240.

[24] Encinar, J.M., González, J.F., Rodríguez, J.J. and Tejedor, A. Biodiesel fuels from vegetable oils: transesterification of cynara cardunculus l. oils with ethanol. Energy Fuels 2002; 16: 443–450.

[25] Sivaprakasam, S. and Saravanan, C. G. Optimization of the transesterification process for biodiesel production and use of biodiesel in a compression ignition engine. Energy and Fuels 2007; 21 (5): 2998–3003.

[26] Meng, X., Chen, G. and Wang, Y. Biodiesel production from waste cooking oil via alkali catalyst and its engine test. Fuel Processing Technology 2008; 89: 851–857.

[27] Ahn, E., Koncar, M., Mittelbach, M. and Marr, R. A low-waste process for the production of biodiesel. Separation Science and Technology 1995; 30 (7–9): 2021–2033.

[28] Ryan, R., Dodge, L.G. and Callahan, T.J. The effects of vegetable oil properties on injection and combustion in two different diesel engines. Journal of American Oil Chemical Society 1984; 61: 1610–1620.

[29] Alcantara, R., Amores, J., Canoira, L., Fidalgo, E., Franco, M. J. and Navarro, A. Catalytic production of biodiesel from soy-bean oil used frying oil and tallow. Biomass and Bioenergy 2000; 18: 515–527.

[30] Srivastava, A.R. and Prasad, R. Triglycerides-based diesel fuels. Renew. Sustain. Energy Reviews 2000; 4: 111–133.

[31] Sharma Y.C., Singh B, Korstad J. Latest developments on application of heterogenous basic catalysts for an efficient and eco friendly synthesis of biodiesel: A review. Fuel 2011; 90:1309–1324.

[32] Cao, P., Tremblay A.Y., Dube,′ M.A. and Morse K. Effect of membrane pore size on the performance of a membrane reactor for biodiesel production. Ind Eng Chem Res 2007; 46(1):52–58.

[33] Ngamcharussrivichai C, Nunthasanti P, Tanachai S, Bunyakiat K. Biodiesel production through transesterification over natural calciums. Fuel Process Technol 2010; 91(11):1409–1415.

[34] Babajide O., Petrik L., Musyoka N., Amigun B., Ameer F. Use of coal fly ash as a catalyst in the production of biodiesel. Petrol Coal 2010; 52(4):261–272.

[35] Babajide O., Petrik L., Musyoka N., Ameer F. Novel zeolite Na-X synthesized from fly ash as a heterogeneous catalyst in biodiesel production. Catalysis Today 2012;190: 54–60.

[36] Babajide O., Petrik, L., Amigun, B. and Ameer, F. Low-cost feedstock conversion to biodiesel via ultrasound technology. Energies 2010; 3: 1691–1703.

[37] Gogate P. R. and Pandit A. B. A review and assessment of hydrodynamic cavitations as a technology for the future. Ultrasonic Sonochemistry 2005; 12: 21–27.

[38] Ji J., Wang J., Li Y., Yu Y. and Xu Z. Preparation of biodiesel with the help of ultrasonic and hydrodynamic cavitation. Ultrasonics 2006; 44: 411–414.

[39] Stavarache C., Vinatoru M., Bandow H. and Maeda Y. Ultrasonically driven continuous process for vegetable oil transesterification. Ultrasonic Sonochemistry 2007;14: 413–417.

[40] Refaat A.A., Attia N.K., Sibak H.A., Sheltawy S.T. and Eldiwani G.I. Production optimization and quality assessment of biodiesel from waste vegetable oil. Int. J. Environ. Sci. Tech. 2008; 5: 75–82.

[41] Lifka J. and Ondruschka, B. Influence of mass transfer on the production of biodiesel. Chemical Engineering and Technology 2004; 27: 1156–1159.

[42] Thanh, L. T., Okitsu, K., Sadanaga, Y., Takenaka, N., Maeda, Y. and Bandow, H. Ultrasound assisted production of biodiesel fuel from vegetable oils in a small scale circulation process. Bioresource Technology 2010;101:639–645.

[43] Kumari V., Shah S. and Gupta M.N. Preparation of biodiesel by lipasecatalyzed transesterification of high free fatty acid containing oil from Madhuca indica. Energy and Fuels 2007; 21: 368–372.

[44] Qiu Z., Zhao L. and Weatherley L. Process intensification technologies in continuous biodiesel production. Chemical Engineering and Processing 2010; 49: 323–330.

[45] Martin H. Heat and mass transfer between impinging gas jets and solid surfaces. Adv. Heat Transfer 1977; 13: 1–60.

[46] Jambunathan K., Lai, E., Moss, M. A., and Button, B. L. A review of heat transfer data for single circular jet impingement. Int. J. Heat Fluid Flow 1992; 13:106–115.

[47] Beitelmal A. H., Saad M. A. and Patel C. D., The effect of inclination on the heat transfer between a flat surface and an impinging two- dimensional air jet. Int. J. Heat Fluid Flow 2000; 21:156–163.

[48] Nieuwoudt O. 2010: Jet reactor in biodiesel production available on https://sites.google.com/site/biofuelson.

[49] Cao P, Tremblay AY, Dube′ MA, Morse K. Effect of membrane pore size on the performance of a membrane reactor for biodiesel production. Ind Eng Chem Res 2007;46 (1):52–58.

[50] Dube M, Tremblay A, Liu J. Biodiesel production using a membrane reactor. Bioresour Technol 2007; 98 (3):639–647.

[51] Aransiola E.F., Ojumu T.V., Oyekola, O.O., Madzimbamuto, T.F., Ikhu-Omoregbe D.I.O. A Review of current technology for biodiesel production: State of the art. Biomass and Bioenergy 2014;61: 276–297.

[52] Ranchod N., Grewan S., Nedambale N., Matambo T., Low M. and Harding K. Biodiesel production using a membrane reactor. Chemical Technology 2013: 21–25.

[53] Dimian AC, Rothenberg G. Process for manufacturing acid esters through reactive distillation. EP Patent 2,154,226; 2010.

[54] Kiss AA, Dimian AC, Rothenberg G. Biodiesel by catalytic reactive distillation powered by metal oxides. Energy Fuel 2008; 22(1):598–604.

[55] Kiss AA. Heat-integrated reactive distillation process for synthesis of fatty esters. Fuel Process Technol 2011; 92(7):1288–1296.

[56] Cossio-Vargas, E., Hernandez, S., Segovia-Hernandez, J.G. and Cano-Rodriguez, M.I. Simulation study of the production of biodiesel using feedstock mixtures of fatty acids in complex reactive distillation columns. Energy 2011; 36 (11):6289–6297.

[57] Novel integrated reactive separation process for FAME synthesis available on members.chello.nl/a.kiss2/projects.html (accessed on 31 October 2014).

[58] Dimian AC, Bildea CS, Omota F, Kiss AA. Innovative process for fatty acid esters by dual reactive distillation. Comput Chem Eng 2009; 33(3):743–750.

[59] Sinkala T. 2009. Development of next Generation biofuels in Africa. http://www.jatropha.pro/PDF%20bestanden/Development%20of%20next%20generation%20biofuels%20in%20Africa%20%28Zambia0.pdf (accessed on 10 October 2014).

[60] Eastern Cape Business 2013 by Global Africa Network http://issuu.com/globalafricanetwork/docs/easterncapebusiness (accessed on 24 June 2014).

http://dx.doi.org/10.5772/59859

[61] Ribeiro D. and Matavel, N. 2009. Jatropha—a socio-economic pitfall for Mozambique. Justian Ambiental & Uniao Nacional de Camponesses. Report for SWISS AID.

[62] Shumba E., Roberntz P. and Kuona, M. (2011a). Assessment of sugarcane out-grower schemes for bio-fuel production in Zambia and Zimbabwe.WWF, Harare, Zimbabwe.

[63] Makkar, H.P.S., Becker, K., Sporer, F. and Wink, M. Studies on nutritive potential and toxic constituents of different provenances of *Jatropha curcas*. Journal of Agriculture and Food Chemistry 1997; 45: 3152–3157.

[64] Shumba E, Roberntz P, Mawire B, Moyo N, Sibanda M, Masuka M. 2011. Community level production and utilization of jatropha feedstock in Malawi, Zambia and Zimbabwe. WWF, Harare, Zimbabwe

[65] Thurmond, W., 2008: Biodiesel 2020: A Global Market Survey. 2nd Edition. Emerging Markets Online http://www.emerging-markets.com accessed on 7 July 2010.

[66] Radford, T. (2014) Grass Is Greener For Biofuels Future. GLOBE Foundation http://www.environmental-expert.com//news/grass-is-greener-for-biofuels-future-430785?utm_source=News_Energy_12062014&utm_medium=email&utm_campaign=newsletter&utm_content=normtextlink (accessed on 12 June 2014).

[67] Nigam, S.P. and Singh, A. Production of liquid bio fuels from renewable resources. Progress in Energy and Combustion Science 2010; 37:52–68.

Chapter 4

Prospects for the Production of Biodiesel in Pakistan

Mushtaq Ahmad, Hammad Ahmad Jan,
Shazia Sultana, Muhammad Zafar,
Muhammad Aqeel Ashraf and Kifayat Ullah

Additional information is available at the end of the chapter

http://dx.doi.org/10.5772/59318

1. Introduction

The use of renewable energy resources (biodiesel) to generate power is receiving attention around the world, and for Pakistan, it can address the current and upcoming energy stresses of the country. Pakistan is facing a severe economic crisis owed to an unceasingly rising gap between energy demand and energy supply. The scarcity in power and gas supply has already frozen a number of industrial sectors such as textile, small and medium enterprises,and local transport. It is common anxiety in today's world that fossil fuels will be exhausted soon. The price of energy is rising unceasingly and is predicted to be at its peak by 2050. The fossil fuel sources are decreasing in Pakistan, the result of which is the import of about 8.1 million tons at approximately US$ 9.4 billion per annum. Thus, renewable and sustainable energy resources, such as biodiesel needs to be maintained so that a sustainable energy mix could be achieved to confirm energy security. In the ambit of this justification, augmenting the scarce energy resources in Pakistan through intense coupling of the various biodiesel sources can effectively address the shortage and can confirm energy security. Towards this end, the advancement achieved in biodiesel-associated researches in Pakistan are assessed and presented, highlighting ways of attaining the objective set forth by the Government. To this effort, biodiesel as a renewable energy source has been deliberated, to overcome energy crisis, achieve pollution-free environment, economic growth and more importantly,significantly increase the income of our farmers. This study has also identified areas in Pakistan where there are substantial possibility to renovate bio-energy production distribution systems to deliver diverse energy carriers like electricity, industrial and domestic fuel, and gases. Fences are observed over the entire bio-energy range and policy issue and institutional roles and odd jobs are discoursed. Now the responsibility is upon the officialdoms such as of the Alternative Energy Development Board (AEDB) and Pakistan State Oil (PSO) to connect the research results from several native universities and to develop a full-scale biodiesel economy in Pakistan.

In the past, mankind has investigated a number of energy resources from wood, coal, oil and fossil fuel to nuclear energy. Recently, the public and governmental sensitivities to pollution and energy security have led to the elevation of renewable energy assets. Biofuel is one such reserve that could play a significant role in a more diverse and sustainable energy mixture [1].Biofuel is that source of fuel whose energy is derived from biological carbon fixation. Biofuels comprise fuels derived from biomass conversion, as well as solid biomass, liquid fuels,and various biogases[1]. The fabrication of liquid biofuels has augmented five-fold in the previous two decades due to policy interferences and changing relative energy prices [2]. The increasing production of biofuels is mainly for energy security by lessening reliance on import of fast-depleting fossil fuels and saving considerable amounts of foreign exchange; to mitigate the global warming emissions by reducing the use of fossil fuels; and to improve agricultural development by offering better prices and new jobs [1].

Energy plays a basal role in the socio-economic advancement of a country by providing for daily life needs. For thousands of years,biofuel has been used as an energy source by men. The statistical data documented by the International Energy Agencyclarify that conventional energy resource, such as fossil fuel are still the main sources of energy tailed by coal and gas, contributing about 80% of Total Primary Energy Supply (TPES) [1].Energy sources diversification is vital for energy security, climate change, and sustainable development issues. Additionally, too much dependence on non-renewable energy sources for power generation in the long-term is unfeasible. Consequently, extensive consumption of renewable energy sources such as biofuel, geothermal energy, solar energy and wind energy, is indispensable to overcome the energy crisis [2]. The biofuel potential is impressive and its capability to address the world's energy demand has been extensively recognized. Presently, renewable energy globally is still ruled by the "old" renewable such as hydropower and traditional biofuel that supplies 6% and 9%, respectively, of the world's key energy demand. While the "new" renewable sources such as mini- and micro-hydro, photovoltaic and windenergies afford only 2% of the world's primary energy. In South Asia, households are likely to follow the energy ladder including power sources like dung, crop residue, firewood, kerosene, gobar gas, LPG, and electricity for cooking purposes. Evidence suggests that while it is possible to observe such transition in urban and semi-urban areas, the change is very slow in rural areas. In developing countries, improving power services for poor households is one of the most tenacious challenges to them.

Pakistan is also located in the south Asian region with a total land area of 888,000 km^2. The approximate population of Pakistan is about 173.51 million with an annual growth rate of 2.05%; it is estimated that Pakistan will become the fourth largest nation on earth in terms of population by 2050 (Economic Survey of Pakistan 2010). The draft population policy 2009–2010 foresees a reductionin fertility level from 3.56 (2009) to 3.1 births per woman by the year 2015. With a median age of around 20 years, Pakistan is also a "young" country. It is assessed that there are currently about 104 million Pakistanis under the age of 30 years [1].Pakistan is basically an agriculture-dependent country. About 62% of the country's inhabitants live in rural areas, and is directly or indirectly reliant on agriculture for their income. Pakistan has about 5.17% of land covered by forest out of which 5% is protected. To lessen dependence on natural forests, national forest policy calls for the promotion of alternate energy resources including energy plantations, micro-hydropower generation, bio-gas, solar and wind energies, liquid petroleum gas (LPG), and natural gas for use in critical mountain ecosystems [1].

The key objective for power sector reforms in Pakistan is rural electrification. However, there is no secure government policy for the growth of devolved power supply. Pakistan, for nearly two decades, has been one of the fastest growing power markets in the world though it has a young and growing population, low per capita electricity consumption, rapid urbanization, and strong economic growth. Its economic growth has been significantly hindered by acute energy dearth for five years (Economic Survey of Pakistan 2009).Among the power sectors, the electricity sector in Pakistan consumed 15 million tons of oil in 2008 equaling to 28% of petro-fuels consumed by the country. As our oil reserves were not enough, therefore 71% of the country's oil requirements were imported in2008 [3]. The high consumption of oil in the energy sector not only tightens the economic pressure, but also increases CO_2 emission from the power generation sector. Moreover, high oil prices condensed the quantity of oil purchased, and therefore, electricity could not be supplied as per demand of the economy, which led to demand-supply gap of electricity. Because of this reason, the total supply of electricity altered from a surplus of 1230MW at the end of financial year 2005 to a shortfall of 5885MW at the end of financial year 2010 [4].

This inability of the electricity supply to address increasing demand is mainly attributed to the lower utilization of existing installed generation capacity which is mainly due to increasing oil prices in international markets [4]. The average annual rise of electricity need from 2005–2010 was 8%, and is predicted to continue till 2035. If the growth continues at the same rate, the total need of the country will be 474 GW up to 2050. The power generation plan till 2030 indicates that electricity supply will be growing at an average annual growth rate of 11% till 2030, and 65% of the increase in installed generation capacity will be thermal-based electricity (Pakistan Economic Survey, 2009).

Under such circumstances, much of our dependence on petro-fuel will not only cause an increment in environmental pollutants but will also result in high electricity prices due to the increasing prices of fossil fuels, and could make the power sector vulnerable to international price vacillations of fossil fuels. In these circumstances, the share of renewable energy in the power sector in Pakistan was less than 1% till 2010 [3]. Therefore, for this reason, it is imperative that Pakistan exploits domestically existing alternative energy sources for power generation. However, to tap renewable energy resources in Pakistan, details on the potential of these energy sources considering commercially available and most promising technologies need to be evaluated and quantified.

It is stated by a careful evaluation that the energy requirements of the country will increase to three times up to 2050 and the capacity to deliver this needis not very encouraging. Consequently, it becomes necessary to tap the substitute and renewable resources for energy [5].Biofuel experts recognize the land of Pakistan as rich in natural species of plants that can be used for biodiesel feedstock like *Pongamiapinnata, Cannabus, Jutropha* and *Ricinuscommu-*

nis. The seeds of such plants are rich in oil that can firmly produce biodiesel in Pakistan. The cultivation of these plants is likely to create at least one job for every acre of planted trees; the overall influence on agricultural employment alone could be huge. Besides, these plants have the gift to nurture marginal, waste or arid land. Pakistan has huge areas of such deprived quality land (more than 80 million acres) ideal for the farming of energy crops, so planting *Pongamiapinnata, Cannabus,Jutropha*and *Ricinuscommunis* would not prevent lands from cultivating vital food crops. To synthesize biodiesel, besides vegetable oil, alcohol (ethanol or methanol) is also required. This is lavishly accessible in Pakistan courtesy of sugar industry. Another source of ethanol synthesis is natural gas. The more we cultivate fuel crops for alternate fuel source, the better it is for the environment and for our economy.

Biodiesel is a renewable form of fuel used in automobiles and is generally extracted from plant seeds or from its other parts (e.g. flowers).Basically, Pakistan is an agricultural country and there are enough resources to be utilized for the production of biodiesel. Unfortunately, no systematic investigations have been done on biodiesel technology due to lack of education, confidence, and interaction between our industries and research institutions [6].The Government of Pakistan has laid down various schemes to harness indigenous renewable sources of energy (biodiesel). In Quaid-i-Azam University Islamabad, a lot of work has also been done on the production of biodiesel from plants.

2. Energy and poverty

Meanwhile energy plays an important role in the daily lives of humans, and poverty deprives people of chances for a better existence; and it is not surprising that there are manifold links between energy and poverty. Poverty means starvation, absence of medical treatment, and deprived access to rudimentary facilities such as electricity and water supply. It means being unable to send children to school, and often needing them to work instead. Eradicating poverty, being the most unrelenting priority, is the realization of man's elementary desires, which include nutriment, housing, water supply and hygiene, and other amenities that will advance their standards of living, such as adequate health care, education, and better transport [7].

Even though energy is not in itself a rudimentary necessity, it is essential as a critical input for providing indispensable human desires. The availability of modern energy facilities can contribute to poverty mitigation by (1) improving living standards through better lighting, access to cleaner cooking fuels, and safe drinking water, and (2) improving effective transfer of services such as reliable space- and water-heating, lighting, refrigeration of vaccines and other medicines, and sterilization of equipment in health centers. The provision of economical and good-quality lighting also allows students to extend study hours and to improve their employment prospects [7].

2.1. Biodiesel synthesis technology

Synthesis involves three reactions, whereby triglyceride is converted successively to diglyceride, monoglyceride, and glycerol, consuming one mole of alcohol in each step and liberating

http://dx.doi.org/10.5772/59318

one mole of ester [8]. The final biodiesel composition depends on the initial feedstock, as well as on the reaction conversions and process separation efficiencies. The thermo-physical properties depend on factors such as chain length, branching, and degree of saturation [9]. The reaction of transesterification proceeds in the presence of a suitable catalyst. When raw materials (oils or fats) have a high percentage of free fatty acids or water, the alkali catalyst will react with the free fatty acids to form soaps. The water can hydrolyze the triglycerides into diglycerides and form more free fatty acids. Both of the above reactions are undesirable and reduce the yield of the biodiesel product. In this situation, the acidic materials should be pre-treated to inhibit the saponification reaction [10].

2.2. Biodiesel: A solution to energy crisis in Pakistan

In 2003, Jeffrey Dukes, a biologist, estimated that the fossil fuels we use in a calendar year were the result of decay of organic matter "containing 44 × 1018 grams of carbon, which is more than 400 times the net primary productivity of the planet's current biota". In simple English, this means that each, year we routinely use four centuries' worth of assets of natural biota (plants and animals) [11]. Currently, Pakistan is in the clutch of a serious energy crisis that is distressing all zones of the economy and the different parts of the society. The way circumstances are currently positioned, explanations or solutions for the problem are scarce. It is time to alter attitudes and life styles at the national level, which should be initiated by those in authority and then followed by all sectors of the society, all of whom have the right to electricity. At best, there could be some short, medium, and long-term solutions to the crisis but they want instant planning and implementation with a massive investment. Former leaders of the republic have not succeeded in resolving the energy crisis, and the problem continued to persist.

To grow the economy of the country, energy is the key source. Pakistan has to face a major energy crisis in *natural gas, power* and *oil and this is the main reason for the* delay in the country's economic growth (which is already in intricacy). The high price of the import of oil, building of giant dams and indeterminate local sanctuary environment attached with non-existence of national covenant to physique dams are probably to adjournment swift resolution of energy crises. This energy discrepancy will lead to inflation. Likewise subventions of billions of rupees have to be dragged out of the "Water and Power Development Authority (WAPDA)" to bear loses. This would seriously disturb the national exchequer.

Unremitting and economy supply of energy is essential for sustainable economic progression. Currently, the role of renewable and sustainable sources of energy in Pakistan is inadequate, to say the least, and key actions are needed to make it a noteworthy player in the country's energy supply mix [5]. It is highlighted that the fabrication of biodiesel is a prerequisite to sustainable growth and will lessen reliance on imported fuel. The core deduction that was derived from the study was that if indigenous vegetation yielding inedible oil is cultivated on massive unproductive areas of the country, the feedstock cost could be reduced and biodiesel could become a solution to Pakistan's ailing energy crisis. About 70% of Pakistan's overall geographical area lay coarse, thus providing an opportunity to cultivate inedible oil-yielding vegetation that in turn is converted to biodiesel. Thus, it is necessary to develop plant-based biodiesel productions in Pakistan, which will be valuable for the improving socio-economic settings of the country [12].

2.3. National biodiesel program

To use biodiesel as substitute energy source in Pakistan, the AEDB has verbalized policy recommendations, the primary aims of which are to minimize the bill of imported fuel of the country, address the demand of raw material for biodiesel which will be the prime goods for biodiesel fabrication, and promote a pollution-free environment. Also on 14th Feb. 2008, the Economic Coordination Committee (ECC) of the National Cabinet has permitted the strategy for the use of biodiesel as an alternate energy source in its conference. Striking points of the dogma are as follows:

1. AEDB shall be the primary coordinating and facilitating body for the National Biodiesel Program.
2. Gradual introduction of biodiesel fuel blends with petroleum diesel so as to achieve a minimum share of 5% by volume of the total diesel consumption in the country by the year 2015 and 10%by 2025.
3. The Ministry of Petroleum & Natural Resources shall come up with the fuel quality standards for B-100 and blends up to B-20.
4. Oil Marketing Companies (OMCs) are to purchase biodiesel (B-100) from biodiesel manufactures; and sell this biodiesel blended with petroleum diesel (starting with B-5) at their points of sale [13].

2.4. Biodiesel feasibility for Pakistan

In general, vegetable oil is treated with either methanol or ethanol to synthesize biodiesel. The main reason for the use of methanol worldwide is its low price. The main source of methanol is coal, and in Pakistan, the coal reserve value is about 180 billion tons and is the 5th largest in world. At present, ethanol production in Pakistan is also high (300,000 tons of cane per day) because currently, 76 sugar mills are operational [14]. The main source of ethanol is the molasses from sugar cane which is actually the by-product. There are 21 distillery units in Pakistan with a capacity to process 2 million tons of molasses to produce 400,000 tons of ethanol; therefore, it means that excess ethanol can be either for gasohol purpose or for biodiesel production [15]. The production capacity of these 21 units is about 400,000 tons and the country's need plus export is up to 80,200 tons and yet, there is still leftover ethanol, which is about 318,000 tons [15]. Thus, in terms of ethanol requirements for synthesis of biodiesel, the stock is sufficient to increase production.

Conventionally, NaOH is used as a catalyst during biodiesel synthesis which is produced in enough quantity to cover the country's requirements, and production can be easily augmented because of massive reserves of NaCl [14].As it is well known, Pakistan is an agricultural country with70% of its population working in the farm fields, and so the Soil Survey of Pakistan surveyed and classified its soil types into 79 major classes, the details of which are given in

http://dx.doi.org/10.5772/59318

Table 1 (Soil Survey of Pakistan).In the figure, land use for agriculture which is spread throughout the country is shown by yellow color (Soil Survey of Pakistan). In the country, 28 million hectares of land is unused and this is due to water scarcity, high temperature, and soil salinity (Fig. 1). Unfortunately, despite the rich land, Pakistan, which is known as an agriculture-based country, has to rely on imports for edible oil, wheat, and milk [14].

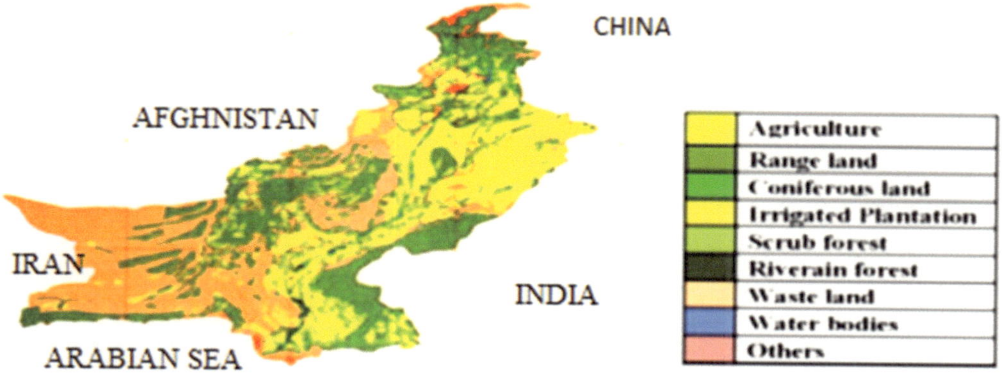

Figure 1. Land use for agriculture in Pakistan

From Jatropha seed, India successfully produces biodiesel. In Pakistan, Jatropha plants can also be cultured very easily and grow well, especially in saline soil with less quantity of water and can also withstand high temperature [14]. A one-hectare field of Jatropha can yield up to two tons of biodiesel fuel per year. If Pakistan makes use of all uncultured land for biodiesel manufacturing, then Pakistan will be able to yield 56 million tons of biodiesel in a calendar year, while the current necessity of fuel is about 8.5 million tons [14]. In short, land is available for cultivation of energy crops. Pakistan's energy requirement is rising and a 10% increase (including for power and transport sectors) is perceived per annum. Consequently, biodiesel assignment is feasible and has a very promising future in Pakistan; raw materials for biodiesel fabrication are accessible, and more importantly, the Government of Pakistan is very serious in energy-generation programs [14].

The organic material resulting from biological organisms (plants and animals) is called biomass. Bio-energy can be defined as energy obtained from biological and renewable sources (biodiesel); it may be processed or converted in the form of heat or transformed into electricity for distribution. Biomass is a portable feedstock that can be easily transmuted into biofuels for the production of bio-energy, manufactured straight or indirectly from biomass. Biofuels are either solid (fuel wood, charcoal, wood pellets, briquettes, etc.) or liquid (bioethanol, biodiesel). Now with the evolving advancement in bio-energy using more recent technology, biomass energy can be divided into traditional biomass and modern bio-energy. Traditional biomass is the chief font of energy used in evolving countries mainly for food preparation and warming at the home level, typically using three-stone stoves, or in some areas improved cooking stoves. The energy source of this type is present in the form of wood-fuel (including fuel wood and charcoal), crop residues, and animal dung and is often processed and used by women and

S/No.	Type of Land Use	Area in 1000ha
1	Agriculture	21,733
2	Range Land	25,475
3	Coniferous Forest	1,353
4	Irrigated Land	80
5	Scrub Forest	196
6	Riverains Forest	239
7	Waste Lands	28,501
8	water bodies	1,274
9	Others	159
	Total	79,610

Table 1. Different lands used by Pakistan

children on everyday basis. On the other hand, modern bio-energy is used generally for the generation of electrical energy or transport power. Liquescent biofuels for transport such as ethanol and biodiesel are examples of emerging energy substitutes [1].

About 62% of Pakistan's residents have restricted access to commercial energy; only the traditional methods of using wood, animal waste and crop waste for home fuel needs are available to them because they are located in the rural areas. Effectiveness of use is very squat and most of the latent is lost because of non-scientific conventional technologies. Therefore, it is essential to develop modern bio-energy technologies so that renewable capitals of energy may serve to supplement the long-term energy requirements of Pakistan to a momentous level [1].

2.5. National biodiesel program

To use biodiesel as substitute energy source in Pakistan, Alternative Energy Developed Board has verbalized Policy Recommendations, the primary aims of which are to minimize the bill of imported fuel of the country, the demand of raw factual of biodiesel which will be the prime goods for biodiesel fabrication and a pollution free environment. Also on 14th Feb 2008 the Economic Coordination Committee (ECC) of the National Cabinet has permitted the Strategy for usage of biodiesel as an alternate energy source in its conference. Striking points of the dogma are as follows:

1. AEDB shall be apex coordinating and facilitating body for the National Biodiesel Program.

2. Gradual introduction of Biodiesel fuel blends with petroleum diesel so as to achieve as minimum share of 5% by volume of the total diesel consumption in the country by the year 2015 and 10%by 2025.

3. Ministry of Petroleum & Natural Resources shall come with the fuel quality standards for B-100 and blends up to B-20.

4. Oil Marketing Companies (OMCs) to purchase Biodiesel (B-100) from Biodiesel manufactures; and sell this Biodiesel blended with Petroleum diesel (starting with B-5) at their points of sale [16].

2.6. Research and development on biodiesel in various Pakistani institutions

On biodiesel technology, numerous projects have been initiated at different universities and industries in Pakistan. Unfortunately, all efforts have been conducted individually, and apparently, there was minimal knowledge sharing between and among institutions. This work is an effort at assembling all the figures produced by numerous national organizations and to present it in an articulate form for the assistance of the future [7].

2.7. Impact of Jatropha and Pongame on Pakistan's biodiesel plans

The efforts done by Pakistan in rearing and harnessing Jatropha and pongame plants for biodiesel production is nothing short of praiseworthy. Research and progress reporting are also ongoing and findings are continually studied and evaluated [7].

2.8. Status of Jatropha cultivation for biodiesel production

In Sindh, the presence of Jatropha has long been reported [17]. It is locally identified as Karanga, RatanJothor or Jamal Ghota and was used to treat several diseases in villages and towns [18]. Through imported seeds from a number of countries, Jatropha has been mostly cultured on small scale by private tycoons [17]. Their plantations have increased from about 2 acres in 2005 to more than 400 acres in 2008 as shown in Fig. 2. This increase in Jatropha farming was primarily due to an aggressive campaign started by the AEDB [17]. In the private sector, numerous organizations are interested in cultivating Jatropha nurseries at several locations in Sindh, Punjab, and Baluchistan. These nurseries have become the root for a quantity of Jatropha ranches in the same areas. In these ranches, the normal age of a plant may range from several weeks to about 18 months [17].

In 2005, roughly 2 acres were cultivated for Jatropha farming [17]. However, after that, the private sector started mounting nurseries for further cultivation. In 2006, more than 10,000 saplings were provided by nursery proprietors to several growers in Sindh and Punjab for transplantation into the farms. Similarly in 2007, these nurseries presented about 50,000 saplings for transplantation to various growers in Sindh and in Baluchistan [17]. However, because of glorified interest, cognizance, and conceivable economic gains in cultivating Jatropha, agrarians in Sindh presented more concentration, and because of the large-scale accessibility of saplings in the nurseries, during 2008, more than 200,000 saplings were provided by several nurseries in Sindh for transplantation [7].

Seeing this interest, in 2008, Pakistan State Oil (PSO) also brought seeds for developing nurseries and for cultivating about 20,000 Jatropha plants in their own farmhouse [17].

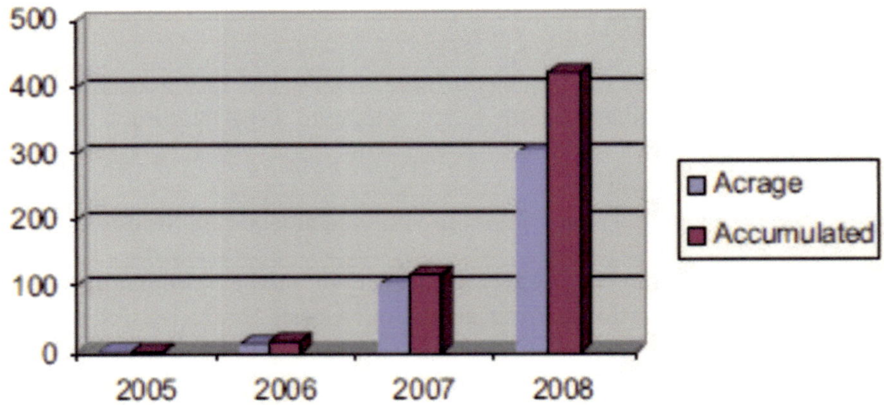

Figure 2. Growth of cultivation of Jatropha in Pakistan

Currently, the PSO has about 10,000–20,000 saplings for such transplantation [19]. Upon calculation, within the quarter of 2009, their nurseries located adjacent to Karachi already had more than 200,000 saplings and were prepared to offer additional saplings if requested. The growth projections of Jatropha cultivation, up to the last of calendar year 2014, are shown in Fig. 3 [17].

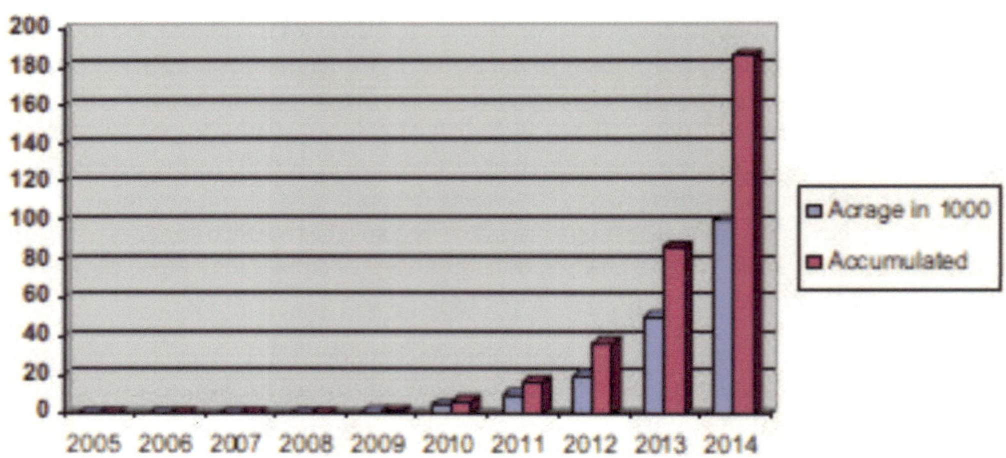

Figure 3. Projected growth of Jatropha cultivation by PSO

PSO spearheaded an experimental project centered on harnessing *Jatropha curcas* plants from its private farmhouses for the synthesis of biodiesel on commercial scale. The predicted outcomes for Pakistan are the following [20]:

- Six million plants will be cultivated for a greener environment.

- Five hundred planters will be rented to accomplish the cultivation of plants on a 5,000- acre land.
- Twenty-four million kilograms of seeds will be produced per year from this plantation.
- A total of 7.2 million liters of biodiesel (costing roughly PKR 345 million @ PKR 48/liter of fossil diesel) will be manufactured/year which is equal to 17 MT of biodiesel/day.

Similarly, other interested investors such as Karachi's Forest Department together with the Pakistan Army have efficaciously planted Jatropha plants in several areas of Sindh [16]. So far, the Forest Department succeeded to nurture 3000 samplings on a trial basis in Malir Cantonment in 2010 for the cultivation of the Jatropha seeds which were supplied by PS [17]. Likewise, the Pakistan Agricultural Research Council (PARC) and a Canadian company, KijaniEnergy, are also interested in developing large-scale cultivation of Jatropha for biodiesel production on marginal lands [21]. In 2009, Kijani Energy capitalized approximately US$ 150 million, which has resulted in the use of 200,000 acres of land in Cholistan, Umerkot, Tharparker, Khairpur, and Sanghar for the purpose of Jatropha cultivation [22].

The advantages and benefits of such deeds can be multifarious. The culturing of Jatropha is likely to produce at least one vacancy for each acre of planted samples; thus, the overall influence on agricultural employment alone can be gigantic [17]. The charge of earning of seeds, cost of synthesis, tax policies made by government, consumption of by-products, oil cake, and other Jatropha surplus continue to influence the cost of biodiesel production. While considering the aspect of expenses, it is indispensable to recognize the rural occupation generation, energy sanctuary, carbon swapping issues, and savings of external exchange. Overall, job opportunities will be created from plantation, seed gathering, extraction of oil, biodiesel manufacturing, and local scattering. Job opportunities generated from plantation and seed collection alone are appraised to be 40 men days/ha/year [7].

2.9. Biodiesel research status by consuming indigenous Jatropha oil

The *Jatropha curcas* plant is a drought-resilient crop that grows profound taproot surface roots permitting it to counterattack and control soil erosion. It yields approximately 2–4 kg/seed/tree/year. The oil yields of *Jatropha curcas* is estimated to be 1590 kg/ha. The chief fatty acids in *Jatropha curcas* seed oil are the oleic, linoleic, palmitic, and stearic acids [23].

For biodiesel fabrication, expenses to be incurred may vary because of variation in location and labor duties, land procurement, and policies in place. However, Silitonga et al. [23] presented a solid case for biodiesel manufacturing from Jatropha. According to them, if 500 hands work on a farmland of 1,500,000 ha, an estimated amount of 2,250,000 liters of oil can be produced. Taking into account the labor charges related to the area of attention, the estimated proceeds can be calculated. For Jatropha cultivation, the cost is negligible because it does not require crop rotation or expensive fertilizers. This partly explains the growing number of organizations in Pakistan that have been involved in the manufacture and testing of biodiesel from *Jatropha curcas* oil. The aim of all this work is to produce/extract high-oil content from the seeds (ca 30% to 40%) as well as to eliminate the presence of anti-nutritional chemicals in the oil that tend to make it inedible [7].

2.10. Research work in universities

The universities that testified results for the synthesis of biodiesel through transesterification of *Jatropha curcas* oil were the University of Agriculture in Government College University, Faisalabad [16], Quaid-i-Azam University in Islamabad [6], and the NED University in Karachi [24]. On the other hand, PSO itself also has partnered with many other universities in Pakistan for the same purpose [20].

Biodiesel fuel properties for Jatropha oil assessed by PSO and unconnectedly at Faisalabad are shown in Table 2.

Parameters	High speed Diesel (PSO)	B10 (PSO)	B100 (PSO)	B100 (Faisalabad)	Test Method
Density at 20 C/cm^3 (lb/in^3)	0.83(0.03)	0.8522	0.8816	0.88	ASTM D 1298
Kinematic viscosity mm^2/s (in^2/s)	2.73 (0.0042)	4.19	4.38	4.8	ASTM D 445
Cetane index	46	53	47	NE	ASTM D 976
Flash point C	37 (310)	90	140	188	ASTM D 93
Calorifis value Btu/lb	19528	19233	17162	NE	ASTM D 420

Table 2. Please send caption

NED University succeeded in experimenting on PSO's biodiesel in a single cylinder four-stroke compression ignition engine (Rotronics) and established that its discharge profile was better than that of diesel, canola oil biodiesel, and castor oil biodiesel [24]. The researchers at NED University also determined that biodiesel from *Jatropha curcas* seeds was inexpensive to manufacture than indigenous castor and taramira oil biodiesels (Table 3) [7].

S/No.	Biodiesel	Cost/L (PKR)
1	Jatropha	94.549
2	Castor	162.708
3	Taramira	277.004

Table 3. Cost of biodiesel synthesis/L in PKR from different non-edible feedstock

2.11. Pakistan state oil research

By using the state-of-the-art trans-esterification unit, PSO has effectively transformed Jatropha oil obtained from its own farms into biodiesel [20]. Engine performance and discharge analysis of PSO's B10 were recently spearheaded by NED University [7]. There results plainly illustrate that Jatropha biodiesel has minimum influence on the environment in contrast to other indigenous biodiesel oils as well as fossil diesel, but its engine efficiency is slightly inferior to its mineral complements. This is because of its lesser greasy value than fossil diesel. This concern can be solved if further investigation is led by Pakistani institutions to create an upsurge in the calorific value of Jatropha oil biodiesel [7].

2.12. Status of biodiesel research using indigenous Pongame oil

The Pongame plant grows well in humid and subtropical habitats and is planted in those zones with an annual rainfall of between 500 and 2500 mm. Pongame is a possible raw material for biodiesel manufacturing in Pakistan. The oil-yielding capacity of its seeds is about 35%. They can flourish in saline soil, in water logged area, and in slightly icy ground; it can also grow on diverse sorts of soil, thus, it is not a problem for highly uncultivated land in Pakistan to be finally used for biodiesel production [7].

In Quaid-e-Azam University (QAU), pongame oil has been effectively transesterified into biodiesel through catalytic transesterification using NaOH as catalyst. As a result, a maximum 90% of crude oil was converted to biodiesel. A comparable, but much effective effort was also made by other researchers in the NED University in Karachi (unreported). The fuel properties of the biodiesel samples produced at both institutions are given in Table 4. The biodiesel synthesis at NED had improved properties compared with the one manufactured at QAU. In addition, its flash point was higher, signifying that it was safer to stock. However, its ignitibility was less than the biodiesel manufactured at QAU [7].

Properties	ASTM Test Method	ASTM Limits	HSD	Pongame B100 (NED)	Pongame B100 (QAU)
Density	ASTM D 1298	0.875-0.900	0.85	0.88	0.92
Viscosity	AASTM D 445	1.9-6.0	2.8	3.9	7.53
Flash point	ASTM D 93	93 C Min	68	152	90
Sulphur contents %	ASTM D 5453	0.0015 Max	2.4	0.001	0.0084
Cetane index	ASTM D 613	47 Min	46.2	58	53
Total acid value	ASTM D 974	0.50 Max	NA	0.2	NA

Table 4. A comparison of the fuel properties of biodiesel from Pongame at QAU and at NED University

The efficiency experiment of biodiesel was taken at the test ground for internal combustion engines in the engines laboratory of QAU and a road test run of a Toyota car (2D) belonging to the AEDB, Islamabad was completed with positive results [25]. The advantage of Pongame biodiesel is that it has slighter discharge of pollutants than Jatropha and had comparable brake power and torque with Jatropha at the same engine speeds.

2.13. Harnessing indigenous resources for biodiesel production—AEDB's pioneering work and contribution

The Government of Pakistan created the AEDB in 2003, with the purpose of endorsing and facilitating the harnessing of renewable energy capitals in the country [26]. The policy for the development of renewable energy was made and issued in 2006. Thus far, AEDB has been able to stimulate different organizations and universities to initiate research work on biodiesel technology and other renewable energy resources. The government also took the task of promoting biodiesel technology through a National Awareness Program via AEDB. The organization has also been successfully nurturing Jatropha in Karachi with the assistance of various stakeholders [27].

2.14. Collaboration with local universities

In 2009, the AEDB has effectively completed their first research work of biodiesel resources. The potential oil resources which were recognized for use in biodiesel synthesis were composed of *Pongamia pinnata*, rapeseed, and castor bean. The organization fruitfully validated B10 and B20 biodiesel fuels from these indigenous resources in vehicles running through petrodiesel. The AEDB also successfully established a research laboratory at Quaid-i-Azam University Islamabad and a fuel-testing laboratory at the University of Engineering and Technology, Taxila [28]. The electrification of a village in inner Sindh province (Goth Umar Din) followed suit. For this, a close policy was made, where the manufacturer of the alternative energy was also the consumer. The villagers were growing the required seeds for the vegetable oil and producing biodiesel through transesterification, and also operated the installed generator set to generate electricity for their village [7].

2.15. Commercial projects

The commercial level fabrication of biodiesel was started by Clean Power (Pvt.) Ltd. along with AEDB by setting up a 400 liters/day refinery. This project has diverse goals that include plantation of crops in waste areas, cultivation of Pongamia and Jatropha plants, and the use of waste vegetable oil. The company also worked with Pakistan Railways to cultivate Pongame seeds in several areas of the country [7].

2.16. Formulation of biodiesel policy recommendation

In 2008, the AEDB also verbalized a policy for biodiesel which encompassed the succeeding major proposals [29]:

a. Introduction of 5% biodiesel blended fuel by 2015 and 10% by 2025 in Pakistan.

b. Oil marketing companies were to buy B (100) biodiesel from biodiesel fabrications and market the biodiesel blended fuel (B-5) at their points of sale.

c. Oil gas regulatory authority was to regulate the pricing mechanism of various blends of biodiesel.

d. All imprinted plants machinery, equipment, and specific items used in the production of biodiesel were to be exempted from customs duty, income tax, and sales tax.

After the endorsement of Policy Recommendations for the use of biodiesel as a substitute fuel, SRO474(I)/2008, for the exemption of taxes and duties on biodiesel-associated paraphernalia, machines, and other specific items was issued by the Federal Board of Revenue (FBR),

Government of Pakistan. In answer to that, the AEDB received endorsement for setting up a 10,000 t/annum biodiesel fabrication capability as part of the Government's viability study for starting a B5-use countrywide program by 2015. The additional advantage of the above-mentioned policy was the inauguration of Pakistan's principal commercial biodiesel fabrication ability (amounting to 18,000 t/annum of fuel) under the sponsorships of M/s Eco-Friendly Fuels Private Ltd. and AEDB [27].

2.17. The possibility of harnessing indigenous algae for biodiesel production

The production of biodiesel from algal cell lipase is not a new concept or technology. It has been carefully studied by numerous investigative groups that have deliberated its potential for the energy market worldwide [7]. A number of officialdoms in Pakistan have used this idea to do primary research biodiesel synthesis from algae based upon preliminary screening studies of numerous indigenous species. Two of them are in Karachi, one being the Pakistan Council for Scientific and Industrial Research and the other being the Department of Biochemistry, University of Karachi [7]. Unfortunately, the oil yield of the algae grown in the laboratories of both institutes was not sufficient to meet the standard defined by Chisti [30].

A Pakistani researcher at Mie University of Japan has recently claimed that the nation could benefit by harnessing its 27–28 million acre saline lands for algal farming [31]. This has also been confirmed by researchers in Malaysia [32]. Considering that about 40% of algal biomass consists of lipids from which oil can be extracted for producing biodiesel, the researcher mentioned that Pakistan should follow the example of other countries that are running similar projects of reclaiming saline lands and producing sustainable biofuels. He also remarked that the Pakistan Technology Board, an organization of Ministry of Science and Technology responsible for identifying and promoting key technologies in Pakistan, had already taken some initiatives to promote innovative research approaches towards biofuel production [31]. Other researchers have identified four strains of algae suitable for cultivation in Pakistan's deserts [33]:

a. *Haematococcus pluvialis.*

b. *Microcoleus vaginatus.*

c. *Chlamydomonas perigranulata.*

d. *Synechocystis.*

So far, there has been quite a clamor on commencing a large-scale algal biodiesel project in Pakistan, but very little has been done to date. No commercial reports are available in the scanty literature either. One of the main reasons may be due to the high costs associated with farming algae on a large scale as reported in a recent article in Dawn [34]. If a cost-effective method of producing algae on both saline lands and sewage networks is developed, algal biodiesel could become a major success in Pakistan [7].

2.18. Comparison between fossil-diesel and biodiesel

Biodiesel as fuel can compete with the existing fossil-diesel fuel. Biodiesel is compared with fossil-diesel in a number of categories such as energy content, hazardous material rating, health and social impacts, and engine performance rating.

2.19. Environmental drawbacks of petro-fuel

The main users of diesel and petrol are vehicles and are the main cause of environmental degradation. Old models of diesel engine are main contributors to air pollution. In Pakistan, many cities have high air pollution as compared with World Health Organization (WHO) standards, and in the past 20 years, the amount of SO_2 increased to approximately 23 folds. The losses due to air pollution in terms of health care are approximately about 500 million dollars per year while the diseases due to air pollution are increasing (Khwaja and Khan 2004). Asthma and lung diseases are caused by SO_2 and it also causes acid rain. Pollutants like CO, CO_2, Ozone, NO_x, and many volatile organic matters are upsetting air quality to dangerous levels in major cities of Pakistan.

2.20. Environmental benefits of biodiesel

Diesel engines are the high pollutant-emission sources (79%) (AEDB, Government of Pakistan). Its pollutants consist of many of organic and inorganic compounds. These particles have hundreds of poisonous chemicals on their surfaces, such as mutagens and carcinogens. Using biodiesel can greatly reduce the emission of pollutants such as SO_x, PM, but not NO_x.

Biodiesel as fuel can also be helpful in decreasing the rate of global warming by reducing the discharge of greenhouse gases. A significant decline has been observed in smog-forming pollutants. It emits much less noxious pollutants compared with petro-diesel as shown in Table 5. The data of B20 and B100 are provided for overall view.

Smog-Producing Pollutants	**B100**	**B20**
Unburned hydrocarbons (HC)	67% Reduction	14% Reduction
Carbon monoxide (CO)	48% Reduction	10% Reduction
Particle matter (PM)	47% Reduction	10% Reduction
Sulphur (SO_x)	100% Reduction	20% Reduction
Nitrogen oxide (NO_x)	10% Reduction	2% Reduction

Table 5. Reduction rate of smog-forming pollutants

Other noxious pollutants, polycyclic aromatic hydrocarbons (PAH) and NPAH (nitrated PAH), also show a great reduction (Table 6).

Smog-Producing Pollutants	B100	B20
Polycyclic aromatic hydrocarbons (PAH)	80% Reduction W/B100	13% Reduction w/B20
NPAH (nitrated PAH)	90% Reduction w/B100	50% Reduction w/B20
(Feasibility paper 4)		

Table 6. Other toxic-emission reduction

2.21. Hazardous rating comparison

Biodiesel in nature is non-hazardous when compared with fossil-diesel. Its flash point is high, therefore, it is safe to use. Biodiesel is biodegradable and 95% of it can be degraded in 28 days. Biodiesel in blended form, for example, as B20 can degrade faster than normal petro-diesel. Hazardous rating comparison between petro-diesel and biodiesel is shown in Table 7 [14].

Properties	Biodiesel	Fossil-Fuel (Diesel)
Biodegradability	Readily biodegradable, about 3 times faster than fossil fuel	Slow biodegradability
Flash point	150 C	51.7 C
Toxicity	Essentially non-toxic	Highly toxic
Spill hazard	Biodiesel is safe with no dangerous chemicals	Dangerous and toxic

Table 7. Hazardous rating comparison of biodiesel vs. fossil fuel (diesel)

2.22. Social benefits

Pakistan is an agriculture country with 70% of its population relying on agriculture for their livelihood. They can uplift their standards of living by producing oil seed generating crops in their field. Small-scale biodiesel production facility can be designed with a small investment and will be helpful in providing biodiesel fuel for their agriculture machinery. Those areas of land which are not being used due to water shortage or soil salinity can now be used for oil crops. The Government pays a huge bill for its crude oil import, thus, giving no benefit to its local economy. If farmers are able to generate energy crops then this will be beneficial not only to themselves but also to the country's overall economy. In Europe, biodiesel plants are being developed by large conglomerate entities. The produced fuel is then transported back to widely dispersed distribution depots. Germany is a successful example and is increasing its production capacities [14].

2.23. Engine performance

The diesel-engine industry has the right to decide whether biodiesel is good or not. After going through a number of experimental tests, scientists agreed upon the better or comparable performance of biodiesel fuel [14].

3. Recommendations

Based on the above analyses, the government of Pakistan may wish to:

- Create a national mission or a central government organization that is dedicated to biodiesel development: It would encourage entrepreneurs to adopt this new technology by working with the small enterprise development authority, to create business models and pre-feasibility studies for biodiesel production facilities and feedstock farms. This organization would also develop an industrial standard for biodiesel specifications.
- Create and empower environmental groups to solicit support in favor of widespread adoption of alternative fuels.
- Work in conjunction with other states to create a more uniform distribution of biodiesel feedstock production.
- To grow plants producing biodiesel in long roadsides to yield raw material for biodiesel and also to benefit the environment.
- Extend the project of biodiesel: There is need to establish pilot projects to commercialize biodiesel and set up its supply chain. The project may be extended step wise like conversion vehicle fleets of designated departments to run on biodiesel.

4. Conclusion

From the reviewed articles and publications, it is clear that the energy need of Pakistan can be attained and that the present energy crisis can be stopped. To achieve this, the practice of indigenous renewable energy sources of Pakistan is highly significant. However, further research and development on renewable energies are needed to create an upturn in their effective consumption. With effective usage of biodiesel in manufacturing energy, Pakistan can encounter a variability of energy requirements, counting producing electricity, providing energy for kitchen proposes to homes, fueling vehicles and providing energy to industries, but there is still a necessity for more thoughtful and broad research to encourage the renewable energy technologies coherent biodiesel policies by the Government essential to been involuntary with more emphasis devoted to supporting the initial local research efforts. In addition, the AEDB and PSO's initiatives to involve various universities all over the country should be harnessed so that effective solutions can be found in meeting the requirement of blending 10% of biodiesel with mineral diesel by 2025.

Author details

Mushtaq Ahmad[1*], Hammad Ahmad Jan[1], Shazia Sultana[1], Muhammad Zafar[1], Muhammad Aqeel Ashraf[2] and Kifayat Ullah[3]

*Address all correspondence to: mushtaqflora@hotmail.com

1 Biofuel Laboratory, Department of Plant Sciences, Quaid-i-Azam University Islamabad, Pakistan

2 Geology Department, Faculty of Sciences, University of Malaya, Kuala Lumpur, Malaysia

3 Department of Biosciences, COMSATS Institute of Information Technology, Park Road, ChakShahzad, Islamabad, Pakistan

References

[1] Bhutto, A. W., A. A. Bazmib and G. Zahedib. 2011. "Greener energy: Issues and challenges for Pakistan—Biomass energy prospective." *Renewable and Sustainable Energy Reviews*.15: 3207– 3219.

[2] Pode, R. 2010. "Addressing India's energy security and options for decreasing energy dependency."*Renewable and Sustainable Energy Reviews*. 14.

[3] HDIP, Pakistan Energy Yearbook, 2009. Hydrocarbon Development Institute of Pakistan (HDIP), Islamabad.

[4] NEPRA, State of Industry Report 2011.National Electric Power Regulatory Authority (NEPRA), Ministry of Water and Power Development, Government of Pakistan, Islamabad.

[5] Amjid, S. S., M. Q. Bilal. M. S. Nazir and A. Hussain. 2011. "Biogas, renewable energy resource for Pakistan." *Renewable and Sustainable Energy Reviews*. 15: 2833–2837.

[6] Ahmad, M., M. A. Khan. M. Zafar. S. Sultana and S. Gulzar. 2007. "Indigenous plants based biodiesel resources in Pakistan." *Ethnobotanical Leaflet*. 11: 224–230.

[7] Chakrabarti, M. H. M. Ali. J. N. Usmani. N. A. Khan. D. B. Hasan. Md. S. Islam. A. A. A. Raman. R. Yusoff and M. F. Irfan. 2012. "Status of biodiesel research and development in Pakistan." *Renewable and Sustainable Energy Reviews*. 16: 4396–4405.

[8] Stephenson, A. L., J. S. Dennis and S. A. Scott. 2008. "Improving the sustainability of the production of biodiesel from oilseed rape in the UK." *Process Safety and Environmental Protection* 86: 427–440.

[9] Saraf, S. and B. Thomas. 2007. "Influence of feedstock and process chemistry on biodiesel quality." *Process Safety and Environmental Protection*.360–385.

[10] Leung, D. Y., C. X. Wu and M. K. H. Leung. 2010. "A review on biodiesel production using catalyzed transesterification." *Applied Energy*. 87: 1083–1095.

[11] Monbiot, G. 2005. The most destructive crop on earth is no solution to the energy crisis. By promoting biodiesel as a substitute, we have missed the fact that it is worse than the fossil-fuel burning it replaces. The Guardian, Tuesday 6 December 2005.

[12] Ahmad, M., M. A. Khan. M. Zafar and S. Sultana. 2007. "Systematic studies of Oil yielding Plants and their applications as a Bio-diesel resource in Pakistan." CSF-HEC workshop on promoting industry academia linkages for developing the knowledge-based enterprise sector in Pakistan.

[13] Ali, S. I. 2009. "CDGK to grow jatropha plants." /http://www.nation.com.pk/pakistan-news-newspaper-daily-english-online/Regional/Karachi/14-Jul-2009/CDGK-to-grow-Jatropha-plants

[14] Khan, N. A., and H. el Dessouky. 2009. "Prospect of biodiesel in Pakistan." *Renewable and Sustainable Energy Reviews*. 13: 1576–1583.

[15] Ministry of Industries, Production and Special Initiatives, Pakistan.

[16] Ali, S. I. 2009. http://biodieselpk.blogspot.com/2010/03/cdgk-to-grow-jatropha-plants.htm.

[17] Khan. N. A., and J. N. Usmani. 2011. "Status of jatropha cultivation for biodiesel production in Pakistan." *Personal communication*.

[18] Soomro, R., and R. A. Memon. 2007. "Establishment of callus and suspension culture in *Jatrophacurcas*." *Pakistan Journal of Botany*. 39: 2431–2441.

[19] CEO AEDB Visits PSO's Biodiesel Pilot Nursery. 2009. /http://www.psopk.com/media/vol10_no5_jan_mar09/ ceo.php

[20] Progress Reporton PSOs' Biodiesel Initiative for the promotion of Biodiesel in Pakistan. 2011. /http://www.psopk.com/products_services/pdf/bio_diesel.pdf

[21] Plan for mass jatropha cultivation. 2009. /http://www.thebioenergysite.com/ news/3596/plan-for-mass-jatropha-cultivation

[22] Canadian firm to produce bio-diesel. 2009. /http://archives.dawn.com/archives/172385

[23] Silitonga, A. S., A. E. Atabania. T. M. I. Mahlia. H. H. Masjuki. I. A. Badruddin. and S. A. Mekhilefe. 2011. "Review on prospect of *Jatrophacurcas* for biodiesel in Indonesia." *Renewable and Sustainable Energy Reviews*. 15: 3733–3756. Soil Survey of Pakistan, www.pakistan.gov.pk/divisions/food-division.

[24] Chakrabarti, M. H., J. N. Usmani and M. A. Ali. 2010. "Techno-economic evaluation of two non-edible vegetable oil based biodiesel in Pakistan." *NED University Journal of Research*. 7(1): 43–54.

[25] Ahmad, M., and M. Zafar. 2005. "*Pongamiapinnata* (Sukh Chain) as a biodiesel resource in Pakistan." Two day seminar on Indigenous development of innovative technologies in the field of renewable energy.Best Western Hotel Islamabad.

[26] Alternative Energy Development Board. 2011. /www.aedb.org

[27] National Biodiesel Program. Alternative energy development board, ministry of water and power, government of Pakistan. 2011. /http://www.aedb.org/bioprogram.htm

[28] National Biodiesel Program.2009. /http://www.research.org.pk/Databank/NATIONAL% 20BIODIESEL /20PROGRAMME.pdf

[29] Siddiqui, M. A. 2007. "Biodiesel as alternative and indigenous source of energy for every village."/http://www.powerasia.com.pk/icaep2007/presentations/1st_Session/Bio_Diesel_as_Alternative_&_Indigenous.pdf

[30] Chisti, Y. 2007. "Biodiesel from microalgae."*Biotechnology Advances*. 25(3): 294–306.

[31] Pakistan's 27 mac resa line land fit for biofuel production. 2011. /http://etcgreen. com/industry-news/biofuel/pakistan-s-27m-acre-saline-land-fit-for-bio fuel-production

[32] Liaquat. A. M., M. A. Kalam. H. H. Masjuki and M. H. Jayed. 2010. "Potential emissions reduction in road transport sector using biofuel in developing countries." *Atmospheric Environment*. 44: 3869–3877.

[33] Production of biodiesel from algae. 2011. /http://www.scribd.com/doc/29870128/ Production-of-Bio-Diesel-From-Algae

[34] High costs seen hampering use of algae as biofuel. 2011. /http://www.dawn.com/2011/06/17/high-costs-seen-hampering-use-of-algae-as-biofuel.html

Chapter 5

Production and Use of Evolving Corn-Based Fuel Ethanol Coproducts in the U.S.

Kurt A. Rosentrater

Additional information is available at the end of the chapter

http://dx.doi.org/10.5772/59951

1. Introduction

We face many challenges in our society, due especially to growing population pressures and increased economic mobility. These can result in increased demands for food, clothing, housing, and consumer goods. Additionally, there has been a growing need for energy during the last several decades, which historically has been met primarily by use of fossil fuels. In the U.S., transportation fuels generally account for about 1/3 of all energy consumed. Of this, about 90% comes from fossil sources. Between 1/2 and 2/3 of the total U.S. demand for petroleum has been met by imports during the last 30 years [1]. Many argue that this scenario is not sustainable in the long run, and other energy alternatives are needed. During 2005-2010, the U.S. experienced some of the highest growth rates ever seen in the domestic biofuels industry.

The U.S. biofuels industry has recently grown as a response to increasing energy needs and energy prices. Bio-based fuels can theoretically be manufactured from many biological materials. Biochemical conversion of carbohydrates into ethanol and lipids into biodiesel are the two most common routes at this point in time. This includes most cereal grains (e.g., corn, barley, wheat, sorghum, rice, etc.), oilseeds (e.g., soybeans, sunflowers, flax, rapeseed, and others), native prairie grasses (including miscanthus, switchgrass, prairie cord grass, reed canary grass, and other grasses), agricultural residues (including corn stover, rice husks, wheat straw, and other crop residues), algae, municipal solid wastes, food processing wastes, and other biological wastes and substrates [2, 3, 4, 5, 6, 7, 8, 9, 10]. Right now, however, the most heavily used feedstock for biofuel production in the U.S. is corn grain. Industrial-scale alcohol manufacturing from corn starch is readily accomplished, and at a low cost (generally about $1/gal), which is considerably lower than other biological material conversions in the U.S.

In recent years, the corn-based fuel ethanol industry in the U.S. has reached a scale which can impact the nation's supply of transportation fuels. Only during the last decade, however, has this industry become visible to the average citizen. This has been due, in part, to the growing demand for domestic transportation fuels for national security, escalating prices at the fuel pump, positive economic effects throughout rural America, as well as questions and controversies surrounding the production and use of corn ethanol. While fuel is the main aim of the ethanol industry, nonfermentable coproducts are also generated during manufacturing. The primary coproduct is distillers dried grains with solubles (DDGS) (Figure 1), which is the focus of this chapter.

Figure 1. Corn-based Distillers Dried Grains with Solubles (DDGS) is currently the most common coproduct available from U.S. fuel ethanol plants (Photo courtesy of Rosentrater).

To help meet the increasing demand for transportation fuels, and to meet federal U.S. mandates, the number of ethanol plants has rapidly increased in recent years, as has the quantity of fuel ethanol and coproducts produced (Figure 2). As of 2015, the RFS (Renewable Fuel Standard) mandates the use of 15 billion gal/y (56.8 billion L/y) of corn-based ethanol in the U.S. [13]. At the beginning of 2015, there were 213 fuel ethanol plants in the U.S., which produced nearly 56.8 billion L/y (15.0 billion gal/y) [12, 15]. As the biofuels industry continues to evolve, cellulosic and other bio-based fuels will gain prevalence in coming years.

Most new ethanol plants have been dry-grind facilities (Figure 3), which will be discussed subsequently. As production volume has increased, the processing residues, known collectively as "distillers grains" – have increased concomitantly (Figure 2). Between 35 and 40 million metric tonnes (t) of distillers grains (both wet and dry) will be produced each year by the U.S. fuel ethanol industry over the next several years. In recent years there has actually been a slight decline in distillers grains production, due to evolving processes which now extract oil from the coproduct streams. This will be discussed at a later point in the chapter.

As the industry has grown over the last few decades, the consumption of corn has grown as well (Figure 4). In recent years, over 30% of the U.S. corn crop is used to produce fuel ethanol.

When examining these data, however, it is important to note several key points: most corn in the U.S. is not human food-grade corn, exports have been relatively constant over time, there has only been a slight decline in the corn used for animal feed, but the overall quantity of corn which is produced by U.S. farmers has been greatly increasing over time. The corn used to manufacture fuel ethanol has arisen from the growing U.S. supply of corn. Furthermore, and more importantly, the corn which has been used to produce ethanol (instead of animal feed) is generally replaced by DDGS and other ethanol coproducts in these animal feeds. Ethanol coproducts (primarily DDGS and DWG) are key to the long-term viability of the fuel ethanol industry as well as the various livestock industries. Thus feed and fuel can be produced simultaneously, contrary to what may be discussed in the media.

2. Objectives

The aims of this chapter are multiple: 1) to review fuel ethanol and coproduct manufacturing in the U.S.; 2) to discuss the importance of distillers coproducts to both the biofuel and the livestock industries; and 3) to illustrate how coproducts are changing and potential uses are expanding as the ethanol industry evolves.

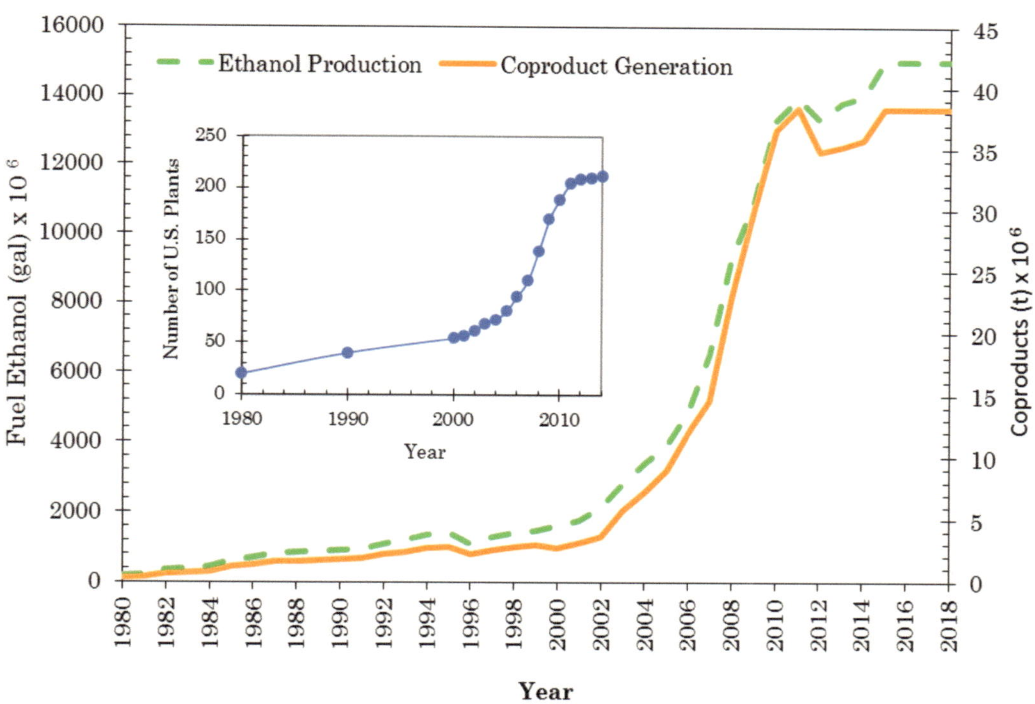

Figure 2. U.S. fuel ethanol (gal) and DDGS (t) production over time; RFS denotes levels mandated by the Renewable Fuel Standard. Inset shows number of U.S. ethanol plants over time [12, 13, 14, 15].

Figure 3. U.S. dry grind corn-to-ethanol manufacturing plants. A. 450 x 10^6 L/y plant. B. 80 x 10^6 L/y plant (Photos courtesy of Rosentrater).

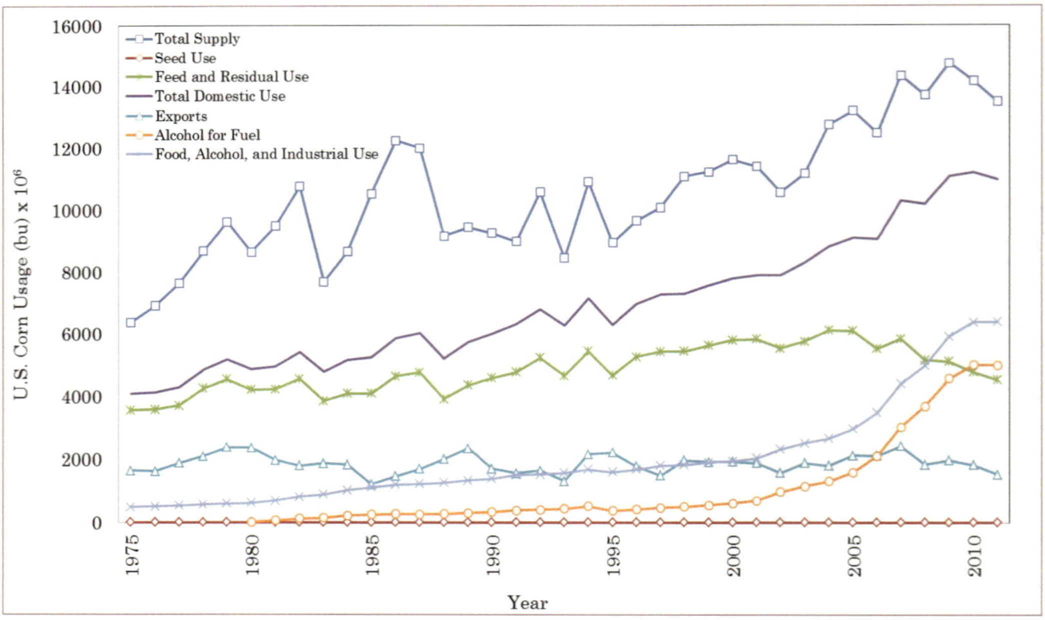

Figure 4. U.S. corn production (bu) and consumption according to major categories of use (adapted from [16]).

http://dx.doi.org/10.5772/59951

3. Manufacturing Processes

Corn ethanol is produced by two main commercial approaches. These include dry grind processing and wet milling. Over the last 20 years, many new fuel ethanol plants have been constructed in the U.S. (Figure 2). During this time, the industry has grown, evolved, and matured. Dry grind processing has become the most popular method for ethanol manufacturing for a variety of reasons, including lower capital and operational costs, lower energy use, and lower water consumption. In fact, most new ethanol plants in recent years have been dry grind facilities, and comprise approximately 80% of ethanol facilities in the U.S. [13].

Dry grind processing (Figure 5) consists of multiple unit operations. Primary steps include grain receiving, cleaning, storage, grinding, cooking, liquefaction, saccharification and fermentation, solids separation, distillation, ethanol storage and loadout, whole stillage centrifugation, coproduct drying, coproduct storage, and loadout. Additionally, ethanol plants entail heat recovery, waste management, CO_2 extraction, dust control, and facility instrumentation systems. Specific details will vary, though, as each ethanol plant is somewhat unique in design and operation. Oil extraction is a relatively new technology for creating additional coproduct streams, and will be discussed later.

Grinding, cooking, and liquefying increase surface area and allow the corn starch to be broken down into dextrins by enzymes (e.g., alpha-amylase) during saccharification, and subsequently into glucose molecules (using gluco-amylase), which are then consumed during fermentation by yeast (*Saccharomyces cerevisiae*). After fermentation, ethanol is removed from water and nonfermentable materials (e.g., proteins, fibres, oils, minerals) by separation and distillation. Dewatering results in wet cake (e.g., suspended solids) and soluble solid-laden water, evaporation of the soluble solids in the water stream, mixing the condensed solubles with the wet cake, and then drying this mixture can result in a variety of coproducts. These are known as "distillers grains", and can be wet or dry, and may or may not have condensed soluble materials added into the mix. Distillers Dried Grains with Solubles (DDGS) is the most common coproduct. It is typically dried to about 10% moisture content or less, to ensure proper shelf life and reduce flowability problems. Distillers grains are generally sold to local livestock producers. It is also transported by truck and rail to livestock producers throughout the nation. DDGS has increasingly been exported to overseas markets in recent years. Distillers Wet Grains (or DWG) is popular with beef and dairy producers near ethanol plants, due to better digestibility and lower price (because it is not dried). In fact, it has been estimated that, nationwide, approximately 25% to 30% of distillers grains sales are for DWG. But, because the moisture contents are generally greater than 50 to 60%, their shelf life is very limited (less than one week, generally), especially in summer months, and shipping large quantities of water is expensive. DDGS is still the most prevalent type of distillers coproduct in the marketplace.

Dry grind ethanol manufacturing generally results in three key products. These include fuel ethanol, the primary end product; residual nonfermentable corn kernel components, which are further processed and then sold as distillers grains (or coproducts); and carbon dioxide. A rule of thumb used in industry is that each 1 kg of corn processed will produce approximately 1/3 kg of each of the products. Another general rule is that each bushel of corn processed (about

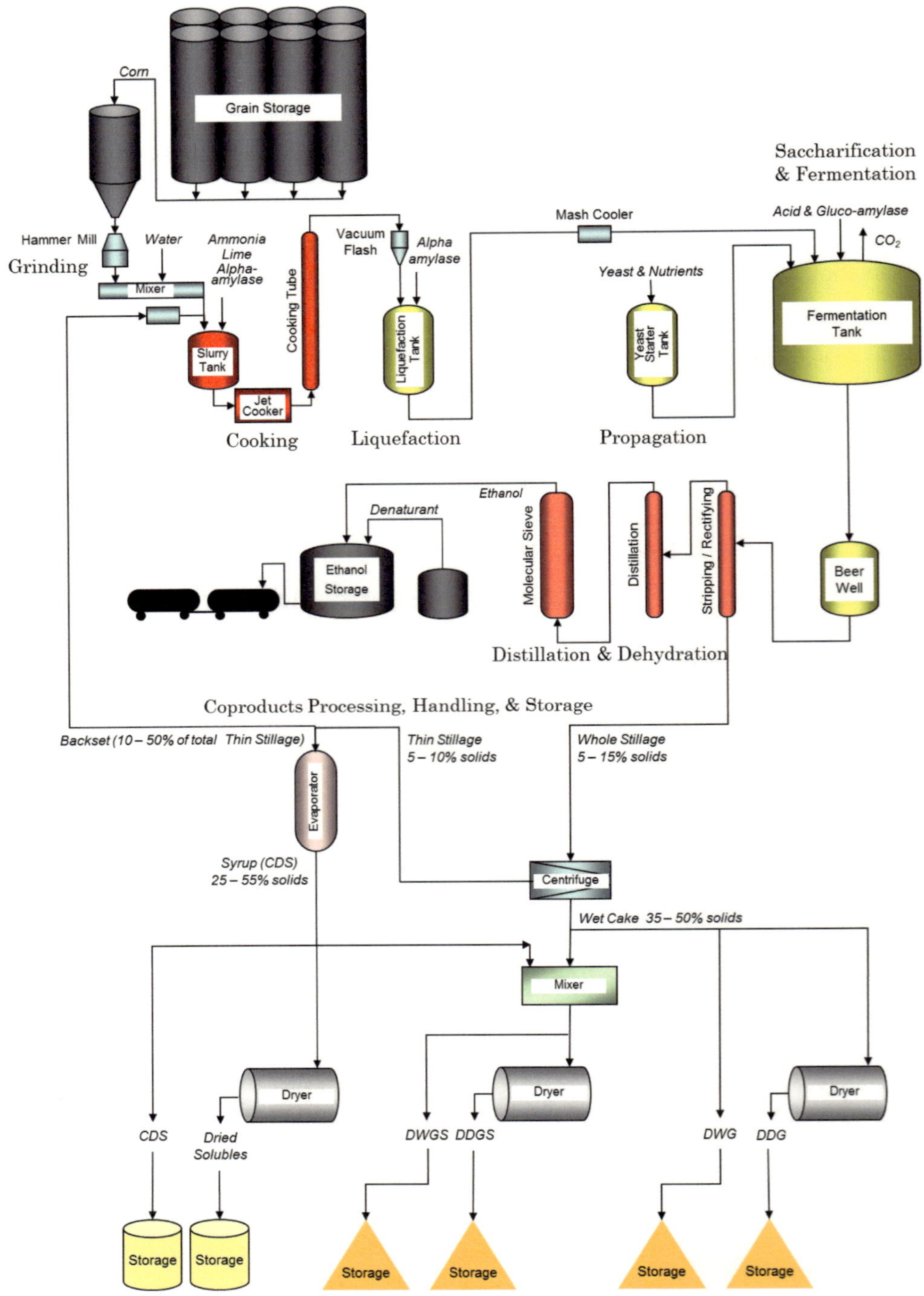

Figure 5. Flow chart of typical corn dry grind fuel ethanol and coproducts processing operations.

http://dx.doi.org/10.5772/59951

56 lb; 25.4 kg) will yield almost 2.9 gal (11.0 L) of ethanol (in reality, it will range from about 2.7 to 2.8 gal/bu, depending on the efficiency of the plant), 18 lb (8.2 kg) of distillers grains, and 18 lb (8.2 kg) of carbon dioxide (which is a byproduct of the yeast metabolism). These will vary from plant to plant due to production practices, equipment used, residence times, process temperatures, concentrations, maintenance schedules, equipment cleanliness, environmental conditions, as well as the composition and quality of the raw corn, the location where the corn was grown, and the growing season that produced that corn.

During fermentation the yeast's metabolic processes which convert glucose into ethanol result in carbon dioxide as a byproduct. This can be captured, compressed, and sold to gas markets such as beverage or dry ice manufacturers. This is especially attractive to ethanol plants near metropolitan areas. Most of the time it is just released to the atmosphere because logistics make CO_2 sales impractical. Additional detailed information on ethanol and DDGS manufacturing can be found in [17, 18].

4. Importance of Coproducts

Coproducts are important to the ethanol industry for a number of reasons. First and foremost, coproducts are additional sources of revenue to ethanol plants. Since these coproducts are primarily used as animal feed ingredients, monitoring and maintaining the consistency of coproduct compositions is critical to sales and utilization. DDGS from most modern U.S. fuel ethanol plants typically contains about 30% protein, 10% fat, at least 40% neutral detergent fibre, and up to 12% starch (of course, the lower the starch the better, as this is indicative of conversion efficiency) [19, 20]. DDGS composition can vary somewhat between plants. Within a single plant over time, however, DDGS is much less variable than amongst plants. Table 1 illustrates composition of DDGS from five ethanol plants in South Dakota, USA. Protein levels ranged from 28.3 to 31.8%, fat ranged between 9.4 and 11.0%, and ash ranged from 4.1 to 13.3%.

Plant	Crude Protein (%)	Crude Lipid (%)	NDF (%)	ADF (%)	Starch (%)	Ash (%)
1	28.33	10.76	31.84	15.56	11.82	13.27
2	30.65	9.75	39.90	15.21	9.81	12.84
3	28.70	10.98	38.46	17.89	11.59	11.52
4	30.65	9.40	36.73	15.28	9.05	4.13
5	31.78	9.50	38.88	17.24	10.05	4.48

Table 1. Composition (% db) of DDGS from ethanol plants in South Dakota [21].

It is instructive to examine composition differences in DDGS from plants throughout the U.S. For example, DDGS from 49 plants from 12 states were analysed for proximate composition (Table 2) and amino acid profiles (Table 3) [22]. On average, dry matter ranged from 87.9% to

90.6%, protein ranged from 29.4% to 32.6%, fat ranged from 9.6% to 12.8%, crude fibre ranged from 6.7% to 9.3%, and ash ranged from 4.2% to 6.6%. It appeared that geographic location of the ethanol plants didn't really play a role for any of the nutrients, as no nutrients seemed to be significantly affected by location.

Some ethanol plants are implementing new fractionation systems to produce new coproduct streams [13], and this is beginning to alter the nutrient composition of the distillers coproducts in the market place. More will be discussed later in this chapter.

State	Plants Sampled	Dry Matter (%)	Crude Protein (%)	Crude Lipid (%)	Crude Fibre (%)	Ash (%)
Minnesota	12	89.03	30.70	11.73	6.96	6.63
Illinois	6	89.72	29.98	11.48	7.26	5.60
Indiana	2	90.55	29.40	12.80	8.07	5.86
Iowa	7	88.92	31.23	10.27	7.57	5.76
Kentucky	3	90.57	29.43	9.77	9.28	4.47
Michigan	1	89.60	32.60	11.00	7.37	6.06
Missouri	2	87.90	30.45	10.25	7.17	5.39
Nebraska	4	89.02	30.40	11.35	8.13	4.23
New York	1	88.21	30.00	9.60	7.87	4.55
North Dakota	4	89.21	31.75	11.70	6.89	6.32
South Dakota	4	88.61	31.80	11.53	6.65	4.78
Wisconsin	3	89.68	31.70	11.63	7.59	5.77
Overall Average	49 (Total)	89.25	30.79	11.09	7.57	5.45

Table 2. Typical proximate composition (% db) of DDGS (averages of samples from 49 ethanol plants) [22].

State	Plants Sampled	Arginine (%)	Histidine (%)	Isoleucine (%)	Leucine (%)	Lysine (%)	Methionine (%)
Minnesota	12	1.39	0.84	1.20	3.63	0.99	0.61
Illinois	6	1.37	0.82	1.15	3.45	0.94	0.63
Indiana	2	1.19	0.79	1.08	3.28	0.85	0.60
Iowa	7	1.34	0.86	1.20	3.63	0.95	0.61
Kentucky	3	1.35	0.79	1.09	3.33	0.89	0.66
Michigan	1	1.28	0.86	1.18	3.67	0.87	0.71
Missouri	2	1.35	0.83	1.18	3.68	0.89	0.73
Nebraska	4	1.46	0.88	1.18	3.61	1.05	0.65
New York	1	1.46	0.85	1.21	3.64	1.04	0.61
North Dakota	4	1.37	0.88	1.24	3.76	0.97	0.65

http://dx.doi.org/10.5772/59951

State	Plants Sampled	Arginine (%)	Histidine (%)	Isoleucine (%)	Leucine (%)	Lysine (%)	Methionine (%)
South Dakota	4	1.47	0.87	1.22	3.70	1.08	0.62
Wisconsin	3	1.45	0.86	1.24	3.75	1.07	0.59
Overall Average	49	1.37	0.84	1.18	3.59	0.96	0.64

State	Plants Sampled	Phenylalanine (%)	Threonine (%)	Tryptophan (%)	Valine (%)	Tyrosine (%)
Minnesota	12	1.59	1.17	0.24	1.62	1.20
Illinois	6	1.51	1.11	0.22	1.52	1.22
Indiana	2	1.45	1.04	0.21	1.44	-
Iowa	7	1.57	1.14	0.25	1.60	-
Kentucky	3	1.48	1.09	0.26	1.43	-
Michigan	1	1.52	1.15	0.25	1.57	-
Missouri	2	1.53	1.15	0.24	1.58	-
Nebraska	4	1.58	1.15	0.26	1.58	1.14
New York	1	1.63	1.11	0.20	1.59	1.19
North Dakota	4	1.62	1.19	0.25	1.67	-
South Dakota	4	1.67	1.19	0.23	1.63	1.35
Wisconsin	3	1.65	1.14	0.22	1.64	1.25
Overall Average	49	1.56	1.13	0.24	1.57	1.22

Table 3. Typical amino acid levels (% db) of DDGS (averages of samples from 49 ethanol plants) [22].

The U.S. ethanol industry's primary market for distillers grains has historically been as a commodity livestock feed ingredient. Most often this has been in the form of DDGS, and in recent years in the form of DWG. All other ethanol coproducts have historically been sold at much lower levels; some of these other coproducts are not produced at some ethanol plants). Using ethanol coproducts for livestock feed or feed supplements have become effective methods for using these materials. Coproducts contain appropriate nutrients and they are highly digestible (depending on the species). Furthermore, use of coproducts in animal feeds (in place of corn grain) will actually help offset corn which has been used for ethanol production (the so-called food vs. fuel debate). In fact, it has been shown that DDGS can replace corn in livestock diets on a 1:1 up to a 1:1.2 level, depending on the species. The majority (over 80%) of U.S. distillers coproducts are used in beef and dairy feeds, because ruminants can use high levels of fibre. As feed ingredient prices have increased in recent years, coupled with increasing knowledge about how to effectively use these feed ingredients, ethanol coproduct use in swine and poultry diets have increased in recent years [22]. Many feeding trials have been conducted on coproducts in livestock diets over the years, for both monogastric and ruminant feeds (many of these studies are fully described in Liu and Rosentrater, 2011). Depending on the diet composition used, all livestock species have been shown to thrive at 10% DDGS inclusion, and most can tolerate levels up to and even greater than 20%.

DDGS use in livestock diets has continued to increase over the years. Various predictions of peak potential DDGS use in domestic U.S. beef, dairy, swine, and poultry markets have estimated that between 40 and 60 million t could be used in the U.S. each year, depending upon inclusion rates, age, etc., for each species [23, 24, 25]. Around the world, the need for protein-based animal feeds continues to grow, and DDGS has become a global commodity. Of the 23 million t of DDGS produced in 2008 [13], 4.5 million t were exported to international markets [26]; this accounted for nearly 20% of U.S. DDGS production that year (Figure 6). And the potential for global exports is projected to increase for the foreseeable future, and will likely hover near about 25% of all DDGS production in coming years [25, 27]. In recent years, China has become the dominant global importer of DDGS. Extensive information about the use of DDGS in livestock diets can be found in [18] as well as [28].

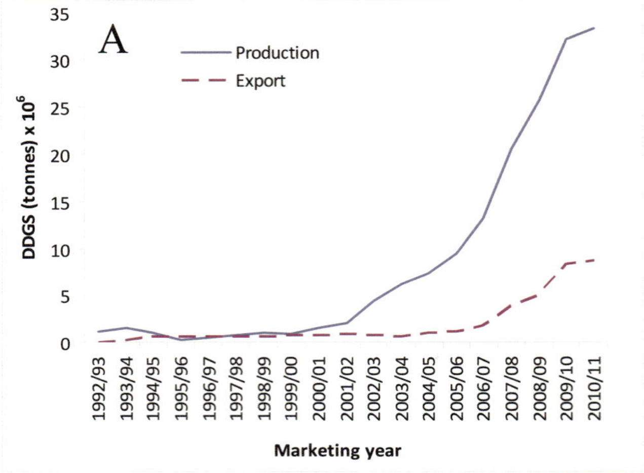

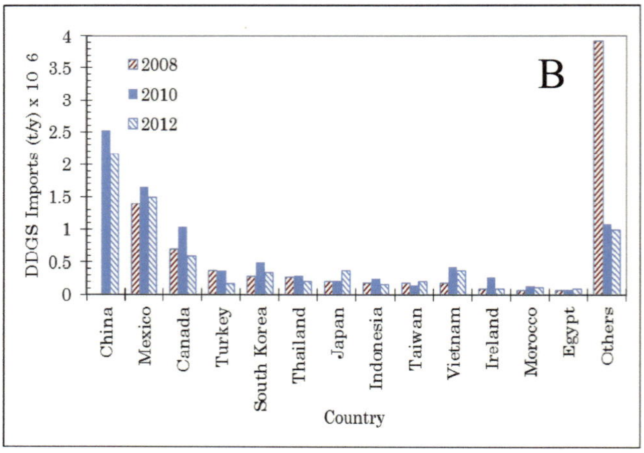

Figure 6. A. DDGS exports from the U.S. over time. B. Countries who imported DDGS from the U.S. in 2008, 2010, 2012 (adapted from [27, 29]).

http://dx.doi.org/10.5772/59951

The sale of nonfermentable coproducts is critical to the fuel ethanol industry as a source of revenue; and these materials have also become important feed ingredients to the livestock industry over the last decade. Sales of dry and wet distillers coproducts generally translates into 10 to 20% of an ethanol plant's total revenues, and can even be as high as 40% (depending on the economics). These materials really are "coproducts", not "byproducts" or "waste materials". In fact, many plants recognize this and promote their simultaneous production of animal feed and biofuel.

In recent years, the market price of DDGS has ranged from approximately $50/t (in the early 2000s) to more than $300/t (2012-2013) (Figure 7). The prices of corn and DDGS have generally paralleled each other fairly well over the years (Figure 8A). This trend occurs due, in large part, to the fact that DDGS is often used to replace corn in livestock diets. In the last decade, DDGS has increasingly been used as a soybean meal replacement also. Because soybean meal has a higher protein content, DDGS is often sold at a lower price compared to either corn or soybean meal (Figure 8A). This has been true volumetrically as well as per unit protein (Figure 8B). In the last few years DDGS has actually been sold at more than 100% the value of corn. This is frequently due to external impacts on the marketplace, including international exports.

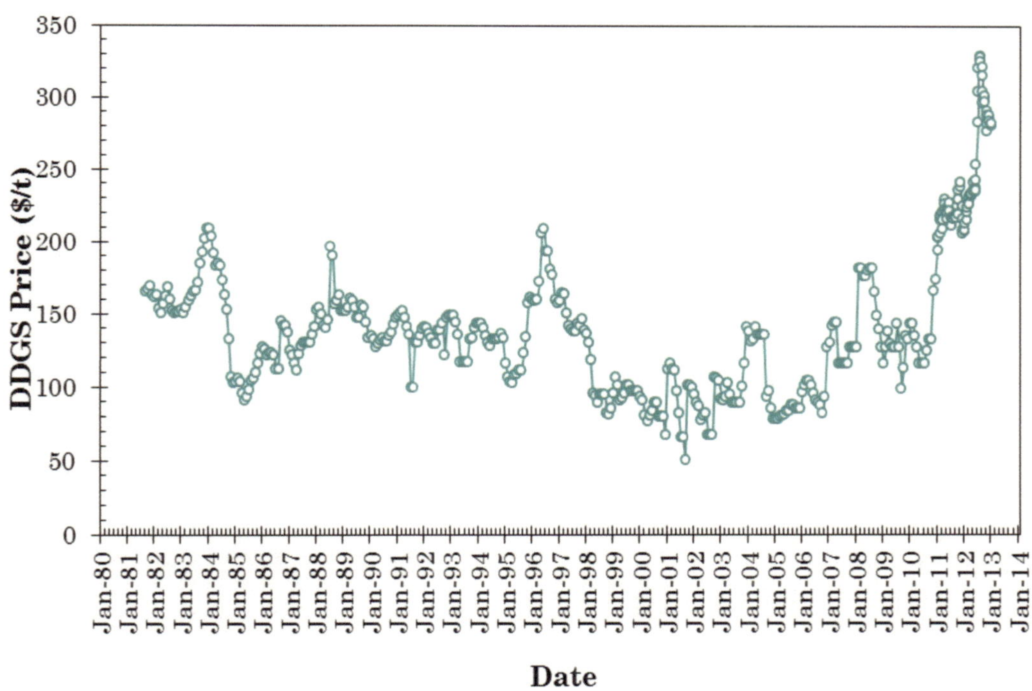

Figure 7. DDGS sales price over time (monthly averages) (adapted from [16, 30]).

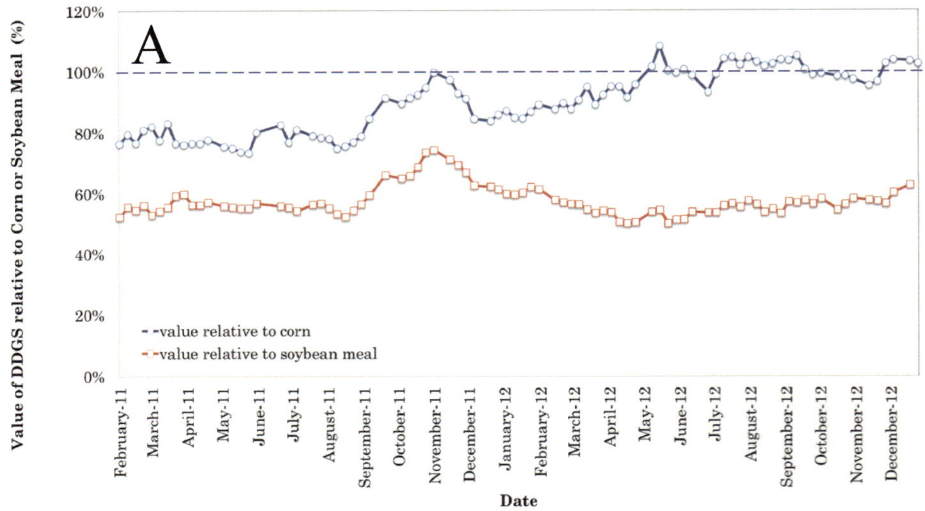

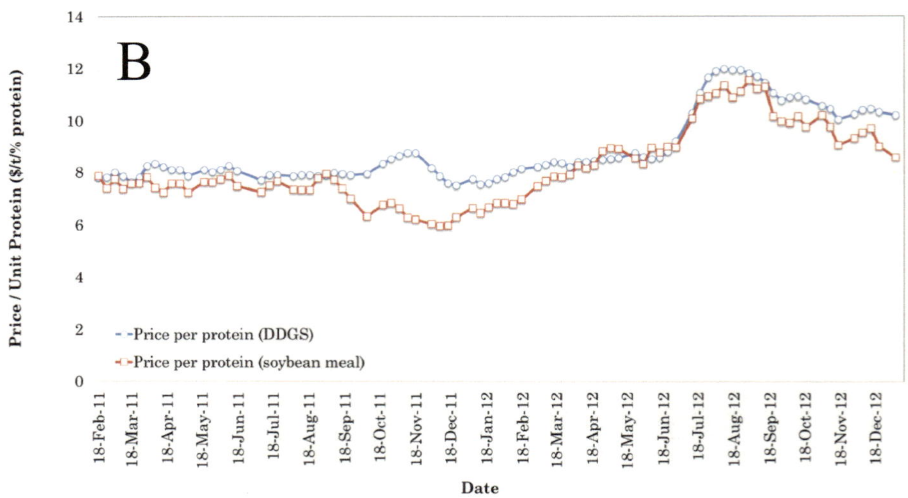

Figure 8. A. Example comparisons of DDGS, soybean meal (SBM), and corn sales prices and their relative comparisons. B. Cost comparisons on a per unit protein basis (adapted from [30]).

5. Coproduct Evolution

Even though the corn ethanol industry is maturing, there continue to be efforts to develop new, valued-added materials from the corn kernels as well as from the coproduct materials. When these research efforts are commercialised, they will result in more products from the corn kernel itself (an approach known as upstream fractionation) and the distillers grains (known as downstream fractionation). These types of fractionation approaches can result in

http://dx.doi.org/10.5772/59951

the separation of components of high, medium, and low value (Figure 9). For example, several mechanical and chemical approaches have been developed to remove protein, fibre, or oil components from the endosperm (which contains the starch). This type of separation will allow a highly-concentrated starch substrate to be introduced to the fermentation process, and will allow the other corn kernel components to be used for human food or other high-value applications. [31] provided an extensive discussion regarding various pre-fermentation fractionation approaches. On the other hand, post-fermentation fractionation techniques have also been examined. For example, [32] used a combination of air classification and sieving to separate fibre particles from DDGS. All of these approaches, if implemented commercially, will alter the chemical composition and digestibility of the resulting DDGS.

Many plants have recently begun adding capabilities to concentrate nutrient streams such as oil, protein, and fibre into specific fractions, which can then be used for targeted markets and specific uses. For example, one company in Iowa is now separating fibre from the DDGS and using it as a feedstock for cellulosic ethanol production. Additionally, many companies have begun removing oil from the whole stillage and/or CDS streams (Figure 10). This oil, which is officially known as Distillers Oil, Feed Grade (Figure 11), can readily be converted into biodiesel or animal feed ingredients, but they cannot be used for food grade corn oil, because they are too degraded. In fact, more than 85% of U.S. ethanol plants are now removing oil, because the economics are so favourable. Of note, in 2010 almost no ethanol plants were extracting oil...the rapid increase has solely been due to added value streams for the ethanol plants. On the horizon is concentrated corn proteins, which can be used for high-value animal feeds (such as aquaculture or pet foods), or other feed applications which require high protein levels (such as monogastrics or younger animals).

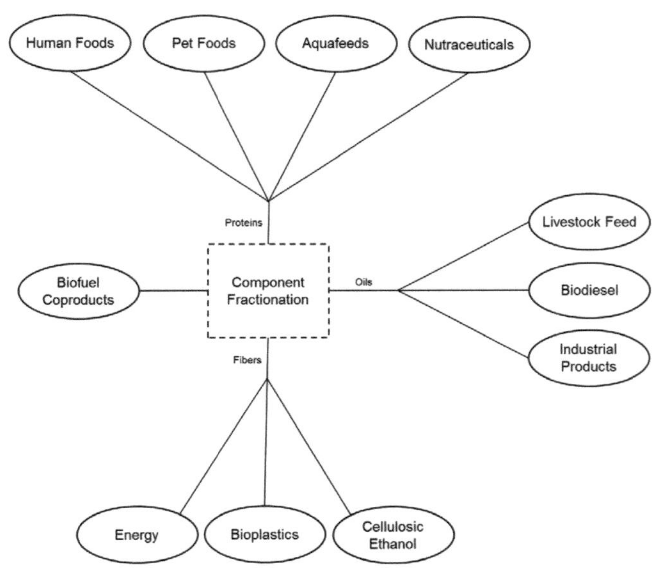

Figure 9. Fractionation of DDGS into individual chemical components (or at least concentrating them) offers the opportunity for new value-added uses and new revenue streams.

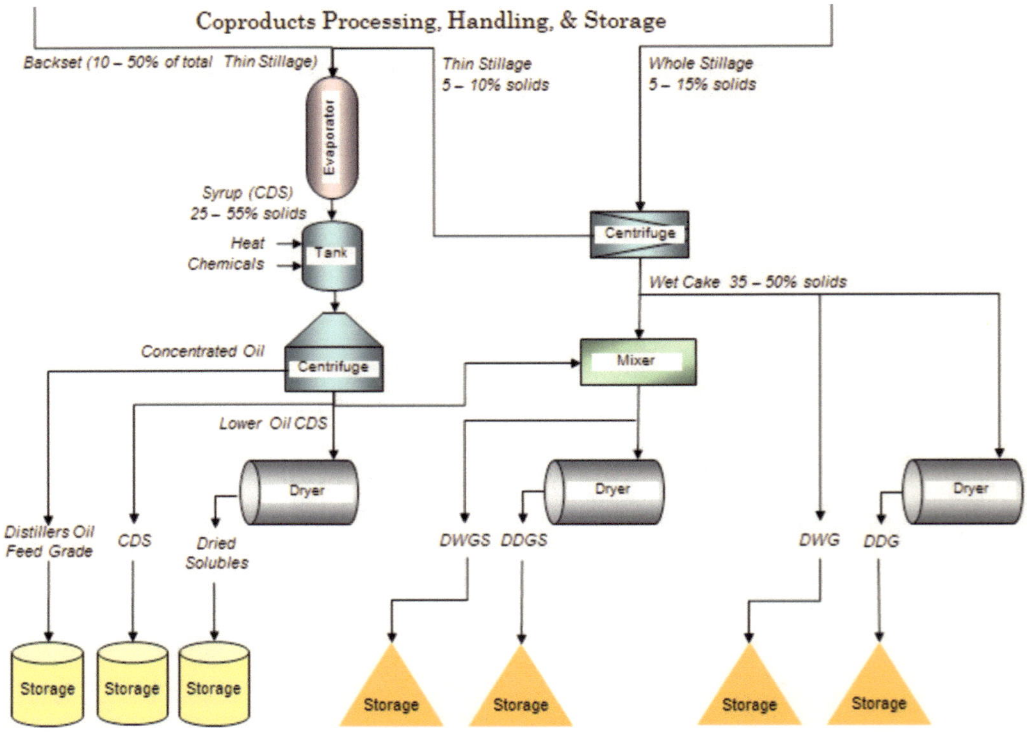

Figure 10. Oil is now removed at many ethanol plants via centrifugation of condensed distillers solubles, after a heating step and additional chemicals are added, which allow the oil to be removed without forming an emulsion.

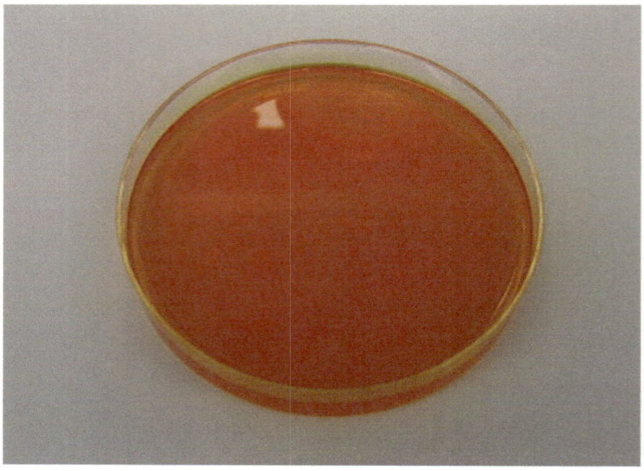

Figure 11. Distillers Oil, Feed Grade, is being extracted from nearly 85% of all U.S. ethanol plants in 2014 (Photo courtesy of Rosentrater).

As these process modifications are developed, tested, and implemented at commercial facilities, improvements in coproducts will be realized and increasingly used in the market-

http://dx.doi.org/10.5772/59951

place. These new products will require extensive investigation in order to determine how to optimally use them and to quantify their values in the marketplace.

6. Conclusions

The fuel ethanol industry in the U.S. has grown exponentially during the last decade in response to government mandates and due to increased demand for alternative fuels. This has become especially pronounced as the price of gasoline has drastically fluctuated, and consumers have realized that fuel prices are problematic. Additionally, energy has become an issue of national security. Corn-based ethanol is not the entire solution to transportation fuel needs. But it is clearly a key component to addressing energy needs. Corn ethanol is seen by many as a transition to other bio-based fuels in the long run; but this industrial sector will continue to play a key role in the bioeconomy, as it is a proven approach to large-scale industrial bioprocessing. And as the industry grows, coproducts will become increasingly important for economic and environmental sustainability. One way to improve sustainability is to diversify coproducts as well as integrate systems (Figure 12, for example), where materials and energy cycle and recycle. For example, upstream outputs become downstream inputs for various components of a biorefinery factory, animal operation, energy production (i.e., heat, electricity, steam, etc.), feedstock operation, and other systems. A closed loop system would be the ultimate scenario. By integrating these various components, and developing a diversified portfolio (beyond just ethanol and distillers grains) the biorefinery will not only produce fuel, but also fertilizer, feed, food, industrial products, energy, and more importantly, can be self-sustaining.

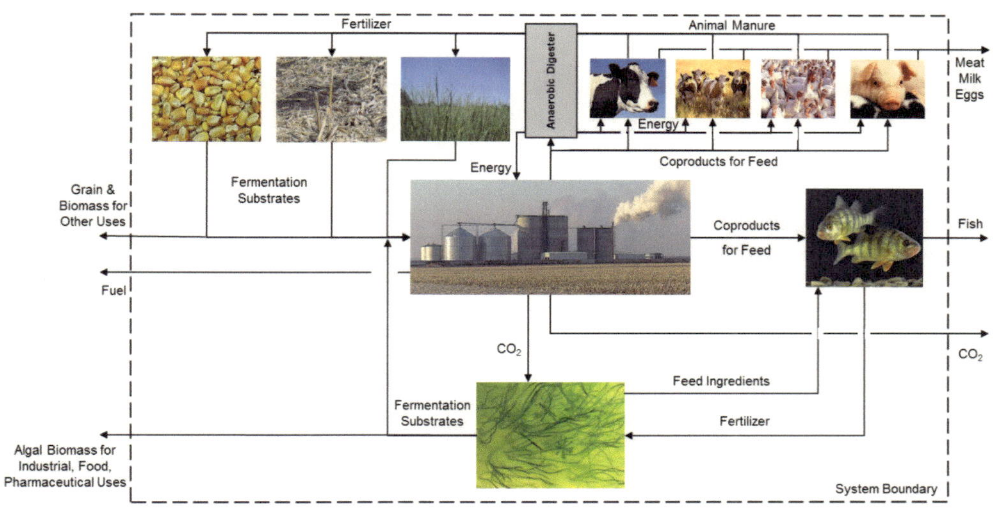

Figure 12. Coproducts will continue to play a key role as the biofuels industry evolves and becomes more fully integrated. This figure illustrates one concept of combining fuel production with animal and plant production systems.

List of Abbreviations

CDS: Condensed distillers solubles

DDG: Distillers dried grains

DDGS: Distillers dried grains with solubles

DWG: Distillers wet grains

DWGS: Distillers wet grains with solubles

RFS: Renewable Fuel Standard

SBM: Soybean meal

Author details

Kurt A. Rosentrater

Address all correspondence to: karosent@iastate.edu

Iowa State University, Ames, Iowa, USA

References

[1] U.S. EIA. (2011). Annual Energy Review. Energy Information Administration, U.S. Department of Energy: Washington, D.C. Available online: www.eia.doe.gov/emeu/aer/.

[2] Agrawal, R., N. R. Singh, F. H. Ribeiro, and W. N. Delgass. (2007). Sustainable fuel for the transportation sector. Proceedings of the National Academy of Sciences 104(12): 4828-4833.

[3] Alexander, C. and C. Hurt. (2007). Biofuels and their impact on food prices. Bioenergy ID-346-W. Department of Agricultural Economics, Purdue University: West Lafayette, IN.

[4] Cassman, K. G. (2007). Climate change, biofuels, and global food security. Environmental Research Letters 2(011002): DOI # 10.1088/1748-9326/2/1/011002.

[5] Cassman, K. G. and A. J. Liska. (2007). Food and fuel for all: realistic or foolish? Biofuels, Bioproducts and Biorefining 1(1): 18-23.

[6] Cassman, K. G., V. Eidman, and E. Simpson. (2006). Convergence of agriculture and energy: implications for research policy, QTA2006-3. Council for Agricultural Science and Technology: Ames, IA.

[7] Dale, B. E. (2007). Thinking clearly about biofuels: ending the irrelevant 'net energy' debate and developing better performance metrics for alternative fuels. Biofuels, Bioproducts, and Biorefining 1: 14-17.

[8] De La Torre Ugarte, D. G., M. E. Walsh, H. Shapouri, and S. P. Slinsky. (2000). The economic impacts of bioenergy crop production on U.S. agriculture, Agricultural Economic Report 816. USDA Office of the Chief Economist, U.S. Department of Agriculture: Washington, D.C.

[9] Dewulf, J., H. Van Langenhove, and B. Van De Velde. (2005). Energy-based efficiency and renewability assessment of biofuel production. Environmental Science and Technology 39(10): 3878-3882.

[10] Lynd, L. R. and M. Q. Wang. (2004). A product-nonspecific framework for evaluating the potential of biomass-based products to displace fossil fuels. Journal of Industrial Ecology 7(3-4): 17-32.

[11] Urbanchuk, J. M. (2009). Contribution of the Ethanol Industry to the Economy of the United States. LECG: Wayne, PA.

[12] RFA. (2011). Biorefinery locations. Renewable Fuels Association: Washington, D.C. Available online: www.ethanolrfa.org.

[13] RFA. (2009a). Growing Innovation. 2009 Ethanol Industry Outlook. Renewable Fuels Association. Washington, D.C. Available at: www.ethanolrfa.org.

[14] RFA. (2009b). Industry resources: co-products. Renewable Fuels Association: Washington, D.C. Available online: www.ethanolrfa.org.

[15] RFA. (2014). Falling walls and rising tides: 2014 annual industry outlook. Renewable Fuels Association: Washington, D.C. Available online: www.ethanolrfa.org.

[16] ERS. (2011). Feed Grains Database: Yearbook Tables. Economic Research Service, U.S. Department of Agriculture: Washington, D.C. Available online: www.ers.usda.gov/data/feedgrains/.

[17] Ingledew, W. M., D. R. Kelsall, G. D. Austin, and C. Kluhspies. (2009). The Alcohol Textbook, 5th Edition. W. M. Ingledew, D. R. Kelsall, G. D. Austin, and C. Kluhspies, ed. Nottingham University Press: Nottingham, UK.

[18] Liu, K. and K. A. Rosentrater. (2011). Distillers Grains: Production, Properties, and Utilization. Boca Raton, FL: CRC Press.

[19] Rosentrater, K. A. (2007). Ethanol processing coproducts – a review of some current constraints and potential directions. International Sugar Journal 109(1307): 1-12.

[20] Rosentrater, K. A. and K. Muthukumarappan. (2006). Corn ethanol coproducts: generation, properties, and future prospects. International Sugar Journal 108(1295): 648-657.

[21] Bhadra, R., K. A. Rosentrater, and K. Muthukumarappan. (2009). Cross-sectional staining and the surface properties of DDGS and their influence on flowability. Cereal Chemistry 86(4): 410-420.

[22] UMN. (2011). The value and use of distillers grains by-products in livestock and poultry feeds. University of Minnesota: Minneapolis, MN. Available online: www.ddgs.umn.edu/.

[23] Staff, C. H. (2005). Question and answer. Biofuels Journal 3(4): 26-27.

[24] Cooper, G. (2006). A brief, encouraging look at 'theoretical' distillers grains markets. Distillers Grains Quarterly 1(1): 14-17.

[25] U.S. Grains Council. (2007). An Independent Review of US Grains Council Efforts to Promote DDGS Exports. U.S. Grains Council: Washington, D.C. Available online: www.grains.org/ddgs-information.

[26] FAS. (2009). Foreign Agricultural Service, U. S. Department of Agriculture: Washington, D.C. Available online: www.fas.usda.gov/.

[27] U.S. Grains Council. (2014). Exports of DDGS in 2013 / 2014. Presented at 18th Annual Distillers Grains Symposium, Irving, TX, USA, May 15, 2014.

[28] FAO. (2011). Opportunities and Challenges in Utilizing Co-products of the Biofuel Industry. Rome, Italy: Food and Agriculture Organization of the United Nations.

[29] Hoffman, L. and A. Baker. (2010). Market issues and prospects for U.S. distillers' grains: supply, use, and price relationships. Report FDS-10k-01. United States Department of Agriculture, Economic Research Service: Washington, D.C. Available online: www.ers.usda.gov.

[30] DTN. (2014). DTN Weekly Distillers Grains Update. Available online: www.dtnprogressivefarmer.com.

[31] Singh, V. and D. B. Johnston. (2009). Fractionation technologies for dry-grind corn processing. Pages 193-207 in: The Alcohol Textbook, 5th Ed. M. W. Ingledew, D. R. Kelsall, G. D. Austin, and C. Kluhspies, ed. Nottingham University Press: Nottingham, UK.

[32] Srinivasan, R., R. A. Moreau, K. D. Rausch, R. L. Belyea, M. D. Tumbleson, and V. Singh. (2005). Separation of fiber from distillers dried grains with solubles (DDGS) using sieving and elutriation. Cereal Chemistry 82: 528-533.

Section 2

Biofuels Second Generation- Advancement

Chapter 6

Microalgal Biofuel

Nitin Raut, Talal Al-Balushi, Surendra Panwar,
R.S. Vaidya and G.B. Shinde

Additional information is available at the end of the chapter

http://dx.doi.org/10.5772/59821

1. Introduction

Transportation is base of economies and all other of developments of any countries, it fulfills the requirement of society. On the other hand transportation is nothing without energy/ petroleum. Petroleum, based fuel account for 97% of transportation energy. Without petroleum shipping of many important good like food and other, driving from one point to other can't be possible. This petroleum is finite and it will be finished in few years and there are serious environmental concerns fuel like hazardous emission causing global warming and air problem also. These problems can be solved if such a sensible fuel will take place on existing fuel and reduces petroleum consumption and thereby reduce emission of hazardous from fuel.

Biofuels are such a sensible fuels and it is generated from biological material. According to use of biomass as a feedstock for biofuels generation, it is classified in three generation. First-generation biofuels are produced from biomass that is edible (sugarcane and corn). It is economical viable but it is responsible for increase in food prices in poor countries. Second-generation biofuels are produced from non-food crops or it is produced from a generally less expensive biomass such as animal, forest, agriculture or municipal wastes.Third-generation biofuels are produced from extracting oil of algae. Algae have been found to have incredible production level compared to other oil seed crops like sunflower, soybean rapeseed. Table 1 below shows a comparison of oil yield for various oilseed crops.

As table 1 shows that algae is capable of producing biodiesel on a large scale compared to other oilseed crops on less area.

Plant	Gallons of Oil/Acre	lb. Oil per Acre
Algae	700	6757
Coconut	285	2070
Jatropha	201	1460
Rapeseed	126	915
Peanut	112	815
Sunflower	99	720
Soybean	62	450

Table 1. Average Production for various Oil Crops

2. World's scenario

The concept of using algae as feedstock's for biofuels was already being discussed 50 years ago but a concerted effort began with the oil crisis in the 1970s. In this series Japan and United State focused on research programs. Main focus of The United State Department of Energy's was production of biodiesel from microalgae (1978-1996), which is known as the Aquatic Species Program (ASP). Japan Government also financed some large research project, but none of them has proven economical on a large scale, due to mainly the production methods used to grow and harvest the algae.

Notwithstanding the technical challenges, the availability of suitable land, in terms of soil type, elevation and slope, in suitable climates (incident radiation, temperature, precipitation/ evaporation balances and severe weather), and the geographical nearness of this land to appropriate water and CO_2 inputs and possibly nearness to markets or transportation infrastructure may impose physical and economic limits to the contribution that algal biofuel can make to the world's future transportation fuel needs. For example, very few large CO_2 emissions sources are in close proximity to regions identified as being most suitable for year round, large scale open pond production systems. In fact, there is an absence of data that could be used in defining limits of production. Land use, land suitability and resource spatial mapping data compiled for the purpose of assessing the geographic potential of algal biofuels does not exist. Claims that algal biofuels could completely replace all petroleum derived transport fuels or even provide a significant contribution to liquid fuels on simple assessment seem improbable, but can be neither supported nor refuted. There is a need to develop this information.

There are as yet no pilot (>100 mt algal biomass/yr) photosynthetic algal biofuels production plants operating in the U.S. The few pre-pilot-scale (e.g. >10 mt) plants have operated for less than a year, with only rather smaller operations of a few hundred square meters operating for two or more years (e.g. *Seambiotic in Israel, Aurora Biofuels* in *Florida,* for example). As mentioned above, *Solazyme* is the front runner with the largest confirmed production of algal lipids for

http://dx.doi.org/10.5772/59821

energy customers to date, using a closed heterotrophic process and genetically modified algae. Three fairly advanced developers who are or will be breaking ground on the next scale demonstrations (20-200 acres) within the next year are *Phycal, Cellana, Sapphire,* and *General Atomics*. All use open pond designs and natural strains. The main interest in microalgae stems from its potential productivity on a per acre-year basis. Claims of current and future relative productivity levels range from 1000 to 5000-plus gallons per acre per year and are summarized in Table 10.

Actual productivity numbers, like other agricultural crops and industrial processes, are highly dependent on the specific site and production process used. At least one company has demonstrated actual productivity in its proprietary process of at least 1400 gal/acre/ year in 2010 for a non-optimized small experimental site in a warm-weather location and estimates productivity could be doubled in the next demonstration at the multi-acre scale. These demonstrated results and model for the next phase were validated by an independent federal agency and review team through the U.S. Pacific Command's Green initiative for Fuels Transition (GIFTPAC) interagency working group, under the leadership of the U.S. Pacific Command Energy Office, J81 Joint Innovation and Experimentation Division, Resources and Assessment Directorate. It is important to remember that these productivity numbers are only for the oil; algae organisms range from 10% of their body mass in oil and up, so for each gallon of fuel produced, a significant proportion of protein and carbohydrates are produced as well. *Cellana Co.* in *Hawaii* (a joint venture of *Shell Oil Co.* and *H.R. Biopetroleum, Inc.*) has operated a pre-pilot plant of between one and two acres to grow diatoms using the *Mera Pharmaceuticals* ponds at the *Natural Energy Laboratory of Hawaii Authority* (NELHA) near *Kona, Hawaii*. The technology was based on prior experience with production of Haematococcus pluvialis biomass by *Aquasearch Co.* in *Hawaii*. Its neighbor at *NELHA*, *Cyanotech*, is one of the traditional nutraceutical companies mentioned above; *Cyanotech* sold $7 million worth of algae-derived astaxanthin in 2009.

Sapphire Energy Co. of *San Diego* was awarded over $100 million in U.S. government grants and loans and is breaking ground on a 300-acre demonstration pilot plant in New Mexico. Sapphire Energy initially announced that it would produce algae oil with oil-excreting genetically modified algae (GMA), but now intends to follow the standard model of growing unmodified algae with naturally high oil content. *Phycal* of Ohio was awarded over $50 million in Department of Energy carbon recycling funds to develop a pilot plant on *Oahu, Hawaii*. *General Atomics*, in *San Diego*, received about $30 million from the U.S. *Department of Defense, Defense Advanced Research Projects Agency*, (*DARPA*) in 2008 to develop a low-cost ($3/gallon initially, $1/gallon later) process for microalgae oil production in an 18-month R&D effort to be followed by a demonstration of this technology over a further 18 months in Hawaii, *Texas*, and *California*. The economic analysis and underlying assumptions on which current projections of $3/gallon oil are based are proprietary-however they include significant animal feed co-product credits.

The contribution of algal biofuels to future liquid transportation fuel supply is assessed against the *US Energy Information Agency* growth projections. By 2030, oil consumption is expected to increase to ca. 6.2 TL yr^{-1}(106 million bbl d^{-1}) with66 % of this growth likely to occur in non-OECD countries in *Asia*. Transportation fuel use is expected to grow slightly to ca. 56% of total oil production. Over the same time period, biofuels will maintain a relatively steady share of unconventional liquid fuel production and grow to between 277 GL/yr & 416 GL/yr (4.8 to 7.2 million Bbl/d· or 8.0% to 12.0% of the liquid transportation fuel supply). The EIA uses a figure of ca. 340 GL/yr as a reference case for total biofuel production in 2030.

A 5% contribution of algal biofuels to total biofuels supply by 2030 would require the construction of 170 100 ML facilities. When the technical uncertainty is considered it seems unlikely that the first large scale plant would be commissioned before the middle of the coming decade, and even this would be ambitious. Approaches that rely on molecular biology to achieve breakthroughs, e.g., the partnership between *Synthetic Genomics Inc.* and *ExxonMobil Corp.*, are promising but will likely take more than a decade to reach commercial viability. Assuming success in the first commercial venture and accelerated rates of adoption beyond 2015-2020, 170 100 ML facilities could conceivably be operational by 2030 as this rate of construction is lower than the recent development rate of ethanol plants in the *US* and *Brazil*. The forty-plus companies tackling the concept of algae production on a large scale for energy use have begun to differentiate into market niches, generally according to their founding technical expertise and physical location.

Companies where the founding members had deep pharmaceutical or bioengineering expertise tend to build their business models around proprietary genetically modified organisms and closed systems. Examples include *Synthetic Genomics, Solazyme, LS9, Targeted Growth, Inc., Amyris, Heliae Development*, and *Algenol*. Companies derived from other industries such as defense, wastewater treatment, and agriculture tend to prefer open pond systems and natural strains. Examples include *General Atomics, SAIC, HR Biopetroleum/Cellana, Aquaflow Bionomics,* and *Phyco Biosciences*.

Companies headquartered in colder latitudes tend to focus on closed algae production systems. Examples include *Solazyme, Amyris Biotechnologies, Algae@Work, Algaedyne,* Heliae, and *Greenfuels Technologies Inc* (now defunct). Companies headquartered in warmer latitudes tend to focus on open-pond photosynthetic systems. Examples include *Sapphire Energy, General Atomics, HR Biopetroleum/Cellana, SAIC*, and *Seambiotic* in *Israel*.

Some companies are pursuing a hybrid approach. One example is *Ohio-based Phycal Inc.*, which plans to use an open-pond system at its *Hawaii* demonstration site to grow out the algae, then put them into a closed heterotrophic for "fattening" prior to harvest. *HR Biopetroleum/Cellana* also uses a hybrid system, where the seedstock are grown in closed photobioreactor systems to reduce contamination and then inoculated into open ponds for bulking up in volume prior to harvest.

Every algae company has at least one other major revenue stream in its business model beyond just lipid production for biofuels markets. That co-product tends to affect its selection of sites, strains, production processes, etc. Some examples include a valuable co-product stream from animal feed (*General Atomics*), human food or nutraceuticals (*Solazyme, LiveFuels*), specialty chemicals (*Amyris*), carbon capture and storage (*Phycal Inc., Algae@Work*), and wastewater treatment (*Aquaflow Bionomics*). Within the closed process market niche is a group of companies that use a non-photosynthetic approach to grow their algae. This "heterotrophic" process involves feeding the microalgae sugar in the absence of light to get them to boost their proportion of oil relative to carbohydrates and proteins. An example is *Solazyme*, which is notable in being the first algae Energy Company to complete commercial sales of algae oil specifically for fuel, by delivering over 20,000 gallons of jet fuel (JP5) and marine diesel (F-76) to the *Defense Logistics Agency*.

3. Classification of biofuels

In three generation of biofuels, First-generation biofuels are the biofuels which are directly related to a biomass that is generally edible. First-generation biofuels are in trend around the world and these are also economically viable, but there are some issues related to this kind of biofuels such as utilization of arable lands which are directly affect food availability in most of poor countries so it leads food versus fuel debate. In some countries where sugar market play vital role in their economy the production of ethanol from sugarcane is facing competition with sugar market and on the other hand where ethanol from corn is also responsible for increasing value of food on the world's market. Some problems are with biodiesel market, which is limited by the price of vegetable oils. So these are some reasons which are leading interest towards second generation biofuels.

Second generation biofuels are also known as advanced biofuels. In this type of biofuels, various types of biomass can be used as a feedstock for manufacturing of biofuels. Biomass is source of organic carbon that is part of carbon cycle so it is available as renewed after completion of carbon cycle and is produced from a generally less expensive biomass such as animal, forest, agriculture or municipal wastes. Generally these biomasses are residual non food parts of crops that are not used for food purpose and food crops can be used as second generation biofuels, if they have already fulfilled their food purpose.

Two transformative technologies for production of second generation biofuels are usually done:

Biochemical: in this modification of the bio-ethanol fermentation process including a pretreatment process and

Thermochemical: in this modification of the bio oil process to produce methanol, fisher – Tropsch diesel or dimethyl ether.

Third generation biofuels are produced from extraction oil of algae. Its production has a very high growth yield and low cost. There are many advantage associated with third generation biofuels production such as fastest growing biomass, less land required compared to agriculture product used in other generation and some environment benefits like it cleans water it uses by removing nutrients & other pollutants, adds oxygen and it consumes CO_2.

4. Algal basics

Algae grow almost everywhere in the world. They live all over the world from the ocean to the desert and from hot springs to snow and ice. Algae are important for aquatic ecosystem because they provide food and shelter to other organism. Algae are also important because they have ability of an aquatic ecosystem to absorb nutrients and heavy metals. Algae use sunlight and chlorophyll to make food. Algae are organisms that are like plants and vegetables. Algae grow in almost any aquatic environment and use light and carbon dioxide (CO_2) to create biomass. Algae range in size from a few micrometers to over 30 m in length.

One most important benefit of algae over conventional crops is algae can be grown under conditions which are unsuitable for conventional crop production.

4.1. Microalgae vs. macroalgae

There are two types of algae, Macroalgae and Microalgae. The word "Macro" means big so microalgae are large in size (in size of inches and greater) and multi cellular. Seaweed is example of largest Macroalgae which can be well over 25 m in length. The largest seaweed, giant kelp, is known to grow as fast as 50 cm/day, and can reach a length up to 80 m. Microalgae cells can double every few hours during their exponential growth period Macroalgae have some advantages. Mainly because of their relatively larger sizes than micro, these can be harvested more easily.

On other side "Micro" means very small (in size of micrometers) and contain one cell so called "Unicellular" organisms. Microalgae are more preferred because of the fact that they grow very quickly and also because they have much higher lipid content than Macroalgae.

The main advantages of using microalgal organisms in a variety of industrial applications are:

- they grow rapidly and have a higher solar conversion efficiency than most terrestrial plants;
- they can be harvested batch-wise or continuously almost all year round;
- algal production facilities can be collocated on otherwise non-productive, non-arable land;
- they can utilize salt and waste water sources that cannot be used by conventional agriculture;
- they can use waste CO_2 sources thereby potentially mitigating the release of GHG into the atmosphere; and,
- they can produce a variety of feedstocks that to generate nontoxic, biodegradable biofuels and valuable co-products.
- biodegradable biofuels and valuable co-products.

4.2. Major compositions of microalgal biomass

Microalgae are one of the best alternatives to traditional forms of biomass for biofuels production, due to its ability to be cultivated on marginal lands, fastest growing biomass, high productivity, and potential to utilize carbon dioxide (CO_2) from various sources.

Microalgal biomass is unicellular organisms which mean they have only one cell. Microalgae biomass contains compounds like protein, carbohydrates, lipids and nucleic acid. The percentages of compounds vary with the type of algae. Under good condition, biomass of green algae can be double in less than 24 hours [20]. Green algae can have huge lipid contents, continuously over 50% [23]. Oil content found in green algae may be different; a comparison of the oil content in algae is shown in table 2. Microalgae are capable of fixing CO_2 in the atmosphere, thus facilitating the reduction of increasing atmospheric CO_2 levels

Species	Oil content (percentage based on dry weight)
Chlorella sp.	28-32
Cylindrotheca sp.	16-37
Nitzschia sp.	45-47
Nannochloropsis sp.	31-68
Schinzochytrium sp.	50-77

Table 2. Oil content of algal species

Many microalgae species can be induced to accumulate substantial quantities of lipids [105] thus contributing to a high oil yield. The average lipid content varies between 1 and 70% but under certain conditions some species can reach 90% of dry weight [20, 74, 75, 109].Table 3 presents both lipid content and lipid and biomass productivities of different marine and freshwater microalgae species, showing significant differences between the various species [74, 95, 96, 109]. As shown in table 2, oil content in microalgae can reach 75% by weight of dry biomass but associated with low productivities (e.g. for Botryococcus braunii). Most common algae (Chlorella, Crypthecodinium, Cylindrotheca, Dunaliella, Isochrysis, Nannochloris, Nannochloropsis, Neochloris, Nitzschia, Phaeodactylum, Porphyridium, Schizochytrium, Tetraselmis) have oil levels between 20 and 50% but higher productivities can be reached.

Chlorella seems to be a good option for biodiesel production. Yet, as other species are so efficient and productive as this one, the selection of the most adequate species needs to take into account other factors, such as for example the ability of microalgae to develop using the nutrients available or under specific environmental conditions. All these parameters should be considered simultaneously in the selection of the most adequate species or strains for biodiesel production.

Also significant is the composition of fatty acids of the different microalgae species, as they can have a significant effect on the characteristics of biodiesel produced. These are composed of saturated and unsaturated fatty acids with 12–22 carbon atoms, some of them of v3 and v6 families. The analysis of seven fresh water microalgae species for the fatty acid compositions shows that all of them synthesized C14:0, C16:0, C18:1, C18:2, and C18:3 fatty acids.

The relative intensity of other individual fatty acids chains is species specific, e.g. C16:4 and C18:4 in *Ankistrodesmus sp.*, C18:4 and C22:6 in *Isochrysis sp.*, C16:2, C16:3 and C20:5 in *Nannochloris sp.*, C16:2, C16:3, and C20:5 in *Nietzsche sp.* Different nutritional and environmental factors, cultivation conditions and growth phases may affect the fatty acid composition. For example, nitrogen deficiency and salt stress induced the accumulation of C18:1 in all treated species and to some extent C20:5 in *B. braunii.*

5. Current usage of microalgae

Microalgae have a specialty that it useful in various ways likes humans use algae as food, for production of useful compounds, as nutrient and fertilizer, wastewaters treatment and other pollutants removal from wastewater, as indicators of environmental change, in space technology, and as laboratory research systems. Microalgae are capable of fixing CO_2 in the atmosphere because when its grow using photosynthesis, also need CO_2, which is waste of various sources and responsible for global warming worldwide.

5.1. Food

Algae are rich in iodine, potassium, iron, magnesium and calcium. Algae are a complete protein with essential amino acids that are involved in major metabolic processes such as energy and enzyme production. Algae contain high amounts of simple and complex carbohydrates which provide the body with a source of additional fuel. The sulfated complex carbohydrates are thought to enhance the immune system's regulatory response. Algae contain an extensive fatty acid profile, including Omega 3 and Omega 6, which also play a key role in the production of energy.

5.2. Nutrient removal and fertilizer

Algae is such a biomass important which take part in treatment of ponds, pollution control and useful as a fertilizer. Nitrogen, phosphorous, potassium is important fertilizers which are requisite for the growth of the plant, knows as Nutrient. Silica, iron and some other is also useful nutrient for growth of an area such as silica is a vital nutrient for growth of diatoms (phytoplanktonic organism) which is important part of various marine food series. Same as iron is responsible for restrain of phytoplankton.

5.3. Wastewater treatment and detoxification

Wastewater may be produced by municipal, agriculture, industrial and other activities. Algae are feasible for treatment of wastewater by removal of nutrient. Algae biomass can be used in wastewater treatment for the removal of bacteria, reduction of both chemical oxygen demand (COD) and biochemical oxygen demand (BOD), removal of N, P, and for the removal of heavy metals.

5.4. CO_2 emissions

Microalgae can generate biomass by absorb CO_2, which is produced at large-scale due to power plant gases. CO_2 utilization capacity around 1.8 tonnes of CO_2 will be utilized per tonnes of algal dry biomass produced, which is varies with algae species. CO_2 is harmful for environment as it leads some serious issue like global warming microalgae protect environment, through photosynthesis metabolism, microalgae absorb CO_2 and release oxygen.

5.5. Biofuels

Algae can be used to convert various types of fuels which depend on both technique and part of cells used. Biodiesel can be extracted from lipid and oily part of the algae biomass using similar process which is used for other vegetable oil. Alternatively or following lipid extraction, the carbohydrate content of algae can be fermented into bioethanol or biobutanol

Heat and electricity can be generated by burning of algae. Some algae can produce H_2 Gas (hydrogen gas) under some specific condition. Microalgae grow quickly and contain high oil content as compared with terrestrial crops [19].

6. Classification of microalgae

Microalgae were among the first life forms on earth [34]. As prefix "micro" mean small, so microalgae are very small in their size (in size of micrometers).Microalgae known as unicellular organisms because it has one cell. Microalgae can make their own energy and this energy is stored in the cell. They are capable of fixing large amounts of carbon dioxide (CO_2) while contributing to approximately 40 percent to 50 percent of the oxygen in the atmosphere thereby helping to support the majority of life on our planet.

Microalgae are highly productive on a global scale, with cell doublings of 1-4 per day. While microalgae make up only 0.2 percent of global biomass generated through photosynthesis, they account for approximately 50 percent of the total global fixed organic carbon [36].

Microalgae, like terrestrial plants, grow and multiply through photosynthesis, a process whereby light energy is converted into chemical energy by —fixing atmospheric CO_2.

Marine and freshwater microalgae species	Lipid content (% dry weight biomass)	Lipid productivity (mg/L/day)	Volumetric productivity of biomass (g/L/day)	Areal productivity of biomass (g/m²/day)
Ankistrodesmus sp.	24.0–31.0	–	–	11.5–17.4
Botryococcus braunii	25.0–75.0	–	0.02	3.0
Chaetoceros muelleri	33.6	21.80	0.07	
Chaetoceros calcitrans	14.6–16.4/39.8	17.6	0.04	–
Chlorella emersonii	25.0–63.0	10.3–50.0	0.036–0.041	0.91–0.97
Chlorella protothecoides	14.6–57.8	1214	2.00–7.70	–
Chlorella sorokiniana	19.0–22.0	44.7	0.23–1.47	–
Chlorella vulgaris	5.0–58.0	11.2–40.0	0.02–0.20	0.57–0.95
Chlorella sp.	10.0–48.0	42.1	0.02–2.5	1.61–16.47/25
Chlorella pyrenoidosa	2.0	–	2.90–3.64	72.5/130
Chlorella	18.0–57.0	18.7	–	3.50–13.90
Chlorococcum sp.	19.3	53.7	0.28	–
Crypthecodinium cohnii	20.0–51.1	–	10	–
Dunaliella salina	6.0–25.0	116.0	0.22–0.34	1.6–3.5/20–38
Dunaliella primolecta	23.1	–	0.09	14
Dunaliella tertiolecta	16.7–71.0	–	0.12	–
Dunaliella sp.	17.5–67.0	33.5	–	–
Ellipsoidion sp.	27.4	47.3	0.17	–
Euglena gracilis	14.0–20.0	–	7.70	–
Haematococcus pluvialis	25.0 – 0.05	–	0.06	10.2–36.4
Isochrysis galbana	7.0–40.0	–	0.32–1.60	–
Isochrysis sp.	7.1–33	37.8	0.08–0.17	–
Monodus subterraneus	16.0	30.4	0.19	–
Monallanthus salina	20.0–22.0	–	0.08	12
Nannochloris sp.	20.0–56.0	60.9–76.5	0.17–0.51	–
Nannochloropsis oculata.	22.7–29.7	84.0–142.0	0.37–0.48	–
Nannochloropsis sp.	12.0–53.0	37.6–90.0	0.17–1.43	1.9–5.3
Neochloris oleoabundans	29.0–65.0	90.0–134.0	–	–
Nitzschia sp.	16.0–47.0	8.8	–	21.6
Oocystis pusilla	10.5	–	–	40.6–45.8

http://dx.doi.org/10.5772/59821

Marine and freshwater microalgae species	Lipid content (% dry weight biomass)	Lipid productivity (mg/L/day)	Volumetric productivity of biomass (g/L/day)	Areal productivity of biomass (g/m²/day)
Pavlova salina	30.9	49.4	0.16	–
Pavlova lutheri	35.5	40.2	0.14	–
Phaeodactylum tricornutum	18.0–57.0	44.8	0.003–1.9	2.4–21
Porphyridium cruentum	9.0–18.8/60.7	34.8	0.36–1.50	25
Scenedesmus obliquus	11.0–55.0	–	0.004–0.74	–
Scenedesmus quadricauda	1.9–18.4	35.1	0.19	–
Scenedesmus sp.	19.6–21.1	40.8–53.9	0.03–0.26	2.43–13.52
Skeletonema sp.	13.3–31.8	27.3	0.09	–
Skeletonema costatum	13.5–51.3	17.4	0.08	–
Spirulina platensis	4.0–16.6	–	0.06–4.3	1.5–14.5/24–51
Spirulina maxima	4.0–9.0	–	0.21–0.25	25
Thalassiosira pseudonana	20.6	17.4	0.08	–
Tetraselmis suecica	8.5–23.0	27.0–36.4	0.12–0.32	19

Table 3. Lipid content and productivities of different microalgae species [111]

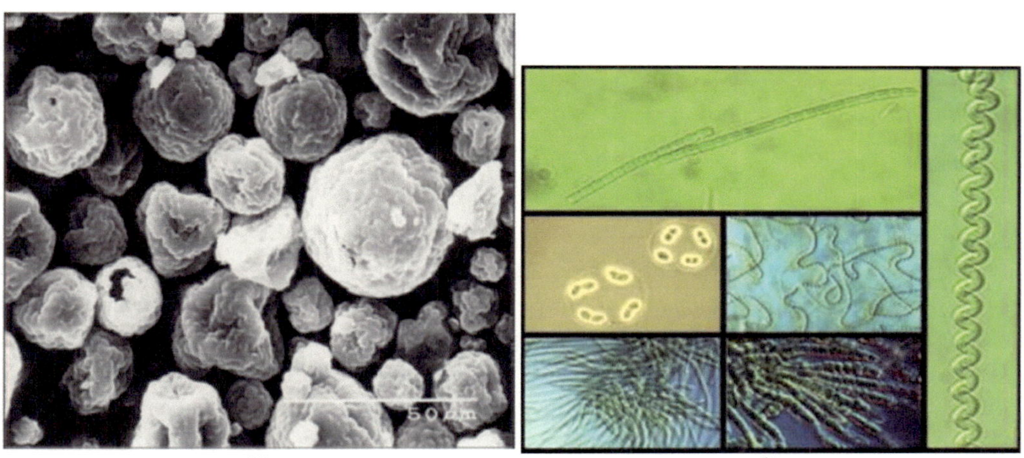

(a) (b)

Figure 1. (a). Scanning electron micrograph of a microalgae (Chlorella)[47]; (b).Cyanobacteria range from simple unicellular organisms to colonies [122]

Over 40,000 separate species of algae have been identified, and that number almost certainly represents a small fraction of the true population (perhaps as high as 10,000,000 different species [55]. Because of the diverse nature of algae, it has been difficult to settle on a universally accepted classification system. For example, some experts will exclude cyanobacteria because of their simple cellular structure relative to other classes of algae. Others will focus on a separation of unicellular (microalgae) and multicellular (macroalgae).

Much of the classification of algae depends upon photosynthetic pigments, whole organism morphology, cellular anatomy and ultrastructure, and metabolism and physiology. The biological divisions that encompass the various classes of algae are;

- Cyanophyta (cyanobacteria)
- Prochlorphyta
- Glaucophyta
- Rhodophyta (red algae)
- Cryptophyta (cryptomonads)
- Chlorophyta (green algae)
- Euglenophyta
- Chloroarachniophyta
- Pyrrophyta (dinoflaggellates), and
- Chromophyta (heterokonts)

Of these classes, those that produce significant amounts of lipids are considered to be of interest for the production of Biofuels. Macroalgae typically require deep bodies of water for growth, and generally are viewed to lack the potential to make a significant contribution to the world's future liquid transportation fuel needs. Notwithstanding this view macroalgae production is increasing and there is interest in the EU and Japan in its use as a feedstock for methane production by anaerobic digestion and ethanol production by saccharification and fermentation.

Most of the algae known to produce more than 20% of their biomass as lipids fall into the divisions *Cryptophyta, Chlorophyta,* and *Chromophtya.* Cryptomonads are biflagellate unicellular algae carrying the photosynthetic pigments chlorophyll a and c, α-carotene and β-carotene giving them the colours green, olive, brown, yellow, red, or blue. They are found in waters ranging from fresh to hypersaline, sometimes in great abundance. *Rhodomonas salina* (also known as *Chroomonas salina)* is a cryptomonad known to produce lipids at high levels.

Chlorophyta or green algae range from unicellular forms to large seaweeds. Their photosynthetic pigments are similar to those in higher plants and include chlorophyll a and b, α-, β-, and γ-carotene, and various xanthophylls. Their cell walls contain cellulose and they often use starch as an energy reserve (attributes of potential feedstocks for ethanol production). *Chlamydomonas reinhardtii,* a chlorophyte, was selected as a model system for the study of plants, and is one of the few algae whose entire gene sequence is known. *C. reinhardtii* can grow autotrophically on a simple medium of inorganic salts and in the presence light and CO_2, but can also grow heterotrophically in total darkness using acetate as a carbon source and O_2.

http://dx.doi.org/10.5772/59821

Several Chlorophytes are known to produce high levels of lipids including *Botryococcus braunii*, *Chlorella vulgaris*, *Neochloris oleoabundans*, and *Nannochloris sp.*The chromophyta contain chlorophyll a and b, α-, β-, and γ-carotenes, zeaxanthin and several other xanthophylls. They comprise many different classes of algae including the *Chrysophyceae* (golden-brown algae), *Bacillariophycease* (diatoms), *Xanthophyceae* (yellow-green algae), *Eustigmatophyceae*, and *Prymnesiophyceae*. Examples of each of these classes are known to produce high levels of lipids including *Ochromonas danica*, *Phaeodactylum tricornutum*, *Nitzschia palea*, *Monallantus salina*, *Nannochloropsis sp.*, and *Isochrysis sp.*

Unlike the other divisions of algae, cyanobacteria or blue green algae is prokaryotic, that is, they lack nuclei and are members of the bacterial kingdom. They contain many different photosynthetic pigments including chlorophyll a and d, phycobilins, β-carotene, zeaxanthin, and other xanthophylls, and phycobilins. Although a *Nostoc* commune has been shown to produce triacylglycerides, cyanobacteria rarely produce more than 20% of their cell weight as lipids, but they will be included in this discussion because they have been shown to accumulate high levels of glycogen (as much as 60% of dry weight) as a storage material, and it is possible to divert the carbon flux from carbohydrate production to lipid production. In addition, cyanobacteria have long-established commercial production methods (mainly for food supplements and nutraceuticals) and genetic techniques have been developed for many different strains.

According to size, color/ pigment, shape, lifecycle and their cellular structure, microalgae are classified in four classes as abundant in below table 4.

6.1. Diatoms (Bacillariophyceae)

Diatoms (Bacillariophyceae) are a type of algae. Mainly diatoms are unicellular but have different in shape such as stars, zigzag, ribbons, fans, spheres, elliptical and triangles when they exists as colonies. Carbon is stored in the form of oil in Diatom. This oil and water current helps them to move within the water to find their food and nutrient. Diatom cells have a unique feature is that, they are enclosed within a cell wall made of silica which is called a frustules. This silica is used to protect the cell.

Microalgae	Known Species	Storage Material	Habitat
Diatoms (Bacillariophyceae)	100,000	Chyrsolaminarin (polymer of carbohydrates)	Freshwater and brackish water
Green Algae (Chlorophyceae)	8000	Starch	Freshwater
Blue Algae (Cyanophyceae)	2000	Starch and TAGs	Different
Golden Algae (Chrysophyceae)	1000	Natural oils & carbohydrates	Freshwater

Table 4. Microalgae Classification

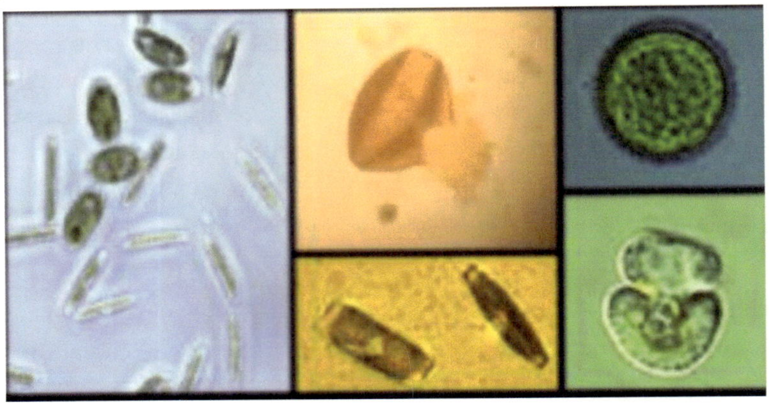

Figure 2. Microalgal Biodiversity (diatoms and green algae) [105]

6.2. Green Algae (Chlorophyceae)

Green Algae (Chlorophyceae) can be unicellular or colonial, generally it found quite abundant in fresh water. They have flagella (tails) attached to each cell, they use these flagella to swim. They include some of the most common species, as well as many members that are important both ecologically and scientifically. There are approximately 350 genera and 2650 living species of chlorophyceans. They come in a wide variety of shapes and forms, including free-swimming unicellular species, colonies, non-flagellate unicells, filaments, and more. They also reproduce in a variety of ways, though all have a haploid life-cycle, in which only the zygote cell is diploid. The zygote will often serve as a resting spore, able to lie dormant though potentially damaging environmental changes such as desiccation.

The Chlorophyceae includes three major groups distinguished primarily by basic differences in the arrangement of their flagellae:

- *Volvocales, Chaetophorales, & Chlorococcales* - together make up more than half of all chlorophyceans. Members of these orders have an offset flagellar arrangement (1 o'clock-7 o'clock).
- *Chlorellales* - Members of this order have opposed flagellae (12 o'clock-6 o'clock), though some have only vestigial flagellae and so have not been definitively associated with this group. Similarities with members of the Chlorococcales make distinctions difficult.
- *Oedogoniales* - Members of this smallest group have a complex multiflagellate crown on their swimming spores. All are filamentous, oogamous, and have net-like chloroplasts.

6.3. Blue Algae (Cyanophyceae)

Blue Algae (Cyanophyceae) grow in both fresh and salt waters of dams, rivers, creeks, reservoir, lakes. Blue Algae are a type of bacteria but due to some ways it act like plant by using to manufacture carbohydrates from carbon dioxide and water and release oxygen, through a process of photosynthesis [105].

http://dx.doi.org/10.5772/59821

Figure 3. (a) Green Algae [112] Figure (b) Blue Green Algae [112]

6.4. Golden Algae (Chrysophyceae)

Golden Algae (Chrysophyceae), similar to diatoms in pigment and biochemical composition, are mostly found in fresh water. A single species "Prymnesium parvum" are referred as Golden The chrysophyceans (golden algae) are heterokontophyte algae with golden chloroplasts. Many chrysophycean algae are unicellular, but colonial or simple multicellular species are also known. The chrysophycean algae are basically autotrophic but there are many mixotrophic & colorless heterotrophic species. Heterotrophic chrysophyceans such as *Spumella and Paraphysomonas* play an important role as lower consumers. The chrysophycean algae mainly inhabit in freshwater, but some species (especially heterotrophs) are common in marine.

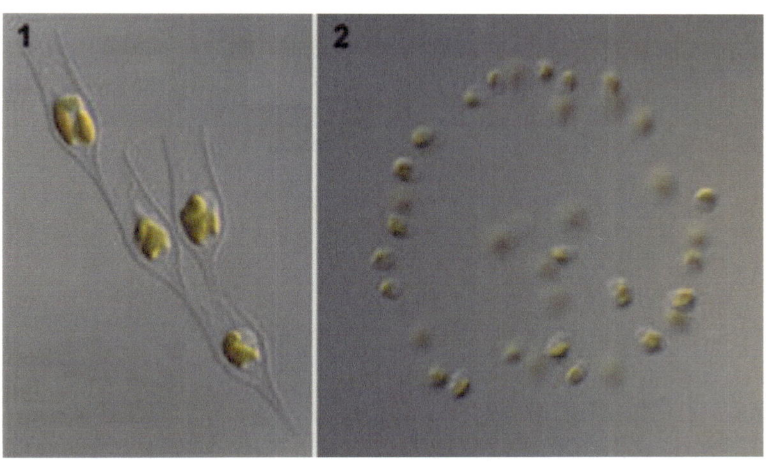

Figure 4. a: Dinobryon b: Uroglena [82]

The cells are naked or covered by scales, lorica or cell wall. The flagellate cell usually possesses two heterodynamic flagella but posterior (a) flagellum is sometimes reduced. Tubular mastigonemes on anterior (b) flagellum possess lateral filaments. Mixotrophic and heterotrophic species engulf particles (e.g. bacteria) through splitted R2 microtubules. Because major photosynthetic carotenoid is fucoxanthin, chrysophycean chloroplasts are golden-yellow in color. Asexual reproduction by means of binary fission, sporogenesis etc. Sexual reproduction has been reported in some species. The chrysophycean algae produce cysts surrounded by siliceous wall, statospore via sexual or asexual reproduction. Statospores form microfossils to be used for paleoenvironmental reconstruction.

7. Microalgae for biodiesel production

Microalgae as compared to conventional crops have high photosynthetic efficiency and therefore potentially high productivity per unit area of plantation. The U.S. Department of Energy's Aquatic Species Program (1978- 1996) focused on biodiesel production from microalgae because biodiesel is a promising fuel product in many ways because it is useful to counter "Energy Security" and "Climate Changes". Day by day the energy required in Transportation sector is increasing and petroleum will not be able to fulfill the all requirements, in that case Microalgae is one of the substitutes to petroleum. Through photosynthesis metabolism, microalgae absorb CO_2 and release oxygen; it will reduce the global warming effect.

7.1. Current biodiesel feedstock

Biodiesel can be produce from various feedstock's which is soybean oil, rapeseed, Jatropha, mustard, jojoba, sunflower, palm oil, coconut hemp, animal fats, sewage sludge and algae. A comparison of feedstock for biodiesel is as below in table 5.

Feedstock	Oil content (% dry weight biomass)	Land use (sqm/year/L of biodiesel)	Productivity of Biodiesel (Liter Biodiesel/Hectare /Year))
Maize	44	56	179
Soybean	18	15	661
Jatropha	28	13	772
Rapeseed	41	10	1014
Sunflower	40	9	1113
Palm oil	36	2	5585
Microalgae (Low Oil Content)	30	0.2	61,091
Microalgae (medium Oil Content)	50	0.1	101,782
Microalgae (high Oil Content)	70	0.1	142,475

Table 5. Feedstock for biodiesel [93]

http://dx.doi.org/10.5772/59821

7.2. Potential of using microalgae as biodiesel feedstock

Microalgae are emerging as a potential high-volume source of lipids for advanced biofuels. While commercial production of microalgae has been established for human nutritional products like *Spirulina, beta carotene,* and *omega-3* fatty acids for at least three decades, the concept of using microalgae as an aquaculture source for energy production on the mega-ton scale meaningful to the petroleum industry has enjoyed a recent resurgence. Over conventional crops, algae can be grown under conditions which are unsuitable for conventional crop production and algae can be grow on the land which is not arable land. Microalgae potential because of the fact that they grow very quickly and live in harsh conditions due to their unicellular structure even Microalgae are able to double their mass within few hours. Microalgae are preferred over Macroalgae because Microalgae have much higher lipid content than Macroalgae.

Microalgae species	Carbohydrates (%)	Proteins (%)	Lipids (%)
Chaetoceros muelleri	11–19	44–65	22–44
Chaetoceros calcitrans	10	58	30
Isochrysis galbana	7–25	30–45	23–30
Chlorella protothecoides	10.62–15.43	10.28–52.64	14.57–55.20
Chlorella sp.	38–40	12–18	28–32
Nannochloropsis sp.	-	-	31–68
Neochloris oleoabundans	-	-	35–54
Schizochytrium sp.	-	-	50–77
Scenedesmus– obliquus	10–17	50–56	12–14
Quadricauda de Scenedesmus	-	47	1.9

Table 6. Chemical composition of biofuel source microalgae

The majority of companies trying to demonstrate commercial production of microalgae for energy and other markets were found within the past six years. The pace of innovation in systems engineering, cultivation techniques, intracellular productivity improvement techniques, and business model development has been extremely rapid. Production and productivity levels have jumped by orders of magnitude each year over the past three to four years, for example from less than 100 verifiable gallons of algae oil produced by the entire industry in 2009 to over 20,000 gallons delivered to customers in 2010.

Table 7, shows that Microalgae (low, medium and high Oil Content) have been found to have incredible production level compared to other oil seed crops.

Species	Oil content	Species, %	Oil content, %
Botryococcus braunii	25–75	Isochrysis sp.	25–33
Chlorella sp.	28–32	**M. Subterraneus**	39.3
Chlorella emersonii	63	**Monallanthus salina**	420
Chlorella minutissima	57	**N. laevis**	69.1
Chlorella protothecoides	23	**Nannochloris sp.**	20–35
Chlorella sorokiniana	22	**Nitzchia sp.**	45–47
Chlorella vulgaris	40, 56.6	**P.incisa**	62
Cylindrotheca	16–37	**Phaeodactylum tricornutum**	20–30
Crypthecodinium cohni	20	**Schizochytrium sp**	. 50–77
Dunaliella primolecta	23	**Tetraselmis sueica**	15–23

Table 7. Oil content in some microalgae

Source	Carbohydrates (%)	Proteins (%)	Lipids (%)
Anabaena cylindrica	25–30	43–56	4–7
Chalmydomonas rheinhardii	17	48	21
Chlorella vulgaris	12–17	51–58	14–22
Dunaliella salina	32	57	6
Porphyidium Cruentum	40–57	28–39	9–14
Spirulina maxima	13–16	60–71	6–7
Baker's yeast	38	39	1
Meat	1	43	34
Milk	38	26	28
Rice	77	8	2
Soya bean	30	37	20

Table 8. Chemical composition of some food source microalgae (% of dry matter) [120]

7.3. Biodiesel production from microalgae biomass

Liquid fuel can be obtained by the process of oil extraction from algae. Hexane is an organic solvent which is used for this process. The hexane removes the oil from the algae. The mixture of hexane and oil is distilled leaving pure algae oil. This technique has significance that solvent is reused for each cycle. Algae fiber, which is remain after this process can be used as fertilizer.

The methodology mostly used for biodiesel production is based on the transesterfication reaction, as follows:

http://dx.doi.org/10.5772/59821

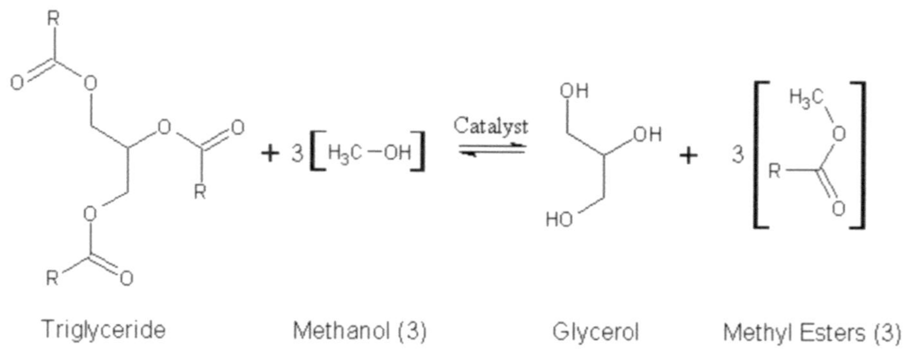

The transesterfication reaction, as above, takes place in the presence of either homogeneous or heterogeneous catalysts (traditional method). Those alternatives can be compared in search for the most efficient method of biodiesel production from microalgae lipids. The biodiesel consists of a biodegradable fuel produced from renewable sources.

Properties	Mcr algal oil Biodiesel	Biodiesel fuel	ASTM biodiesel standard
Density (kg L^{-1})	0.864	0.838	0.86–0.9
Viscosity, (mm2s^{-1}, cSt at 40^{0}C)	5.2	1.9–4.1	3.5–5.0
Flash point , (^{0}C)	115	75	Min 100
Solidifying point (^{0}C)	-12	-50 to 10	-
Cold filter plugging point, (^{0}C)	-11	-30 (Max -6.7)	Summer max 0; winter <15
Acid value, (mg KOH g^{-1})	0.374	Max 0.5	Max 0.5
Heating value , (MJ kg^{-1})	41	40–45	-
H/C ratio	1.81	1.81	-

Table 9. Properties of biodiesel from microalgal oil, biodiesel fuel & ASTM biodiesel standard [120].

The synthesis of this fuel can be accomplished by methodologies such as cracking, esterification or transesterfication using animal fat or vegetable oils. Table 9 shows a comparison of characteristics of biofuels and petro diesel along with ASTM biodiesel standard [118].

7.4. Current limitations for algal biodiesel production

Algae biodiesel is still a new technology because more than 30 years, research and development program was initiated by US Department of Energy but due to lack of funding and comparative low cost of petroleum fuel than algae biodiesel, in 1996 this program was discontinued. Now further research will be required to make algae fuel more viable and efficient then petroleum [120]. Algae biodiesel have also lower stability during regular seasonal temperature because during processing, microalgae differ in polyunsaturated from which is other form of biodiesel and polyunsaturated fats have an ability to retain their fluidity at lower temperature during winter but it will have also lower stability during regular seasonal temperature.

Microalgae species	Carbohydrates (%)	Proteins (%)	Lipids (%)
Chaetoceros muelleri	11–19	44–65	22–44
Chaetoceros calcitrans	10	58	30
Isochrysis galbana	7–25	30–45	23–30
Chlorella protothecoides	10.62–15.43	10.28–52.64	14.57–55.20
Chlorella sp.	38–40	12–18	28–32
Nannochloropsis sp.	-	-	31–68
Neochloris oleoabundans	-	-	35–54
Schizochytrium sp.	-	-	50–77
Scenedesmus– obliquus	10–17	50–56	12–14
Quadricauda de Scenedesmus	-	47	1.9

Table 10. Chemical composition of biofuel source microalgae

8. Technology for growing algae

There are two algae cultivation technologies currently in use for commercial microalgae production and proposed for algal biofuel production (viz. extensive or open ponds, intensive or raceway ponds or closed photobioreactors in many designs and closed fermenter systems).

8.1. Open pond system

Large extensive or open pond systems are currently in use for wastewater treatment and *Dunaliella salina* production. Oxidation ponds in wastewater treatment systems are not in the true sense for algae production as no algae are harvested. Cognis Australia Pty Ltd produce β-carotene from *D. salina* harvested from hypersaline extensive ponds in Hutt Lagoon and Whyalla. The halotolerant *D. salina* dominates naturally in brine at salt concentrations >100 g L^{-1} but grows relatively slowly (producing perhaps not much more than 2.2 t ha^{-1} yr^{-1}). *Hutt Lagoon* has a total pond surface area of ca. 520 ha and Whyalla is ca. 440 ha. In terms of pond surface area, *Hutt Lagoon* and *Whyalla* are among the largest algal production systems in the world. These extensive pond algae production systems have limited mixing, and rely on natural selection and the bounty of nature with minimal intervention.

Open pond system used big shallow pond which is open. This type of pond is easy to construct and operate than close pond system. Shallow pond is constructed to provide large area to algae for exposed to sunlight. Open pond system is use for cultivation of algae especially having high oil content. Both natural and artificial water pond can be used to algae biomass production. Main advantages of open pond system are low operating cost and their simple structure.

Figure 5. Spirulina farms on Hainan Island China & Circular chlorella ponds at Yaeyama on Okinawa Island, Southern Japan [121]

Similarly their many disadvantage are also associated with open pond system are poor productivity, little control over algae production, large evaporative losses, large area required, diffusion of carbon dioxide to the atmosphere and expensive harvesting etc [93].

8.2. Closed ponds

Closed pond system mean which is not open to expose in the air. Control over environment is much better much better than open pond system and it allows more species to grown than other. It is not only more expensive system than open pond system, but also low productivity of biomass.

8.2.1. Photo bioreactor

Photobioreactors are closed systems of transparent tubes, plates, bags or hemispherical domes. Photobioreactors improve yields by protecting productive strains to some extent from contamination, pathogens, and predators, offer the benefits of some temperature control and eliminate climate related impacts of open ponds (viz. rainfall, evaporation, and diurnal and seasonal temperature fluctuations). While better mixing in photobioreactors may provide slight area productivity gains, claims of productivity, which refer to the area or footprint of the growth vessel, can be extremely high when the reactors are configured vertically and are misleading. Vertical photobioreactors must be situated far enough from each other so as to not shade, and consequently the basic limitation on productivity remains the same for both open ponds and closed photobioreactors.

Surface fouling due to bacteria, other organisms, and, in particular, algae, is a major problem with photobioreactors, and cleaning can be a major design and operational problem. Where CO_2 input and O_2 evolution must be optimized for maximum productivity, gas transfer, which is restricted to the surface area of gas liquid interfaces, can limit scalability of photobioreactor designs.

Commercial photobioreactors as shown in figure 6 are in operation at different facilities including the production of *H. pluvialis* in Israel and Hawaii and C. vulgaris in Germany.

Typical plant gate selling prices/production costs are well above \$100/kg from such systems. Consequently, biofuels production based entirely on photobioreactor systems is generally considered unlikely to be commercially viable.

Algae pumped with nutrient rich water through plastic and borosilicate tube, exposed to sunlight called photobioreactor (PBR). Algae biomass produces using carbon dioxide and light by the process of photosynthesis and nutrient from wastewater in artificial environment not in natural environment. Using photobioreactor, algae easily grow on the land which is not arable such as desert and even ocean surface also. PBR is more productive and controlled but more costly and difficult than open pond system.

Table 10 makes a comparison between PBR and ponds for several culture conditions and growth parameters. Comparison of performances achieved by PBRs and open ponds may not be easy, as the evaluation depends on several factors, among which the algal species cultivated and the method adopted to compute productivity. There are three parameters commonly used to evaluate productivity in algae production units: Volumetric productivity (VP): productivity per unit reactor volume (expressed as g/L d).Areal productivity (AP): productivity per unit of ground area occupied by the reactor (expressed as g/m^2 d). Illuminated surface productivity (ISP): productivity per unit of reactor illuminated surface area (expressed as g/m^2 d).

Figure 6. Photobioreactor for large scale algae production [122]

http://dx.doi.org/10.5772/59821

As stated by Richmond [96] despite closed systems offer no advantage in terms of areal productivity, they largely surpass ponds in terms of volumetric productivity (8 times higher) and cell concentration (about 16 times higher). In conclusion, PBR and open ponds should not be viewed as competing technologies, but the real competing technology will be genetic engineering [96].

9. Harvesting of algae

Algae harvesting overcome to get desired algae product that is fuel. Harvesting method use for algae harvesting depends upon type of algae. There is number of algae harvesting method but some of most common is Flocculation, Centrifugation and Microorganism. There are some issues related to algae harvesting that should carried out before harvesting Process to be done such as the water content should be within desired limit, algae must be in paste form before processing. Size of microalgae cells increase by Flocculation so that sedimentation will be easily done with large cells particle. Chemical flocculation and centrifugation is useful in high density algae because using certain chemical such as alum, lime and aluminum sulphate will coagulate and precipate the cell down or float to the surface. This method is very high costly because of the large amount of the chemical used in this process.

Algal harvesting consists of biomass recovery from the culture medium that may contribute to 20–30% of the total biomass production cost [47]. In order to remove large quantities of water and process large algal biomass volumes, a suitable harvesting method may involve one or more steps and be achieved in several physical, chemical, or biological ways, in order to perform the desired solid–liquid separation. Experience has demonstrated that albeit a universal harvesting method does not exist, this is still an active area for research, being possible to develop an appropriate and economical harvesting system for any algal species.

Most common harvesting methods include sedimentation, centrifugation, filtration, ultra-filtration, sometimes with an additional flocculation step or with a combination of floccula-tion–flotation. Flocculation is used to aggregate the microalgal cells to increase the effective particle size and hence ease sedimentation, centrifugal recovery, and filtration [47]. Weissman and Goebel [298] studied four primary harvesting methods for the purpose of biofuels production: microstraining, belt filtering, flotation with float collection, and sedimentation. These methods discriminate on a size and density basis in performing the biomass separation. Microstrainers are an attractive harvesting method because of their mechanical simplicity and availability in large unit sizes. The recent availability of very fine mesh polyester screens has revived interest in their use for microalgae harvesting.

Subsequent studies concluded that it would be necessary to flocculate the cells prior to microstraining. Filter presses operating under pressure or vacuum can be used to recover large quantities of biomass, but for some applications filtration can be relatively slow which may be unsatisfactory. Also filtration is better suited for large microalgae such as Coelastrum probo-scideum and S. platensis but cannot recover organisms with smaller dimensions such Scene-desmus, Dunaliella, or Chlorella [47]. Alternatively, membrane microfiltration and ultra-

Culture systems for microalgae	Closed systems (PBRs)	Open systems(Ponds)
Contamination control	Easy	Difficult
Contamination risk	Reduced	High
Sterility	Achievable	None
Process control	Easy	Difficult
Species control	Easy	Difficult
Mixing	Uniform	Very poor
Operation regime	Batch or semi-continuous	Batch or semi-continuous
Space required	A matter of productivity	PBRs ≅ Ponds
Area/volume ratio	High (20–200 m^{-1})	Low (5–10 m^{-1})
Population (algal cell) density	High	Low
Investment	High	Low
Operation costs	High	Low
Capital/operating costs ponds	Ponds 3–10 times lower cost	PBRs "/> Ponds
Light utilization efficiency	High	Poor
Temperature control	More uniform temperature	Difficult
Productivity	3–5 times more productive	Low
Water losses	Depends upon cooling design	PBRs ≅ Ponds
Evaporation of growth medium	Low	High
Hydrodynamic stress on algae	Low–high	Very low
Gas transfer control	High	Low
CO_2 losses	Depends on pH, alkalinity, etc.	PBRs≅Ponds
O_2 inhibition	Greater problem in PBRs	PBRs "/> Ponds
Biomass concentration	3–5 times in PBRs	PBRs "/> Ponds
Scale-up	Difficult	Difficult

Table 11. A comparison of open and closed large-scale culture systems for microalgae [115].

filtration are other possible alternatives to conventional filtration for recovering algal biomass, which are more suitable for fragile cells and small scale production processes. Furthermore these filtration processes are more expensive especially because of the need for membrane replacement and pumping.

Richmond [96] suggested one main criterion for selecting a proper harvesting procedure, which is the desired product quality. In one hand for low value products, gravity sedimentation may be used, possibly enhanced by flocculation. Sedimentation tanks or settling ponds are also possible, e.g. to recover biomass from sewage-based processes. In other hand for high-value products, to recover high quality algae such as for food or aquaculture applications, it is often recommended to use continuously operating centrifuges that can process large volumes of biomass.

Albeit at considerable cost, centrifuges are suitable to rapidly concentrate any type of micro-organisms, which remain fully contained during recovery. Additionally, these devices can be easily cleaned or sterilized to effectively avoid bacterial contamination or fouling of raw product.

Another basic criterion for selecting the harvesting procedure is its potential to adjust the density or the acceptable level of moisture in the resulting concentrate right to the optimum subsequent process [47, 96]. Gravity sedimented sludge is generally more diluted than centrifugally recovered biomass, which substantially influence the economics of product recovery further downstream. Since costs of thermal drying are much higher than those of mechanical dewatering, in order to reduce the overall production cost, a concentrate with higher solids content is required after harvest to easy biomass dehydration (e.g. in a drum drying).

In this case a combination of methods can also be used, e.g. a pre-concentration with a mechanical dewatering step such as microstrainer, filtration, or centrifugation and then, a postconcentration by means of a screw centrifuge or a thermal drying. After separation from the culture medium algal biomass (5–15% dry weight) must be quickly processed lest it should get spoiled in only a few hours in a hot climate.

Harvesting Process	Relative Cost	Concentrated Solids	Energy Input
Centrifugation	Very High	More than 10 %	High
Chemical Flocculation	High	8-10 %	High
Direct sedimentation	Low	1-3%	Low
Bioflocculation	Low	1-3%	Low
Auto flocculation	Low	1-3%	Low
Microstrainin	Low	2-4%	Medium
filtration	High	2-6%	High

Table 12. Comparison between various Harvesting Techniques [93]

10. Algae oil extraction techniques

The general extraction techniques are mechanical extraction/ cell disruption methods and Solvent extraction coupled with mechanical cell disruption methods. Other novel methods are

Supercritical CO_2 extraction and direct conversion of algal biomass to biodiesel. Extraction methods such as ultrasound and microwave assisted are also employed for oil extraction from vegetable sources. The results indicate that compared with conventional methods these new methods can greatly improve oil extraction with higher efficiency. Extraction times are reduced and yields increased by 50–500% with low or moderate costs and minimal added toxicity. In the case of marine microalgae Crypthecodinium cohnii, ultrasound worked best as the disruption of the tough algal cell wall considerably improved the extraction yield from 4.8% (in Soxhlet) to 25.9%.

10.1. Mechanical extraction/ cell disruption methods

The first and simple extraction method is mechanical cell disruption of algal cells to extract oil without contamination of other chemicals. Mechanical pressing or French pressing of dry algal lumps involves pressurizing the algal biomass to high-pressure, where the cell walls are ruptured to releases the oil similar to oil extraction from seeds or nuts through mechanical pressing. Homogenization through bead or ball milling is a process to disintegrate the alga cells which takes place in a jacketed chamber or vessel. The shear force created by the high velocity beads which moves radially causes the disruption of cells [121]. Cell disruption in this method depends on factors like residence time, cell concentration, chamber volume, bead volume and umber of rotations. All these mechanical cell disruption are usually combined with the solvent extraction to improve the extraction efficiency. Along with the mechanical methods new pretreatment techniques ultra-sonication, microwave also getting attention. In ultra-sonication & microwave pretreatments, the biomass will be treated in a sonication [96]/ microwave [105] chamber prior to solvent extraction.

10.2. Solvent extraction

Solvent extraction is a common practice used to extract oils from the algal biomass and other biomasses. The solvent should be selected based on efficiency, selectivity towards the different classes of lipids and ability of solvent to prevent any possible degradation of lipids. In order to achieve maximum extraction, the linkages between the lipids and other organelles of the algae cells which are connected with van der Waals interactions, hydrogen bonding and covalent bonding should be broken18. The most common solvents used for extraction are n-hexane, chloroform, petroleum ether, methanol, ethanol, isopropanol, dichloromethane and mixture of any of these solvents depending upon method and desired class selection of lipids. The conventional solvent extraction methods are Bligh and dyer, folch [90], Soxhlet extraction. The steps involved in the solvent extraction at micro level were explained by halim et al. When the algal cells interacted with the organic solvents, these solvents penetrate through the cell wall and interact with the selective class of lipids depending upon its dielectric constant to form a solvent- lipid complex. This complex diffuses in to the bulk solvent due to the concentration gradient continues until this process reaches equilibrium [51]. The solvent extraction methods shows a lot of variability depending upon the organic solvent (dielectric constant) used and biological matrix being used in selection of different class of lipids [20, 51]. The cell wall and its composition and solvents dielectric constant could be the reasons for these variable extraction properties of individual methods [51].

These solvent extraction methods have been slightly modified by many researchers to improve the kinetics of the extraction process often called as accelerated solvent extraction (ASE). Kauffmann and christen reviewed these accelerated solvent extraction techniques involving microwave heating and pressurized solvent extraction. In the microwave assisted extraction (MAE) the acceleration is achieved by faster disruption of weak hydrogen bonds the dipole rotation of the molecules caused by electromagnetic radiation. In pressurized solvent extraction (PSE) the higher temperature and pressure accelerates the extraction process as the high temperature accelerates the extraction kinetics, high pressure keeps the solvent in liquid state and forces the solvent to pass through the pores of the matrix thoroughly [68]. When coupled with the cell disruption techniques described earlier the solvent extraction will be very faster and utilizes small amounts of solvents [81]

10.2.1. Hexane solvent method

Algae oil extraction can be done thorough various techniques, hexane extraction is one of them. Hexane, Benzene and ether chemical used as a solvent extraction, in which Benzene and ether is widely used in food industry because of low cost factor. Isolation and oil press/express method are method in which hexane solvent extraction can be used for oil extraction. After oil has been extracted through expeller, remaining pulp can be mixed with cyclohexane chemical to further extraction of the remaining oil content in pulp. When the oil dissolved in the cyclohexane chemical, again pulp s filtered out from the solution and using distillation process oil and cyclohexane can be separated. Using this process more than 95 % of the total oil by the algae can be obtained.

10.3. Supercritical CO_2 extraction

Commercial applications of supercritical CO_2 extraction dates back to early 1990's. Supercritical extraction is being used in food and pharmaceutical industries due to its range of selectivity of compounds, non-toxic nature, and easy separation [87]. The principle behind this technology is, when fluids crosses both critical temperature and critical pressure they attains properties of both gases and liquids. This state of the fluid is called supercritical state of fluid, and it exhibits mass transfer properties of gas and solvent properties of liquid with greater diffusion coefficients [100]. Because of the lower critical point at 31.1°C and 72.9 atm carbon dioxide became preferred fluid for extraction applications. The solvent properties of supercritical fluid can be modified by altering extraction pressure and the extraction temperature. As an example target compounds like pigments, proteins and neutral lipids can be extracted at their respective extraction temperature and pressure, where they interact with the solvents [51, 80]. Due to its high selectivity, lower toxicity, chemical inertness and high purity of the extracted compounds, supercritical CO2 extraction is being used in many pharmaceutical, nutraceutical and food industries worldwide [80].

11. Biodiesel production from algal oil

Biodiesel is a mixture of fatty acid alkyl esters obtained by transesterification (ester exchange reaction) of vegetable oils or animal fats. These lipid feedstocks are composed by 90–98% weight) of triglycerides and small amounts of mono and diglycerides, free fatty acids (1–5%), and residual amounts of phospholipids, phosphatides, carotenes, to copherols, sulphur compounds, and traces of water [17].

Transesterification is a multiple step reaction, including three reversible steps in series, where triglycerides are converted to diglycerides, then diglycerides are converted to monoglycerides, and monoglycerides are then converted to esters (biodiesel) and glycerol (by-product). The overall transesterification reaction is described in Fig. 3 where the radicals R1, R2, R3 represent long chain hydrocarbons, known as fatty acids.

For the transesterification reaction oil or fat and a short chain alcohol (usually methanol) are used as reagents in the presence of a catalyst (usually NaOH). Although the alcohol: oil theoretical molar ratio is 3:1, the molar ratio of 6:1 is generally used to complete the reaction accurately. The relationship between the feedstock mass input and biodiesel mass output is about 1:1, which means that theoretically, 1 kg of oil results in about 1 kg of biodiesel.

A homogeneous or heterogeneous, acid or basic catalyst can be used to enhance the transesterification reaction rate; although for some processes using supercritical fluids (methanol or ethanol) it may not be necessary to use a catalyst [295]. Most common industrial processes use homogeneous alkali catalysts (e.g. NaOH or KOH) in a stirred reactor operating in batch mode.

Recently some improvements were proposed for this process, in particular to be able to operate in continuous mode with reduced reaction time, such as reactors with improved mixing, microwave assisted reaction [44,65], cavitations reactors [43, 44] and ultrasonic reactors [130, 68].

Transesterification is process algae oil must go through to become desired product biodiesel which is required two chemicals (Methanol and Sodium hydroxide) and following steps to be done as mix Methanol and Sodium hydroxide which make sodium methoxide now this sodium methoxide mix with algae oil and allow it to settle for about 8 hours. Now filter biodiesel to 5 microns and drain glycerin. This glycerin is used to make products such as soap and others.

12. Sustainability

Environmental protection will be one of the prominent reasons foe utilization of biomass resources. Microalgal biofuels are more important because it is useful to counter Energy Security and Climate Change problem which is main issue through worldwide. The microalgal Biomass absorbs carbon dioxide during growth, and emits it during combustion. Hence it does not contribute to green house effect. There can be a substantial reduction in the overall carbon dioxide emission as the microalgal biomass is a carbon dioxide neutral fuel. Microalgal biofuels

http://dx.doi.org/10.5772/59821

is also sustainable because away from conventional crops algae biomass can be grow on the land which is not arable land so it not affect food security to anywhere in the world. As discussed above, algae is fastest growing biomass on less land required to agriculture product so there is no any problem associated with row material required for biofuels generation.

Sustainability is the subject of much discussion at international scientific and governmental forums on biofuels. Emerging from this discussion is a consensus that sustainability is of foremost importance as an overarching principle for the development of biomass-to-energy agro-industrial enterprises. While sustainability criteria that are agreeable to all nations are still being expounded, the generally accepted principles of sustainability include that;

- the greenhouse gas balance of the production chain is positive;
- the biomass production is not at the expense of carbon sinks in existing vegetation and soil;
- the biomass production does not endanger the food supply and existing local business activity (i.e. local supply of energy, medicines and building materials);
- the biomass production has no impact on biodiversity (protected or vulnerable biodiversity is not affected or if possible strengthened);
- soil and soil quality are retained or improved;
- ground water and surface water are not depleted and water quality is maintained or improved;
- air quality is maintained or improved; and
- the production and processing of biomass contributes to local prosperity and to the social well being of employees and the local population.

It is self evident that where there is a natural abundance of freshwater, it is likely on arable land (that may be under agriculture and may have multiple competing uses for the water resource), or on land in its natural state with considerable biodiversity value. With few exceptions where the abundance of freshwater is the consequence of human intervention, the water has multiple competing uses.

Consequently, from the perspective of sustainability it seems obvious that algal production systems should target water resources other than freshwater. In fact, the proponents of algal biofuel claim that the production system is superior to biofuels based on terrestrial biomass because it can utilize non-arable land and waste water resources.

While the literature on the sustainability of algal biofuels is sparse, recent analyses appear to dispute the claims of superiority of algal production systems when compared to terrestrial crops.

Clarens *et al.* (2010) compared the environmental life cycle impacts of algal biomass production to corn, switch grass and canola production. The functional unit was 317 GJ of biomass derived energy or the amount of energy consumed by one American citizen in one year (i.e. the study sort to inform on the life cycle impacts associated with the production of 317 GJ of biomass based on the higher heating value of the material on a dry basis). Biomass production was modeled for three locations in the USA, and for algae was based on fresh water and municipal sewerage effluents from conventional activated sludge and biological nitrogen removal treatment plants. Algae production in raceway ponds varied from 0 g m^{-2} d^{-1}(seasonal shut down) to 20 g m $^{-2}$ d^{-1} depending on site location and climate. All four biomass production systems had net positive energy (i.e. more energy produced than consumed in the biomass production). Algae cultivation had better land use and eutrophication LCA outputs than terrestrial crops, but the terrestrial crops were found to have lower energy use, greenhouse gas emissions and water use than algae production based on fresh water or municipal sewerage effluents. When industrial grade CO_2 was used in algal biomass production the system emitted more greenhouse gases (GHG) than it sequestered. Even when flue gas was used, the algal production system consumed more energy and emitted more GHG than the terrestrial plant production systems (mostly as a consequence of high mineral fertilizer use).

Lardon *et al.* (2009) compared the environmental life cycle impacts of microalgae biodiesel production to the impacts of palm, rape and soybean oil biodiesel and petroleum diesel production. The LCA was based on a cradle to combustion' boundary (i.e. all products and processes upstream of fuel combustion in a diesel engine). The functional unit was 1 MJ of fuel in a diesel engine. The study considered four algae biofuel production scenarios, viz. production under nitrogen fertilizer rich and starved conditions and with oil extraction from wet and dry raceway ponds varied from 19.25 g m^{-2} d $^{-1}$ (in the nitrogen starved case) to 24.75 g m^{-2} d $^{-1}$ (in the nitrogen rich case). Of the four algae biofuel production scenarios, only growth under starved nitrogen conditions with oil extraction from wet biomass had a positive net energy. In the three other algal biofuel scenarios, the energy consumed in the production was greater than the energy in the delivered biofuel. These balances assumed 100% recovery of energy from the algae cake residue after oil extraction. Fertilizer (nitrogen) consumption had a far greater impact on cumulative energy demand than drying biomass for extraction. Algae biofuel had better land use and eutrophication LCA outputs than biofuels from the terrestrial crops, but petroleum diesel had better land use and eutrophication impacts than all biofuels. In all other assessed metrics, one or all of the terrestrial crop biofuels had lower LCA impacts than all algal biofuel scenarios (again mostly as a consequence of high mineral fertilizer use).

It should be stressed that these LCA studies are based on hypothetical operating scenarios, not real production systems. The purpose of the studies is to highlight inefficiencies in the production systems that need to be addressed to create sustainable microalgae-to-biofuel enterprises. Nevertheless, these studies created debate in the scientific community and the exchange of comments published in subsequent editions of the journal. Principal among the criticisms form algae biofuel proponents are that the authors of LCA studies that report negative outcomes use too low growth rates and too high mineral fertilizer consumption figures.

In contrast, Christi (2008), a proponent of algal biofuels, provides an opinion in *Trends in Biotechnology* titled *Biodiesel from microalgae beats bioethanol.* The claimed superiority of algal biofuel over sugarcane ethanol is based solely on land use efficiencies. In this article, Christi claims algal biofuel can sustainably and completely replace all petroleum derived transport

http://dx.doi.org/10.5772/59821

fuels, and quotes average annual algal biomass production in tropical regions as high as 1.535 kg m^{-3} d^{-1} in photobioreactors (a productivity/reactor volume measurement). This report has already noted that claims of extremely high growth in vertically configured photobioreactors are misleading. Vertical photobioreactors must be situated far enough from each other so as to not shade, and consequently the basic limitations on land use and productivity remains the same for both open ponds and closed photobioreactors. Christi (2007) had previously claimed very high land use efficiencies in raceway ponds (viz. 136,000 L/ha of oil for algal biomass with an oil mass fraction of 70% and 58,700 L/ha of oil for algal biomass with an oil mass fraction of 30%). Such yields are only achievable with production of greater than 340 days in a year and at a pond productivity of ca. 50 g m^{-2} d^{-1} (unrealistically high at the current state of technology). *Christi* also assumes that CO_2 is available at little or no cost (presumably in these same tropical regions); this is a challengeable assumption. Despite the liberal use of the word 'sustainable', *Christi* provides no other LCA metric than land use efficiency.

Reijnders (2008) in a rejoinder notes that Christi did not consider fossil fuel inputs during the biofuel life cycle, that previous LCA studies on *Dunaliella* and Spirulina production showed little or no net energy benefit, and that by comparison terrestrial plant production systems are characterized by much lower fossil fuel inputs. The studies of *Clarens et al.* and *Lardon et al.* support Reijnders views. It would seem probable that while the assumptions imbedded in hypothetical production scenarios do have significant impacts on LCA outcomes algal biofuel production faces significant challenges to meet sustainability criteria. Limited LCA studies indicate that significant advances need to be made in reducing fossil fuel inputs associated with nutrient use, harvesting and extraction.

Acknowledgements

The above work is part of the research titled "Harnessing clean and green energy via integrated treatment of industrial and domestic waste water" (ORG/EI/10/2009) sponsored by The Research Council, Oman, The authors are indebted to The Research Council, Oman for funding this project and giving an opportunity to realize successful completion. A special acknowledgement of the entire team at The Research Council Oman who helped and co-operated to administer the project for successful completion.

Author details

Nitin Raut[1], Talal Al-Balushi[2], Surendra Panwar[3], R.S. Vaidya[4] and G.B. Shinde[5]

1 Faculty of Engineering, Sohar University Sohar, Oman

2 International Relation Dept. The Research Council-Oman, Oman

3 Faculty of Civil Engineering, MJPR University, Jaipur, Rajasthan, India

4 Dept. of Chemistry, Shri Shivaji Science. College, SGBAU, Amravati, MS, India

5 Dept. of Chemical Engineering, SVIT, Nashik, Maharashtra, India

References

[1] Abdel-Raoufa N, Al-Homaidanb A.A., Ibraheemb I.B.M., Microalgae and wastewater treatment, Saudi Journal of Biological Sciences, Volume 19, Issue 3, July 2012, Pages 257–275

[2] Admon S., Tarre S., Sabbah I., Lahav O., and Green M., "Treatment of pre-settled municipal wastewater using a passively aerated vertical bed." Environ. Eng. pp 707-715, 2005.

[3] Ali M. S., Khan M.R. and Afroz T, "Portable water from industrial wastewater", Chemical Engineering Research Bulletin, Bangladesh, pp 20-23, Vol. 12, 2008.

[4] Azcan N, Danisman A. Microwave assisted transesterification of rapeseed oil. Fuel 2008;87:1781–8.97

[5] Balmant W, Modeling and simulation of anaerobic bioreactors. The Second International Green Energy Conference, IGEC Proceedings, Oshawa, Ont., 25–29 June 2006; 1–12.

[6] Barclay WR. Mixtures of omega-3 and omega-6 highly unsaturated fatty acids from euryhaline microorganisms. US Patent 6977167; 2005.

[7] Becker W, *Microalgae in human and animal nutrition.* Richmond, A., ed. Handbook of microalgal culture. Blackwell, Oxford. pp. 312-351.

[8] Benemann, J.R., Goebel, R.P., Weissman, J.C., Rubenstein, D.C. (1982). *Microalgae as a source of liquid fuels.* Final technical report USDOE ER. Www. osti.gov/ bridge/ product.biblio.jsp?query_ id=0&page=0 &osti_id=6374113

[9] Benemann, J.R., Koopman B.L., Baker, D.C., Goebel, R.P., Oswald, W.J. (1978). *The photosynthesis energy factory: Analysis, synthesis, and demonstration.* DOE contract No. EX-76-C-01-2548: Intertechnology/Solar Corporation.

[10] Benemann, J.R., Pursoff P., Oswald, W.J. (1978). *Engineering Design and Cost Analysis of a Large-Scale Microalgae Biomass System.* NTIS #H CP/T1605-01 UC61.Washington, D.C.: U.S. Department of Energy, pp. 91.

[11] Bhanu V., "New technologies in the paper and pulp industry wastewater treatment", Elsevier, 2011.

[12] Blight EG, Dyer WJ. A rapid method of total lipid extraction and purification. Canadian Journal of Biochemistry and Physiology 1959; 37(8) 911–917.

[13] Boari G., Mancini I.M. and Trulli E, "Technologies for Water and Wastewater Treatment", University degli Study della Basilicata., Italy, Potenza., P.N. 283, 1997.

[14] Borowitzka M. A., *Fats, oil, and hydrocarbons.* Micro-algal Biotechnology. 1988

[15] Bozbas K. Biodiesel as an alternative motor fuel: production and policies in the European Union. Renewable and Sustainable Energy Reviews 2008; 12: 542–52.

[16] Changl I.S., Clench P.L., Jefferson B., and Judd S."Membrane Fouling in Membrane Bioreactors for Wastewater Treatment", Journal of Environmental Engineering, (ASCE), P.N.1-2, 2002.

[17] Chao He, Yunshan Ge, Jianwei Tan, Xiukun Han. Spray properties of alternative fuels: a comparative analysis of biodiesel and diesel. International Journal of Energy Research 2008; 32(14):1329–1338. DOI: 10.1002/er.1440.

[18] Chi, Z., D. Pyle, et al. (2007). "A laboratory study of producing docosahexaenoic acid from biodiesel-waste glycerol by microalgal fermentation." Process Biochemistry 42(11): 1537-1545.

[19] Chisti Y. Biodiesel from microalgae beats bioethanol. Trends in Biotechnology 2008; 26(3):126–131. DOI: 10.1016/j.tibtech.2007.12.002.

[20] Chisti Y. Biodiesel from microalgae. Biotechnology Advances 2007; 25(3):294–306. DOI: 10.1016/j.biotechadv.2007.02.001.

[21] Chisti Y., Do biofuels from microalgae beat biofuels from terrestrial plants? Trends in Biotechnology 08; 26(7):351–352. DOI: 10.1016/j.tibtech.2008.04.002.

[22] Christi, Y. (2008). *Biodiesel from microalgae beats bioethanol.* Trends in Biotechnology. (25), pp.126-131.

[23] Clarens A., Resurreccion, E.P., White, M.A., Colosi, L.M. (2010). *Environmental Life Cycle Comparison of Algae to Other Bioenergy Feedstocks.* Environ. Sci. Technol. (44), 1813-1819.

[24] Cooney M., Young G. and Nagle N., Separation and Purification Reviews, 2009, 38, 291-325.

[25] Cravotto G, Boffa L, Mantegna S, Perego P, Avogadro M, Cintas P. Improved extraction of vegetable oils under high-intensity ultrasound and/or microwaves. Ultrasonics Sonochemistry 2008; 15(5):898–902.

[26] Demirbas A. Biofuels sources, biofuel policy, biofuel economy and global biofuel projections. Energy Conversion and Management 2008; 49(8):2106–2116. DOI: 0.1016/j.enconman.2008.02.020.

[27] Deshmane VG, Gogate PR, Pandit AB. Ultrasound-assisted synthesis of biodiesel from palm fatty acid distillate. Industrial and Engineering ChemistryResearch

[28] Dewulf J, van Langenhove H, van de Velde B. Energy-based efficiency and renewability assessment of biofuel production. Environmental Science and Technology 2005; 39(10):3878–3882. DOI: 10.1021/ es048721b.

[29] Doucha J. and Livansky, K., Applied Microbiology and Biotechnology, 2008, 81.

[30] Dunahay, T. J., Jarvis, E.E., Dais, S.S., Roessler, G. (1996). *Manipulation of microalgal lipid production using genetic engineering.* Appl. Biochem. Biotechnol. (57), pp. 223-231.

[31] Dunahay, T.G., Jarvis, E.E., Roessler, P.G. (1995). *Genetic transformation of the diatoms Cyclotella cryptica and Navicula saprophila.* J. Phycol. (31), pp. 1004-1012.

[32] Energy Information Administration. (2009). *International Energy* Outlook 2009. http://www.eia.doe.gov/oiaf/archive/ieo09/index.html

[33] Falkowski, P.G., Katz, M.E., Knoll, A.H., Quigg, A., Raven, J.A., Schofiled, O., Taylor, F.J.R. (2004). *The evolution of modern eukaryotic phytoplankton.* Science (305), pp. 354-360. Falkowski, P.G., Raven, J.A. (1997). *Aquatic photosynthesis.* Malden, MA: Blackwell Science.

[34] Ferna ´ ndez FGA, Hall DO, Guerrero EC, Rao KK, Grima EM. Outdoor production of Phaeodactylum tricornutum biomass in a helical reactor. Journal of Biotechnology 2003; 103(2):137–152. DOI: 10.1016/ S0168-1656(03)00101-9.

[35] Field, C.B., Behrenfeld, M.J., Randerson, J.T., Falkowski, P. (1998). *Primary production of the biosphere: integrating terrestrial and oceanic components.* Science (281), pp. 237-240.

[36] Florentinus, A., Hamelinck, C., de Lint, S., van Iersel, S. (2008). *Worldwide potential of aquatic biomass.* Utrect, Ecofys.

[37] Fortman JL, Chhabra S, Mukhopadhyay A, Chou H, Lee TS, Steen E, Keasling JD. Biofuel alternatives to ethanol: pumping the microbial well. Trends in Biotechnology 2008; 26(7):375–381. DOI: 10.1016/j.tibtech.2008.03.008.

[38] Fun S., Sing A., Isdepsky M., Borowitzka A. and Bohemian N.R, "Production of biofuels from microalgae, Ausralia, 2011.

[39] Gavrilescu M, Chisti Y. Biotechnology-a sustainable alternative for chemical industry. Biotechnology Advances 2005; 23(7–8):471–499. DOI: 10.1016/ j.biotechadv.2005.03.004.

[40] Gerpen JV. Business management for biodiesel producers. NREL Technical Report, NREL/SR51036342, 2004.

[41] Gnanapragasam G., Senthilkumar M., Arutchelvan V., Velayutham T, and Nagarajan S., " Bio-kinetic analysis on treatment of textile dye wastewater using anaerobic batch reactor", Bioresource Technology, India, PP 627-632, 2010.

[42] Gogate PR, Kabadi AM. A review of applications of cavitation in biochemical engineering/biotechnology. Biochemical Engineering Journal 2009; 44:60–72.

[43] Gogate PR., Cavitational reactors for process intensification of chemical processing applications: a critical review. Chemical Engineering and Processing 2008; 47:515–27.

[44] Gongloff, M. (2008). *VC's Get Clean.* Wall Street Journal online, http://blogs.wsj.com/environmentalcapital/category/alternative-energy/cleantech/venture-capital/. Accessed February 11, 2008.

[45] Gravalos I, Gialamas T, Koutsofitis Z, Kateris D, Xyradakis P, Tsiropoulos Z, Lianos G. Comparison of performance characteristics of agricultural tractor diesel engine operating on home and industrially produced biodiesel. International Journal of Energy Research 2009; 33(12):1048–1058. DOI: 10.1002/er.1533.

[46] Grima EM, Belarbi EH, Fernandeza FGA, Medina AR, Chisti Y. Recovery of microalgal biomass and metabolites: process options and economics. Biotechnology Advances 2003; 20:491–515. DOI: 10.1016/S0734-9750(02)00050-2.

[47] Grima ME, Belarbi EH, Fernandez FGA, Medina AR, Chisti Y. Recovery of microalgal biomass and metabolites: process options and economics. Biotechnology Advances 2003; 20(7-8):491–515.

[48] Groom MJ, Gray EM, Townsend PA. Biofuels and biodiversity: principles for creating better policies for biofuel production. Conservation and Biology 2008; 22(3):602–609. DOI: 10.1111/j.1523-1739. 2007.00879. X.

[49] Grosssman A. R. (2007). *In the Grip of Algal Genomics.* Transgenic Microalgae as Green Cell Factories (Leon, R., Galvan, A., Fernandez, E.). Austin, TX:

[50] Halim R., Danquah M. K. and Webley P. A., Biotechnology Advances, 2012, 30, 709-732.13

[51] Hampannavar U.S and Shivayogimath C.B, "Anaerobic treatment of sugar industry wastewater by Up flow anaerobic sludge blanket reactor at ambient temperature", Karnataka, international journal of environmental science, Volume 1, No 4, 2010.

[52] Hassannia J. H., Vice President, Business Development Diversified Energy Corporation A Project-Based Perspective: Algae Biofuels Economic Viability, 21 April 2009,

[53] http://universe-review.ca

[54] Hu Q, Sommerfeld M, Jarvis E, Ghirardi M, Posewitz M, Seibert M, et al.Microalgal triacylglycerols as feedstocks for biofuels production: perspectives and advances. The Plant Journal 2008; 54:621–39. [23]

[55] Huber GW, Iborra S, Corma A. Synthesis of transportation fuels from biomass: chemistry, catalysts, and engineering. Chemical Review 2006 106:4044–4098. DOI: 10.1021/cr068360d.

[56] Indarti E., Majid M.I.A., "Direct FAME synthesis for rapid total lipid analysis from fish oil and cod liver oil." Journal of Food Composition and Analysis 18(2-3): 161-170.

[57] Iverson S. J., Lang S. L. C. and CoopeM. H. r, Lipids, 2001, 36, 1283-1287.

[58] J.-Y. Lee, C. You S.-Y. Jun, C.-Y. Ahn and H.-M. Oh, Bioresource Technology, 2010, 101, S75-S77.

[59] James C. Ogbonna1, Hitoshi Yoshizawa1 and Hideo Tanaka, "Treatment of high strength organic wastewater by a mixed culture of photosynthetic microorganisms", Japan, pp 277 - 284, 2000.

[60] Kalva A, Sivasankar T, Moholkar VS. Physical mechanism of ultrasoundassisted synthesis of biodiesel. Industrial and Engineering Chemistry Research 2008; 48(1):534–44.

[61] Kalyuzhnyi S. V., Valdez Saucedo J., and Martinez Rodriguez J.," The Anaerobic Treatment of Soft Drink Wastewater in UASB and Hybrid Reactors", Humana Press Inc., Moscow, pp 1-11, 1996.

[62] Kaneko, T., Sato, S., Kotani H., Tanaka, A., Asamizu, E., Nakamura, Y., Miyajima N., Hirosawa M., Sugiura M., Sasamoto S., Kimura, T., Hosouchi T.,

[63] Kaufmann B. and Christen P., Photochemical Analysis, 2002, 13, 105-113.

[64] Kheshgi H.S., Prince, R.C., Marland G. (2000). *The Potential of Biomass Fuels in the Context of Global Climate Change: Focus on Transportation Fuels.* Ann. Rev. Energy Environ. (25), pp. 199-244.

[65] Khotimchenko S. V., Yakovleva I.M. (2005). *Lipid composition of the red algal Tichocarpus crinitus exposed to different levels of photon irradiance.* Phytochem. (66), pp. 73-79.

[66] Krauss C., *Venture Capital Wants to Put Algae in Your Gas Tank.* New York Times online, www.nytimes.com/2007/03/07/business/07algae.html?_r=1&oref=slogin. Accessed February 11, 2008.

[67] Kumari P., Reddy C. R. K. and Jha B., Analytical Biochemistry, 2011, 415, 134-144.

[68] Lahav O., Artzi E., Tarre S., and Green M.., "Ammonium removal using a novel unsaturated flow biological filters with passive aeration". Water Res. 35(2), 397–404, and 2001.

[69] Lardon L., Helias A., Sialve B., Steyer J.P., Bernard, O. (2009). *Life-Cycle Assessment of Biodiesel Production from Microalgae.* Environmental Science & Technology, In Press. http://pubs.acs.org/doi/abs/10.1021/es900705j

[70] Larson E. D., *Technology for Electricity and Fuels from Biomass.* Annual Review of Energy and Environment 18, 567-630. 1993

[71] Lee Y.K.., *Microalgal mass culture systems and methods: Their limitations and potential.* Journal of Applied Phycology. (13), pp. 307-315. (2001)

[72] Li Q., Du W., Liu D., Perspectives of microbial oils for biodiesel production. Applied Microbiology and Biotechnology 2008; 80:749–756. DOI: 10.1007/ s00253-008-1625-9.

[73] Li Y., Horsman M., Wu N., Lan CQ., Dubois-Calero N. Biofuels from microalgae. Biotechnology Progress 2008;24(4):815–20.15

[74] Li Y, Wang B, Wu N, Lan CQ. Effects of nitrogen sources on cell growth and lipid production of Neochloris oleoabundans. Applied Microbiology and Biotechnology 2008; 81(4):629–36.

[75] Liu B., Zhao Z., Biodiesel production by direct methanolysis of oleaginous microbial biomass. Journal of Chemical Technology and Biotechnology 2007; 82(8):775–780. DOI: 10.1002/jctb.1744.

[76] Ma F. R., and Hanna M. A., (1999). "Biodiesel production: a review." Bioresource Technology 70(1): 1-15. 69

[77] Mariano A.B., Ghidini L.F.C., Spray drying of Nannochloropsis oculata microalgae biomass. 20th International Congress of Mechanical Engineering, Proceedings of COBEM, Gramado, RS, Brazil, 15–20 November 2009.

[78] McGinnis, K. M., Dempster R.A., Sommerfeld, M.R. (1997). *Characterization of the growth and lipid content of the diatom Chaetoceros muelleri.* Journal of Applied Phycology. (9), pp. 19-24.

[79] Mendes R. L., Nobre B. P., Cardoso M. T., Pereira A. P. and. Palavra A. F, Inorganica chimica Acta, 2003, 356.

[80] Mercer P. and Armenta R. E., European Journal of Lipid Science and Technology, 2011, 113,539-547.

[81] Miao X, Wu Q. Biodiesel production from heterotrophic microalgal oil. Bioresource Technology 2006; 97(6):841–846. DOI: 10.1016/j.biortech.2005.04.008.

[82] Miao XL, Wu QY. High yield bio-oil production from fast pyrolysis by metabolic controlling of Chlorella protothecoides. Journal of Biotechnology 2004; 110(1):85–93. DOI: 10.1016/j.jbiotec.2004.01.013.

[83] Milne, T.A., Evans, R.J., Nagle, N. (1990). *Catalytic conversion of microalgae and vegetable oils to premium gasoline, with shape-selective zeolites.* Biomass. 21: 219-232.

[84] Netravali AN, Chabba S. Composites get greener. Materials Today 2003; 6(4):22–29.

[85] Oller A, Malato S and Sanchez Perez J A, "Combination of Advance Oxidation Processes and biological treatments for wastewater decontamination", Elsevier, Spain, 2011.

[86] Perrut M., Industrial & Engineering Chemistry Research, 2000, 39.

[87] Pienkos P. T. and Darzins A., The promise and challenges of microalgal-derived biofuels. Biofuels, Bioprod. Bioref. 3: 431-440. doi: 10.1002/bbb.159

[88] Piorreck M., Pohl, P., *Formation of biomass, total protein, chlorophylls, lipids and fatty acids in green and blue-green algae during one growth phase.* Phytochemistry. (23:2), pp. 217-223.

[89] Pratoomyot J, Srivilas P, Noiraksar T. Fatty acids composition of 10 microalgal species. Songklanakarin Journal of Science and Technology 2005; 27(6):1179–87. [19]

[90] Pulz O, Gross W. Valuable products from biotechnology of microalgae. Applied Microbiology Biotechnology 2004; 65(6):635–648. DOI: 10.1007/ s00253-004-1647-x.

[91] Raja R, Hemaiswarya S, Kumar NA, Sridhar S, Rengasamy R. A perspective on the biotechnological potential of microalgae. Critical Reviews in Microbiology 2008; 34(2):77–88. 31.

[92] Rajvanshi S, Sharma M P, Microalgae: A Potential Source of Biodiesel Journal of Sustainable Bioenergy Systems, 2012, 2, 49-59

[93] Reijnders L. Do biofuels from microalgae beat biofuels from terrestrial plants? Trends in Biotechnology 2008; 26(7):349–350. DOI: 10.1016/j.tibtech.2008.04.001.

[94] Renaud SM, Thinh LV, Parry DL. The gross chemical composition and fatty acid composition of 18 species of tropical Australian microalgae for possible use in mariculture. Aquaculture 1999; 170:147–59.

[95] Richmond A. Handbook of microalgal culture: biotechnology and applied Phycology. Blackwell Science Ltd; 2004.

[96] Rodolfi L, Zittelli GC, Bassi N, Padovani G, Biondi N, Bonini G, et al. Microalgae for oil: strain selection, induction of lipid synthesis and outdoor mass cultivation in a low-cost photobioreactor. Biotechnology and Bioengineering 2009; 102(1):100–12.

[97] Rodrigues JBR, Belli Filho P. Eficie^ncia da microalga Chlorella minutissima no tratamento de residuos de suinocultura enriquecido com ure ´ ia. Biotemas 2004; 17(2):7–26 (in Portuguese).

[98] Roessler, P.G. (1990). *Environmental control of glycerolopid metabolism in microalgae: Commercial implications and future research directions.* Journal of Phycology. (26), pp. 393-399.

[99] Rosenberg JN, Oyler GA, Wilkinson L, Betenbaugh MJ. A green light for engineered algae: redirecting metabolism to fuel a biotechnology revolution. Current Opinion in Biotechnology 2008; 19(5):430–6.

[100] Satyanarayana K. G., Mariano A. B., Vargas J. V. C., A review on microalgae., and Int. J. Energy Res. 2011; 35:291–311 r 2010 John Wiley & Sons, Ltd. DOI: 10.1002/ 307

[101] Schenk P.M., Hall S.R.T., Stephens E, Marx UC, Mussgnug JH, Posten C, et al. Second generation biofuels: high-efficiency microalgae for biodiesel production. Bioenergy Research 2008; 1:20–43.

[102] Schneider D., "Grow your Own? Would the Wide Spread Adoption of Biomass-Derived Transportation Fuels Really Help the Environment," American Scientist, Vol. 94, pp. 408, 409.

[103] Scragg A.H., Growth of microalgae with increased calorific values in a tubular bioreactor. Biomass and Bioenergy 2002; 23(1):67–73. DOI: 10.1016/S0961-9534(02)00028-4.

[104] Sheehan J., Dunahay T., Benemann J., Roessle P., A Look Back at the U.S. Department of Energy's Aquatic Species Program—Biodiesel from Algae, National Renewable Energy Laboratory, NREL/TP-580-24190

[105] Sheehan, J., Cambric V., Duffield, J., Grabowski, M., Shapouri H. (1998). *Life Cycle Inventory of Biodiesel and Petroleum Diesel for Use in an Urban Bus.*

[106] Shi X.M., Zhang X.W., Chen F. Heterotrophic production of biomass and lutein by Chlorella protothecoides on various nitrogen sources. Enzyme Microbiology and Technology 2000; 27(3–5):312–318. DOI: 10.1016/S0141-0229(00)00208-8.

[107] Smith S., (2007). *The Algae Solution.* American Radioworks online, www. noventivc.co m / news / npr08.17.07.htm. Accessed August 17, 2007.

[108] Spolaore P, Joannis-Cassan C, Duran E, Isambert A. Commercial applications of microalgae. Journal of Bioscience and Bioengineering 2006; 101(2): 87–96.

[109] Spolaore, P., Joannis-Cassan, C., Duran, E., Isambert, A. (2006). *Commercial applications of microalgae.* Journal of Bioscience and Bioengineering. (101), pp. 87-96.

[110] T.M. Mata et al. / Renewable and Sustainable Energy Reviews 14 (2010) 217–232 230

[111] Takagi M.K., Yoshida T., Effect of salt concentration on intracellular accumulation of lipids and triacylglyceride in marine microalgae Dunaliella cells Journal of Bioscience and Bioengineering 2006; 101(3):223–226. DOI: 10.1263/jbb.101.223.

[112] Travieso L., Pello N., Beni F., Sanchez E., Borja R., O'Farrill N., Weiland P., BIOALGAL reactor: preliminary studies for heavy metals removal. Biochemical Engineering Journal 2002; 12:87–91. DOI: 10.1016/S1369-703X (02)00045-1.

[113] Warabi Y., Kusdiana D., Saka S., Reactivity of triglycerides and fatty acids of rapeseed oil in supercritical alcohols. Bioresource Technology 2004; 91(3): 283–7.

[114] Ward O.P., Singh A., (2005). *Omega-3/6 fatty acids: alternative sources of production.* Process Biochemistry. (86), pp. 3627-3652.

[115] Weissman J.C., Goebel R.P., Design and analysis of microalgal open pond systems for the purpose of producing fuels: a subcontract report. US DOE- SERI; 1987.

[116] Wen Z.Y., Chen F., Heterotrophic production of eicosapentaenoic acid by microalgae. Biotechnology Advances 2003; 21(4):273–294. DOI: 10.1016/S0734- 9750(03)00051-X.

[117] Wiyarno B., Yunus R. M., in Journal of Applied Sciences, 2011, vol. 11,pp. 3607-3612.

[118] Xiao L., Mjos S. A., and Haugsgjerd B. O., Journal of Food Composition and Analysis, 2012, 25.

[119] Xu H, Miao X, Wu Q. High quality biodiesel production from a microalga Chlorella protothecoides by heterotrophic growth in fermenters. Journal of Biotechnology 2006; 126(4):499–507. DOI: 10.1016/j.jbiotec.2006.05.002.

[120] Zhang F., Endo T., Kitagawa R., Kabeya H., Hirotsu T., Synthesis and characterization of a novel blend of polypropylene with Chlorella. Journal of Materials Chemistry 2000; 10(12):2666–2672. DOI: 10.1039 b004489j.

[121] http://www.algaeindustrymagazine.com/special-report-spirulina-part-5-development-of-a-spirulina-industry-production/

[122] http://www.suprabio.eu/suprabio-consortium/igv-gmbh-igv-biotech/

Chapter 7

An Overview of Bioethanol Production From Algae

Didem Özçimen and Benan İnan

Additional information is available at the end of the chapter

http://dx.doi.org/10.5772/59305

1. Introduction

The depletion of fossil fuels which meet most of our energy requirements in near future and the pollutants from fossil fuels necessitates the usage of alternative renewable energy sources extensively. In this context, biomass is considered as an important alternative energy source to fossil fuels. Biodiesel and bioethanol produced from biomass sources are one of the best alternatives for petroleum-based fuels and recently, they are commonly used for transportation in many countries. Bioethanol is the most produced biofuel in the world and especially in Brazil and the United States two main producing countries with 62% of the world production. Large scale manufacture of ethanol as fuel is performed from sugar cane in Brazil, while it is produced from corn as a raw material in the United States [1]. Bioethanol production of 2013 in the countries is given in Table 1 [2].

Country	Production
USA	50,274
BRAZIL	23,690
EUROPE	5,182.38
CHINA	2,630.88
INDIA	2,060.1
CANADA	1,976.94
OTHER	2,748.06

Table 1. Bioethanol production amounts of countries in 2013 (million liter) [2]

Bioethanol is basically produced from first or second generation feedstocks. First generation bioethanol is produced from some cereals and lugumes such as corn, sugar beet, wheat and barley used for also food sources. Sugars which are obtained from first generation feedstock such as sugar cane, molasses, sugar beet and fruits can be fermented via yeast directly. Advantages of these raw materials are high sugar yields and low conversion cost. Their disadvantage is their production in just certain periods of the year. While 25 gallons of ethanol produced from an average of 1 ton sugar beet, 20 gallons of ethanol is produced from 1 ton of sweet sorghum stalk yearly. However their production is more expensive than that produced from sugar cane due to its energy and chemical inputs [3].

Usage of this first generation feedstock for bioethanol production leads to various discussions about increasing food prices and occupation of agricultural land. These problems are solved partially by using second generation feedstocks lignocellulosic materials such as waste or forest residues. Second generation feedstocks have some advantages over first generation feedstocks due to not being used as food source and less land requirement. However their harvesting, purification and various pre-treatment needs made their production quite challenging and not economical. Algae which are the third generation feedstock for biofuels are an alternative for the first and second generation feedstocks due to their productivity, easily cultivation and convenient harvesting time [4-6]. Recently, they are mostly utilized for biodiesel production because of their high lipid content. On the other hand, they have cellulosic structure and large amounts of carbohydrate embedded in, so they can be also utilized for bioethanol production directly or with the remains which is obtained after oil extraction. Since bioethanol production from conventional feedstock is considered for emitting more green-house gases than fossil fuels in consequence of the production steps and applications during the process, algal bioethanol production can overcome these problems. In comparison with conventional feedstocks, algal production areas don't occupy agricultural lands and they needn't any fertilizer for cultivation. With these advantages and significant carbohydrate content, higher ethanol yields are obtained from algae. In table 2, ethanol yield values from different feedstocks including first and second generations are given [7].

Feedstock	Ethanol yield (gal/acre)	Ethanol yield (L/ha)
Corn stover	112–150	1,050–1,400
Wheat	277	2,590
Cassava	354	3,310
Sweet sorghum	326–435	3,050–4,070
Corn	370–430	3,460–4,020
Sugar beet	536–714	5,010–6,680
Switch grass	1,150	10,760
Algae	5,000–15,000	46,760–140,290

Table 2. Ethanol yield values from different feedstocks [7]

Although it depends on the raw material which is used, ethanol production have three main steps: to obtain fermentable sugars, conversion of sugars to ethanol via fermentation process and distillation and purification of produced ethanol. In this chapter, these steps are presented in detail with their alternatives. All literature studies on the subject are reviewed, discussed and also new approach to pre-treatment methods of raw materials to produce bioethanol is presented.

2. Algae as bioethanol feedstock

Algae are simple organisms containing chlorophyll and they use light for photosynthesis. Algae can grow phototrophically or heterotrophically. Phototrophic algae convert carbondioxide in atmosphere to nutrients such as carbohydrate. Conversely, heterotrophic algae continue their development by utilizing organic carbon sources [8]. Algae can grow in every season and everywhere such as salty waters, fresh waters, lakes, deserts and marginal fields etc. However for their cultivation, generally open systems like ponds and photobioreactors as closed systems are used. Open ponds are the most used cultivation systems in industry. They are more preferable than other systems due to having low investment and operation costs. On the other hand difficult control of cultivation conditions and contamination risk are the main disadvantages of the open systems. Besides being cheap and low energy need, their cleaning also can be done easily. Although, open tanks have low cost and easy operation, parameters like light intensity, temperature, pH and dissolved oxygen concentration cannot be controlled easily. Most produced algae species in open systems are *Spirulina, Chlorella* and *Dunaliella* [9].In comparison with open systems, photobioreactors have very high photosynthetic efficiency. Thus, photobioreactors ensures high biomass yield. Though they are expensive, they are preferred for specific algal production. Algal production which is controlled in terms of parameters like light, pH, carbon dioxide etc., can be achieved and also contamination risk is not seen mostly in photobioreactors. Since they are closed systems, evaporation doesn't occurred and they enable production of special biochemical materials. Although there are many types of photobioreactors, most commonly systems are vertical and horizontal tubular columns and flat-type photobioreactors [10]. These photobioreactors which are made of glass or plastic, can be designed in shapes of horizontal vertical, conical or curved etc [11,12].

Algae are classified as microalgae and macroalgae. Microalgae as their name implies, are prokaryotic or eukaryotic photosynthetic microorganisms. They can survive in hard conditions with their unicellular or simple colony structures [13]. Because of being photosynthetic organism, they can produce high amount of lipid, protein and carbohydrate in a short time. Besides biodiesel and bioethanol there are lots of high value products and sub-products produced from microalgae such as biogas [14, 15], biobutanol, acetone [16], Omega 3 oil [17], eicosapentaenoic acid [18], livestock feed [19], pharmaceuticals and cosmetics [20, 21]. Especially sub-products are preferred for economic support of main process [22]. Chemical composition of microalgae can change according to the cultivation type and cultivation conditions. They can have rich or balanced composition of protein, lipid and carbohydrate amounts. Microalgae especially get attention due to have high lipid content [23]. Many species of microalgae accumulate a significant amount of lipids in their structure and can provide high oil yield. Their average lipid content can change between 1-70%, but this ratio can reach up to 90% of dry weight under certain conditions [13]. Macroalgae or seaweed are plants which are adapted to the marine life, often located in coastal areas. They are classified as brown seaweeds, red seaweeds and green seaweeds according to their pigments [24]. Due to have high photosynthesis capability, they have sufficient carbon source for usage in biorefinery. On the contrary of their appearances, their features of morphologic and physiological and chemical compositions are different from terrestrial plants [24]. Unlike the structure of the lignocellulosic biomass of microalgae, they comprise substances such as carrageenan, laminaran, mannitol, alginate which are used in various sectors [25]. They are separated from microalgae with having low lipid content and different from lignocellulosic material with having less or no lignin in their structure [6].

Microalgae stand out as biodiesel feedstock with the ability of lipid production and high photosynthetic efficiency. As for macroalgae, they are utilized for biogas or bioethanol production with their carbohydrates [26]. First studies as algal biofuels are focused on biodiesel production. However, there is a potential for carbohydrates in the structure of algae which can be utilized for ethanol production after various hydrolysis processes. Algal cells in the water don't need structural biopolymers such as hemicellulose and lignin which are necessary for terrestrial plants [4]. This simplifies the process of bioethanol production. Marine algae can produce high amount of carbohydrate every year. Also it is expected that algae will meet the demand of biofuel feedstock due to harvest in a short time than other biofuel raw materials [27]. Microalgae which have high amount of starch such as *Chlorella, Dunaliella, Chlamydomonas, Scenedesmus* are very useful for bioethanol production. In addition to that, microalgae don't need energy consumption for distribution and transportation of molecules like starch. Like microalgae, macroalgae are also raw materials that can be used in ethanol production. Absence of lignin or having less lignin in the structure, simplifies the hydrolysis stages [4,28]. Although it changes with the algal species, they have various amounts of heteropolysaccharides in their structures. Whereas red algae contain carrageenan and agar, brown algae have laminaran and mannitol in their structure [6].

3. Algal polysaccharides

Ethanol production from algae is based on fermentation of algal polysaccharides which are starch, sugar and cellulose. For microalgae, their carbohydrate content (mostly starch) can be reached to 70% under specific conditions [29]. Microalgal cell walls are divided into inner cell wall layer and outer cell wall layer. Outer cell layer can be trilaminar outer layer and thin outer monolayer. Also there can be no outer layers as well [30]. Outer cell walls of microalgae contain certain polysaccharides such as pectin, agar and alginate. However their composition can be vary from species to species [30]. On the contrary, inner cell walls of microalgae constitute mostly cellulose, hemicellulose and other materials [30]. Due to have cellulose in their cell walls and starch, microalgae are considered as a feedstock for production of bioethanol [31]. Most of their cell wall polysaccharides and starch can be fermented for bioethanol production [32].

Similarly, carbohydrate content of macroalgae is found 25-50% in the green algae, 30-60% in the red algae and 30-50% in the brown algae. Macroalgae species which have the highest polysaccharide content are *Ascophyllum* (42-70%), *Porphyra* (40-76%) and *Palmaria* (38-74%). High carbohydrate content of algal species are presented in Table 3 [32]. Polysaccharides in the cell wall of macroalgae are composed of cellulose and hemicelluloses. Cellulose and hemicelluloses content of macroalgae compose 2-10% of dry weight. Lignin is only exists in *Ulva* species and it constitutes 3% of dry weight [27]. Differently from microalgae, alginate, mannitol, glucan and laminarin are the most abundant polymers in macroalgal structure [42]. Alginates are

Algal species	Carbohydrate content (%)	Ref.
C. vulgaris	55.0	[33]
Chlamydomonas reinhardtii UTEX 90	60.0	[34]
Chlorococum sp.	32.5	[35]
S. obliquus CNW-N	51.8	[36]
Tetraselmis sp.CS-362	26.0	[37]
Ulva lactuca	55-60	[38]
Ascophyllum	42-70	[39]
Porphyra	40-76	[40]
Palmaria	38-74	[41]

Table 3. Carbohydrate content of algal species

polymers which are obtained from cell walls of various brown algae. They consist of mannuronic acid and L-gluronic acid monomers and they are extracted from cell walls by using sodium carbonate. Although they are usually used as stabilizer in pharmaceutical industry, they also used in paper and adhesive manufacture, oil, photography and textile industries [43-45]. Caragenan is another polysaccharide which is obtained from red algae. It is used as stabilizer in food, textile and pharmaceutical industry. Agar is also acquired from red algae. Like caragenans and alginates, it is extracted with hot water and used as stabilizer and gelling agent. 90% of produced agar is utilized in the food industry, the remaining 10% is used in microbiological and biotechnological field [44,46]. Mannitol which is a structure in brown algae is a sugar alcohol, especially found in *Laminaria* and *Ecklonia*. Mannitol content of macroalgae can change with season and environmental conditions. Mannitol can be used in pharmaceutical, paint, leather and paper manufacture. In addition to that, mannitol can be utilized in food industry as a coating material [27]. Laminarin is a polysaccharide which helps the immune system by increasing the B cells, provides protection against infection by bacterial pathogens and severe irradiation. Another polysaccharide from macroalgae is ulvan. It is mainly presented in *Ulva sp.* and it is source of sugars for production of fine chemicals [27].

4. Pre-treatment technologies for biomass

The most important difficulties encountered in the production of bioethanol are the pre-treatment of biomass. The objectives of an effective pre-treatment are obtaining sugars directly or later by hydrolysis, preventing lost or degradation of obtained sugars, limiting the toxic materials which inhibit the ethanol production, reducing energy requirement for process and minimizing the production cost. There are four pre-treatment techniques including physical, chemical, physicochemical and biological pre-treatments that are applied to biomass [1]. Pretreatment process is the step that forms the significant part of the cost of ethanol production. Although there is no technique that can be considered as the best option, researches and developments are carried on to reduce cost and improve performance [3].

4.1. Physical pre-treatments

4.1.1. Mechanical comminution

Chipping, grinding and milling are the most used techniques for mechanical comminution. Comminutions improve the efficiency of the process for the next steps by reducing the polymerization degree and increase the specific surface by reducing cellulose cristallinity. Energy that is need in the process depends on the initial and final dimensions of particles, moisture content and structure of the raw material [1,47]. In order to assist enzymatic hydrolysis of lignocellulosic materials various milling techniques can be used. For instance, pre-treatment of rice straw with wet disk milling gave higher hydrolysis yields than usual dry milling [48].

4.1.2. Pyrolysis

Pyrolysis is an endothermic process which is a reaction needs low energy input and treats biomass over the temperature of 300°C and degrades cellulose to char and gaseous products like CO and H_2. When the char is washed with water or diluted acid, remaining solution contains sufficient amount of carbon source to support microbial growth for the production of bioethanol. Approximately 55% of biomass weight is lost in the washing step [1]. It is reported in a study that Fan et al. [49] have performed 80-85% conversion of cellulose to reducing sugars.

4.1.3. Microwave oven pre-treatment

Microwave oven pre-treatment is a simple method with short reaction time, high heating efficiency and low energy input. Thermal and non-thermal effects which are generated by microwaves in a liquid medium are used in this technique. The heat generated in biomass results in a polar bond vibration. This causes an explosion between the particles and degradation of lignocellulosic structure. Asetic acid is released from lignocellulosic material and an acidic medium is occurred for hydrolysis [50]. Ooshima et al. [51] investigated the effect of microwave pre-treatment on rice straw and baggase and it was found that an improvement in total reducing sugar production. In recent years, microwave pre-treatments are carried out with various chemical reagents and their potential are investigated. In the studies of alkali microwave pre-treatment, NaOH provides higher reducing sugar yields on switchgrass and coastal bermudagrass in comparison with other alkaline reagents such as Na_2CO_3, $Ca(OH)_2$ [52,53]. Also for pre-treatment of rice straw and its hulls, this technique made cellulose more accessible to enzymes.

4.2. Physicochemical pre-treatments

4.2.1. Steam explosion method

Steam explosion method is a technique that provides accessibility on the biomass for degradation of cellulose. This method comprise the heating of biomass under high pressure steam (20–50 bar, 160-270 °C) for a few minutes, then reaction is stopped when the pressure conditions arrive to the atmospheric conditions. Diffusion of the steam into the lignocellulosic matrix leads to the dispersion of fibers. No catalyst is used during the applied method. Levulinic acid, xylitol and alcohols are obtained after the degradation of biomass [54,55]. Many types of biomass such as poplar [56], eucalyptus [57], olive residues [58], corn stover [59], wheat straw [60], sugarcane bagasse [61], grasses [62] have been pre-treated with steam explosion method efficiently.

4.2.2. Liquid hot water method

Liquid hot water method treats biomass by using water which is kept in a liquid state under high pressure and temperature for 15 minutes without adding any chemical or catalyst. Instead of steam explosion method, this technique does not need rapid pressure drop or expansion. Pressure is used to prevent evaporation and to stabilize the water in this method [60]. Although it provides the release of hemisellulosic sugars as oligomers, it causes the formation of little amounts of undesirable components which inhibit microbial growth such as carboxylic acid, furfural [63]. Since there is no need for chemicals, it is an environmental and economic method [64]. It is reported that liquid hot water method improves the enzymatic hydrolysis by removing 80% of hemicelluloses when it is pre-treated corn stover, sugarcane bagasse and wheat straw [65].

4.2.3. Ammonia Ffiber Explosion (AFEX)

Ammonia fiber explosion (AFEX) is a method that liquid ammonia and steam explosion are carried out together. In this method, biomass which has 15-30% moisture content is treated with liquid ammonia at a loading ratio of 1–2 kg NH_3/kg dry biomass. To acquire appropriate temperature, pressure over 12 atm is required. Whereas being an easy method and have short reaction time, it is not effective on raw materials that contain high lignin content [54]. Ammonia has effects such as shredding biomass fibers, partially decrystallization of cellulose and destroying carbohydrate attachments [65]. Although sugars are not released directly with this method, it enhances polymers (hemicellulose and cellulose) to be attacked enzymatically. Thus, low amount of enzyme is enough for enzymatic hydrolysis after AFEX. In order to improve the process economically, ammonia must be recover after the pre-treatment. Ammonia loading, temperature, high pressure, moisture content of biomass, and residence time are the basic parameters which effect AFEX process. Up to 90% cellulose and hemicelluloses conversions can be acquired with this technique [3].

4.2.4. CO_2 explosion

CO_2 explosion is similar to AFEX method. However this method has low process cost due to need low temperature. Also formation of inhibitors in the steam explosion is not occurred in this technique. In addition to that, its conversion yields are very high compared to steam explosion [50,66].

4.2.5. Wet oxidation

Wet oxidation method is based on the treatment of biomass with water and air or oxygen as a catalyst over the temperature of 120 °C. Although solubility of hemicellulose and lignin are increased with this method, free hemicelluloses molecules do not hydrolyze. Whereas sugar monomers are formed in steam explosion and dilute acid pre-treatment, sugar which released in wet oxidation method are oligomers [67,68]. In a study performed by Pederson [69] et al. 40% glucose yield was obtained for wet oxidation of wheat straw.

4.3. Chemical pre-treatments

Chemical pre-treatments include dilute acid, alkaline, ammonia, organic solvent pre-treatments and methods that use other chemicals. These processes are easy to perform and also good conversion yields are achieved in a short time [1].

4.3.1. Acid pre-treatment

Acid pre-treatments are methods that acid is used as catalyst to make cellulose more accessible to the enzymes. These processes are divided into two groups as using concentrated acid or diluted acid. Using concentrated acid is less preferable than dilute acid because of forming high amount of inhibiting components and causing corrosion in the equipments [68]. Generally sulphuric acid, hydrochloric acid, nitric acid and phosphoric acid are used in these pre-treatments. Dilute acid are applied at moderate temperatures to convert lignocellulosic structures to soluble sugars [54]. Nowadays biomass is pre-treated with dilute sulphuric mostly to hydrolyze hemicelluloses and facilitate enzymatic hydrolysis [70]. Dilute sulphuric acid hydrolyzes biomass to hemicelluloses, and then hydrolyzes to xylose and other sugar and break xylose down to furfural. Furfural which is a toxic component in ethanol production process, is recovered by distillation [54]. Miranda et al. have investigated the effect of acid pre-treatments with the concentrations between 0.05-10 N, and have obtained the highest sugar yield under the condition of 2 N acid pre-treatment. In their experiments, 2 N to 10 N acid pre-treatments, it is reported that a decrease have been observed in sugar yields [71]. Larsson et al. also mentioned that in an experiment about acid pre-treatment of soft wood, a decrease in ethanol yields have been observed with an increasing acid concentration. In addition to this, it is indicated that formic acid which is a toxic molecule, is presented in the media and inhibits the fermentation [72].

4.3.2. Alkaline pre-treatment

These processes are carried out at low temperature and pressure compared to other techniques. Unlike acid pre-treatments, lignin can be removed without major effects on the other components. However there are limitations such as transformation of some alkaline to unrecoverable salts. In addition to that, solubility of hemicelluloses and cellulose are less in this pre-treatment compared to solubility in acid pre-treatment [73]. Alkaline pre-treatment reduces the lignin and hemicelluloses content of biomass and improves the surface area and helps water molecules for breaking bonds between hemicelluloses and lignin [54]. The most used catalysts in this method are sodium hydroxide, potassium hydroxide, calcium hydroxide and ammonia [74]. Effects of alkaline pre-treatments are varies according to biomass. In an alkaline pre-treatment of coastal bermudagrass, reducing sugar yields are decrease with an increasing alkaline concentration [75]. However, Wang et al. reported that under the conditions of increasing alkaline concentrations, glucose yields were increased [76]. Like dilute acid pre-treatments, dilute alkaline pre-treatments also can form inhibitory by-products such as furfural, hydroxymethylfurfural and formic acid [77]

4.3.3. *Organosolv pre-treatment*

Organosolv pre-treatment is a process that uses organic solvents such as methanol, ethanol, acetone, ethylene glycol. Catalysts are also can be added to the process along with solvents. Hydrochloric acid, sulphuric acid, sodium hydroxide and ammonia are the catalysts used in the process. Besides bonds of lignin and hemicellulose can be broken, pure and high quality lignin can be obtained as a by-product [78]. Removal of lignin improves the surface area and provides accessibility of enzymes to cellulose. After the pre-treatment, cellulosic fibers, solid lignin and liquid solution of hemicellulose sugars are obtained. This method has some disadvantages like oxidation, volatilization and creating high risk in process at high pressure. Also solvents must be recovered due to formation of significant amounts of furfural and soluble phenols and to reduce operation cost [50,67].

4.4. Biological pre-treatments

Compared to the above methods applied to the production of bioethanol, using fungi in pre-treatments is considered environmentally friendly because of not using chemicals, less energy input, not required reactors that resistant to corrosion and pressure, and minimum inhibitor formation [79]. Fungi which are used in biological pre-treatments are generally brown, white and soft mold. These fungi can be degrade lignin, hemicelluloses and cellulose partially. Despite of its advantages, long process time, large production are and need of control continuously for growth of microorganisms ensue as disadvantages for commercial productions[50].

Enzymatic hydrolysis is the step of hydrolysis of cellulose by specific cellulase enzymes. Obtained products after hydrolysis are reducing sugars that include glucose. Cost of the enzymatic hydrolysis are less than acid or alkaline hydrolysis due to reaction is carried out under mild conditions (4.8 pH, temperature of 45-50 °C) [50]. Cellulase enzymes that are used in hydrolysis can be produced by bacteria and fungi. These microorganisms can be aerobic, anaerobic, mesofilic or thermophilic. Bacteria which produce cellulase can be exemplify as *Clostridium, Cellulomonas, Bacillus, Thermomonospora, Ruminococcus, Bacteriodes, Erwinia, Acetovibrio, Microbispora* and *Streptomyces. Trichoderma, Penicillium, Fusarium, Phanerochaete, Humicola* and *Schizophillum sp.* are identified as cellulase produced fungi among the fungi [1]. Although there are anaerobic bacteria which produce cellulase with high specific activity, these bacteria are not suitable for commercial productions. Cellulase enzymes consist of mixture of endoglucanase, exoglucanase and b-glucosidase. While endoglucanase attacks the regions where cellulose fibers have low crystallinity, exoglucanase removes the cellulose units from released chains with the effect of endoglucanase and then degrades the molecule. B-glucosidase hydrolyzes the cellulose units and enables the formation of glucose [64]. Enzymatic hydrolysis can be affected by certain factors which are enzyme-related and substrate-related factors. Substrate-related factors have a directly influence on enzymatic hydrolysis. These factors are connected to each other and effect the enzymatic conversion. These factors can be defined as *degree of polymerization and crystallinity of cellulose, accessibility of the substrate, lignin and hemicelluloses content and pore size.*

Hydrolysis rates of biomass depend on the degree of polymerization and crystallinity of cellulose. Degree of polymerization is related to crystallinity. Cellulase enzymes can hydrolyze the crystalline structure of cellulose. Endoglucanase enzymes decrease polymerization degree of cellulosic component by cutting the internal sites of cellulose chains in the enzymatic hydrolysis [80]. Accessibility of the substrate is another main factor effect hydrolysis rate. The rate of hydrolysis increases with increasing substrate accessibility because of being surface area more available for enzymatic attack [80]. Lignin and hemicellulose are complex structures to hydrolyze in lignocellulosic materials. Due to have a role like cement, lignin acts as physical barrier and prevents the digestible parts of cellulose to hydrolyze and it becomes very difficult for enzymes to access cellulose. For this reason, they reduce the efficiency of hydrolysis. Removal of hemicellulose enhances the pore size and provides accessibility to cellulose for enzymes in order to perform hydrolysis efficiently [81,82]. Pore size of the substrate is one of the limiting factors in enzymatic hydrolysis process. In many lignocellulosic material, external area of the biomass is smaller than internal area and this situation causes cellulase enzymes to entrap in the pores of the material. In order to increase hydrolysis rate, porosity of the biomass should be increased [83].

5. Fermentation

Fermentation is a process that based on disciplines of chemistry, biochemistry and microbiology and which fermentable sugars are converted to ethanol by microorganisms [84]. Process consists of conversion of glucose to alcohol and carbon dioxide:

$$C_6H_{12}O_6 \rightarrow 2C_2H_5OH+2CO_2 \quad (1)$$

In this process 0.51 kg bioethanol and 0.49 kg carbon dioxide are obtained from per kg of glucose in theory maximum yield. However practically, microorganisms also use glucose for their growth, the actual yield is less than 100% [85]. Microorganisms used in fermentation are utilized from 6-carbon sugars in ethanol production. Therefore, cellulosic biomass which have high amount of glucose are the materials that have easiest conversion capability. One of the most effective yeast which produces bioethanol is *Saccharomyces cerevisiae*. Besides having high bioethanol production yields, it has a resistance to high bioethanol concentration and inhibitor components which can be occurred after acid hydrolization of lignocellulosic biomass. Because reaction occurs under anaerobic conditions, oxygen molecules must be removed with nitrogen gas as a swept gas. Yeast and fungi can tolerate 3.5-5.0 pH ranges [86]. *S.cerevisiae* has high

http://dx.doi.org/10.5772/59305

osmotic resistance and can tolerate low pH levels like 4.0. *Zymomonas* stands out with rapid bioethanol production and high productivity compared to other traditional yeasts. However *Z.mobilis* cannot tolerate the toxic effects of asetic acid and various phenolic compounds in the lignocellulosic hydrolysate [87]. Bioethanol yields of microorganisms are depend on temperature, pH level, alcohol tolerance, osmotic tolerance, resistance for inhibitors, growth rate and genetic stability [54]. Fermentation processes generally are carried out with two basic processes as *simultaneous saccharification and fermentation* and *separate hydrolysis and fermentation*, yet new production processes have been developed [1].

5.1. Separate Hydrolysis and Fermentation (SHF)

Enzymatic hydrolysis is performed separately from fermentation in this process. Liquid which comes from hydrolysis reactor first converted to ethanol in a reactor that glucose fermented in, and then ethanol is distilled and remained unconverted ksilose is converted to ethanol in a second reactor. Advantage of the process is performing reactions in optimum conditions. On the other hand, usage of different reactors is increasing the cost. Also glucose and cellulose units that obtained after hydrolysis, inhibit activity of the enzyme and decrease hydrolysis rate [3,54].

5.2. Simultaneous Saccharification and Fermentation (SSF)

In this process, pre-treatment and enzymatic hydrolysis steps are carried out with fermentation step in the same reactor. It is very efficient when dilute acid or hot water at high temperature is applied in the process. High bioethanol yields can be achieved with SSF process. Also inhibiton of enzyme activity is very low due to fermenting glucose and cellulose units in the same media by yeast. Therefore, this process needs low amount of enzyme. In addition to that, process cost is reduced because of the reactions are carried out in one reactor. As a disadvantage, temperatures differences between saccharification and fermentation cause various effects in growth of microorganisms. *Saccharomyces* cultures are used in pH of 4.5 and temperature of 37 °C this process [3,54,88].

5.3. Simultaneous Saccharification and Co-Fermentation (SSCF) & Separate Hydrolysis and Co-Fermentation (SHCF)

Saccharomyces cerevisiae which used in fermentation cannot convert carbohydrates like pentos under moderate conditions and this causes impurity for biomass and decreases bioethanol yield. In order to overcome this, recombinant yeasts can be used to convert residues such as pentose to ethanol. In SSCF, recombinant yeasts and cellulase enzyme complex are fed to the same vessel to convert biomass to ethanol. This system is generally the same as SSF process. SCHF process is a combination of SSCF and SHF. In this process, fermentation and hydrolysis are carried out in different vessel. This process can produce ethanol with high productivity in comparison with SHF process [88].

Due to their simple structure and being a new raw material for bioethanol production, most of these pre-treatment techniques have not applied to algal biomass yet, and just few studies have been found in literature which is presented in Table 4.

Algae	Classification	Pre-treatment	Fermenting organism, process type and process time	Yield (%)	Ref.
Chlorococum sp.	Microalgae	Supercritical CO_2 lipid extraction	*Saccharomyces bayanus*, SHF, 60 h	38.30	[28]
Chlorococcum infusionum	Microalgae	0.75% (w/v) NaOH pre-treatment in120°C for 30 min	*Saccharomyces cerevisiae*, SHF, 72 h	26.00	[89]
Chlamydomonas reinhardtii UTEX 90	Microalgae	3% H_2SO_4 pre-treatment in 110°C for 30 min	*Saccharomyces cerevisiae* S288C, SHF, 24 h	29.10	[90]
Chlamydomonas reinhardtii UTEX 90	Microalgae	a-amylase (90 °C, 30 min) and glucoamylase (55 °C, 30 min) enzymatic hydrolysis	*Saccharomyces cerevisiae S288C*, SSF, 40 h	23.50	[91]
Chlorella vulgaris	Microalgae	3% H_2SO_4 pre-treatment in 110 °C, for 105 min	*Escherichia coli* SJL2526, SHF, 24 h	40.00	[92]
Schizochytrium sp	Microalgae	Hydrothermal degradation and a-enzymatic hydrolysis	*Escherichia coli KO11*, SSF, 72 h	5.51	[93]
Chlorococcum humicola	Microalgae	H_2SO_4 pre-treatment	*Saccharomyces cerevisiae*, SHF, 50 h	48	[94]
Chlamydomonas reinhardtii cw15	Microalgae	12 N H_2SO_4 pre-treatment	*Saccharomyces cerevisiae*, SHF, 48 h	44	[95]
Kappaphycus alvarezii	Macroalgae	0.9 N H_2SO_4 pre-treatment in 120 °C, for 60 min	*Saccharomyces cerevisiae NCIM 455*,SHF, 96 h	15.4	[96]
Kappaphycus alvarezii	Macroalgae	0.2% H_2SO_4 pre-treatment in 130 °C, for 15 min	*Saccharomyces cerevisiae* SHF, 4h	1.31	[97]
Gracilaria salicornia	Macroalgae	2% H_2SO_4 pre-treatment in 120 °C for 30 min and enzymatic hydrolysis with cellulose in 40 °C	*Escherichia coli KO11*, SHF, 48 h	7.90	[98]
Gelidium elegans	Macroalgae	Meicelase pre-treatment in 50 °C for 120 h	*Saccharomyces cerevisiae IAM 4178*, SHF, 48 h	36.7	[99]
Sargassum sagamianum	Macroalgae	Thermal liquefaction in 15 MPa 200 °C for 15 min	*Pichia stipitis CBS 7126*, SHF, 48 h	10.0	[100]
Laminaria japonica	Macroalgae	0.1 N HCl acid pre-treatment in 121 °C for 15 min and enzymatic hydrolysis with Celluclast 1.5 L, Viscozyme L	*Escherichia coli KO11*, SSF, 72 h	16.1	[101]
Laminaria hyperborea	Macroalgae	Disruption and washing in water in pH 2 65 °C	*Pichia angophorae*, SHF, 48 h	0.86	[102]
Saccharina latissima	Macroalgae	Shredding and saccharification pre-treatment	*Saccharomyces cerevisiae Ethanol Red*, SSF, 48 h	0.47	[103]
Laminaria digitata	Macroalgae	Shredding and saccharification pre-treatment	*Pichia angophorae*, SSF, 96 h	13.2	[104]
Laminaria japonica	Macroalgae	0.1 M H_2SO_4 acid pre-treatment in 121 °C for 1 h and enzymatic hydrolysis with cellulose and cellobiase	*Saccharomyces cerevisiae*, SHF, 36 h	11.3	[105]
Laminaria japonica	Macroalgae	Milling and otoclave in 120 °C for 15 min	*Pichia stipitis* KCTC7228	2.9	[106]
Gracilaria verrucosa	Macroalgae	Enzymatic hydrolysis with cellulose and β–glucosidase	*Saccharomyces cerevisiae*, SHF, 48 h	43	[107]
Gelidium corneum	Macroalgae	0.5–1% acid pre-treatment in 121°C for 30 and 60 min	*Saccharomyces cerevisiae*, SHF, 96 h	5.8	[108]
Sargassum spp.	Macroalgae	3.4-4.6% H_2SO_4 acid pre-treatment and enzymatic hydrolysis	*Saccharomyces cerevisiae*, SHF, 48 h	65	[109]
Ulva lactuca	Macroalgae	2 N acid pre-treatment	*Saccharomyces cerevisiae*, SSF, 48 h	24.48	[38]

Table 4. Studies of ethanol production from micro and macroalgae

6. Distillation and purification

A distillation process is necessary for separation of ethanol from mixture and purification of ethanol after fermentation process. Process is performed simply with boiling ethanol-water mixture. Because of boiling point of water (100°C) is higher than boiling point of ethanol (78 °C), ethanol vaporized before water [110]. However, due to being an azeotrop mixture, high amount of energy is used for distillation [42]. In order to separate azeotrop mixtures, an agent which changes the azeotrop structure must be added to the mixture. Added substance changes the volatility of mixture by effecting the molecular attractions in the mixture. Various separation agents such as benzene, pentane, cyclohexane, hexane, acetone, and diethyl ether can be used in this process [111]. Distillation column which has two streams as top and bottom, separates most of the bioethanol from the mixture. While top stream is rich in bioethanol, bottom stream is rich in water. 37% bioethanol then concentrated in rectifying column to approach concentration of 95% [78]. Product which is remained in the bottom is fed to stripping column in order to remove excess water [112]. Mostly in plants, recovery of bioethanol in distillation columns is fixed to be 99.6% due to decrease bioethanol loss [54].

7. Conclusion

Today, demand for fossil fuels cannot be met with current reserves and increasing oil prices with economical and political crisis and effects of global warming are led countries to use renewable energy sources. Algae as third generation feedstock have a great potential because of their characteristics. Different valuable products can be obtained from algae such as biodiesel, bioethanol, biogas, pharmaceuticals and nutraceuticals. Nowadays algae are mostly utilized for biodiesel production due to their high lipid content. However algae have also high carbohydrate content that cannot be ignored. Thus they can be utilized for bioethanol production directly or with the remains which are obtained after oil extraction. In this study, potential of algae as a bioethanol feedstock, important steps of bioethanol production especially pre-treatment techniques have been mentioned. In production sections, pre-treatment techniques and fermentation processes are explained in details. Recently, bioethanol production from algae is very new technology and open to development. Innovative and efficient fermentation processes and pre-treatment techniques are needed to make ethanol production preferable. In conclusion, algae will with their huge potential will outclass the first and second generation feedstocks and lots of improvements for usage of it will carried out in the future.

Author details

Didem Özçimen* and Benan İnan

*Address all correspondence to: ozcimen@yildiz.edu.tr

YıldızTechnical University, Faculty of Chemical and Metallurgical Engineering, Bioengineering Department, Istanbul, Turkey

References

[1] Hill J, Nelson E, Tilman D, Polasky S, Tiffany D. Environmental, economic, and energetic costs and benefits of biodiesel and ethanol biofuels. Proceedings of the National Academy of Sciences 2006; 103(30): 11206-11210.

[2] Renewable fuels association,http://ethanolrfa.org/pages/World-Fuel-Ethanol-Production, 20 June 2014.

[3] Sarkar N, Ghosh S. K, Bannerjee S, Aikat K. Bioethanol production from agricultural wastes: An overview. Renewable Energy 2012; 37(1): 19-27.

[4] John R. P, Anisha G. S, Nampoothiri K. M, Pandey A. Micro and macroalgal biomass: a renewable source for bioethanol. Bioresource Technology, 2011; 102(1): 186-193.

[5] Nigam P. S, Singh A. Production of liquid biofuels from renewable resources. Progress in Energy and Combustion Science, 2011;37(1): 52-68.

[6] Daroch M, Geng S, Wang G. Recent advances in liquid biofuel production from algal feedstocks. Applied Energy, 2013;102: 1371-1381.

[7] Chaudhary L, Pradhan P, Soni N, Singh P, Tiwari A. Algae as a Feedstock for Bioethanol Production: New Entrance in Biofuel World, Int.J. ChemTech Res.2014;6(2),pp 1381-1389.

[8] Wen Z, Chen, F.Heterotrophic production of eicosapentaenoic acid by microalgae. Biotechnology Advances, 2003;21: 273–294.

[9] Vieira Costa, J.A, de Morais, M.G. An Open Pond System for Microalgal Cultivation" Biofuels From Algae, Chapter 1, In Pandey, A., Lee., D, Chisti, Y. Rsoccol, C., (ed.) USA. 2014;pp.1-22.

[10] Yen H, Hu I, Chen C, Chang J. Design of Photobioreactors for Algal Cultivation", Biofuels From Algae, Chapter 2, In Pandey, A., Lee., D, Chisti, Y. Ve Rsoccol, C., USA. (Ed.) 2014; pp.23-46

[11] Bitoga J.P, Lee I.-B, Lee C.-G, Kim K.-S, Hwang H-S, Hong S.-W, Seo I.-H, Kwon K.-S, Mostafa E. Application of computational fluid dynamics for modeling and designing photobioreactors for microalgae production, Computers and Electronics in Agriculture, 2011;76: 131–147.

[12] Ugwu C.U, Aoyagi H, Uchiyama H. Photobioreactors for mass cultivation of algae, Bioresource Technology, 2008;99: 4021–4028.

[13] Mata T. M, Martins A. A, Caetano N. S. Microalgae for biodiesel production and other applications: a review, Renewable and Sustainable Energy Reviews, 2010;14(1): 217-232.

[14] Vergara-Fernandez A, Vargas G, Alarcon N, Velasco A. Evaluation of marine algae as a source of biogas in a two-stage anaerobic reactor system. Biomass and Bioenergy 2008;32:338–44.

[15] Yen H.W, Brune, D.E. Anaerobic co-digestion of algal sludge and waste paper to produce methane. Bioresource Technology 2007;98, 130–134.

[16] Nakas JPS, Parkinson CM, Coonley CE, Tanenbaum SW. System development for linked-fermentation production of solvents from algal biomass. Applied and Environmental Microbiology 1983;46:1017–23.

[17] Belarbi EH, Molina E, Y. A process for high yield and scaleable recovery of high purity eicosapentaenoic acid esters from microalgae and fish oil. Process Biochemistry 2000;35:951–69.

[18] Cheng-Wu Z, Zmora O, Kopel R, Richmond A. An industrial-size flat plate glass reactor for mass production of Nannochloropsis sp. (Eustigmatophyceae). Aquaculture 2001;195:35–49.

[19] Besada V, Andrade JM, Schultze F, Gonzalez JJ. Heavy metals in edible seaweeds commercialised for human consumption. Journal of Marine Systems 2009;75:305–13.

[20] Spolaore P, Joannis-Cassan C, Duran E, Isambert A. Commercial applications of microalgae. Journal of Bioscience and Bioengineering 2006;101: 87-96.

[21] Olaizola M. Commercial development of microalgal biotechnology: from the test tube to the marketplace. Biomolecular Engineering 2003;20: 459-466.

[22] Demirbas, M. F., Biofuels from algae for sustainable development", Applied Energy, 2011;88(10): 3473-3480.

[23] Burton T, Lyons H, Lerat, Stanley M, Rasmussen M. B. A review of the potential of marine algae as a source of biofuel in Ireland, Dublin: Sustainable Energy Ireland-SEI.2009.

[24] Jung K. A, Lim S. R, Kim Y, Park J. M. Potentials of macroalgae as feedstocks for biorefinery, Bioresource technology, 2013;135: 182-190.

[25] Micro-and Macro-algae: Utility for Industrial Applications: Outputs from the EPOBIO Project, September 2007. CPL Press, 2007.

[26] Özçimen D, Gülyurt M.Ö, İnan B. Algal Biorefinery for Biodiesel Production", Biodiesel-Feedstocks, Production and Applications, Prof. Zhen Fang (Ed.), ISBN: 978-953-51-0910-5, InTech; 2012.p

[27] Kraan S. Algal polysaccharides, novel applications and Outlook, Carbohydrates–comprehensive studies on glycobiology and glycotechnology. 2012;ISBN, 978-953.

[28] Harun R, Danquah M. K, Forde G. M. Microalgal biomass as a fermentation feedstock for bioethanol production, Journal of Chemical Technology and Biotechnology, 2010;85(2): 199-203.

[29] Bra′nyikova′, I., Mars ˇa′lkova′, B., et al.,. Microalgae—novel highly efficient starch producers. Biotechnol. Bioeng.2011;108 (4), 766–776.

[30] Yamada T, Sakaguchi K. Comparative studies on Chlorella cell walls – induction of protoplast formation, Arch. Microbiol. 1982;132 10–13.

[31] Metting F.B. Biodiversity and application of microalgae, J. Ind. Microbiol. Biotechnol. 1996;17 477–489..

[32] Hall J, Payne G. Factors controlling the growth of field population of Hydrodictyon reticulatum in New Zealand. Journal of Applied Phycology, 1997; 9: 229-236.

[33] Brennan L, Owende P. Biofuels from microalgae – a review of technologies for production, processing, and extractions of biofuels and co-products, Renew. Sust. Energ. Rev. 2010;14,557–577.

[34] Hirano A, Ueda R, Hirayama S, Ogushi Y. CO2 fixation and ethanol production with microalgal photosynthesis and intracellular anaerobic fermentation, Energy 1997;22 137–142

[35] Ike N. Toda, N Tsuji, K. Hirata, K Miyamoto. Hydrogen photoproduction from CO2-fixing microalgal biomass: application of halotolerant photosynthetic bacteria, J. Ferment. Bioeng. 1997;84, 606–609

[36] Ho S.H, Chen C.Y, Chang J.S. Effect of light intensity and nitrogen starvation on CO2 fixation and lipid/carbohydrate production of an indigenous microalga Scenedesmus obliquus CNW-N, Bioresour. Technol. 2012;244–252.

[37] Brown M.R, McCausland M.A, Kowalski K. The nutritional value of four Australian microalgal strains fed to Pacific oyster Crassostrea gigas spat, Aquaculture 1998; 165 281–293.

[38] İnan B. Utilization of algae for bioethanol production. MSc. Thesis. Yıldız Technical University;2014.

[39] Becker E.W. Microalgae: Biotechnology and Microbiology. Cambridge University Press. 1994.

[40] Jensen A. Present and future needs for algae and algal products. Hydrobiologia 1993; 260–261, 15–23.

[41] Ross A.B, Jones J.M, Kubacki M.L, Bridgeman T. Classification of macroalgae as fuel and its thermochemical behaviour. Bioresour. Technol. 2008; 99, 6494–6504.

[42] Wargacki A.J, Leonard E., et al., An engineered microbial platform for direct biofuel production from brown macroalgae. Science 2012;335 (6066), 308–313.

[43] Carlsson S.A, Beilen J.B, Möller R, Clayton D. Micro and Macro Algae: Utility for Industrial" 2007.

[44] Korkmaz E, Selçuk G, Çayırgan F, Ayrım A. Biyoteknolojide Mikroalgler, Bilim ve Teknik, August, 2003.

[45] Radmer R.J. Algal diversity and commercial algal products. Bioscience 1996;46 (4), 263.

[46] Morrissey J, Kraan S, et al., A guide to commercially important seaweeds on the Irish coast. Bord Iascaigh Mhara/Irish Sea Fisheries Board, Dun Laoghaire, Co, Dublin, Ireland.2001

[47] Karimi K, Shafiei M, Kumar R. Progress in Physical and Chemical Pretreatment of Lignocellulosic Biomass", Biofuel Technologies Recent Developments, Springer-Verlag Berlin Heidelberg, 2013;pp.53-96.

[48] Hideno A, Inoue H, Tsukahara K, Fujimoto S, Minowa T, Inoue S. et al.,. Wet disk milling pretreatment without sulfuric acid for enzymatic hydrolysis of rice straw. Bioresour. Technol. 2009;100, 2706–2711.

[49] Fan LT, Gharpuray MM, Lee YH. Cellulose hydrolysis biotechnology monographs. Berlin: Springer; 1987. p. 57.

[50] Tabil L, Adapa P. Kashaninejad M. Biomass feedstock pre-processing–part 1: pretreatment, Biofuel's engineering process technology 2011.

[51] Ooshima H, Aso K, Harano Y, Yamamoto T. Microwave treatment of cellulosic materials for their enzymatic hydrolysis. Biotechnology Letters 1984;6(5):289e94..

[52] Hu Z, Wen Z, Enhancing enzymatic digestibility of switchgrass by microwave-assisted alkali pretreatment.Biochem. Eng. J. 2008;38, 369–378.

[53] Keshwani D.R, Cheng J.J, Microwave-based alkali pretreatment of switchgrass and coastal bermudagrass for bioethanol production. Biotechnol. Prog. 2010;26, 644–652.

[54] Balat M, Balat H. Öz C, Progress in bioethanol processing, Progress in energy and combustion science, 2008;34(5): 551-573.

[55] Tomás-Pejó E, Alvira P, Ballesteros M, Negro M.J. Pretreatment Technologies for Lignocellulose-to-Bioethanol Conversion", Biofuels, In:Ashok Pandey, Christian Larroche, Steven C. Ricke, Claude-Gilles Dussap ve Edgard Gnansounou (Ed.)2011;pp. 149-176.

[56] Oliva J.M, Sa ′ez F, Ballesteros I, Gonza ′lez A, Negro M.J, Manzanares P, et al., Effect of lignocellulosic degradation compounds from steam explosion pretreatment on

ethanol fermentation by thermotolerant yeast Kluyveromyces marxianus. Appl. Biochem. Biotechnol. 2003;105, 141–154.

[57] Ballesteros M, Oliva J.M, Negro M.J, Manzanares P, Ballesteros I. Ethanol fromlignocellulosicmaterials by a simultaneous saccharification and fermentation process (SFS) with Kluyveromyces marxianus CECT 10875. Process Biochem. 2004;39, 1843–1848.

[58] Cara C, Ruiz E, Ballesteros I, Negro M.J, Castro E. Enhanced enzymatic hydrolysis of olive tree wood by steam explosion and alkaline peroxide delignification. Process Biochem. 2006;41, 423–429.

[59] Yang M, Li W, Liu B, Li Q, Xing J. High-concentration sugars production from corn stover based on combined pretreatments and fed-batch process. Bioresour. Technol. 2010;101, 4884–4888.

[60] Ballesteros I, Negro M, Oliva J.M, Caban ˜as, A, Manzanares P, Ballesteros M. Ethanol production from steam-explosion pretreated wheat straw. Appl. Biochem. Biotechnol. 2006;130, 496–508.

[61] Martin C, Galbe M, Wahlbom C.F, Hahn-Ha °gerdal, B, Jonsson L.J. Ethanol production from enzymatic hydrolysates of sugarcane bagasse using recombinant xylose-utilising Saccharomyces cerevisiae. Enzyme Microb. Technol. 2002;31, 274–282.

[62] Viola E, Cardinale M, Santarcangelo R, Villone A, Zimbardi F. Ethanol from eel grass via steam explosion and enzymatic hydrolysis. Biomass Bioenerg. 2008;32, 613–618.

[63] Mabee W. E, Gregg D. J, Arato C, Berlin A, Bura R, Gilkes N.,... Saddler, J. N., Updates on softwood-to-ethanol process development", Applied Biochemistry and Biotechnology, 2006;129(1-3): 55-70.

[64] Potumarthi R, Baadhe R.R, Bhattacharya S. Fermentable Sugars from Lignocellulosic Biomass: Technical Challenges. Biofuel Technologies Recent Developments, Springer-Verlag Berlin Heidelberg, 20013;pp.3-28.

[65] Taherzadeh M. J, Karimi, K. Pretreatment of lignocellulosic wastes to improve ethanol and biogas production: a review, International journal of molecular sciences, 2008;9(9): 1621-1651.

[66] Chundawat S. P, Venkatesh B, Dale B. E. Effect of particle size based separation of milled corn stover on AFEX pretreatment and enzymatic digestibility", Biotechnology and bioengineering, 2007;96(2): 219-231.

[67] Prasad S, Singh A, Joshi H. C.bEthanol as an alternative fuel from agricultural, industrial and urban residues, Resources, Conservation and Recycling, 2007;50(1): 1-39.

[68] Martin C, Klinke H. B, Thomsen A. B. Wet oxidation as a pretreatment method for enhancing the enzymatic convertibility of sugarcane bagasse, Enzyme and Microbial Technology, 2007;40(3): 426-432.

[69] Pedersen M, Meyer AS. Influence of substrate particle size and wet oxidation on physical surface structures and enzymatic hydrolysis of wheat straw. Biotechnology Progress 2009;25(2):399e408.

[70] Lee YY, Iyer P, Torget RW. Dilute-acid hydrolysis of lignocellulosic biomass. Adv Biochem Eng/Biotechnol 1999;65:93–115.

[71] Miranda J. R, Passarinho P. C, Gouveia, L. Pre-treatment optimization of Scenedesmus obliquus microalga for bioethanol production, Bioresource technology, 2012;104: 342-348.

[72] Larsson S, Palmqvist E, Hahn-Hägerdal B, Tengborg C, Stenberg K, Zacchi G, Nilvebrant N. O. The generation of fermentation inhibitors during dilute acid hydrolysis of softwood, Enzyme and Microbial Technology,1999;24(3): 151-159.

[73] Carvalheiro F, Duarte L. C, Gírio F. M. "Hemicellulose biorefineries: a review on biomass pretreatments 2008

[74] Wan C, Zhou Y, Li Y. Liquid hot water and alkaline pretreatment of soybean straw for improving cellulose digestibility", Bioresource technology,2011;102(10): 6254-6259.

[75] Wang Z, Keshwani D. R, Redding A. P, Cheng J. J. Sodium hydroxide pretreatment and enzymatic hydrolysis of coastal Bermuda grass, Bioresource technology, 2010;101(10): 3583-3585.

[76] Wang Z, Li R, Xu J, Marita J. M, Hatfield R. D, Qu R, Cheng J. J. Sodium hydroxide pretreatment of genetically modified switchgrass for improved enzymatic release of sugars" Bioresource technology, 2012;110: 364-370.

[77] Vincent M, Pometto III A.L, van Leeuwen J. Ethanol production via simultaneous saccharification and fermentation of sodium hydroxide treated corn stover using Phanerochaete chrysosporium and Gloeophyllum trabeum, Bioresource Technology, 2014;158:1–6.

[78] Mesa L, González E, Cara C, González M, Castro E, Mussatto S. I. The effect of organosolv pretreatment variables on enzymatic hydrolysis of sugarcane bagasse, Chemical Engineering Journal, 2011;168(3): 1157-1162.

[79] Keller F. A, Hamilton J. E, Nguyen Q. A. Microbial pretreatment of biomass, Applied biochemistry and biotechnology, 2003;105(1-3): 27-41.

[80] Allan M. B, Arnon D. I. Studies on nitrogen-fixing blue-green algae. I. Growth and nitrogen fixation by Anabaena cylindrica Lemm. Plant Physiol. 1955; 30: 366–372.

[81] Mansfield S.D, Mooney C, Saddler J.N, Substrate and enzyme characteristics that limit cellulose hydrolysis. Biotechnol. Prog. 1999;15, 804–816.

[82] Chang V.S, Holtzapple M. Fundamentals factors affecting biomass reactivity.Appl. Biochem. Biotechnol. 2000;84–86, 5–37..

[83] Chandra R.P, Bura R, Mabee W.E, Berlin A, Pan X, Saddler J.N. Substrate pretreatment: the key to effective enzymatic hydrolysis of lignocellulosics?Adv. Biochem. Eng. Biotechnol. 2007;108, 67–93.

[84] Almodares A, Hadi M. R. Production of bioethanol from sweet sorghum: A review, African Journal of Agricultural Research, 2009;4(9): 772-780.

[85] Balat M, Balat H. Recent trends in global production and utilization of bio-ethanol fuel", Applied energy, 2009;86(11): 2273-2282.

[86] Demirbaş A. Ethanol from cellulosic biomass resources, International journal of green energy, 2004;1(1): 79-87.

[87] Doran-Peterson J, Cook D. M, Brandon S. K. Microbial conversion of sugars from plant biomass to lactic acid or ethanol, The Plant Journal, 2008;54(4): 582-592.

[88] Harun R, Liu B, Danquah M. K. Analysis of process configurations for bioethanol production from microalgal biomass, Progress in Biomass and Bioenergy Production, Intech Science, Technology & Medicine, Croatia. ISBN, 2011;978-953.

[89] Harun R, Jason WSY, Cherrington T, Danquah MK. Exploring alkaline pre-treatment of microalgal biomass for bioethanol production", Appl Energy, 2011;88:3464–7.

[90] Nguyen MT, Choi SP, Lee J, Lee JH, Sim SJ. Hydrothermal acid pretreatment of Chlamydomonas reinhardtii biomass for ethanol production, J Microbiol Biotechnol, 2009;19:161–6.

[91] Choi SP, Nguyen MT, Sim SJ, Enzymatic pretreatment of Chlamydomonas reinhardtii biomass for ethanol production, Bioresour Technol, 2010;101:5330–6.

[92] Lee S, Oh Y, Kim D, Kwon D, Lee C, Lee J. Converting carbohydrates extracted from-marine algae into ethanol using various ethanolic Escherichia coli strains, Appl Biochem Biotechnol,2011;164:878–88.

[93] Kim J, Um B-H, Kim T. Bioethanol production from micro-algae, Schizocytrium sp., using hydrothermal treatment and biological conversion, Korean J Chem Eng,2012; 29:209–14.

[94] Harun R, Danquah M. K. Influence of acid pre-treatment on microalgal biomass for bioethanol production, Process Biochemistry, 2011;46(1): 304-309.

[95] Scholz M. J, Riley M. R, Cuello J. L. Acid hydrolysis and fermentation of microalgal starches to ethanol by the yeastSaccharomyces cerevisiae, Biomass and Bioenergy, 2013;48: 59-65.

[96] Khambhaty Y, Mody K, Gandhi M. R, Thampy S, Maiti P, Brahmbhatt H.,... Ghosh, P. K. Kappaphycus alvarezii as a source of bioethanol", Bioresource technology, 2012;103(1): 180-185.

[97] Meinita M. D. N, Kang J. Y, Jeong G. T, Koo H. M, Park S. M, Hong, Y. K. Bioethanol production from the acid hydrolysate of the carrageenophyte Kappaphycus alvarezii (cottonii), Journal of Applied Phycology, 2012;24(4): 857-862.

[98] Wang X, Liu X, Wang G. Two-stage Hydrolysis of Invasive Algal Feedstock for Ethanol Fermentation, Journal of integrative plant biology, 2011;53(3): 246-252.

[99] Yanagisawa M, Nakamura K, Ariga O, Nakasaki K. Production of high concentrations of bioethanol from seaweeds that contain easily hydrolyzable polysaccharides, Process Biochemistry, 2011;46(11): 2111-2116.

[100] Lee H. Y, Jung K. H, Yeon J. H. Repeated-batch operation of surface-aerated fermentor for bioethanol production from the hydrolysate of seaweed Sargassum sagamianum, Journal of microbiology and biotechnology,2011;21(3): 323-331

[101] Kim N. J, Li H, Jung K, Chang H. N, Lee P. C. Ethanol production from marine algal hydrolysates using Escherichia coli KO11, Bioresource technology, 2011;102(16): 7466-7469.

[102] Horn S. J, Aasen I. M, Østgaard K. Ethanol production from seaweed extract, Journal of Industrial Microbiology and Biotechnology, 2000;25(5): 249-254.

[103] Adams J. M, Gallagher J. A, Donnison I. S. Fermentation study on Saccharina latissima for bioethanol production considering variable pre-treatments, Journal of applied Phycology, 2009;21(5): 569-574.

[104] Adams J. M. M, Toop T. A, Donnison I. S, Gallagher J. A. Seasonal variation in Laminaria digitata and its impact on biochemical conversion routes to biofuels, Bioresource technology, 2011;102(21): 9976-9984.

[105] Ge L, Wang P, Mou H. Study on saccharification techniques of seaweed wastes for the transformation of ethanol, Renewable energy, 2011;36(1): 84-89.

[106] Lee S. M, Lee, J. H. Ethanol fermentation for main sugar components of brown-algae using various yeasts", Journal of Industrial and Engineering Chemistry, 2012;18(1): 16-18.

[107] Kumar S, Gupta R, Kumar G, Sahoo D, Kuhad R. C. Bioethanol production from Gracilaria verrucosa, a red alga, in a biorefinery approach", Bioresource technology, 2013;135: 150-156.

[108] Yoon M. H, Lee Y. W, Lee C. H, Seo Y. B. Simultaneous production of bio-ethanol and bleached pulp from red algae, Bioresource technology, 2012;126: 198-201.

[109] Borines M. G, de Leon R. L, Cuello J. L. Bioethanol production from the macroalgae Sargassum spp, Bioresource technology, 2013;138: 22-29.

[110] Limayem A, Ricke S. C.Lignocellulosic biomass for bioethanol production: current perspectives, potential issues and future prospects, Progress in Energy and Combustion Science, 2012;38(4): 449-467,

[111] Kumar S, Singh N, Prasad R. Anhydrous ethanol: A renewable source of energy, Renewable and Sustainable Energy Reviews, 2010;14(7): 1830-1844.

[112] Kwiatkowski JR, McAloon AJ, Taylor F, Johnston DB. Modeling the process and costs of fuel ethanol production by the corn dry-grind process. Ind Crops Products 2006;23:288–96.

Chapter 8

Sub and Supercritical Fluid Technologies for the Production of Renewable (Bio) Transportation Fuels

Harvind Kumar Reddy, Tapaswy Muppaneni and Shuguang Deng

Additional information is available at the end of the chapter

http://dx.doi.org/10.5772/59818

1. Introduction

1.1. Biofuels and feed stocks

The limitation on the availability of petroleum based fuels and environmental concerns have made humans look at alternative fuels such as biofuels, wind energy, solar energy, hydro-electric power, hydrogen, nuclear energy and so on. Among all these alternative energy sources, biofuels are one of the potential alternative energy sources that can fulfill a part of the current energy demand. The concept of biofuels dates back to the 19th century when researchers tried to use vegetable oil as a fuel for diesel engines. But the problems associated with the usage of oils directly as fuel and the availability of fossil fuels made the concept of vegetable oil fuels uneconomical. Due to rapid depletion of petroleum based fuels, ever increasing consumption of fuels in developing nations and environmental concerns led scientists and governments to develop fuels based on oils or lipids. These oils or lipids cannot be used directly as liquid fuels but can be processed and/or upgraded to be used as a blend or as a direct substitute for the regular fuel. Biodiesel, green diesel, bio-jet fuel or green jet fuel and green gasoline are examples of renewable transportation fuels. Vegetable and plant oils, animal fats, and waste cooking oil from food processing industries have been used as a primary feedstock for the production of renewable transportation fuels. Along with these sources, micro algae have also attained significant attention as a source of oil/lipids, due to high productivity of oil compared to plants in a given period of time and area [1]. Apart from the higher lipids, algal biomass is being used as a source for the production of proteins, amino acids and carbohydrates [2, 3]. Cultivation of algal biomass in waste water treatment plants which contribute to the cleaning of water and production of fuels. In spite of having more oil yield than plant crops, algae feedstock has its own problems of processing the biomass such as harvesting, drying and extraction of oil to produce biofuels. Various processing technologies are available to produce these bio fuels from different feed stock materials. This chapter focuses on the production of biodiesel from oils/fats and wet algal biomass through supercritical alcohol transesterification, novel methods for extraction of oil/lipids from wet algal biomass, liquefaction of whole algal biomass through hydrothermal extraction and liquefaction (HT E&L) and catalytic hydrothermolysis to produce regular hydrocarbon fuels from oils using supercritical water. Before discussing the biofuels production, the sub and super critical technologies will be discussed.

1.2. Sub and supercritical conditions

As shown in figure 1, the four phases of a pure material or compound can be observed at different temperature and pressure conditions. When a compound is heated above its boiling point and below its critical point under pressure, it is called a subcritical fluid and when a compound is heated above its critical point is called as supercritical fluid. The sub critical and supercritical fluids possess different physical-chemical properties compared to their properties at normal conditions. They are compressibility (like gases) due to reduced densities, increased polarity due to reduced dielectric constants and they have catalytic properties attained by variations in ion dissociation constants. Above the critical point the particular material obtains gas like densities, liquid like solvating properties and intermediate mass transfer kinetics [4]. By varying temperature and pressure, these enhanced capabilities of the sub and supercritical fluids are being used for environmentally benign selective separations, catalytic reactions for production or purification of various products. The commonly known supercritical fluids are water, CO_2, ethanol, methanol, ethane, methane etc. In this chapter the utilization of water, methanol, and ethanol to produce various kinds of biofuels or fuel intermediates will be discussed.

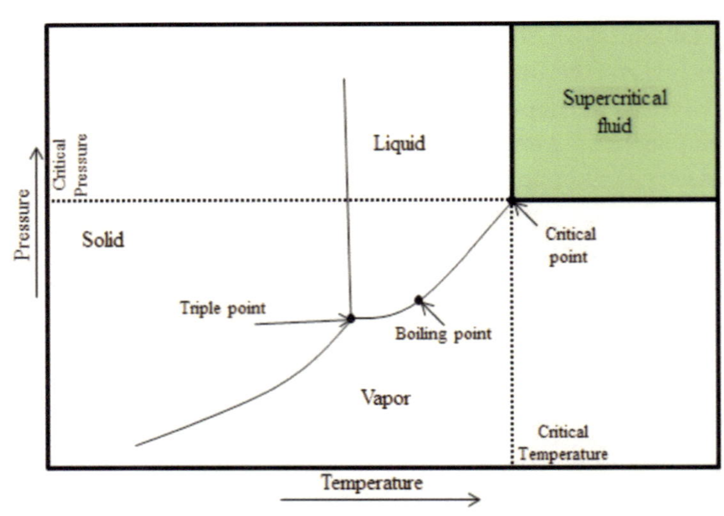

Figure 1. Phase diagram of a pure compound

http://dx.doi.org/10.5772/59818

2. Supercritical alcohol process for the production of biodiesel from oils/fats/lipids

Biodiesel is one of the first generation biofuel developed that is being used in present day transportation. Biodiesel is not a new source of alternative fuel, it has a long history. When Rudolph diesel invented the diesel engine, he also suggested that pure vegetable oils can be used as a fuel in those engines. Three decades later basic research has started to use modified vegetable oil as a fuel in the diesel engine. This modified vegetable oil can be called as biodiesel. It took almost a century after the invention of diesel engine to start extensive research on biodiesel and its use as a fuel. Biodiesel is a fuel derived from biomass such as vegetable oil, animal fat, algae or other renewable resources which consist of long chain alkyl esters. Biodiesel is a nontoxic, renewable, biodegradable, and eco-friendly fuel. Biodiesel produces lower emissions compared to that of regular petroleum based fuels. Biodiesel usage in the place of regular diesel fuel can reduce emissions such as SOx, CO, particulate matter and hydrocarbons in the exhaust gas and it is better than regular diesel fuel in terms of sulfur content, flash point, aromatic content, and cetane number. Biodiesel does not contribute to a net rise in the level of carbon dioxide in the atmosphere and has the capability of minimizing the intensity of the greenhouse effect. Biodiesel is more promising fuel because of its renewability, energy security and the high energy content consistent with that of petroleum based fuels. Biodiesel can be used as a fuel blend or as a substitute and will have similar properties to that of regular diesel. Several countries around the world have made it mandatory to sell regular diesel fuel with a blend biodiesel and gasoline with a blend of ethanol in order of environmental concerns to address. The blend concentration varies and can be denoted by different notations such as B100 (pure biodiesel), B50 (50% biodiesel, 50% petroleum diesel), B20 (20% biodiesel, 80% petroleum diesel), B10, B5 etc.

There are different processes to make biodiesel from renewable feedstock. These include but are not limited to pyrolysis, micro emulsions, dilution, catalytic cracking, and transesterification. Pyrolysis is a thermo chemical process that decomposes organic material in the absence of oxygen. In this process, the biomass will be converted into bio oil which is similar to crude oil. This oil will be further converted to small chain hydrocarbons via hydro treating and hydrocracking and then used as transportation fuels. Micro emulsions are isotropic mixtures of oil, water and a surfactant; which can be blended with petroleum diesel fuels, solvents such as alcohols and can be used as transportation fuels. Other methods like dilution and catalytic cracking and transesterification can also be used for the production of biodiesel.

Among all these processes, transesterification is one of the most economic and to produce biodiesel simplest way. Transesterification or alcoholysis is a process in which the triglycerides present in the oils chemically react with alcohol to produce alkyl esters with or without the aid of a catalyst. Alcohols like methanol, ethanol, propanol, butanol and amyl alcohol can be used for transesterification. When the transesterification occurs in the presence of methanol, it is known as methanolysis and the esters formed are known as fatty acid methyl esters. Ethyl esters, propyl esters and butyl esters will be produced when their respective alcohols are used in the transesterification process. On an industrial scale, methanol a 'refinery residue' is the primary alcohol used for the production of biodiesel. Ethanol is an agricultural product which is renewable, non-toxic, eco-friendly and can also be used for biodiesel synthesis. Figure 2 shows a simple mechanism of transesterification reaction with ethanol as alcohol.

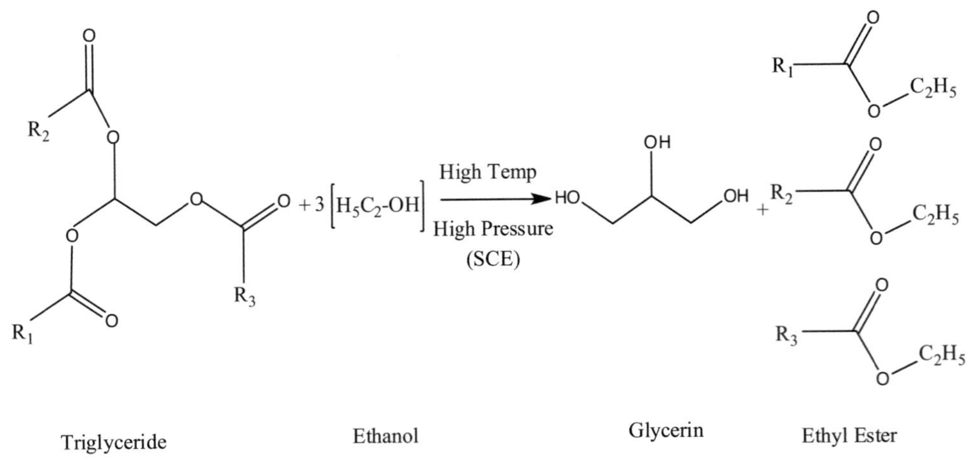

Figure 2. Transesterification reaction: R1, R2 and R3 are long chain hydrocarbons which may be same or different

Biodiesel can be produced by various transesterification methods using alkali, acid or enzyme catalysts or by advanced methods such as microwave irradiation and ultrasonic transesterification. The alkali process gives a high purity, high yield biodiesel in a short span of reaction time but is not suitable for oils with high free fatty acid (FFA) content, for these oils, acid esterification followed by alkali transesterification can be employed to reduce the high FFA content and to improve the biodiesel yield. However, the longer reaction time and low catalyst recovery are problems in this process. Enzyme catalytic transesterification requires longer reaction times. All the methods mentioned above have their limitation and challenges such as longer reaction time, lower reaction rate, and weak catalytic activity.

To overcome these limitations, non-catalytic transesterification can be implemented to produce biodiesel under supercritical alcohol conditions [5]. Under supercritical conditions, the intermolecular hydrogen bonding in the alcohol molecule will be significantly decreased. As a result, the polarity and dielectric constant of alcohol are reduced allowing it to act as a free monomer. Alcohol at supercritical conditions can solvate the triglycerides to form a single phase of oil/alcohol mixture and yield fatty acid alkyl ester and diglycerides. Diglyceride is further transesterified to form ethyl ester and monoglyceride, which in the last step is then converted into alkyl ester and glycerol. Vegetable oils which include edible oils such as palm oil, sunflower oil, rice bran oil, rapeseed oil and non-edible oils such as jatropha oil, paradise oil, and pongamia oil can be used in biodiesel production. Waste cooking oil, algae and animal fats such as lard, tallow, yellow grease are also potential feedstock in the production of biodiesel. Table 1 shows the fatty acid profiles of some common biodiesel feed stocks collected from literature from Balat et al.,[6] and others. The fatty acid profiles of the same (particular) feed stocks may vary due to its cultivation conditions and extraction methods.

http://dx.doi.org/10.5772/59818

Feed stock	C16:0	C16:1	C18:0	C18:1	C18:2	C18:3	others
Sunflower oil	6.4	0.1	2.9	17.7	72.9	0	0
Palm oil	42.6	0.3	4.4	40.5	10.1	0.2	1.1
Soybean oil	13.9	0.3	2.1	23.2	56.2	4.3	0
Rapeseed oil	3.5	0	0.9	64.1	223	8.2	0
Tallow [7]	26.2	1.9	33.67	30.1	0.7	0	0
Lard [8]	24.7	2.9	13.1	42.48	13.64	1.16	2
Waste cooking oil [9]	6.8	0.4	3.7	22.8	65.2	0.1	0

Table 1. Fatty acid profiles of various biodiesel feed stocks

The catalytic transesterification processes requires a lower amount of alcohol (1:9 oil to alcohol ratio), and mild temperatures (60°C) for the production of biodiesel. Alkali catalysts like potassium hydroxide, sodium hydroxide, sodium methoxide, potassium methoxide and acid catalysts like hydrochloric acid, phosphoric acid, and sulfuric acid can be used as catalysts in catalytic transesterification but catalyst separation, free fatty acid and water interference in the reaction, glycerol separation, and energy intensive are disadvantages. The use of different feed stocks, different alcohols greatly vary the processing conditions; the complete conversion may not be achieved with such changes in the process, reaction times could reach hours or days and separation of product becomes much more challenging.

In a non-catalytic supercritical alcohol process, the transesterification of triglycerides and the alkyl esterification of fatty acids will occur simultaneously with a shorter reaction time and reduced the energy consumption due to the simplified separation and purification steps. This process does not require any pre-treatment of the feed stock regardless of its fatty acid composition and profile. In non-catalytic supercritical transesterification the oil to alcohol ration varies between 1:40-45 depending upon the feed stocks fatty acid profile, 290-350°C temperature, and reaction pressure above saturation pressure [5, 10]. Introduction of co-solvent into the reaction mixture decreases the critical point of alcohol, increases the mutual solubility of the oil and alcohol at lower reaction temperatures and accelerates the reaction rate under supercritical alcohol conditions [11]. Normally methanol and ethanol are being used as alcohol to produce biodiesel. But longer chain alcohols like 1-butanol, 1-propanol and 1-octanol could also be used to produce biodiesel. The critical conditions of these alcohols are presented in Table 2.

Alcohol	Critical Temperature (°C)	Critical Pressure (bar)
Methanol	240	78.6
Ethanol	243	63.1
1-Propanol	264	50.3
1-Butanol	290	43.5
1-Ocatanol	386	26.5

Table 2. Transesterification alcohols and their critical conditions

The major influencing factors on the yields of biodiesel are type of alcohol, reaction temperature, oil to alcohol ratio, reaction time and pressure. The critical temperatures increase with increase in chain length or molecular weight of the alcohol. At the same temperature, the acidity of longer chain alcohols tends to decrease resulting in slower reactivity or slightly more reaction time than the short chain alcohols. However the cold flow properties of the biodiesel produced with long chain alcohols are better than the biodiesel produced with short chain alcohols [12]. All these factors influence the selection of alcohol, as it affects both the economics and energetics of the process. The yield of biodiesel increases with the increase in reaction temperature above the critical conditions of the alcohols. Beyond the optimum temperature, the yield may start decreasing due to degradation of fatty acids at higher temperatures. Usually this also depends on the fatty acid profiles; as poly unsaturated fatty acids (PUFAs) are thermally unstable at higher temperatures. Feed stocks having more PUFAs may give higher yields at slightly lower temperatures than the feed stocks having less PUFAs [13]. The usual optimum reaction temperature ranges between 290-350°C, which also depending on the other reaction parameters.

Feed stock	FAME	FAEE	FABE
Camelina oil [15] Reaction temperature: 310°C Oil to alcohol molar ratio: 1:40	Yield: 91% Reaction time:24 min.	Yield: 85% Reaction time:28 min.	Yield: 84% Reaction time:45 min.
Rapeseed oil [12] temperature: 300°C Oil to alcohol molar ratio: 1:42	Yield: 99% Reaction time:15 min.	Yield: 94% Reaction time:45 min.	Yield: 86% Reaction time:45 min.
Rapeseed oil [12] temperature: 350°C Oil to alcohol molar ratio: 1:42	Yield: 94% Reaction time:4 min.	Yield: 94% Reaction time:10 min.	Yield: 89% Reaction time:10 min.

Table 3. Variation in the yields of biodiesel with different alcohols

As mentioned earlier the long chain alcohols need higher reaction temperatures to get higher yields of biodiesel than the short chain alcohols. The molar ratios of oil to alcohol vary for different feed stocks with different alcohols. This usually ranges between 1:40-45 at optimum reaction temperature. A lower amount of alcohol negatively affects the yields as the reverse transesterification reaction tries to go backwards. On the other hand more alcohol also reduces the yields by changing the critical point of the mixture to higher temperatures, where the optimum temperature of the reaction is not sufficient to perform the forward reaction. This also imposes another economic barrier as this extra alcohol requires more energy to heat, and will need to be recycled after separation process [14]. Supercritical alcohol processing is very

http://dx.doi.org/10.5772/59818

fast compared to conventional transesterification. The typical processing times of supercritical processes are 5-30 minutes depending upon the type of alcohol and reaction temperature. The reaction pressure also slightly increases the yield of biodiesel above its saturation pressure at a particular reaction temperature. But it is always a better option to use the lowest possible pressures; as high pressure demands more energy and capital investment. Variation in the yields of biodiesel and reaction times with respect to alcohol is shown in Table 3.

3. Supercritical water process for the production of green fuels from oils/fats/lipids

The catalytic hydrothermolysis (CH) is another process which produces regular hydrocarbon transportation fuels from oils or fats. The supercritical water performs the hydrolysis of vegetable oils to produce biocrude oil. This biocrude oil consists of a wide range of compounds such as straight chain, branched chain and cyclic hydrocarbons (alkanes, alkenes and aromatics etc.,) and their distribution varies depending upon the processing conditions and feed stock. An increase in temperature towards critical point the causes oils/fats become miscible with water around 300-330°C [16].

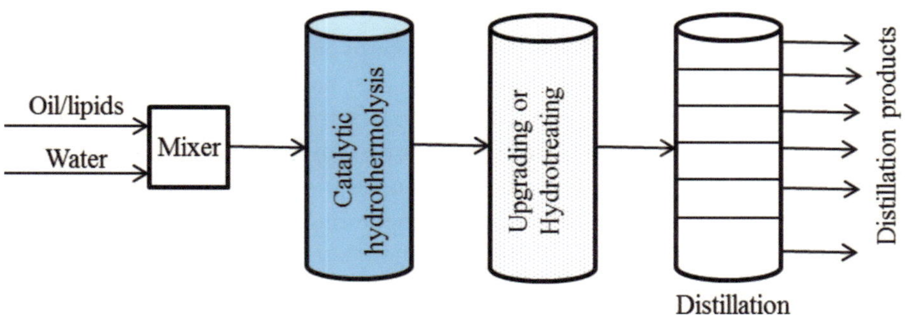

Figure 3. Catalytic hydrothermolysis process

During this process, triglycerides undergo hydrolysis reaction to form free fatty acids and glycerol. Compared to thermal cracking and pyrolysis, the formation of gaseous products will be reduced to the minimum in CH process. At higher temperatures, decarboxylation and cyclization reactions of fatty acids and glycerol produce alkanes, alkenes, carbon dioxide and water. The aqueous phase contains small amounts of smaller hydrocarbons and glycerol. Use of external catalyst such as KOH and NaOH enhances the production of alkanes. Use of metal oxide catalysts favors the production of alkenes. [17]. The biocrude oil produced during this process must be hydrotreated or reformed to meet commercial fuel standards. The upgrading process is discussed later in the chapter. Catalytic hydrothermolysis is performed to produce jet fuel and diesel range hydrocarbons. Oil and water are processed at 9:1 (vol. ratio) and between 450-475°C. The resulting biocrude is then upgraded with commercial nickel catalyst to produce jet fuel and diesel range hydrocarbons [18]. The schematic of the CH process is shown in Figure 3. The hydrogen utilization during the CH process can be reduced compared to direct catalytic cracking of oils. The fatty acid profile of feed stock, water to oil ratio, rate of heating and reaction pressure determines the final product properties.

4. Supercritical alcohol process for the production of biodiesel from wet algal biomass

The utilization of vegetable oils to produce biodiesel has resulted in stress on domestic markets and often disrupted the production capacities to lower levels of the operating plants [19]. Other feed stocks like waste cooking oil and animal fats are not sufficient to meet commercial demands and to make biofuels at profitable scales. Researchers identified microalgae as an alternative crop to produce oils in larger volumes, with smaller areas of land and in shorter periods of time. The active research on algae has started in the 1970s due to the oil crisis. Micro-algae are single cell plants which grow in most of the marine environments around the globe. They can be cultivated under autographic and heterotrophic conditions depending upon the species. Due to their faster photosynthesis they grow much faster and consume more CO_2 when compared to oil producing energy crops. Algae can be harvested in cycles of 6-14 days depending on the strategies and cultivating conditions.

Microalgae are being cultivated in open race way ponds and in closed photo bioreactors. Open raceway ponds are much cheaper to operate, but very hard to control the conditions within the pond. Open ponds are more vulnerable the atmospheric conditions and other invading species which greatly effect both quantity and quality of algal biomass. On the other hand, photo bioreactors provide a very controlled environment which helps to produce biomass with better quality and quantity than open ponds. But operating costs of photo bioreactors are very high, prohibiting it for the use in biofuels production. Many research institutions and private corporations have developed the best suitable systems for their needs. After cultivation biomass can be harvested with techniques such as centrifugation, flocculation and hydro cyclones etc. The biomass content or water content in the biomass varies for different systems. The extraction of oil is the most energy intensive step in algal biofuels production. The drying step that occurs prior to the solvent extraction of oil consumes nearly 90MJ of energy which is nearly 85% of the total energy consumed to produce 1 kg of biodiesel. Techniques like Supercritical CO_2 extraction, pyrolysis and gasification also need dry biomass. Due to this these methods are also not suitable for the production of biofuels with algal biomass. Biodiesel could be produced directly from the algal biomass by using supercritical alcohol transesterification process.

The direct conversion of wet algal biomass into biodiesel is demonstrated with *Nannochloropsis sp.* with supercritical methanol and ethanol as conversion media without using catalyst. The FAMEs and FAEEs can be produced with free fatty acids and triglycerides present in the algal biomass. As the temperature rises from normal room temperature to supercritical conditions, the reduced dielectric constant provides enhanced extraction capabilities to the alcohols to open/break the cell walls and to extract lipids. The cells structure disintegrates into small particles at higher temperatures due to enormous pressure, providing complete conversion of lipids into biodiesel. After reaching the critical points of the alcohols, the transesterification reaction takes place. At higher pressures the cell structure will be destroyed and provides more access to the lipids for transesterification reaction. The cell structures of the biomass before and after conversion are presented in Figure 3. The algal cells containing large globules of lipids (indicated by arrows) along with other cellular organelles (Figure 3a) are completely destroyed and disintegrated (figure 3b) into a network of pieces [20].

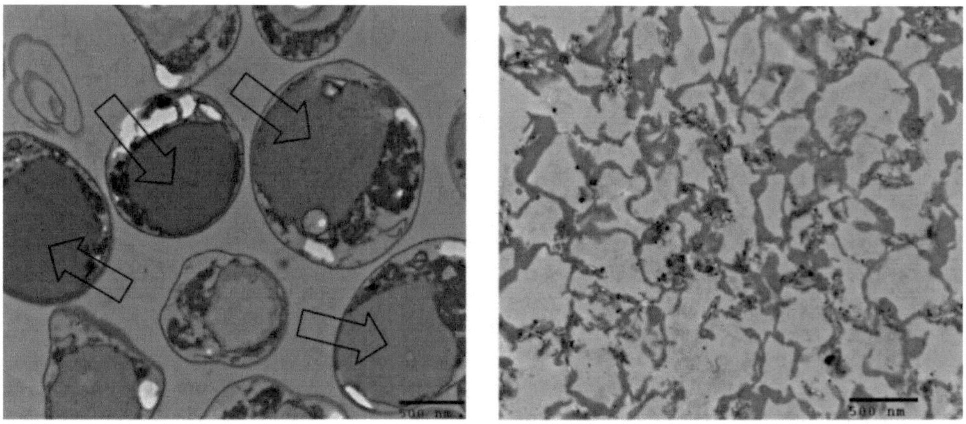

(a) (b)

Figure 4. TEM images of algal cells before (a) and after (b) in supercritical ethanol conversion

The major influencing parameters of the direct conversion are reaction temperature, algae to alcohol ratio (wt.:vol.) and reaction time. The optimum reaction conditions for both methanol and ethanol are presented in Table 4. The algal biomass used in the experiments has 50% (supercritical methanol) and 52% (supercritical ethanol) of total lipids on ash free dry weight basis. As the temperature increases the extraction takes place below critical point, and transesterification starts from critical point. The maximum yields observed are 84% and 67% at 255°C and 265°C respectively. The short chain containing methanol has produced more biodiesel than ethanol similar to the vegetable oil transesterification. Same amount of alcohol may have been another reason for the lower yields with ethanol. When compared to the supercritical alcohol transesterification of vegetable oils and fats, the reaction temperatures of the direct conversion methods of algae are very low. This is due to the difference in fatty acid profile of algal biomass. The algal biomass used in these studies has more unsaturated (~40-45%) and polyunsaturated fatty acids (PUFAs) (~10%); which are thermally unstable and causes the reduction of biodiesel yields at higher temperatures. The decomposition of PUFAs was observed at higher temperatures above the optimum reaction temperature in both studies [21, 22].

Algal biomass: ***Nannochloropsis sp.***	**Yield (on the basis of total lipids)**	**Reaction temperature (°C)**	**Algae to alcohol ratio (wt./vol.)**	**Reaction time (min.)**
FAME (methanol)[22]	84%	255	1:9	25
FAEE (ethanol)[21]	67%	265	1:9	20

Table 4. Reaction conditions for maximum yields of biodiesel with algal biomass

When compared to biodiesel production using vegetable oils by supercritical alcohol process, nearly 2-3 times more alcohol is needed for algal biomass conversion. More energy is required for the separation of the extra alcohol, making the process more energy intensive. The production of biodiesel directly from the wet algal biomass is possible; but supercritical processing of expensive feedstock like algae demands complex infrastructure and higher energy, making production of biofuels less profitable. During this process valuable byproducts like polyunsaturated fatty acid ethyl esters are lost in order to maintain the fuel properties.

5. Subcritical water processing of wet algal biomass to produce green fuels

Recently, many researchers have identified, and studied the importance of byproducts in the production of algal biofuels. Algae are being used to produce various kinds of natural products like protein, polyunsaturated fatty acids (PUFAs), vitamins, carbohydrates, and dietary fibers. Extraction and separation of these compounds is a key factor for the commercialization of algal biofuels [23, 24]. The recovery of these valuable products cannot be achieved using direct conversion methods due to severe processing conditions and degradation of these compounds. New novel methods called hydrothermal extraction and liquefaction (HT E&L) or subcritical water extraction or liquefaction can be used for the recovery of valuable products along with energy dense biocrude oil, and bio-char from wet algal biomass. Water attains these selective extraction and liquefaction properties due to increased temperature and pressure. The dielectric constant of water reduces slowly with increasing temperatures, which provides more solvating power to the water. This solvating power varies with the process parameters which include temperature, pressure, solids loading and extraction time. This technique has been used for many selective extraction applications such as bioactive components from *Haematococcus pluvialis* microalga[25], nutraceutical compounds from citrus pomaces[26], and oils from coriander seeds[27]. Typical extraction temperature will vary between 120-250°C, and changes along with pressure variation for a particular target compound. Another major advantage of this method is higher extraction efficiencies can be achieved when compared to conventional solvent extraction.

Similar to the above given examples, lipids from wet algal biomass can be extracted. In this process water present in the harvested algal biomass itself acts as the solvent; which also eliminates the drying step in the process and achieves maximum extraction of oil. This selective extraction of lipids is demonstrated with the wet algal biomass of *nannochloropsis salina.* The reaction temperature, solids (biomass) loading and reaction time are studied to extract the maximal amount of lipids from the biomass. The RSM (response surface methodology) analysis of extraction results indicated that the increase in temperature causes increase in the yield of crude extract and lipids with in the crude extract up to 217°C. Due to increased solvating power or reduced dielectric constant; further rise in temperature caused a reduction in crude extract due to secondary reactions which caused redistribution of crude extract into other product fractions like the water phase or gas phase. The biomass loading is another crucial parameter which affects the extraction efficiency. The amount of crude extract is increased with more diluted biomass and a maximum is found to be at 5%; meaning less biomass loading provides more solvent for the extraction of target compounds. But at the same time the energy consumption should also be considered; more water in the biomass requires more energy to reach the target processing conditions. Due to this, optimum biomass loading was increased to 7.5%, which slightly affects the extraction efficiency. The reaction time was found to be very when compared large to solvent extraction techniques, and maximum extraction was achieved within 25 min. Through this subcritical water extraction method nearly 60-70% of the lipids present in the algal biomass can be extracted. Along with the lipids, the water fraction contains sugars, sugar alcohols, and PUFAs in algae oil which have commercial value in the pharmaceutical and food industry; Additionally the bio-char contains proteins (~45% by wt.) and 24.9 MJ/kg of energy making it another valuable byproduct [28]. The biodiesel produced with subcritical water extraction with thermal energy recovery (60%) and utilization of bio-char to produce biogas (methane) consumes only 28.23 MJ of energy for 1kg biodiesel [29]. The process can be used only for algal biomass which has more lipids and for the production of biodiesel. In order to produce biofuels from algal biomass having fewer lipids, higher hydrothermal processing conditions should be used.

Further increase in temperature of the hydrothermal extraction process, results in hydrothermal liquefaction due to the enhanced reaction capabilities of water. The density of water decreases as the temperature and pressure rise towards the critical point and drops drastically after the critical point; here water medium attains gas like densities and liquid like solvent properties. The ability of water in these conditions to provide H^+ or OH- ions (varying ion product (K_w) of water) is useful in performing acid or base catalyzed reactions without using external catalysts [30-32]. During the hydrothermal liquefaction, the macromolecules present in the biomass are subjected to hydrolysis, which degrades them into smaller molecules. During this process, the oxygen present in the biomass will be removed by dehydration in the form of water, and by decarboxylation in the form of carbon dioxide. During the liquefaction process energy dense biocrude oil, aqueous phase with water soluble compounds, bio-char and gaseous product will be produced [31, 33]. The subsequent intermediate reactions of the hydrolyzed or extracted compounds determine the yields of product fractions, which are influenced by processing conditions.

Numerous studies on hydrothermal liquefaction of algae are available in literature. Few examples of these studies are hydrothermal liquefaction of algal biomass *nannochloropsis sp.*, [34], *chaetomorpha linum*[35], *chlorella pyrenoidosa*[36], or *spirulina plantesis*[37]. The typical processing temperature used for hydrothermal liquefaction ranges between 250-370°C. The yields and properties of the products vary with reaction temperature. The yield of energy dense biocrude oil increases with rise in temperature up to 350°C. Lower yields can be observed at lower temperatures due to a decrease in hydrolysis of the biochemical compounds and repolymerization. Above 350°C, hydrothermal gasification becomes more dominant and contributes to the production of more gaseous products rather than biocrude oil. The typical yields of biocrude oil yields range between 20-60% on the ash free dry weight basis (AFDW). As the temperature rises the water becomes more reactive than an extraction solvent due to its increased ion dissociation constant. The ion dissociation constant at 250°C is 1000 times greater than at the room temperature [38]. Because more protons and hydroxide ions present in these conditions hydrolysis becomes more active reaction. The hydrolysis reaction causes the degradation of basic chemical compounds, and when accompanied by reactions like repolymerization produces bio-crude oil. The yields differ with process conditions, and biochemical composition of the algal biomass. The contribution of lipids, proteins and carbohydrates to the yield of biocrude oil yield are in the order of lipids>proteins>carbohydrates [39]. The biocrude oil yields of commonly used algal biomass are presented in Table 5. The high heating value (H.H.V) of biocrude oil varies in the range of 32-39 MJ/kg. The reaction temperature greatly affects the quality of biocrude oil. At lower processing temperature, lipids contribute more to the yield of biocrude oil which contains more energy or higher H.H.V. At higher temperatures, due to the contribution of protein derived compounds and increased nitrogen in biocrude oil the H.H.Vs of biocrude oil reduce to lower levels. The variation reaction pressure beyond saturation has little to no effect on the product distribution in HTL of algae [36]. Even though most of the studies have used residence times between 20-120 minutes in batch (reactor) mode operation, 2-5 min. of reaction time is sufficient to get better biocrude oil yields in a continuous flow system [40].

Algal Biomass	**Biochemical composition**			**Reaction conditions**	**Biocrude yield (% of AFDW)**
	Lipids	Protein	Carbohydrates	Temperature, solids loading and Reaction time	
Chlorella sp.[40]	4	60	25	350°C, 10% and 3 min.	42
Chlorella sp. [41]	60	9	13	300°C, 20% and 90 min.	66
Nannochloropsis sp. [41]	14	52.4	5	300°C, 15% and 60 min.	48.4
Nannochloropsis salina [42]	12	37	33	310°C, 25% and 30 min.	46
Spirulina platensis [42]	6	60	19	350°C, 25% and 30 min.	38
Nannochloropsis sp. [34]	14	59	20	300°C, and 10 min.	50
Nannochloropsis sp. [43]	28	52	12	350°C, 21% and 60 min.	43
Spirulina platensis [44]	11	49	31	350°C, 20% and 60 min.	39
Scenedesmus sp.[45]	13	56	25	300°C, 20% and 30 min.	45
Spirulina sp.[45]	5	64	20	300°C, 20% and 30 min.	31
Dunaliella tertiolecta [46]	20	63	15	340°C, and 5 min.	41
Dunaliella tertiolecta [47]	22	24	46	300°C, 10% and 60 min.	30
Chlorella sorokiniana [47]	4	30	54	300°C, 10% and 60 min.	18

Table 5. Yields of biocrude oil produced with various strains of algal biomass

The other product fractions are bio-char, aqueous phase and gaseous fraction. The bio-char yield decreases with an increase in temperatures, as the metabolites of the biomass converts into other products. The range of bio-char yields can be found between 10-70% depending on the processing temperature and time. Bio-char produced at lower temperatures tend to have higher H.H.V due to hydrothermal carbonization. The H.H.V of bio-char can be around 18-22 MJ/kg at lower temperatures and reduces to 8-10 MJ/kg at higher processing temperatures above 300°C. Due to extraction or conversion of biochemical compounds present in the biomass at higher temperatures, H.H.Vs of the bio-char are reduced compared to the H.H.V of the original biomass and bio-char produced at lower temperatures [20].

The aqueous phase of the HTL is another valuable product fraction which contains essential nutrients (NH_3-N and PO_4^{3-}), amino acids and carbohydrates depending on the processing temperature. The hydrolysis of proteins results in the formation of amino acids, and the deamination (further hydrolysis) of these amino acids produces the ammoniacal nitrogen. Usually the amount of ammoniacal nitrogen increases with an increase in temperature, because the amino acids decompose rapidly at higher temperatures. At the same time, the concentration of amino acids decreases due to rapid conversion with an increase in HTL temperatures. The other valuable nutrient phosphate behaves in a different way; the amount of phosphate in water phase decreases with an increase in temperature and deposits in the bio-char fraction. The concentrations of these nutrients can be as low as 400 ppm for NH_3-N, 6 ppm for PO_4^{3-} and as high as 6300 ppm for NH_3-N, 3000 ppm for PO_4^{3-}. However, these concentrations vary greatly with the biochemical composition of the biomass and the processing temperature [20, 48, 49]. Recycling these nutrients back to cultivation is a very crucial step as it can save of fresh on addition nutrient supply and reduce overall cost. Along with the nutrients, the aqueous phase also contains valuable carbohydrates including polysaccharides, monosaccharides, sugar alcohols, amino acids and glycerol. At milder temperatures most of the carbohydrates present in the algal biomass are extracted into the aqueous phase. With an increase in HTL temperature, these polysaccharides hydrolyze to yield simple sugars and derivative compounds. At higher processing temperatures these compounds start degrading or converting into other product fractions due to secondary and tertiary reactions. A Typical optimum point for the extraction of polysaccharides is around 160°C [50]. Similar to the extraction of lipids, polysaccharides from algal biomass can be extracted at lower temperatures, and the remaining biomass can be used to produce biocrude oil. The recoverable quantity of polysaccharides and biocrude oil depends on the original biochemical compositions of the biomass.

As mentioned earlier, separation and recovery of these valuable amino acids, nutrients, and carbohydrates is vital for the production of algal biofuels. This concept of algal bio refinery with sequential HTL (SEHTL) or hydrothermal extraction and liquefaction (HT E&L) is shown in Figure 4. The processing strategies for each strain of biomass vary as they possess different biochemical composition. The separation of these compounds provides a much needed economic life line and also improves the quality of biocrude oil. The removal of amino acids and nitrogen based compounds at lower temperatures reduces the quantity of nitrogen in the biocrude oil produced in the second step. When compared to cellulosic ethanol production,

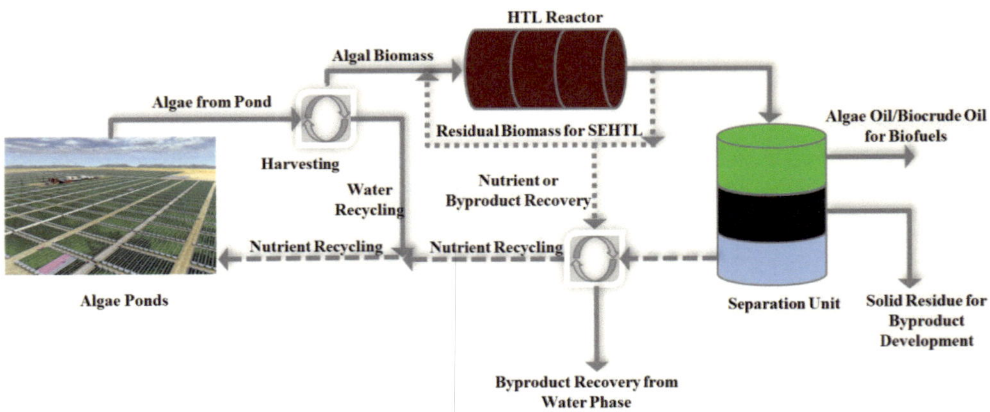

Figure 5. Algal biorefinery with HTL process

HTL of algal biomass has better energy return on investment (EROI) and lower emissions can be achieved [41].

The biocrude oil produced in the HTL process contains nitrogen (4-9 wt.%) and oxygen (2-7wt. %). The biocrude oil produced with oils/fats in the CH process is easy to upgrade compared to biocrude oils produced with algae due to the absence of nitrogenous compounds. Due to the presence of nitrogen and oxygen, processing the biocrude oil becomes slightly complex. Two strategies can be followed to produce hydrocarbon fuels from biocrude oil. The first one is, co-processing the biocrude in existing petroleum refineries by diluting the crude petroleum up to the permissible levels of nitrogen and oxygen. The second option is direct processing of biocrude oil with suitable catalysts. These catalysts include metal oxides of Ni, Co, Mo, Pt and W supported on γ-Al_2O_3, SiO_2, zeolites and carbon. Commercialization of these catalysts for processing biocrude oil may take some more time. More research and development is needed to optimize both the HTL of algal biomass and suitable catalysts for biocrude oils with varying properties. Compared to conventional jet fuel, the biojet fuel produced from algal biomass with HTL and upgrading can reduce life cycle greenhouse gas emissions by 76% [51].

6. Conclusions

The sub and supercritical processing methods seems promising for the production of renewable transportation fuels from various feed stocks. Supercritical alcohol processing provides much better path to produce biodiesel from cheap feed stocks such as animal fats and waste cooking oils. These feed stocks need very intensive pretreatments to produce biodiesel through catalytic transesterification. They can be directly processed using the supercritical alcohol method. At the same time the price of feed stock also plays a vital role to determining the sustainability of this process. Catalytic hydrothermolysis is being used by many corporate companies to demonstrate the possibility of producing high grade jet fuels and naval distillates. This technology is ready to deploy during certain critical political or environmental scenarios. Hydrothermal extraction and liquefaction is acquiring more interest in the research domain due to its ability to produce a wide variety of products. HT E & L processes are identified as the best possible techniques for the production of biofuels from algal biofuels. The current research is more focused on cultivation and recovery of byproducts from algal biomass though sequential HTL. The algae biofuel research is still in the research and development stage and may take a couple of years to be commercialized.

Author details

Harvind Kumar Reddy, Tapaswy Muppaneni and Shuguang Deng*

*Address all correspondence to: sdeng@nmsu.edu

Chemical and Materials Engineering Department, New Mexico State University, Las Cruces, NM, USA

References

[1] Chisti Y. Biodiesel from microalgae. Biotechnology Advances. 2007;25:294-306.

[2] Singh J, Cu S. Commercialization potential of microalgae for biofuels production. Renewable & Sustainable Energy Reviews. 2010;14:2596-610.

[3] Brennan L, Owende P. Biofuels from microalgae-A review of technologies for production, processing, and extractions of biofuels and co-products. Renewable & Sustainable Energy Reviews. 2010;14:557-77.

[4] Eckert CA. Supercritical fluids as solvents for chemical and materials processing. Nature. 1996;383:313-8.

[5] Saka S, Kusdiana D. Biodiesel fuel from rapeseed oil as prepared in supercritical methanol. Fuel. 2001;80:225-31.

[6] Balat M. Biodiesel fuel production from vegetable oils via supercritical ethonol transesterification. Energy Sources Part a-Recovery Utilization and Environmental Effects. 2008;30:429-40.

[7] Chakraborty R, Gupta AK, Chowdhury R. Conversion of slaughterhouse and poultry farm animal fats and wastes to biodiesel: Parametric sensitivity and fuel quality assessment. Renewable and Sustainable Energy Reviews. 2014;29:120-34.

[8] Hong IK, Lee JR, Lee SB. Fuel properties of canola oil and lard biodiesel blends: Higher heating value, oxidative stability, and kinematic viscosity. Journal of Industrial and Engineering Chemistry. 2014.

[9] Demirbas A. Biodiesel from waste cooking oil via base-catalytic and supercritical methanol transesterification. Energy Conversion and Management. 2009;50:923-7.

[10] Muppaneni T, Reddy HK, Patil PD, Dailey P, Aday C, Deng S. Ethanolysis of camelina oil under supercritical condition with hexane as a co-solvent. Applied Energy. 2012;94:84-8.

[11] Muppaneni T, Reddy HK, Ponnusamy S, Patil PD, Sun YQ, Dailey P, et al. Optimization of biodiesel production from palm oil under supercritical ethanol conditions using hexane as co-solvent: A response surface methodology approach. Fuel. 2013;107:633-40.

[12] Warabi Y, Kusdiana D, Saka S. Biodiesel fuel from vegetable oil by various supercritical alcohols. Proceedings of the Twenty-Fifth Symposium on Biotechnology for Fuels and Chemicals Held May 4–7, 2003, in Breckenridge, CO: Springer; 2004. p. 793-801.

[13] Imahara H, Minami E, Hari S, Saka S. Thermal stability of biodiesel in supercritical methanol. Fuel. 2008;87:1-6.

[14] Tan KT, Gui MM, Lee KT, Mohamed AR. Supercritical Alcohol Technology in Biodiesel Production: A Comparative Study between Methanol and Ethanol. Energy Sources Part a-Recovery Utilization and Environmental Effects. 2011;33:156-63.

[15] Sun Y, Reddy HK, Muppaneni T, Ponnusamy S, Patil PD, Li C, et al. A comparative study of direct transesterification of camelina oil under supercritical methanol, ethanol and 1-butanol conditions. Fuel. 2014.

[16] King J, Holliday R, List G. Hydrolysis of soybean oil. in a subcritical water flow reactor. Green Chemistry. 1999;1:261-4.

[17] Watanabe M, Iida T, Inomata H. Decomposition of a long chain saturated fatty acid with some additives in hot compressed water. Energy Conversion and Management. 2006;47.

[18] Li L, Coppola E, Rine J, Miller JL, Walker D. Catalytic Hydrothermal Conversion of Triglycerides to Non-ester Biofuels. Energ Fuel. 2010;24:1305-15.

[19] Schenk PM, Thomas-Hall SR, Stephens E, Marx UC, Mussgnug JH, Posten C, et al. Second Generation Biofuels: High-Efficiency Microalgae for Biodiesel Production. Bioenergy Research. 2008;1:20-43.

[20] Reddy HK. Production of Transportation Fuels from Wet Algae. Chemical Engineering. Las Cruces: New Mexico State University; 2014. p. 182.

[21] Reddy HK, Muppaneni T, Patil PD, Ponnusamy S, Cooke P, Schaub T, et al. Direct conversion of wet algae to crude biodiesel under supercritical ethanol conditions. Fuel. 2014;115:720-6.

[22] Patil PD, Gude VG, Mannarswamy A, Deng S, Cooke P, Munson-McGee S, et al. Optimization of direct conversion of wet algae to biodiesel under supercritical methanol conditions. Bioresource Technology. 2011;102:118-22.

[23] Spolaore P, Joannis-Cassan C, Duran E, Isambert A. Commercial applications of microalgae. J Biosci Bioeng. 2006;101:87-96.

[24] Wijffels RH, Barbosa MJ, Eppink MHM. Microalgae for the production of bulk chemicals and biofuels. Biofuels Bioproducts & Biorefining-Biofpr. 2010;4:287-95.

[25] Rodriguez-Meizoso I, Jaime L, Santoyo S, Senorans FJ, Cifuentes A, Ibanez E. Subcritical water extraction and characterization of bioactive compounds from Haematococcus pluvialis microalga. Journal of Pharmaceutical and Biomedical Analysis. 2010;51:456-63.

[26] Kim J-W, Nagaoka T, Ishida Y, Hasegawa T, Kitagawa K, Lee S-C. Subcritical Water Extraction of Nutraceutical Compounds from Citrus Pomaces. Separation Science and Technology. 2009;44:2598-608.

[27] Eikani MH, Golmohammad F, Rowshanzamir S. Subcritical water extraction of essential oils from coriander seeds (Coriandrum sativum L.). Journal of Food Engineering. 2007;80:735-40.

[28] Reddy HK, Muppaneni T, Sun Y, Li Y, Ponnusamy S, Patil PD, et al. Subcritical water extraction of lipids from wet algae for biodiesel production. Fuel. 2014;133:73-81.

[29] Ponnusamy S, Reddy HK, Muppaneni T, Downes CM, Deng S. Life cycle assessment of biodiesel production from algal bio-crude oils extracted under subcritical water conditions. Bioresource Technology. 2014;170:454-61.

[30] Savage PE. A perspective on catalysis in sub- and supercritical water. Journal of Supercritical Fluids. 2009;47:407-14.

[31] Peterson AA, Vogel F, Lachance RP, Froeling M, Antal MJ, Jr., Tester JW. Thermochemical biofuel production in hydrothermal media: A review of sub- and supercritical water technologies. Energy & Environmental Science. 2008;1:32-65.

[32] Akiya N, Savage PE. Roles of water for chemical reactions in high-temperature water. Chemical Reviews. 2002;102:2725-50.

[33] Bobleter O. HYDROTHERMAL DEGRADATION OF POLYMERS DERIVED FROM PLANTS. Progress in Polymer Science. 1994;19.

[34] Valdez PJ, Nelson MC, Wang HY, Lin XN, Savage PE. Hydrothermal liquefaction of Nannochloropsis sp.: Systematic study of process variables and analysis of the product fractions. Biomass and Bioenergy. 2012;46:317-31.

[35] Aresta M, Dibenedetto A, Carone M, Colonna T, Fragale C. Production of biodiesel from macroalgae by supercritical CO2 extraction and thermochemical liquefaction. Environmental Chemistry Letters. 2005;3:136-9.

[36] Yu G, Zhang Y, Schideman L, Funk TL, Wang Z. Hydrothermal liquefaction of low lipid content microalgae into bio-crude oil. Transactions of the Asabe. 2011;54:239-46.

[37] Toor SS, Reddy H, Deng S, Hoffmann J, Spangsmark D, Madsen LB, et al. Hydrothermal liquefaction of Spirulina and Nannochloropsis salina under subcritical and supercritical water conditions. Bioresource Technology. 2013;131:413-9.

[38] Okajima I, Sako T. Energy conversion of biomass with supercritical and subcritical water using large-scale plants. Journal of Bioscience and Bioengineering. 2014;117:1-9.

[39] Biller P, Ross AB. Potential yields and properties of oil from the hydrothermal liquefaction of microalgae with different biochemical content. Bioresource Technology. 2011;102.

[40] Jazrawi C, Biller P, Ross AB, Montoya A, Maschmeyer T, Haynes BS. Pilot plant testing of continuous hydrothermal liquefaction of microalgae. Algal Research. 2013;2:268-77.

[41] Li H, Liu Z, Zhang Y, Li B, Lu H, Duan N, et al. Conversion efficiency and oil quality of low-lipid high-protein and high-lipid low-protein microalgae via hydrothermal liquefaction. Bioresource Technology. 2014;154:322-9.

[42] Toor SS, Reddy H, Deng S, Hoffmann J, Spangsmark D, Madsen LB, et al. Hydrothermal liquefaction of Spirulina and Nannochloropsis salina under subcritical and supercritical water conditions. Bioresource Technology. 2013;131:413-9.

[43] Brown TM, Duan P, Savage PE. Hydrothermal Liquefaction and Gasification of Nannochloropsis sp. Energ Fuel. 2010;24:3639-46.

[44] Jena U, Das KC, Kastner JR. Comparison of the effects of Na2CO3, Ca3(PO4)2, and NiO catalysts on the thermochemical liquefaction of microalga Spirulina platensis. Applied Energy. 2012;98:368-75.

[45] Vardon DR, Sharma BK, Blazina GV, Rajagopalan K, Strathmann TJ. Thermochemical conversion of raw and defatted algal biomass via hydrothermal liquefaction and slow pyrolysis. Bioresource Technology. 2012;109:178-87.

[46] Minowa T, Yokoyama S-y, Kishimoto M, Okakura T. Oil production from algal cells of Dunaliella tertiolecta by direct thermochemical liquefaction. Fuel. 1995;74:1735-8.

[47] Reddy HK, Muppaneni T, Rastegary J, Shirazi SA, Ghassemi A, Deng S. ASI: Hydrothermal extraction and characterization of bio-crude oils from wet chlorella sorokiniana and dunaliella tertiolecta. Environ Prog Sustainable Energy. 2013;32:910-5.

[48] Alba LG, Torri C, Samori C, van der Spek J, Fabbri D, Kersten SRA, et al. Hydrothermal Treatment (HIT) of Microalgae: Evaluation of the Process As Conversion Method in an Algae Biorefinery Concept. Energ Fuel. 2012;26:642-57.

http://dx.doi.org/10.5772/59818

[49] Biller P, Ross AB, Skill SC, Lea-Langton A, Balasundaram B, Hall C, et al. Nutrient recycling of aqueous phase for microalgae cultivation from the hydrothermal liquefaction process. Algal Research. 2012;1:70-6.

[50] Chakraborty M, Miao C, McDonald A, Chen S. Concomitant extraction of bio-oil and value added polysaccharides from Chlorella sorokiniana using a unique sequential hydrothermal extraction technology. Fuel. 2012;95:63-70.

[51] Fortier MOP, Roberts GW, Stagg-Williams SM, Sturm BSM. Life cycle assessment of bio-jet fuel from hydrothermal liquefaction of microalgae. Applied Energy. 2014;122:73-82.

Section 3

Alternative Fuels

Chapter 9

Overview of Obtaining Alternative Fuels in The Co-liquefaction Processes with Biomass and Coal in Malaysia

Mohd Azlan Mohd Ishak, Siti Nur Ain Mohd Hassan, Siti Nurlia Ali, Mohd Fauzi Abdullah, Asnida Yanti Ani, Nur Nasulhah Kasim, Ali H. Jawad, Wan Izhan Nawawi Wan Ismail and Khudzir Ismail

Additional information is available at the end of the chapter

http://dx.doi.org/10.5772/59362

1. Introduction

The extensive utilization of nonrenewable fossil fuels sources such as petroleum, coal and others has imposed a serious problem of energy depletion beyond the 21st century. Currently, our environment has been severely polluted and the instant effect such as climate change, acid rain, greenhouse effect, etc were the results of these problems. Therefore, it is crucial to balance between exploitation of energy resources and environment protection so that development can be sustainable for the human beings. Thus, a new alternative method should be proposed to overcome this problem, especially in the area of renewable energy. In many developing countries, there is an abundance of biomass and agricultural wastes like sawdust, rice and coconut husks, bagasse and animal dung. Unfortunately, these renewable sources were discharged daily due to its properties (wastes) and too expensive to dispose off. For instance, in Malaysia, a huge amount of biomass wastes from oil palm and rubber industries are being produced daily, estimated at not less than 4.5 million metric tonnes effluent per year.

The simplest way for the farmers to discharge the agricultural waste is by burning them. This practice will contribute to air pollution and increase the greenhouse gases effect. To overcome the waste properties of biomass, they are treated in two senses. First, biomass plant matter is used to generate electricity with steam turbines and gasifiers or produce heat, usually by direct combustion. Second, biomass which includes plant or animal matter can be converted into fibers or other industrial chemicals, including biofuels. Vegetative biomass is generally composed of lignin, cellulose and hemicelluloses and varies in composition depending on plant species. Because of the nature of their chemical structures, biomass materials can easily be converted into biofuels and other useful chemicals in comparison to coals and other fossil energy sources. For instance, in Malaysia, rubber trees plantation cover more than 1.7 million hectares all over the country. These plantations produced agricultural waste that includes rubber seed, rubber seed pod, and rubberwood. Each hectare can produce an approximately amount of 150 – 160 kg of seeds. The search for a low-cost raw material with adequate fuel characteristics for biofuels production is an important step toward establishing a successful biofuels industry. Typically, rubber seed oil, which is nonedible, is considered as a prospective feedstock for alternative fuels production.

In recent years, the liquefaction potential of coals has been investigated to increase the yield of coal conversion processes and the quality of liquid fuels obtained from coal. However, the liquefaction process of coal and biomass materials known as co-liquefaction has not been developed in Malaysia. Typically, rubber seed oil, which is nonedible, obtained from rubber seed is considered as a prospective feedstock for alternative fuels production since it has been found to be rich in oil. Thus, these abundant biomass sources are potential candidates for the production of alternative fuels via co-liquefaction process, and the development of this technology becomes significant in both efficient utilization of resources and improvement of ecological environment. Owing to the fact that energy composition properties are rich in coal, poor in oil and little in gas [1], coal liquefaction to produce alternative petroleum in Malaysia is of great significance to energy security. Malaysian low-rank coals are chosen as the starting material in this study due to the fact that these coals are abundant, low grade and have lower energy content because of their low carbon composition. They are lighter (earthier) and have higher moisture levels. Hence, an attempt should be made for co-liquefaction of Malaysian coals and rubber tree wastes (rubber seed, rubber seed pod or rubberwood), especially for the production of alternative fuels or other chemical feedstocks.

2. World energy supply

Since 2004, there was no increase in world's oil supply because of the unavailability of finding new sources and the declining output from older oil fields. The increasing production difficulties mean that the supply of oil will soon begin to decline and that, month by month, the decline will be at an accelerating pace [2]. World energy supply is largely dependent on conventional petroleum products and most of the expected increases in oil demand come from the motorized transport sector, with the largest growth from developing countries. Consequently, the transport sector will become responsible for about one-third of the world's future greenhouse gas (GHG) emission growth and oil prices may reach dramatically high levels [3].

http://dx.doi.org/10.5772/59362

Coincidentally, a growing concern for fossil fuels exhaustion has shown up in the scientific community in the last decade or so. Because of oil, natural gas and coal are finite natural resources; thus the production of these fossil fuels will reach a peak and eventually start to drop. Then, the process of declining will be accelerated because extraction costs will increase after the fossil fuel resources have been consumed. Thus, the implementation of alternative sources becomes more and more important [4].

The depletion of fossil fuels and tighter environmental regulations has forced the world to adopt alternative renewable fuel sources such as hydro, wind and biomass [5], or others. The government expects that an investment through RM9.7 billion is required for the development of the electric utility sector until 2010 [6]. Analysts predicted that the government's aim for 2, 080 MW or 11% of all electricity generated nationwide in 2020 to be sourced from environment-friendly renewable energy and this will be a challenging and long-term task [7]. Hence, due to current demands in electricity and with recent developments in the energy sector, alternative renewable fuels must be recognized as an important energy source for the foreseeable future.

2.1. Biomass

Biomass is a local resource that can contribute to the diversification of energy supply and potentially create employment for cultivation, harvesting, transport and fuel preparation. The importance of biomass can clearly be seen from the action plan as reported by the European Commission. The commission has targeted to increase the usage of biomass from 289, 000 TJ in 2003 to about 628, 000 TJ in 2010, as reported in their 2005 "Biomass Action Plan". The factors that led to almost double the biomass use was due to some important advantages that have been identified. These include (i) less dependence on short-term weather changes, (ii) low costs, (iii) an alternative source of income for farmers and finally (iv) able to promote regional economic structures [8].

Well-managed biomass yields carbon emission-saving fuels when substituted for fossil fuels. Amongst renewable energy sources, biomass appears to be the most important in terms of technical and economic feasibility. Since the content of both nitrogen and sulfur are low in biomass materials, the formation of NOx and SOx gases during combustion or firing the biomass materials is also low. The carbon dioxide released during combustion is equal to the amount of carbon dioxide intake when the plants grow up and this means produce almost no net carbon dioxide emission after combustion [1, 3, 4]. It is therefore today that biomass is considered a major future energy source for development and industry, arousing growing interest worldwide, not only for use in transport. Thus, the utilization and development of biomass energy could solve not only the energy issues, but could surely reduce the main environment problems such as pollution and greenhouse effects. Food crops containing starch/sugar/oil can be processed to produce biofuels where the conversion technologies and markets are readily available. Moreover, with the increasing petroleum prices, the commercial opportunity must be quickly taken into consideration.

Huge amount of biomass wastes in Malaysia, especially from rubber plantation are being produced daily. The use of biomass in direct liquefaction with the aim to ease the petroleum importation has been paid a major concern, especially among researchers. However, several drawbacks that affected the direct liquefaction of biomass are actually attributed to the nature of the biomass itself. These include source of biomass being seasonal and biomass normally having a relatively higher distributing area in comparison to its total amount. Eventually, these factors will directly affect the feedstock supply of biomass and of course limit its large-scale utilization.

Energy from biomass is a large untapped energy source and its direct liquefaction is possible [9-15]. Energy that can be derived from biomass actually exists in all three phases, i.e. (i) gas fuels (synthesis gas from gasification), (ii) liquid fuels (biodiesel, bio-oil and bio-ethanol) and (iii) solid fuels (for use in boiler combustion). However, in terms of energy output, commercially proven technology and versatility in a wide range of processes (biochemical or thermochemical), energy from biomass is proven to be the most feasible short-term solution to substitute fossil fuels in comparison to other underdeveloped technologies, namely, photovoltaics and fuel cells [16].

2.2. Rubber seed oil

Peninsular Malaysia, compromising 12 of the 14 states in the Malaysian federation is among the world's most important rubber growing area. Rubber tree is also grown in Sabah and Sarawak. Altogether, Malaysia produces almost 20% of the world's natural rubber. More than half of Malaysia's rubber comes from thousands of privately owned small landholdings, which are usually about 2 hectares. The rest is grown on big estates owned by various companies; each can cover over a 1000 hectares. Overall, Malaysia has about 1.7 million hectares of rubber plantation [10]. Rubber tree, or its scientific name is Hevea brasiliensis, belongs to the family Euphorbiaceae and was found to be the most economically important member of the genus Hevea. This is mainly due to the milky latex collected on the daily basis by using special knife tapping from its bark around the tree. In Malaysia, the leaves of the rubber tree fall by end of year, usually from December up to February and refoliate quickly after that, followed by flowering and also producing a large volume of seeds. These seeds, however, were left underutilized [11]. Because of environmental issues, researchers in Malaysia have pursued it quite extensively to evaluate the potential usage and applications of rubber tree wastes (mostly rubber seed, rubber seed pod and rubber wood) through various chemicals and biological determinations. Figure 1 shows the typical scenario of rubber tree plantation in Malaysia and Figure 2 shows the components of rubber tree wastes. Rubber seeds have been found to be rich in oil. Its content in the dried kernel varies from 35 to 45%. It is semi-drying and consists of 17 – 22% saturated fatty acids and 17 – 82% unsaturated fatty acids, and is comparable to drying oils commonly used in surface coating [17]. Rubber seed oil has been found to have potential applications in many areas, which include production of biodiesel as fuel for compression engines [18, 19], foaming agent in latex foam, in the synthesis of alkyd resin used in paints and coatings [17] and several other uses [20]. Thus, the potential of rubber seed to be further utilized in other areas such as co-liquefaction with coal to produce alternative fuels and other useful products provides great opportunity. Table 1 shows the properties of rubber seed oil.

Figure 1. Typical rubber tree plantation scenario in Malaysia.

Figure 2. Components of rubber tree wastes: (a) young seeds, (b) ripe seeds, (c) fruit in an opened pod/pericarp and (d) kernel and shell of rubber fruits.

Fatty acid composition (%)	Rubber seed oil
Palmitic acid C16:0	10.2
Stearic acid C18:0	8.7
Oleic acid C18:1	24.6
Linoleic acid C18:2	39.6
Linolenic acid C18:3	16.3
Specific gravity	0.9
Viscosity at 40 °C (mm^2/s)	66.2
Flash point (°C)	198.0
Calorific value (MJ/kg)	37.5
Acid value (mg KOH/g)	34.0

Table 1. Properties of rubber seed oil [18].

3. Energy demand scenario in Malaysia

The Malaysian government policy on energy utilization and consumption has approved to increase the coal base generation of electricity from 11% to about 33% by the year 1995 and 2020, respectively [21]. Moreover, the coal reserves in Malaysia, which are mainly located in the states of Sarawak (70% reserves), Sabah (28%) and other states (2% from Selangor, Perak and Perlis), have total reserves of about 1, 050 million tonnes of various qualities, ranging from lignite to anthracite [21, 22, 23]. However, the only coal deposit being mined in Malaysia is from Kapit area, Sarawak.

Since early 1999, Tenaga Nasional Bhd. (TNB), the major electricity provider and Sarawak Electric Supply Co. (SESCO) have purchased 120, 000 and 400, 000 tonnes per annum (tpa), respectively, from Kapit coal mine [24]. SESCO operates the 100 MW Kapit minemouth coal-fired power station, where two 50 MW units supply electricity to the Sarawak grid. TNB has projected to double to 20 million tpa its coal import once two planned coal-fired plants (total of 1400 MW) are fully commissioned. Moreover, the construction work has started on the 2100 MW Pulau Bunting power station, which will burn 6 million tpa of coal. In order to secure low coal prices and improve the security of coal supply, TNB's long-term plan is to buy 30 to 50% of its annual coal requirement from its Indonesian coal mining subsidiary, TNB Coal International Ltd., which owns the right to mine in five areas in South Kalimantan. Malaysia imported about 2.9 million tonnes of coals in 1999, i.e. 85% were steam coal and 15% anthracite and bituminous coal. This amount of coal was needed to fulfill the requirement for its cement and utility industries. Indonesia, Australia, China and South Africa were the major overseas suppliers [23].

3.1. Malaysian coals

Based on the statistics reported in The Eighth Malaysia Plan by the Department of Minerals and Geosciences, Sarawak; Mukah Balingian (MB) was identified as the second largest coal area in Malaysia with reserves of *ca.* 710 km^2 [6]. However, most of the known coal areas in the states of Sabah and Sarawak, including Mukah Balingian, are not commercially mined due to poor availability of infrastructure and located far inland [6], and most of the coal types are of low rank, i.e. sub-bituminous. Thus, the usage of abundant of coal reserves was not optimized. Low rank coals (brown coal, lignite and lower sub-bituminous coals) are the most abundant fossil resources, but they have not been utilized in a large amount because of their low calorific values. This is mainly due to the presence of a large amount of oxygen functional groups in such coals, namely, carboxylic, -COOH; hydroxyl, -OH; carbonyl, -CO; etc. The traits of these coals are low price, relatively large porosity and high reactivity, which benefit their utilization for direct coal liquefaction (DCL) [25]. Owing to the fact that low rank coal contains less carbon, liquefaction process is one of the best option thus far to utilize them.

3.2. Characteristics of Mukah Balingian coal

From Table 2, it can be seen that MB coal has relatively high oxygen and volatile matter contents. The petrographic analysis of this coal shows a vitrinite reflectance value of 0.40% and thus can be categorized as a low rank coal, i.e. sub-bituminous C rank [26]. Further, previous work by Ismail et al. [27] using high pressure Temperature Programmed Reduction (TPR) on the pyrite-free MB coal also confirms that this coal is of a low rank by observing the organic sulfur distribution in the coal. Moreover, Rodriguez et al. [28] suggested that low rank coals are normally composed of small aromatic clusters and contain many cross-links and functional groups, and thus, are very reactive and undergo fast and extensive bond breaking during liquefaction.

Ultimate analysis		Proximate analysis			
(wt% daf)		(wt% as received)		(wt% db)	
Carbon	63.9	Volatile matter	33.0	Volatile matter	44.7
Hydrogen	5.1	Fixed carbon	35.6	Fixed carbon	51.1
Nitrogen	1.9	Ash content	4.8	Ash content	4.2
Sulfur	0.5	Moisture	26.6	Fuel ratio[2]	1.1
Oxygen[1]	28.6	Fuel ratio[2]	1.1		
H/C ratio	1.0				
O/C ratio	0.3	Calorific value: 24.6 MJ kg^{-1}			
Petrographic analysis (vol. %)					
Vitrinite	60	Vitrinite reflectance = 0.40%			
Liptinite	31				
Inertinite	8				
Mineral matter	1				

daf = dry ash free basis; db = dry basis; [1]calculated by difference;

[2]Fixed carbon/volatile matter.

Table 2. Characteristics of raw Mukah Balingian coal [29].

Moreover, the petrographic analysis shows that the composition of the coal is dominated by vitrinite that constitutes about 60% by volume of the whole rock, with 31% liptinite (formerly known as exinite), 8% inertinite and 1% mineral matter. Pyrite, however, is present in trace amount. The high content of reactive macerals, i.e., vitrinite and liptinite, that contributes up to 91% of the organic matter content, serves as an important characteristics for carbonization and liquefaction processes of coal [30]. Interestingly, this high value of reactive macerals is in accord with the value suggested by Van Krevelen [31], under the reactivity parameter group to characterize coal for direct liquefaction. It is also known that the vitrinite and liptinite can be extracted readily, which makes the extraction yield correlate broadly with its content [31].

3.3. Thermal behavior of Mukah Balingian coal

Another important characteristic of MB coal is the thermal behavior of decomposition under pyrolysis conditions via thermogravimetric analyser (TGA). The pyrolysis of raw MB coal via TGA at a heating rate of 20 °C min^{-1} is shown in Figure 3. The TG curve (Figure 3(a)) of the raw coal shows the weight loss profile with the weight decreasing as temperature is increased from ambient to 1000 °C. The weight loss profile resembles that of a first-order reaction.

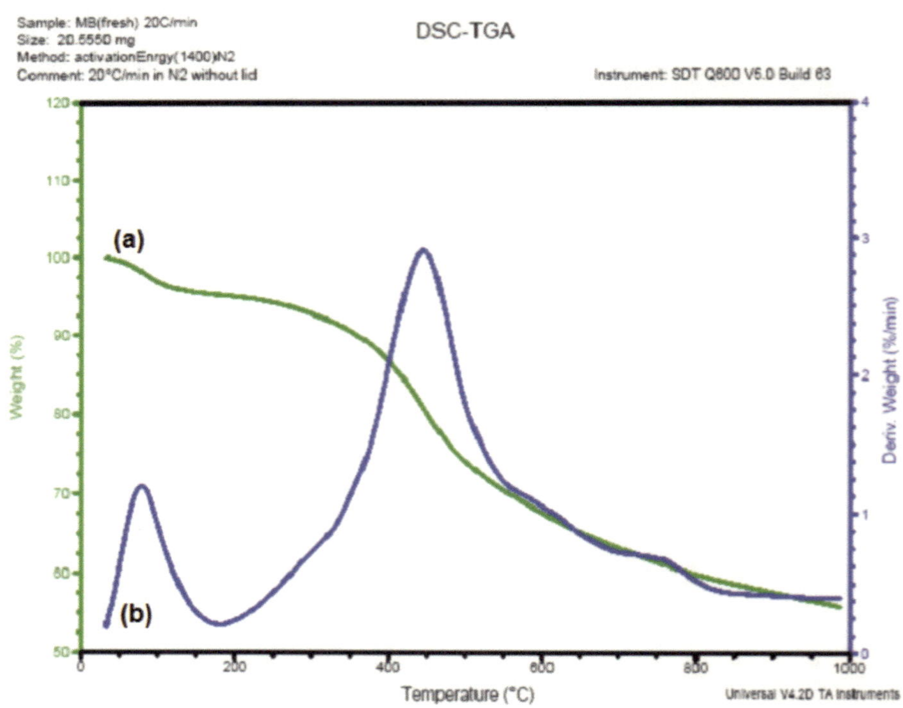

Figure 3. (a) TG and (b) DTG profiles for the thermal decomposition of pyrolysed raw MB coal [29].

Figure 3 (b) shows the differential weight loss (DTG) for the raw MB coal that consists of three main stages. This conforms to previous findings as reported by Probstein and Hicks [32],

Kastanaki et al. [33], Serio et. al. [34] and Radovic et al. [35]. The first-stage pyrolysis, which occurs at temperatures ranging from ambient to 150 °C, involves the dehydration of water and releasing of gas composed of oxides of carbon from the coal. The second-stage pyrolysis, which occurs at temperatures of 200 – 550 °C, however, is due to the release of volatile matter such as hydrocarbon gases, light oils and tars, and is of interest in this study. Finally, the third-stage pyrolysis shows the appearance of some minor curves at temperatures ranging from 550 to 650 °C and from 700 to 800 °C, that were attributed to the release of heavier hydrocarbons and non-condensable gases, mainly hydrogen, and from thermal decomposition of carbonates that are abundant in low rank coals.

Also, from the DTG profile of the raw MB coal, it can be estimated that the softening temperature of this coal is around 350 °C. Merrick [36] suggested that with extraction using liquid solvents, the preferred extraction temperatures lie in the range where the coal starts to decompose thermally, and typically, the extraction is carried out at about 350 – 450 °C. Moreover, Van Krevelen [30] suggested that temperatures around 350 °C were found to be an indicative value for softening temperature in order to characterize coal for direct liquefaction. Thus, it can be suggested that MB coal would be a good feedstock and suitable for liquefaction and/or gasification processes and, hence, optimize the utilization of low rank Malaysian coal.

3.4. Potential of Mukah Balingian coal for liquefaction

In predicting potential conversion of coal, Guyot [37] investigated the relationship between coal properties and liquefaction potential. Guyot introduced a useful correlation called "petrofactor", which derived from the correlations containing terms of factors based on the maceral composition and the maximum reflectance of vitrinite (as shown by Equation 1). A wide rank of coals ranging from brown coal, high-, medium- and low-volatile bituminous coals have been used to obtain potential conversion for all coals, where –2.5 was obtained from the slope of the correlation. The maximum reflectance of vitrinite functions as a rank parameter, whilst the reactive maceral content as the petrological parameter. Usually, the reactive maceral content is considered to be equal to the sum of vitrinite and liptinite. Figure 4 shows the correlation of potential coal conversion with petrofactor as suggested by Guyot [37], and the estimated coal conversion was approximated based on formula in Equation 2.

$$\text{Petrofactor} = 1000 \text{ x} \frac{\text{Maximum reflectance of vitrinite } (\%)}{\text{Total reactive maceral content } (\%)} \quad (1)$$

$$\text{Coal conversion} = 100 - 2.5 \text{ x Petrofactor} \quad (2)$$

The potential conversion of raw MB coal during liquefaction could be estimated using the "Petrofactor" proposed by Guyot [37] (refer Equations 1 and 2). The petrofactor value for the raw MB coal is calculated from Equation 1 and the potential conversion is then estimated using Equation 2. Table 3 shows the results of petrofactor value and estimated coal conversion of

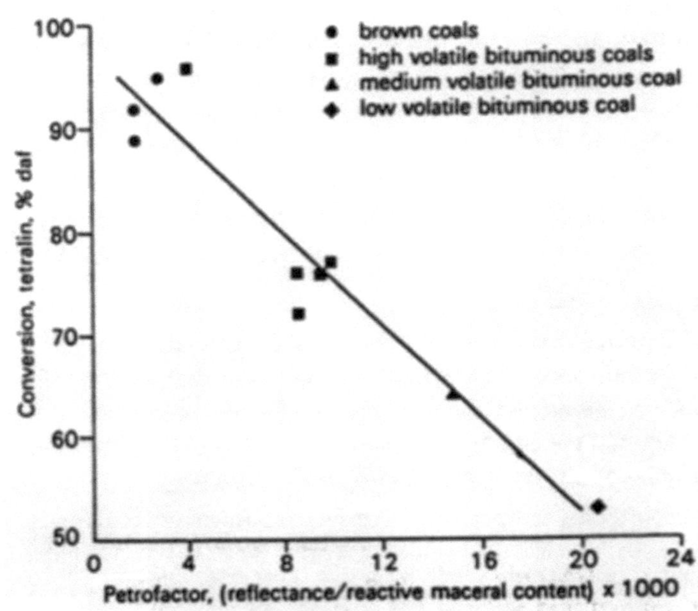

Figure 4. Correlation of coal conversion with petrofactor [37].

raw MB coal. Thus, based on Equation 2, it is estimated that about 89% of raw MB coal would be converted during liquefaction process.

Parameters	**Results**
Maximum reflectance of vitrinite	0.40%
Total reactive maceral content	(Vitrinite + Liptinite) = 91%
Petrofactor	1000 x (0.40%) / (91%) = 4.40
Estimated coal conversion	100 – 2.5 (4.40) = 89%

Table 3. Petrofactor value and estimated conversion of raw MB coal [29].

4. Coal liquefaction

In order to reduce operation cost and capital, and also to increase coal liquefaction efficiency, the liquefaction processes were usually carried out at less severe conditions. However, in most of coal liquefaction, the temperature used was higher than 400 °C and at relatively higher pressures. Thus, many attempts have been made by researchers to establish and develop methods of liquefying coals at lower temperature and pressure. Mukah Balingian, a low rank Malaysian coal, was extensively used in liquefaction study [29]. In the investigation, three different types of pretreatments, i.e. solvent swelling, in-situ solvent soaking and heating and microwave irradiation have successfully been applied to the untreated Mukah Balingian coal prior to liquefaction. The main objective of the pretreatments was to further enhance coal conversion (solid to liquid) and oil yield at less severe liquefaction conditions. The study found that all the applied pretreated coals showed an increase in both conversion and oil yield. A good coal-to-solvent contact and weakened coal structure were found to be equally important factors in determining high coal conversion and oil yield obtained during liquefaction [29].

Because of the increased demands and limited reserves of petroleum, the concern of energy security has made the direct coal liquefaction (DCL) become important in producing alternative transportation fuels. DCL is a technology to convert coal directly into transportation fuels and chemicals, which is conducted at high temperatures and pressures in the presence of solvents [1]. However, because DCL is usually conducted under severe reaction conditions with higher hydrogen consumption, it makes the cost of oil from DCL difficult to compete with that from crude oil, and thus the development of DCL is limited for large-scale applications [25].

On the contrary, biomass such as rubber seed is a cheap and renewable organic energy source; the liquefaction of biomass to convert into alternative transportation fuels has been paid more and more attention. Thus, a new alternative method should be proposed or investigated in order to overcome this crucial global issue.

5. Co-liquefaction of coal and biomass

One of the solution to the above-mentioned situations encountered with coal and biomass is by the co-liquefaction of both of them. This can also maintain the stability supply of the materials. Moreover, one of the advantages of using biomass is that it accelerates the thermal decomposition of coal macromolecular structure by reducing the severe reaction conditions, especially the temperature during liquefaction. By producing smaller molecular fragments, it enables the combination with the large fractions produced from coal decomposition and simultaneously terminate the cross-linking reaction between them (this is to avoid the production of larger molecular weight products). Thus, the yield of low-molecular-weight products will be increased, oil quality improved and eventually, will reduce the yield of the residue during co-liquefaction process [25].

Most of the research works done for alternative energy sources have shown that, in general, co-processing or known as co-liquefaction of coal with biomass-type wastes has a positive effect on the liquefaction yields and product quality [14, 15]. Co-liquefaction of coal with biomass has gained increasing research interest due to the growing concerns over greenhouse gas emissions [28]. Because biomass materials also contain abundant hydrogen, co-liquefaction of coal and biomass waste is one of the best and feasible option to reduce the consumption of hydrogen as well as to avoid severe reaction conditions during co-liquefaction. Till today, however, very limited research work has been carried out in this field.

The mechanism of co-liquefaction of coal and biomass is believed to be a radical process. Thus, the unstable free radical fragments which are formed during pyrolysis of coal and biomass will be stabilized by active hydrogen contributed from biomass to form hydrogenation/lower-molecular-weight products. Researchers anticipated that there is a synergistic effect (synergy is two or more things functioning together to produce a result not independently obtainable) in the co-thermolysis process, and the yield of solid products from co-thermolysis is different from the arithmetic calculated value [1].

Hua *et al.* [38] carried out co-liquefaction of coal and rice straw and believed that there exists a synergistic effect during co-pyrolysis of Shenfu coal and rice straw. Shui *et al.* [39] investigated the co-liquefaction behavior of a sub-bituminous coal and sawdust. They found that the thermolysis of Shenhua coal was accelerated by sawdust and more volatile matter was released from the coal molecular structure during the co-thermolysis process. In another study, Guo *et al.* [25] also investigated the synergistic effect existence in the co-liquefaction of coal and biomass. They found that a positive synergistic effect during the process actually existed. Thus, they concluded that the synergistic effect depends on several factors, i.e. (i) coal rank, (ii) liquefaction conditions and (iii) different liquefaction products.

The liquefaction process of coal and biomass materials, which is known as "Co-liquefaction", has not been developed in Malaysia. The yields and quality (especially H/C ratio) of the liquid products obtained from coal under less severe liquefaction conditions (at lower temperature and pressure) can be improved with co-liquefaction of coal and biomass. Therefore, the cost of oil produced from direct coal liquefaction can be reduced significantly. The process can make full use of hydrogen in biomass, thus decreasing the consumption of hydrogen and moderating the conditions of DCL [1].

Some important parameters for co-liquefaction are the materials used, the design of the reactor, pressure, extraction solvent, temperature, holding time and catalyst used. Hua et al. [38] reported that the rice straw contains 68.3 w/% of volatile matter and resulted in 60.3% of oil at 420 °C. However, rice straw contains high amount of silica. Shui et al. [39] reported that fir sawdust contains 78.2 w/% of votalite matter and results in 55.2% of oil at 420 °C. However, fir occurs in mountains over most of the range. Guo et al. [25] reported that poplar sawdust contains 80.27 w/% of volatile matter and results in 59.19% of oil at 360 °C. However, high tannic acid content is present in poplar. Basic properties of crude rubber seed oil and crude palm oil blend as a potential feedstock for biodiesel production with enhanced cold flow characteristics were studied by Yusup et al. [40] and the inspections determined that the rubber seed oil can be used in the current diesel machines with no alteration required, confirming the adaptability of the produced biodiesel to the current standards. This shows that the characteristic of rubber seed as a biofuel material is more suitable than rice straw, fir sawdust, poplar sawdust and other biomass that contains less oil.

Hua et al. also [38] suggest that the FeS catalyst used has a good catalytic hydrogenation activity on rice straw, but the drawback of this catalyst is that it is low in basicity to reduce carboxylic acid present in the vegetable oil. An alternative for the FeS usage as a catalyst is by using dolomite. Dolomite is a natural rock found abundantly in certain areas of Malaysia and Thailand. Due to its very low cost to produce and being easy to obtain, the main domestic usage of dolomite is currently in the landfill site and in cement manufacturing. $CaCO_3$ and $MgCO_3$ are the major components of dolomite with a small amount of silica and ferrite. In a

http://dx.doi.org/10.5772/59362

simple calcination process at high temperatures, the CO_3 group in pure $CaCO_3$ will decompose to produce CaO (which is highly basic) and MgO [41].

Guo et al. [25] state that the temperature is the most important factor during liquefaction. Shui *et al.* [39] reported that the main pyrolysis temperature range of Shenhua coal (a sub-bituminous coal) (362 – 750 °C) is much higher than that of sawdust (260 – 420 °C), and the releasing rate of volatile matter of Shenhua coal is much lower than that of sawdust. Hua et al. [38] reported that the pyrolysis temperature of Shenhua Coal is in the range of 360 – 750 °C; however, the pyrolysis temperature of biomass is in the range of 250 – 400 °C.

Hence, an attempt should be made for co-liquefaction of low-rank Malaysian coal and biomass wastes such as rubber tree wastes (rubber seed, rubber seed pod or rubberwood) for the production of alternative fuels and other important purposes such as chemical feedstocks. Working on the area of energy has now become the priority in most of laboratories worldwide. In Malaysia particularly, under the New Economic Model (NEM), innovation and research is regarded as crucial factor to propel the industry. The research in this area certainly contributes to the increases in energy sector that part of National Key Economic Area (NKEA). Furthermore, a competitive domestic economy can be created as part of Strategic Reform Initiative (SRI) [42].

6. Biofuels and alternative fuels

Today, the term biofuels mostly refers to ethanol and esterified vegetable oil. Scientifically, a biofuel is a type of fuel whose energy is derived from biological carbon fixation. Biofuels include fuels derived from biomass conversion, as well as solid biomass, liquid fuels and various biogases. It is known that especially agricultural-based alternative fuels have a considerable effect on decreasing net CO_2 emissions [43].

New products such as methanol, dimethyl ether, Fischer-Tropsch (FT) diesel and ethanol from lignocellulosic feedstock are benefiting from R&D programs. The most controversial such energy carrier is first-generation biofuels, i.e. biodiesel and bioethanol from sugar, starch and oil bearing crops or animal fats that in most cases can also be used as food and feed [44]. During the past years, researches on converting lignocellulosic biomass into bioethanol are actively undertaken, aiming to produce the second-generation biofuels, which has no competition with food and is thus sustainable [45]. A review from Akhtar et al. [46] on the process conditions for optimum bio-oil yield in hydrothermal liquefaction of biomass includes various parameters, including temperature, particle size, biomass feedstock, heating rate, density, pressure, residence time and reducing gas/hydrogen donors. In short, they found out that the major parameters that influence yield and composition of bio-oil are temperature, properties of solvent, solvent density and type of biomass.

A number of biofuels for transport are potentially available and are currently being used or investigated at different stages of development worldwide. A study in Australia [47] found out that the crop stubble – the fibrous stalk, leaf and chaff material left after grain (or other products) has been harvested – is an agricultural source of lignocelluloses biomass for second-generation biofuels.

One of other such plants that were used for biofuels production is the once unpopular Jatropha Curcas (JC). JC is a perennial subtropical shrub that produces oil-rich seeds. A study by Jingura [48] on the technical options for optimization of production of JC as a biofuel feedstock in arid and semiarid areas of Zimbabwe proves that JC has been promoted extensively as an energy crop for biodiesel in the tropics.

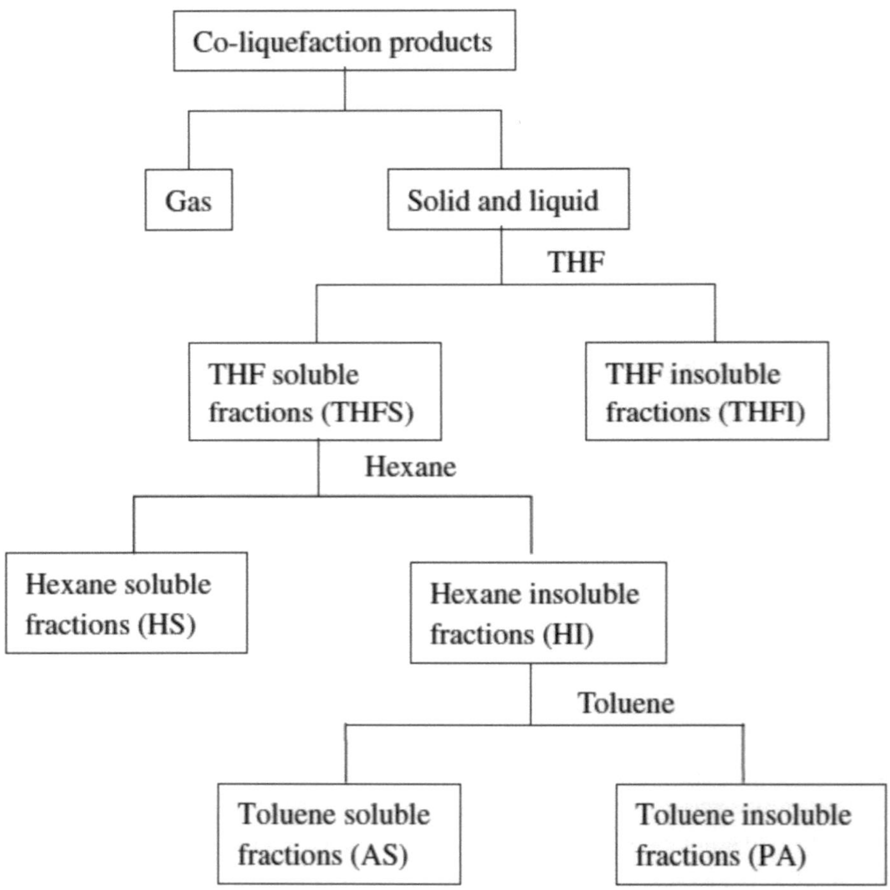

Figure 5. Example of a complete fractionation process of co-liquefaction products.

Apart from the high oil content in the seed, JC was only planted for the fruit. It is not JC seeds that we found more massive when placed on a scale of comparison with the rubber seed. Unlike JC, rubber trees can produce both latex and high oil content seed. According to Kalam et al. [49], the flash point of JC oil is 229 ± 4 °C which is higher than that of rubber seed oil with the

http://dx.doi.org/10.5772/59362

flash point of 198 °C. These properties can be an advantage to the rubber seeds as an alternative biomass in producing bio-oil or biofuel.

There are so many advantages associated with the use of rubber seed to produce biofuels including the large reserve for the production of nonedible products, do not necessarily require cultivation of new plantations, more profitable to farmers, and reinvogorate the economy of local communities. These advantages show great potential of rubber seeds as the biomass for biofuel productions. Several works have been done dealing with the production of biofuel from the rubber seed. A study by Melvin et al. [50] entitled, "A multi-variant approach to optimize process parameters for biodiesel extraction from rubber seed oil" shows that the discarded rubber seed from the hefty rubber plantation of Southern India is considered as the potential source for extracting oil for biodiesel production.

Therefore, alternative fuels, which are also known as nonconventional or advanced fuels, are any substances or materials that have the ability to be utilized as fuels. Commonly available alternative fuels include biodiesel (oil obtained from plant or fruit which was transesterified), bio-alcohols (methanol, ethanol and butanol), fuel cells, batteries, nonfossil methane, hydrogen, nonfossil natural gas, vegetable oil, propane and also other biomass sources. Figure 5 shows the example of a complete fractionation process of co-liquefaction products after liquefaction.

7. Problem statements

Huge amount of biomass wastes in Malaysia, especially from rubber plantations, are being produced daily. The rubber tree produces large volumes of seed, which is underutilized. Rubber seed would be a highly potential candidate for co-liquefaction process as it is absent in silica, easily collected and does not contain tannic acid. In addition, the oil content in the rubber seed is roughly around 40 – 60 %, which is suitable to replace other non-seed biomass. No study has been done on the rubber seed as the material in co-liquefaction process. The existence of synergystic effect between the coal and rice husk, fir sawdust and poplar sawdust was already known. However, the synergistic effect between the rubber seed and Mukah Balingian, a low rank Malaysian coal in the co-liquefaction process, has not yet been evaluated by any research works. And so far, there is no research on co-liquefaction of biomass and coal in Malaysia. It is also a must to investigate the synergistic effect between the coal and rubber seed.

Vegetable oil has the potential to supplement petroleum-based fuels; however, its high FFA, thereby its high viscosity, deters its usage in internal combustion engine directly. Dolomite has received, however, less attention in the basic catalyzed chemical syntheses, in spite of its high basicity, low cost, less toxicity and environmental friendliness. Thus, dolomite as reduction catalyst is a good option to reduce the amount of acid present in the rubber seed oil during the co-liquefaction process. The main pyrolysis temperature range of coal is much higher than that of biomass and the releasing rate of volatile matter of coal is much lower than that of biomass, causing different liquefaction temperature range. The temperature range will be investigated to determine the temperature that can promote the highest conversion of both coal and biomass. From the co-liquefaction process with varied parameters, it is predicted that the percent of product yield will be more than the individual liquefaction of coal and rubber seed alone.

8. Significant contribution to knowledge

It is anticipated that a significant contribution could be made from this co-liquefaction process between rubber seed and coal, i.e.:

i. Contribution in terms of development and improvement of the co-liquefaction technique by using high pressure – high temperature reactor system.

ii. Valuable information of the extracts, i.e. alternative fuels obtained from the co-liquefaction processes would be useful to other areas and researchers.

iii. Provide useful information of Malaysian coal, especially in fuel chemistry area for Coal Bank data.

iv. As an alternative choice to conventional methods of energy consumption in Malaysia.

Acknowledgements

The authors would like to acknowledge the Ministry of Education, Malaysia, Ministry of Science, Technology and Innovation Malaysia and Universiti Teknologi MARA for supporting the research work.

Author details

Mohd Azlan Mohd Ishak[1*], Siti Nur Ain Mohd Hassan[1,2], Siti Nurlia Ali[1,2], Mohd Fauzi Abdullah[1], Asnida Yanti Ani[1], Nur Nasulhah Kasim[1], Ali H. Jawad[1], Wan Izhan Nawawi Wan Ismail[1] and Khudzir Ismail[2]

*Address all correspondence to: azlanishak@perlis.uitm.edu.my

1 Coal and Biomass Energy Research Group, Faculty of Applied Sciences, Universiti Teknologi MARA, Perlis Branch, Arau, Perlis, Malaysia

2 Faculty of Applied Sciences, Universiti Teknologi MARA, Shah Alam, Selangor, Malaysia

References

[1] Zhou, H., Cai, Z., Shui. H., Lei, Z., Wang, Z., Li, H. Co-liquefaction properties of Shenfu coal and rice straw. Journal of Fuel Chemistry and Technology. 2011; 39(10), 721-727.

[2] Douthwaite, R. Degrowth and the supply of money in an energy-scarce world. Ecol. Econ. doi:10.1016/j.ecolecon. 2011.03.020.

[3] The Annual Energy Outlook 2008 with Projections to 2030. National Energy Information Center, Energy Information Administration, Washington DC, 2008.

[4] Chiari, L., Zecca, A. Constraints of fossil fuels depletion on global warming projections. Energy Policy. 2011; 39, 5026-5034.

[5] Guan, C.T. Quaterly Bulletin of the Department of Environment Malaysia (MOSTE). Technical Update: Biomass: Renewable Energy Sources in Malaysia. 2003; 2, 15-16.

[6] The Eight Malaysian Plan. Percetakan Nasional Malaysia Berhad, Kuala Lumpur, 2001.

[7] Yee, L.H. Initiatives to promote renewable energy. The Star Online. 2010 October 15.

[8] Caserini, S., Livio, S., Giugliano, M., Grosso, M., Rigamonti, L. LCA of domestic and centralized biomass combustion: The case of Lombardy (Italy). Biomass and Bioenergy. 2010; 34, 474-482.

[9] Sato, S., Murakata, T., Chiba, T., Tukakosi, M., Oda, S. Effect of lignin addition on coal liquefaction, in International Conference on Coal Science, Tokyo, Japan, New Energy and Industrial Technology Development Organization. 1989; 717-720.

[10] Jung, H., Tierney, J.W., Wender, I. Co-processing of Cellulosic Waste and Coal. Prepr. – Am. Chem. Soc., Div. Fuel Chem. 1993; 38, 880-884.

[11] Anderson, L.L., Tuntawiroon, W., Ding, W.B. Co-liquefaction of Coal and Waste Polymers, in Pajares, J.A., Tascon J.M.D. (Eds), Coal Science: Proceedings of the Eighth International Conference on Coal Science 2. Elsevier, Amsterdam, The Netherlands, 1995; 1515-1518.

[12] Tang, Y., Curtis, C.W. Thermal and Catalytic Co-processing of Waste Tires with Coal. Fuel Pro. Technol. 1996; 46(3), 195-215.

[13] Karaca, F., Bolat, E. Co-processing of a Turkish Lignite with a Cellulosic Waste Material. 1. The Effect of co-processing on liquefaction yields at different reaction temperatures. Fuel Proc. Technol. 2000; 64, 47-55.

[14] Rafiqul, I., Lugang, B., Yan, Y., Li, T. Study on Co-liquefaction of Coal and Bagasse by Factorial Experiment Design Method. Fuel Proc. Technol. 2000; 68, 3-12.

[15] Yuan, X., Cao, H., Li, H., Zeng, G., Tong, J., Wang, L. Quantitative and Qualitative Analysis of Products Formed During Co-liquefaction of Biomass and Synthetic Polymer Mixtures in Sub- and Supercritical Water. Fuel Proc. Technol. 2009; 90, 428-430.

[16] Steven, L., Keat, T. L. Leading global energy and environmental transformation: Unified ASEAN biomass-based bio-energy system incorporating the clean development mechanism. Biomass and Bioenergy. 2010; 35, 2479-2490.

[17] Aigbodion, A. I.; Pillai, C. S. T. Preparation, analysis and applications of rubber seed oil and its derivatives in surface coatings. Prog. Org. Coat. 2000; 38, 187-192.

[18] Ramadhas, A. S., Jayaraj, S., Muraleedharan, C. Characterization and the effect of using rubber seed oil as fuel in the compression engines. Renew. Ener. 2005; 30, 795-803.

[19] Ikwuagwu, O. E., Ononogbo, I. C., Njoku, O. U. Production of biodiesel using rubber seed oil. Ind. Crop. Prod. 2000; 12, 57-62.

[20] Iyayi, A. F., Akpaka, P. O., Ukpeoyibo, U., Balogun, F. E., Momodu, I. O. Rubber seed oil with great potentials. Chem. Tech. J. 2007; 3, 507-516.

[21] The Eighth Malaysia Plan: 2001 – 2005, Unit Perancang Ekonomi (UPEN), Jabatan Perdana Menteri. Percetakan Nasional Malaysia Berhad, Kuala Lumpur, 2001.

[22] Thadeus, J. Complementary roles of natural gas and coal in Malaysia. 2003, http://www.apec-egcfe.fossil.energy.gov.

[23] Wu, J. C. The mineral industry of Malaysia. U.S. Geological Survey Minerals Yearbook. 1999; 14.1-14.7.

[24] http://www.worldcoal.com/coal/Assets/SEAsia.pdf, 2005

[25] Guo, Z., Bai, Z., Bai, J., Wang, Z., Li, W. Co-liquefaction of lignite and sawdust under syngas. Fuel Proc. Technol. 2011; 92, 119-125.

[26] Ishak, M. A. M., Ismail, K., Abdullah, M. F., Kadir, M. O. A., Mohamed, A. R., Abdullah, W. H. Liquefaction studies of low-rank Malaysian coal using high-pressure high-temperature batch-wise reactor. Coal Preparation – A Multinational Journal. 2005; 25(4), 221-237.

[27] Ismail, K., Ishak, M. A. M., Kamal, M. L., Snape, C. E., Brown, S. D., Nawi, M. A. M., Kadir, M. O. A., Ismail, A. F. Determination of organic sulphur forms in some Malaysian and Indonesian coals by high pressure temperature programmed reduction. Advances in Malaysian Energy Research. 1999; 1-7.

[28] De Marco Rodriguez, I., Chomon, M. J., Caballero, B., Arias, P. L., Legarreta, J. A. Liquefaction behaviour of a Spanish sub-bituminous A coal under different conditions of hydrogen availability. Fuel Proc. Technol. 1998; 58, 17-24.

[29] Ishak, M.A.M. Direct liquefaction of pretreated low rank Malaysia coal – towards enhancing conversion and oil yield. LAP Lambert Academic Publishing; 2010.

[30] Van Krevelen, D. W. Coal – Typology, Physics, Chemistry, Constitution. Amsterdam: Elsevier Science Pub.; 1993; 755.

[31] Merrick, D. Coal Combustion and Conversion Technology. Energy Alternative Series. London & Basingstoke: Macmillan Publishers Ltd.; 1984.

[32] Probstein, R. F., Hicks, R. E. Synthetic Fuels. Mc Graw-Hill, Inc.; 1982.

[33] Kastanaki, E., Vamvuka, D., Grammelis, P., Kakaras, E. Thermogravimetric studies of the behavior of Lignite–Biomass blends during devolatilization. Fuel Proc. Technol. 2002; 77-78, 159-166.

[34] Serio, M. A., Hamblen, D. G., Markham, J. R., Solomon, P. R. Kinetics of volatile product evolution in Coal Pyrolysis: Experiment and Theory.Energy & Fuels. 1987; 1, 138-152.

[35] Radovic, L. R., Walker, P. L. Jr., Jenkins, R. G. Importance of carbon active sites in The Gasification of Coal Chars. Fuel. 1983; 62, 849-856.

[36] Merrick, D. Coal Combustion and Conversion Technology. Energy Alternatives Series. London & Basingstoke: Macmillan Publishers Ltd.; 1984, 168-230.

[37] Guyot, R. E. Relationship between Coal Properties and Liquefaction Potential. ACIRL-PR-79-1, North Ryde, NSW, Australian Coal Industry Research Lab. 1978; 28.

[38] Hua Z., Zhen-yi, C., Heng-fu, S., Zhi-ping, L., Zhi-cai, W., Hai-ping, L. Co-liquefaction properties of Shenfu coal and rice straw. Journal of Fuel and Technology. 2011; 39(10), 721-727.

[39] Shui, H., Shan, C., Cai, Z., Wang, Z., Lei, Z., Ren, S., Pan, C., Li, H. Co-liquefaction behavior of a sub-bituminous coal and sawdust. Energy. 2011; 36, 6645-6650.

[40] Yusup, S., Khan, M. Basic properties of crude rubber seed oil and crude palm oil blend as a potential feedstock for biodiesel production with enhanced cold flow characteristics. Biomass and Bioenergy. 2010; 34, 1523-1526.

[41] Ngamcharussrivichai, C., Wiwatnimit, W., Wangnoi, S. Modified dolomites as catalysts for palm kernel oil transesterification. Journal of Molecular Catalysis A: Chemical. 2007; 276, 24-33.

[42] The Tenth Malaysian Plan. Percetakan Nasional Malaysia Berhad, Kuala Lumpur.; 2011.

[43] Karabektas, M. et al. Effects of the blends containing low ratios of alternative fuels on the performance and emission characteristics of a diesel engine. Fuel, doi:10.1016/j.fuel.2011.04.036, 2011.

[44] Gasparatos, A., Stromberg, P., Takeuchi, K. Biofuels, ecosystem services and human wellbeing: Putting biofuels in the ecosystem services narrative. Agriculture, Ecosystem and Environment. 2011; 142, 111-128.

[45] Lin, L., Yan, R., Liu, Y., Jiang, W. In-depth investigation of enzymatic hydrolysis of biomass wastes based on three major components: Cellulose, hemicellulose and lignin. Bioresource Technology. 2010; 101, 8217-8223.

[46] Akhtar, J., Amin N. A. S. A review on process conditions for optimum bio-oil yield in hydrothermal liquefaction of biomass. Renewable and Sustainable Energy Reviews. 2011; 15, 1615-1624.

[47] Rodriguez, L. C., May, B., Herr, A., Farine, D., O'Connell, D. Biofuel excision and the viability of ethanol production in the Green Triangle, Australia. Energy Policy. 2011; 39, 1951-1957.

[48] Jingura, R. M. Technical options for optimization for optimization of production of Jatropha as a biofuel feedstock in arid and semi-arid areas of Zimbabwe. Biomass and Bioenergy. 2011; 35, 2127-2131.

[49] Kalam, M. A., Ahamed, J. U., Masjuki, H. H. Land availability of Jatropha production in Malaysia. Renewable and Sustainable Energy Reviews. 2012; 16, 3999-4007.

[50] Melvin, J. D. F., Edwin, Raj R., Durga, Prasad B., Robert, Kennedy Z., Mohammed, Ibrahim A. A multi-variant approach to optimize process parameters for biodiesel extraction from rubber seed oil. Applied Energy. 2011; 88, 2056-2063.

Chapter 10

A Study of Polystyrene Solubility in Biodiesel

Koji Yamane and Kiyoshi Kawasaki

Additional information is available at the end of the chapter

http://dx.doi.org/10.5772/59112

1. Introduction

Waste plastics, such as polystyrene (PS), polypropylene (PP), and polyethylene (PE), are usually disposed together and so require time for sorting. Polystyrene articles (e.g., Styrofoam, food trays, and packing materials) occupy a large volume at a very low weight. This results in inefficient transportation of waste plastic. In addition, subjecting waste plastics to pyrolysis, which involves conversion of plastic to oil requires energy and is expensive. It would be useful to reduce waste plastic volume and to recover the energy of waste polystyrene recycling by a simple process. In this chapter, a simple process is proposed to obtain liquid fuel dissolved polystyrene in biodiesel at lower energy costs.

As shown in Figure 1, waste polystyrene such as expanded polystyrene is selectively dissolved in biodiesel derived from waste cooking oil, and biodiesel-dissolved polystyrene is utilized as diesel fuel for a cogeneration system. This chapter describes the potential for selective dissolution of polystyrene in biodiesel, which consists of fatty acid methyl esters (FAME), by using Hansen solubility parameters and for properties and combustion characteristics as diesel fuel by using fuel ignition analyzer.

The results obtained by Hansen solubility parameters indicated that methyl oleate, the main component of biodiesel, provides selective solubility for polystyrene. Results of experiments that examined solubility properties revealed that the kinematic viscosity increased with polystyrene concentration in FAME derived from soybean cooking oil. The cause of the increase in kinematic viscosity was due to the presence of a polymer with a molecular weight of up to approximately 10^7.

On examining the use of dissolved polystyrene as fuel for diesel engines, experiments showed that the cetane number decreased with an increase in polystyrene concentration; however, the cycle-to-cycle fluctuations in combustion pressure and ignition timing when polystyrene dissolved in FAME was used were less than those found for neat FAME. Polystyrene dissolved in FAME crystallized and precipitated in the gel upon addition of hydrocarbons such as engine lubricant oil. Therefore, the use of FAME-dissolved polystyrene as diesel fuel requires attention to tribology.

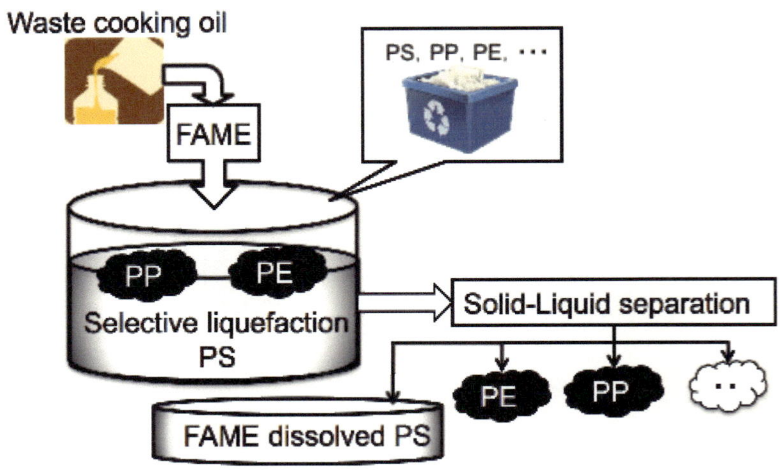

Figure 1. Energy recovery using biodiesel fuel (FAME) derived from waste cooking oil selective-dissolved waste polystyrene (PS).

Biodiesel consists of fatty acid methyl esters (FAME) and can be produced from a great variety of feedstocks including vegetable oil (e.g., soybean, palm, rapeseed oil) and animal fats, as well as waste cooking oils (e.g., used frying oils). FAME has been used as alternative fuel of diesel fuel. Many research projects have been carried out and have been published in books and journals [1-3]. Most of all are concerning to fuel quality, combustion and exhaust emission characteristics, which regards the utilization with the diesel engines.

Effective use of waste plastics is important in establishing a recycling-oriented society. FAME can dissolve plastics and rubbers, which are used as fuel system parts of diesel engines, resulting in the need to replace these parts. If a method to utilize FAME as a solvent to dissolve plastics can be developed, waste plastics could be recovered and could be utilized as liquid fuel.

Several studies on solubility report that *n*-alkenes and di-*n*-alkyl are better solvents for the low and medium molecular weight samples of polystyrene than the corresponding *n*-alkanes [4]. Also, it is shown that certain food items have been shown to be incompatible with the expanded polystyrene (EPS) used for the manufacture of food containers. Citronella, limonene and terpinene, which are constituents of many flavor oils, are excellent solvents for polystyrene [5].

Studies of solubility for volume reduction and waste management of polystyrene recycling have been conducted earlier [6,7]. Solubility values of extrude poly-styrene(XPS) in several solvents such as benzene, toluene, xylene, tetrahydrofuran, chloroform, 1,3-butanediol, 2-butanol, linalool, geranoil, *d*-limonene, *p*-cymene, terpinene, phellandrene, terpineol, metha-

http://dx.doi.org/10.5772/59112

nol, eucalyptus, cinnamaldehyde, nitrobenzene, N,N-dimethylformamide, and water have been determined. The solubility of the polymer in the mentioned solvents at different temperatures has been investigated. The solvent can be easily recycled by distillation.

Polystyrene dissolved in FAME has a greater heating value as fuel than that of FAME alone, so that a reduction of the fuel consumption per output and the increase in the use of diesel and boiler cogeneration can be expected. The diesel combustion characteristics of soybean oil FAME-dissolved packing peanuts has been investigated by Kuzhiyil et al. [8]. The results showed that engine power increased initially with the polystyrene concentration and then decreased at concentrations greater than 5%. The decline in engine power at high polystyrene concentrations could be caused by the poor spray atomization and deteriorated combustion efficiency due to the high viscosity of polystyrene mixtures.

Studies on the use of soy methyl ester-polystyrene (SME-PS) blends to increase durability of concrete have been conducted [9,10]. Experimental results show that SME-PS reduces water absorption of concrete (by up to 75%), protects from damage caused by freezing and thawing (reduces damage by 66%, reduces chloride ingress depth by up to 50%, and slows the rate of the alkali-silica reaction by 50%.

The method for solubilizing polystyrene in a fatty acid ester has been granted as a US patent [11]. The purpose of the invention is to produce commodity material useful for the production of polymers from polystyrene dissolved in fatty acid esters. This patent also provides the method for disposing of polystyrene solubilized in a fatty acid ester. A fatty acid ester composed of a variety of alkyl and alkene esters is used and heated to temperatures between 100°C and 180°C.

Expanded polystyrene (EPS) is used for packaging material to cushion appliances and containers of seafood and agricultural products. Extruded polystyrene (XPS) board is used as insulating material in building. The production of EPS in the world came up to 10 Mt in 2010. Presently, various EPS organizations from more than 25 countries around the world have subscribed to the international agreement on recycling of EPS. In Japan, EPS of 1.1 Mt was produced and 88% of waste EPS was recovered by means of material recycling and energy recovery in 2010; however, the EPS trays with food residue were not recovered. Almost all were sent to landfill sites or were incinerated. EPS and XPS are low density, high volume materials, which consume a significant amount of space in a landfill. The transport efficiency of waste polystyrene is poor because its volume-to-mass ratio is large. Reducing the volume of waste EPS and XPS will reduce CO_2 emissions and decrease the cost of energy per volume due to the transportation of waste plastic.

2. Experimental

2.1. Solubility parameter determination

Solubility parameters can be used to easily identify solvents for polymers. Many successful studies have used methods based on the solubility parameter [12-14]. The term "solubility parameter" was first used by Hildebrand and Scott [15]. The theory relates the energy of mixing to the energy of vaporization of the pure component. This theory was developed for mixing of nonpolar substances. However, many solvents and polymers in common use are polar compounds. Hansen divided the polar portion into a dipole-dipole contribution and hydrogen-bonding contribution, both of which could be determined through solubility experiments with polymers [16]. The Hansen solubility parameter (HSP) separates the solubility energy into three parts: dispersion (δ_D), polarity(δ_P), and hydrogen bonding(δ_H). The HSP concept can be described as "like vectors dissolve like vectors." For the vector "likeness," the HSP distance (*Ra*), which is the distance between the solvent and center of the polymer solubility sphere, was used. *Ra* is calculated using Eq.1:

$$R_a = \sqrt{4(d_{D1} - d_{D2})^2 + (d_{P1} - d_{P2})^2 + (d_{H1} - d_{H2})^2} \tag{1}$$

where subscript 1 represents the solvent, and subscript 2 represents the polymer.

This equation was developed from plots of experimental data where the constant "4" was found convenient and correctly represented the solubility data as a sphere encompassing the good solvents. When the scale for dispersion (δ_D) parameter is doubled, in comparison with other two parameters essentially spherical, rather than spheroidal, regions of solubility are found [16].

To determine if the parameters for the solvent and polymer are within an acceptable range, a value called the interaction radius (R_0) of the polymer is applied to the substance being dissolved. Good solvents are within R_0, and poor ones are outside it. A simple composite affinity parameter, relative energy distance (*RED*), has been defined as:

$$RED = R_a / R_0 \tag{2}$$

Good solvents will have *RED* less than 1.0. Poor solvents will have increasingly higher *RED* values.

In this study, the solubility of polymer in methyl ester, methanol, and fatty acid was determined by HSPiP (Hansen solubility parameters in practice) software [17]. Fatty acids are source materials of FAME production by esterification process. Oleic acid methyl ester (methyl oleate) is a main component of FAME derived from waste cooking oil, because a typical raw material for cooking oil is the rapeseed oil and soybean oil. Tables 1 and 2 show the HSP from the dataset in HSPiP for methyl oleate, methanol, and oleic acid as solvents, and polystyrene (PS), polypropylene (PP), and polyethylene (PE) as polymers. The HSP generally in use for liquids have all been calculated at 25°C. In this study, the dataset at 25°C was used.

http://dx.doi.org/10.5772/59112

Solvent*	Dispersion δ_D MPa$^{1/2}$	Polarity δ_P MPa$^{1/2}$	Hydrogen bonding δ_H MPa$^{1/2}$
Methyl Oleate	16.2	3.8	4.5
Oleic Acid	16	2.8	6.2
Methanol	14.7	12.3	22.3

* data from the dataset in HSPiP [17]

Table 1. Dispersion δ_D, polarity δ_P, and hydrogen bonding δ_H of methyl oleate, methanol, and oleic acid as solvents.

Polymer*	Dispersion δ_D MPa$^{1/2}$	Polarity δ_P MPa$^{1/2}$	Hydrogen bonding δ_H MPa$^{1/2}$	Interaction radius R_0 MPa$^{1/2}$
PS	18.5	4.5	2.9	5.3
PE	16	0.8	2.8	3.2
PP	18	0	1	6

* data from the dataset in HSPiP [17]

Table 2. Dispersion δ_D, polarity δ_P, and hydrogen bonding δ_H and interaction radius of polystyrene(PS), polypropylene(PP), and polyethylene(PE) as polymer.

2.2. Measurement of the solubility and fuel properties

Expanded polystyrene (EPS) and food trays (PSP), cut to pieces, were stirred slowly into FAME and dissolved at room temperature. FAME was prepared by the batch-type production equipment using an alkaline catalyst method [3]. Commercial soybean cooking oil was used as the raw material to produce biodiesel, soybean oil methylester (SME).

To determine the dissolved molecular weight of the polystyrene, polystyrene standards with an average molecular weight (MW) of 4000 and 50,000 were also dissolved in FAME. The polystyrene molecular weight distribution in FAME was measured by gel permeation chromatography (GPC).

To clarify the fuel characteristics as diesel fuel, kinematic viscosity was measured according to JIS K2283, the ignition quality as diesel fuel was analyzed by fuel ignition analyzer (FIA) through constant-volume combustion (Fueltech, FIA-100 ver3). Diesel combustion property was evaluated by an ignition delay. Figure 2 shows the configuration of the FIA. In the experiments, conducted under constant pressure of 2.0MPa and initial temperatures of 450°C, the fuel was injected and the ignition delay was measured. Ignition delay was defined as the time difference between the fuel injection start time and the time at which combustion pressure was 0.02MPa greater than the initial pressure in the chamber, as shown in Figure 3.

The cetane number (*CN*) is used often for estimating ignition quality. In this study, the CN value of FAME-dissolved polystyrene was estimated by calibration of the *CN* obtained from a mixture hexadecane (CN=100) and heptamethylnonane (CN=15). For practical purposes, the FIA cetane number(CN_{FIA}) was determined by Eq.3.

$$CN_{FIA} = 1413 \cdot t^{-\frac{1}{0.69}} \tag{3}$$

where τ denotes the ignition delay [ms].

Furthermore, the relationship between CN of FAME using FIA and CN value using the CFR standard institutional organization engine test, CN_{CFR}, is described by the following relation [18]:

$$CN_{CFR} = CN_{FIA} + 22 \tag{4}$$

One of the important characteristics of diesel fuel is the carbon residue (CR). The CR is a characteristic value related to the amount of carbon deposits stored inside the engine, carbon deposits for petroleum-based fuels in general are measured using a sample condensed to 10% in volume. For FAME, the high-temperature heating process under condensation results in thermal decomposition of FAME components. In addition, the chemical structure of the fatty acid methyl ester component is changed. For this reason, the CR value for FAME was measured using a sample without condensation in this study.

The heating value of fuel is related to fuel economy and engine power. In this study, the higher heating value of fuel was measured using an automatic bomb calorimeter (Shimadzu, CA-4PJ).

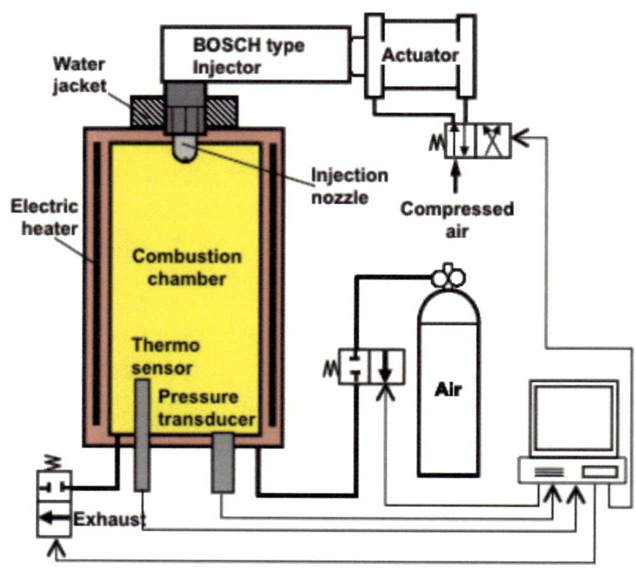

Figure 2. Configuration of fuel ignition quality analyzer (FIA).

http://dx.doi.org/10.5772/59112

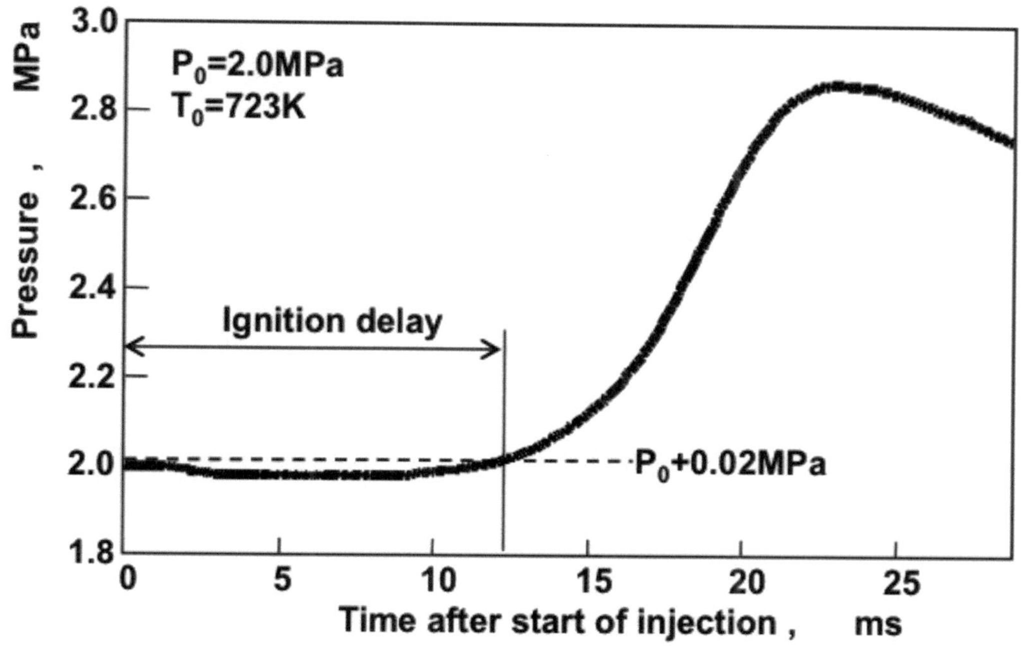

Figure 3. Definition of ignition delay by FIA.

3. Results and discussion

3.1. Solubility of polymers

Figure 4 represents data of the polymers as mesh spheres, and the solvents as dots. Figure 5 also represents the data as two-dimensional planes. Table 3 shows the results calculated from the HSP analysis. In Figures 4 and 5, the scale in coordinate of dispersion(δ_D) is expressed in HSPiP as twice as large as those in coordinates of polarity(δ_P) and hydrogen bonding(δ_H), due to the coefficient "4" of dispersion component in Eq.1.

In Figures 4 and 5, it seems that the methyl oleate is inside of polystyrene (PS) and polypropylene(PP) spheres, methanol and oleate acid are outside of them. From *RED* value in Table 3, the combination of PS and methyl oleate had a *RED* value less than 1.0. The PP and polyethylene (PE) combinations with methyl oleate had *RED* values slightly higher than 1.0. This indicates that methyl oleate is significantly outside the PE and PP spheres. The *RED* value of methanol and oleic acid are significantly greater than 1.0. Therefore, only polystyrene can be considered to be sufficiently dissolved in methyl oleate. The results also show that FAME will selectively dissolve the polystyrene in the form of mixed waste plastic.

Polymer	Solvent		
	Methyl Oleate	Oleic Acid	Methanol
PS	0.93	1.17	4.20
PE	1.08	1.23	7.12
PP	1.05	1.15	40.6

Table 3. RED numbers computed by HSPiP.

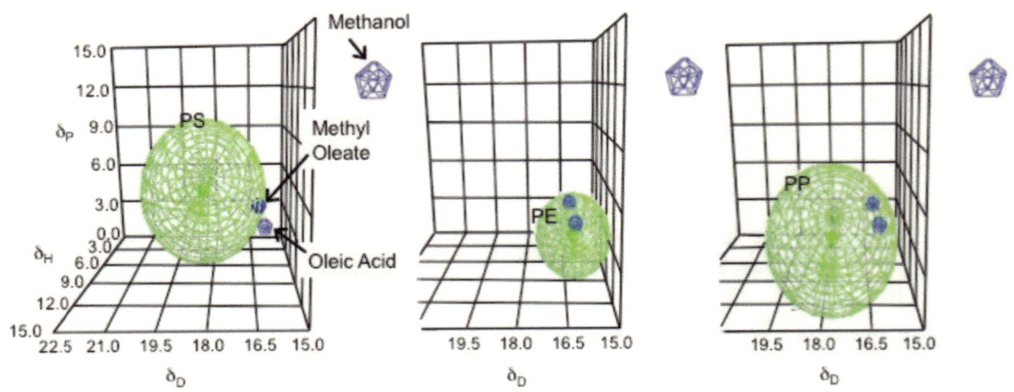

Figure 4. plot of the HSP sphere for polymers and solvents.

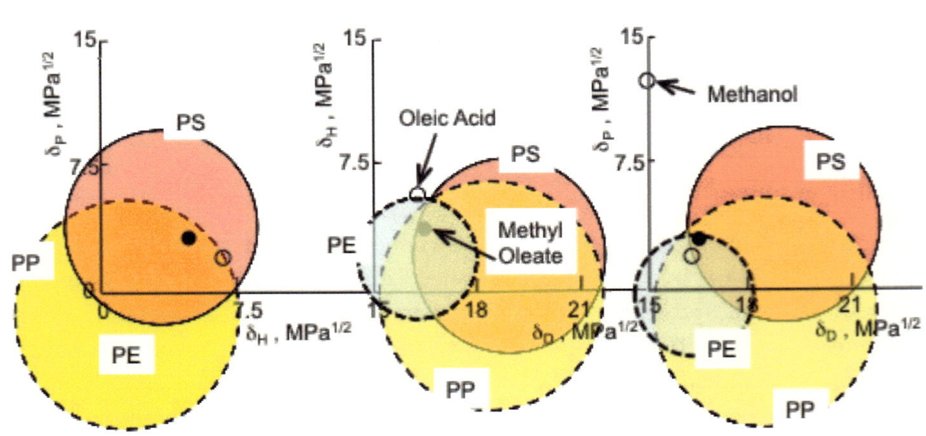

Figure 5. plot of the HSP for polymers and solvents.

3.2. Characteristics of FAME-dissolved polystyrene

To examine solubility, both the food trays (polystyrene paper; PSP) and the expanded polystyrene(EPS) were completely dissolved in FAME, which was completely transparent

http://dx.doi.org/10.5772/59112

Figure 6. Photos of FAME dissolved polystyrene.

after the dissolution, as shown in Figure 6. Figure 7 shows the relation between kinematic viscosity and mass concentration of PS dissolved in FAME. The figure also shows results obtained by another researcher [8] and the dissolved polystyrene standards (average molecular weight of 4000 and 50,000). Kinematic viscosity increased with increasing PSP and EPS concentrations. For EPS, kinematic viscosity increased exponentially, revealing a high kinematic viscosity such as that of heavy oil at a concentration of 9%(m/m) EPS in FAME. In contrast, the kinematic viscosity of the PS standard in FAME was less than those of EPS and PSP. In addition, the kinematic viscosity increased with an increase in average molecular weight, suggesting that an increase in kinematic viscosity is related to the degree of polymerization and molecular weight of the dissolved polystyrene.

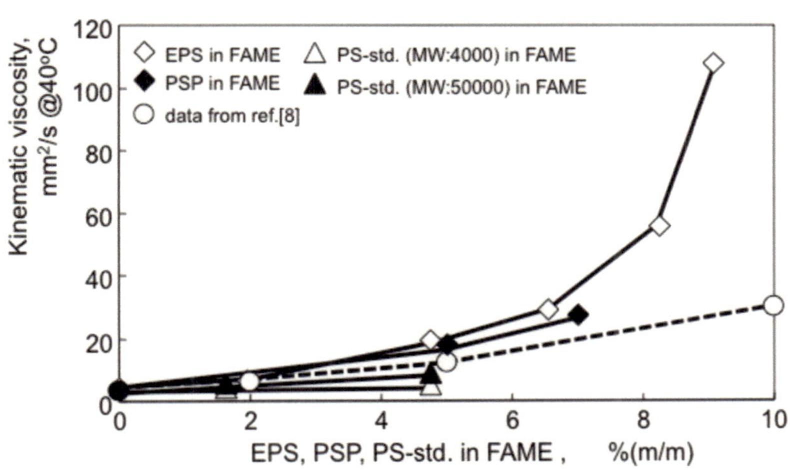

Figure 7. Relationship between kinematic viscosity and mass concentration of the polystyrene dissolved in FAME.

The molecular weight distribution of polystyrene in FAME was measured by GPC. Figure 8 shows the experimental results for the polystyrene (PS) standard with an average molecular weight of 50,000 dissolved in FAME and 5%(m/m) EPS in FAME. The molecular weight peak was similar to that of the polystyrene standard; however, the EPS molecular weights in FAME were distributed across a wide range compared to the range of the PS standard, and indicated a compound with a molecular weight greater than 10^7. These results suggest that the existence of a very large polymer causes an increase in kinematic viscosity.

Next, the rate of volume reduction of polystyrene and PSP caused by dissolution in FAME was investigated. The solvent *n*-hexane—a non-polar solvent—was added to crystallize the PS. The crystallized PS was filtered and the mass was measured to determine specific volume. The specific volume of PS in FAME was 1.38×10^{-3} m^3/kg, indicating that the specific volume of 82.6×10^{-3} m^3/kg for PS before dissolution was reduced. Therefore, dissolution of PSP in FAME can reduce the volume of waste plastic. Also, in case of EPS and XPS which show high expansion ratio, the volume will be extremely reduced by dissolution in FAME.

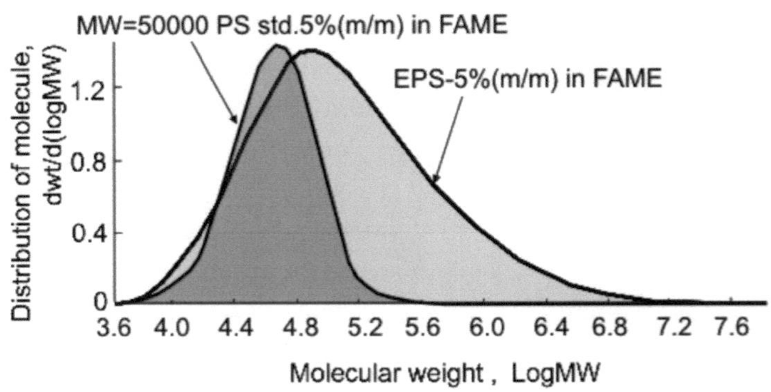

Figure 8. Distribution of polystyrene molecular weight in FAME.

3.3. Fuel characteristics

Figure 9 shows a series of combustion pressures obtained by the FIA fuel ignitability tester. The cycle-to-cycle fluctuation in combustion pressure for EPS dissolved in FAME is less than that for neat FAME.

Figure 10 shows the changes in ignition delay and cetane number (CN_{FIA}) against the dissolved EPS concentration. The ignition delay gradually increased with increasing EPS concentration. For this reason, the CN_{FIA} value also decreases. This is caused by the increased kinematic viscosity and the suppression of fuel spray atomization upon dissolution of EPS. Furthermore, generally the petroleum-based fuels are known to possess poor ignitability at higher concentrations of aromatic hydrocarbons. The raw material for styrene monomers is represented by the aromatic chemical formula of C_8H_8. For this reason, the ignition delay gradually increases with PS concentration.

http://dx.doi.org/10.5772/59112

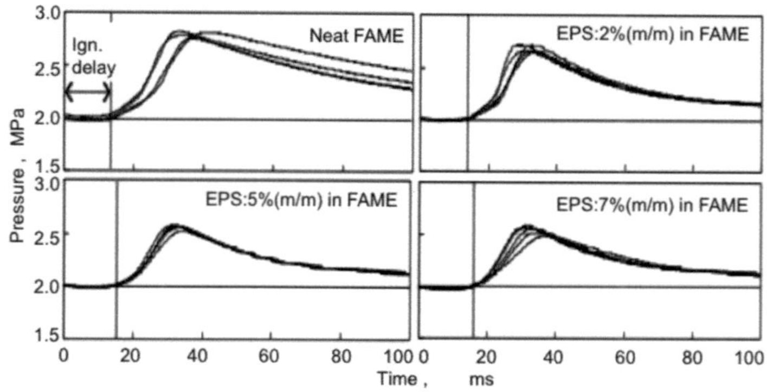

Figure 9. Courses of combustion pressure obtained by FIA fuel ignitability tester.

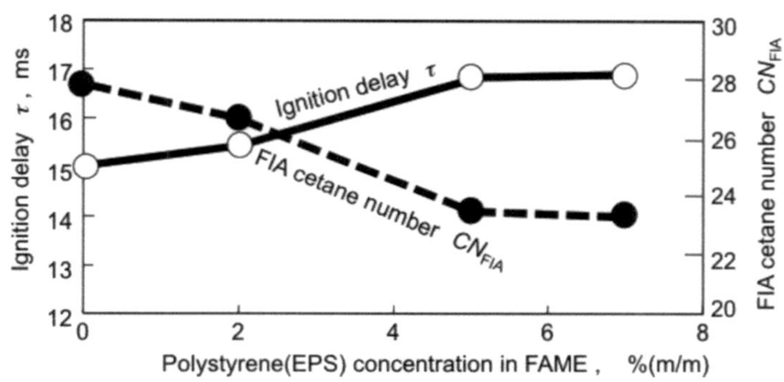

Figure 10. Changes of ignition delay and cetane number (CN_{FIA}) against the dissolved polystyrene (EPS) concentration.

Figure 11 shows the relation between calorific value and EPS concentration. Initially, the dissolution of EPS in FAME was expected to result in a sufficiently high heating value. However, the increase at a lower heating value was greater than that at the higher heating value, because the oxygen content in the original FAME of approximately 10%(m/m) is reduced by dissolution of EPS.

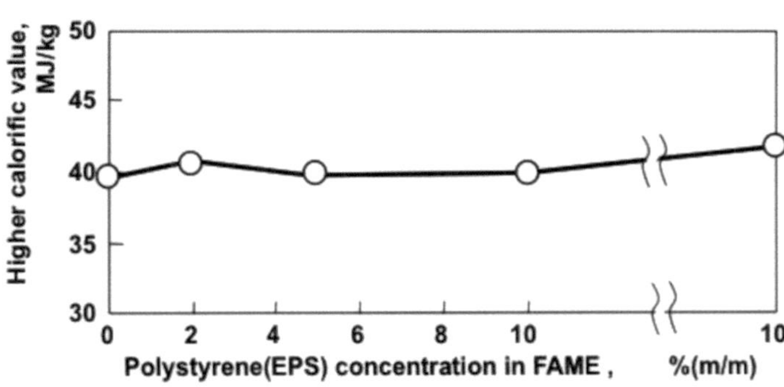

Figure 11. Effect of polystyrene (EPS) concentration in FAME on higher calorific value.

Properties	FAME(SME)	EPS-2%(m/m) in FAME(SME)	EPS-5%(m/m) in FAME(SME)
Density kg/L@15°C	0.886	0.888	0.894
Kinematic viscosity mm^2/s@40°C	4.16	6.46	19.5
Higher heating value MJ/kg	39.6	40.8	39.8
Cetane Number CN_{FIA}	28	27	23.5
100%-Carbon residue %(m/m)	0.012	0.042	0.078

Table 4. Comparison of density, kinematic viscosity, higher heating value, cetane number(CN_{FIA}), carbon residue.

Table 4 shows the density, kinematic viscosity, higher heating value, cetane number (CN_{FIA}), and carbon residue without sample condensation. The carbon residue at 5%(m/m) dissolved EPS was twice that at 2%(m/m) dissolved EPS. When this was used as diesel fuel, the amount of carbon deposit in the combustion chamber increased. This deposit may affect the fuel injection system and fuel spray atomization.

Figures 12 and 13 show the effect of fatty acid components of FAME on kinematic viscosity and carbon residue. As shown in Figure 12, the kinematic viscosity increases with an increase in concentration of EPS, and the fatty acid methyl ester with higher carbon number shows higher kinematic viscosity. Also, the carbon residue at methyl oleate shows higher value than methyl palmitate and methyl laurate at all concentrations of EPS. From the results, fuel properties of FAME dissolved EPS may be improved by changing fatty acid composition in FAME. In other words, to use FAME dissolve EPS as fuel, FAMEs with short and middle length of carbon chains as fatty acid component will be better than fatty acid methyl ester with the long length chains.

3.4. Diesel engine performance

This section describes the engine performance and problem of diesel generator fuelled with FAME dissolved EPS. The experiment was carried out by using small diesel engine generator. Fuel consumption was measured by a burette-installed fuel line at various engine loads. And the brake thermal efficiency was calculated. Table 5 shows the main specification of engine generator used in experiment. Figure 14 shows the result obtained by engine test. From Figure 14, the thermal efficiency at the FAME dissolved EPS-5% shows higher value than that at neat FAME at rated engine output. This might be caused by lower cycle-to-cycle fluctuation in combustion as shown in Figure 9 in case of EPS-5%.

In general, the lubricating oil in the diesel engine was mixed with some injected fuel, a process called "oil dilution by fuel." Then, a mixing test that involved mixing of regular diesel lubricant hydrocarbons and FAME-dissolved EPS was conducted. Figure 15 shows the photos of mixture with regular diesel lubricant hydrocarbon and with lubricant oil derived from castor oil. These photos show that, immediately after mixing, the dissolved EPS crystallizes in the gel and precipitates for both cases. Thus, the use of FAME- dissolved polystyrene as diesel fuel requires the prevention of precipitation and deposition of polystyrene.

	Diesel engine generator
Type	Yanmar L40A Direct-injection Single cylinder Air-cooled
Bore x Stroke	68 mm x 55 mm
Displacement	199 cm^3
Compression ratio	20
Injection nozzle	YDLL-P type 4holes-0.22mm 150 degree of spray angle
Rated engine output	2.4kW/3,600rpm

Table 5. Main specification of diesel generator.

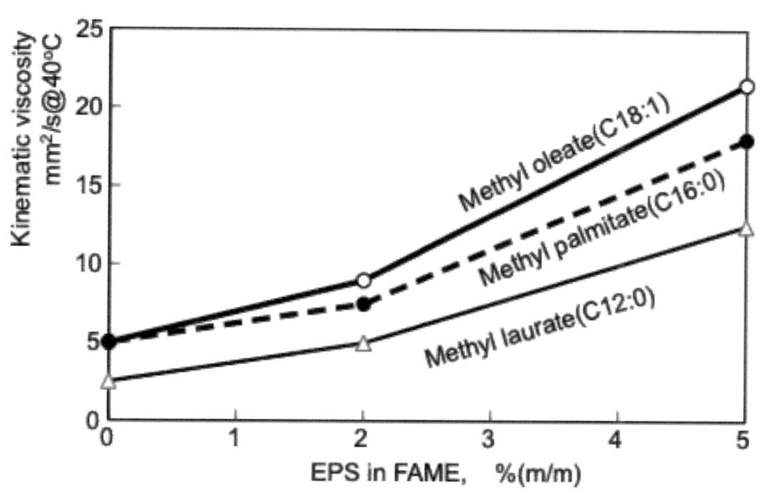

Figure 12. Effect of polystyrene(EPS) concentration in FAME on kinematic viscosity.

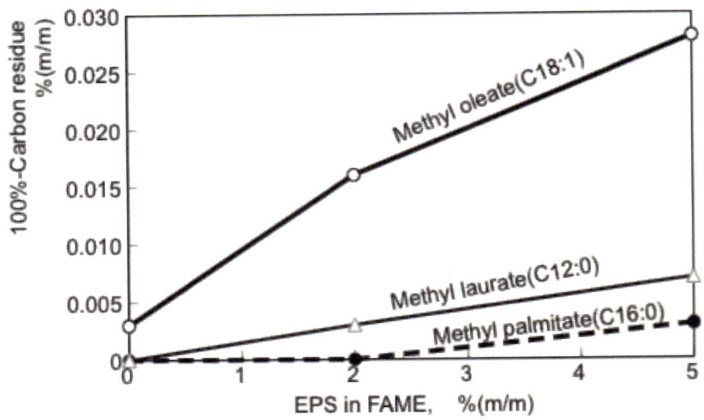

Figure 13. Effect of polystyrene(EPS) concentration in FAME on carbon residue.

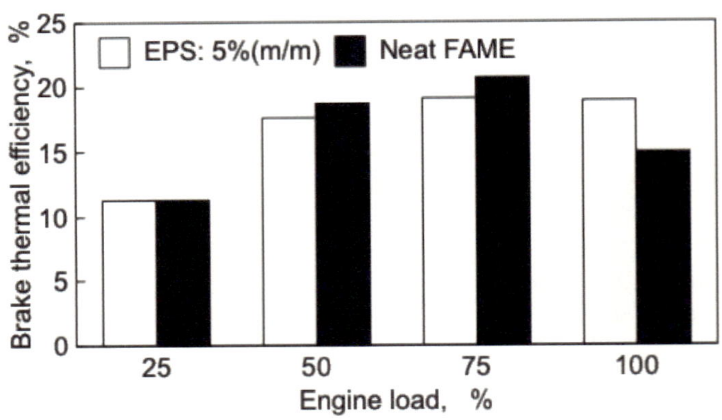

Figure 14. Brake thermal efficiency vs. engine load of diesel engine generator.

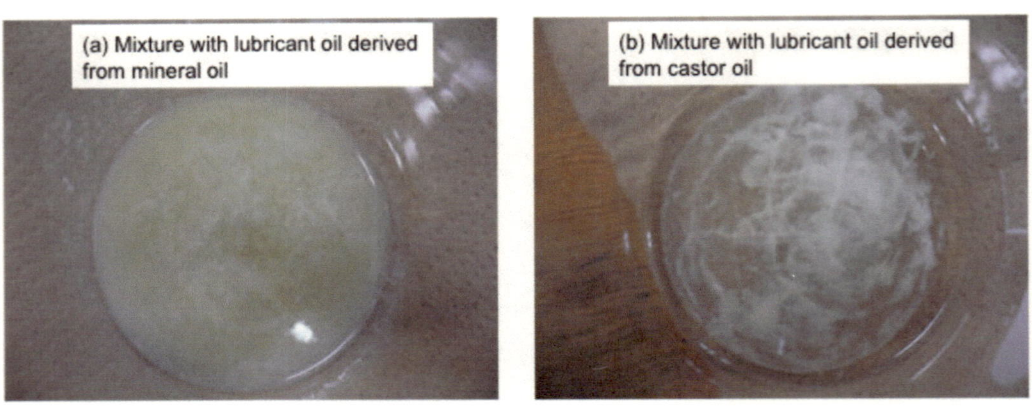

Figure 15. Photo of mixture of FAME-dissolved EPS and engine lubricant oil.

4. Conclusions

1. The potential for dissolution of polystyrene in fatty acid methyl esters (FAME) was estimated using Hansen solubility parameters. Results indicated that FAME provides selective solubility for polystyrene.
2. Results of experiments that examined solubility properties revealed that the kinematic viscosity increased with polystyrene (EPS) concentration in FAME. The cause of the increase in kinematic viscosity was due to the presence of a polymer with a molecular weight of up to approximately 10^7.
3. The volume of polystyrene (PSP) can be reduced by approximately 1/60 by dissolution in FAME.
4. When examining the use of dissolved EPS as fuel for diesel engines, experiments showed that cetane number decreased with an increase in EPS concentration; however, the cycle-to-cycle fluctuation in combustion pressure and ignition timing when using EPS dissolved in FAME were less than those found for neat FAME.
5. EPS dissolved in FAME crystallized and precipitated in the gel upon addition of hydrocarbons such as engine lubricant oil. Therefore, the use of FAME-dissolved polystyrene as diesel fuel requires attention to tribology.

Acknowledgements

We would like to express our appreciation to Mr. Ryuta Onishi, graduate student from the University of Shiga Prefecture, for his assistance with the experiments.

Author details

Koji Yamane* and Kiyoshi Kawasaki

Department of Mechanical Systems Engineering, The University of Shiga Prefecture, Hassaka Hikone Shiga, Japan

References

[1] Knothe G, Krahl J, Gerpen VJ. The biodiesel handbook, 2nd ed. Urbana IL: AOCS Publishing;2010.

[2] Yamane K, Kawasaki K. Potential of existing emission control technologies for diesel engines fuelled with biodiesel. FISITA Trans.2006;F2006P015T:1-12.

[3] Yamane K, Kawasaki K, Sone K, Hara T, Prakoso T. Oxidation stability of biodiesel and its effects on diesel combustion and emission characteristics. Int. J. Engine Research. 2007;8-3:307-19.

[4] Powers PO. Solubility of polystyrene. Ind. Eng. Chem. 1949;41-10:2213-17.

[5] Monte WC, Landau-West D. Solubility of polystyrene in certain vegetable oils, essential oils and their constituents. Journal of Food Science.1982;47-6:1832-35.

[6] Garcia MT, Gracia I, Duque G, de-Lucas A, Rodriguez JF. Study of the solubility and stability of polystyrene wastes in a dissolution recycling process. Waste Management. 2009;29-6:1814-18.

[7] Garcia MT, Duque G, Gracia I, de-Lucas A, Rodriguez JF. Recycling extruded polystyrene by dissolution with suitable solvents. J. Material Cycles and Waste Management. 2009;11-1:2-5.

[8] Kuzhiyil N, Kong SC. Energy recovery from waste plastics by using blends of biodiesel and polystyrene in diesel engines. Energy & Fuels. 2009;23:3246-53.

[9] Golias MR. The use of soy methyl ester-polystyrene sealants and internal curing to enhance concrete durability. Purdue University. e-Pubs. 2010;AAT 1490651:138.

[10] Coates KC, Mohtar S, Tao B, Weiss WJ. Can soy methyl esters reduce fluid transport and improve durability of concrete? Transportation Research Board. 2009;2113:22-30.

[11] Kapila S, Flanigen VJ, Maples MF, Mills MW. Fatty acid esters to dissolved polystyrene for production of commodity material useful for production of polymers. US Patent, 8 Jan 2002;6337413B1.

[12] Belmares M, Blanco M, Goddard-III WA, Ross RB, Caldwell G, Chou SH, Pham J, Olofson PM, Thomas C. Hildebrand and Hansen solubility parameters from Molecular Dynamics with applications to electronic nose polymer sensors. J Comp Chem. 2004;25-15:1814-26.

[13] Koenhen DM, Smolders CA. The determination of solubility parameters of solvents and polymers by means of correlations with other physical quantities. J. App. Polymer Sci. 1975;19-4:1163-79.

[14] Archer WL. Hansen solubility parameters for selected cellulose ether derivatives and their use in the pharmaceutical industry. Drag Dev Ind Pharm. 1992;18-5:599-616.

[15] Hildebrand J, Scott RL. The solubility of nonelectrolytes, 3rd ed. Dover Pub Inc, NY; 1964.

[16] Hansen CM. Hansen solubility parameters. A Users Handbook. 2nd ed. Boca Raton FL: CRC-Press;2007.

[17] Abott S, Hansen CM, Yamamoto H. Hansen solubility parameters in practice(HSPiP)-complete software, data and examples, 3rd ed. v.3.1.20. Dec.2010.

[18] Kawasaki K, Yamane K, Ikawa T. Regression equation for predicting the cetane number of biodiesel fuel based on fuel composition and properties. SAE paper series. 2011;2011-01-1941:1234-39.

Section 4

Quality and Exploitation

Chapter 11

Impact of the Biofuels Burning on Particle Emissions from the Vehicular Exhaust

Lílian Lefol Nani Guarieiro and Aline Lefol Nani Guarieiro

Additional information is available at the end of the chapter

http://dx.doi.org/10.5772/60110

1. Introduction

To help address continuing environmental and public health degradation, most governments adhere to international energy policies with regard to the use of biofuels. The increased environmental concerns, depletion of petroleum, public health and socioeconomic challenges have driven research to develop alternative fuels from renewable resources that are cheaper and environmentally acceptable.

In the last years, the use of alternative fuels has been shown to reduce emissions without negative effects on engine performance. There are also some studies that demonstrate an increase in particle emission with the use of biofuels. However, the literature reviewed here found noticeable decrease in particle emissions and smoke opacity with the use of biofuels. Indeed, many studies have been conducted to characterize and better understand biofuel proprieties related to particle emission.

The particles emitted by combustion of vehicle engines can be distributed in different ranges of size and number of particles. The particle size distributed in the range covering the nano-particles (particles smaller or equal to 100 nm) has a greater toxic potential compared with a larger particle. Studies suggest that exposure to nanoparticles causes serious damage to health, such as lung inflammation, asthma, chronic obstruction of arteries and lungs, cell death, obstruction and accumulation in the olfactory bulb, access to brain damage, tumor necrosis, oxidative stress, neural effects, heart problems, and even death.

Particles emitted from combustion of vehicle engines can also be of different chemical and toxicological proprieties. The chemical composition of particle emission is directly related to the burned fuel and vehicular exhaust condition, and may affect its toxicity. Studies suggest that the higher the organic carbon composition is, the higher damage of particle emitted will be to health.

Conscious about this subject, there is now considerable research and investment on the kind of fuel and biofuel to develop and promote fuels able to minimize the toxicity and particle emissions from vehicular exhausts. Thus, the viability of using biofuels, its impact on public health and environment, and the impact of biodiesel burning on particle emissions from vehicular exhaust, can be better understood.

2. Origin and characterization of particles

The particles present in the atmosphere have diverse origins from multiples sources. Particles consist of a conglomerate of solid particles with variable sizes and physical-chemical proprieties presenting a toxicity level dependent on its size and chemical composition [1]. According EPA, particle or particle matter (PM) is a complex mixture of extremely small particles and liquid droplets. Particle pollution is made up of a number of components, including acids (such as nitrates and sulfates), organic chemicals, metals, and soil or dust particles [2].

There are many sources of PM. An air pollutant can originate from natural processes, like forest fires and wind erosion, and from human activities, like agricultural practices, smokestacks, vehicular emissions, and construction. Examples include dust, dirt, soot, soil, and smoke [2]. In this chapter, we will focus on vehicular sources of particles and the impacts of biofuel burning in this concentration and size distribution.

The particle size is an important propriety and is related to its inhalable potential causing the particle to get into the human respiratory system and then causing health problems [3]. Furthermore, particle size is one of the most important parameters in determining the atmospheric lifetime of particles, which is a key consideration in assessing health effect information because of its relationship to exposure. The US Environmental Protection Agency (EPA) have concerned about particles that are 10 micrometers in diameter or smaller because those are the particles that generally pass through the throat and nose and enter the lungs. Once inhaled, these particles can affect the heart and lungs and cause serious health problems. The EPA groups particle pollution into categories, these categories are based on studies that show a relationship between adverse health effects and the concentration of fine particles in the atmosphere:

"Inhalable coarse particles," such as those found near roadways and dusty industries, are larger than 2.5 micrometers and smaller than 10 micrometers in diameter (PM_{10}). "Fine particles," such as those found in smoke and haze, are 2.5 micrometers in diameter and smaller ($PM_{2.5}$). These particles can be directly emitted from sources such as forest fires, or they can form when gases emitted from power plants, industries and automobiles react in the air. The "Nanoparticles," such as those found close to combustion systems, are 100 nanometers in diameter and smaller. These particles can be directly emitted from sources such as vehicle combustions or any combustion source. Nanoparticles have been hardly studied by the scientific community to better understand its role and relationship with human health. Nanoparticles are also known as ultrafine particles.

The multi-modal distribution of particles based on diameter has long been recognized since the 1980. Notwithstanding, particles display a consistent multi-modal distribution over several physical metrics such as volume and mass; specific distributions may vary over place, conditions, and time because of different sources, atmospheric conditions, and topography [4]. Based on particle size and formation mechanism, particles can be classified into three fundamental modes: nuclei, fine, and coarse modes and particles can be observed in an idealized mass distribution of these modes (Figure 1). High temperature vapor, coagulation and condensation processes, aggregate formation, mechanical processes, precipitation washout and sedimentation process could contribute to the particle formation as seen on this figure.

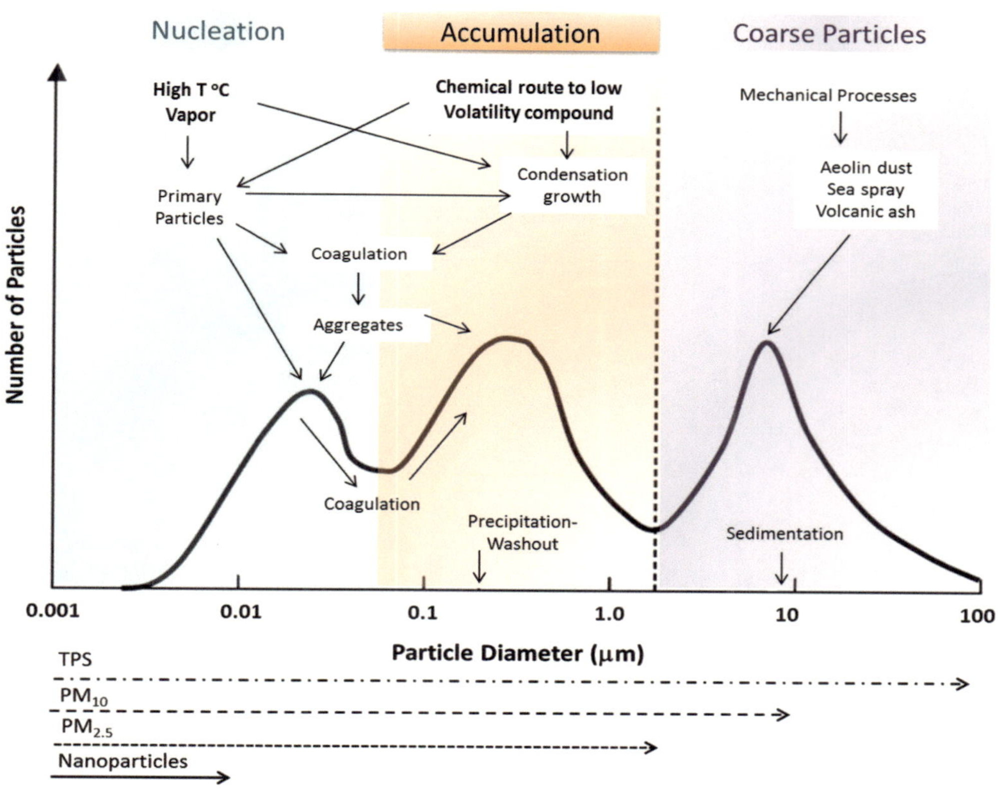

Figure 1. Particle size distribution, formation process, and multi-modal distribution [Adapted from EPA].

Nuclei-mode particles range in diameter from about 5 to 50 nm. They usually consist of particles formed from volatile precursors as exhaust mixes with air during dilution and cooling process, it consists of metallic compounds, elemental carbon, and semi-volatile organic and sulfur compounds. The accumulation mode ranges in size from roughly 30 to 500 nm. They consist mainly of carbonaceous agglomerates that have survived the combustion process, most of the mass, composed primarily of carbonaceous agglomerates and adsorbed materials. The coarse mode consists of particles larger than about 1 μm. These relatively large particles are formed by natural material and re-entrainment of particulate matter, which has been deposited on cylinder and exhaust system surfaces. Also shown in Figure 1 are size range definitions for atmospheric particles: coarse particle PM_{10} (diameter < 10 μm), fine particles $PM_{2.5}$ (diameter < 2.5 μm), and nanoparticles (diameter < 100 nm) [5-7].

The profile for the three modes can change with the characteristics of the emitting source. Figure 2 shows a typical diesel particle matter size distribution weighted by number, surface area, and particle mass; it also shows the alveolar deposition curve [8,9]. As particles increase in size, the deposition efficiency decreases. The most difficult thing about the measurement of engine exhaust size distributions is that most of the nanoparticles emitted by current engines are not formed in the engine itself, but instead are formed from gas phase precursors as the exhaust dilutes and cools. This gas-to-particle conversion process involves homogeneous nucleation, adsorption, and absorption, and is highly nonlinear.

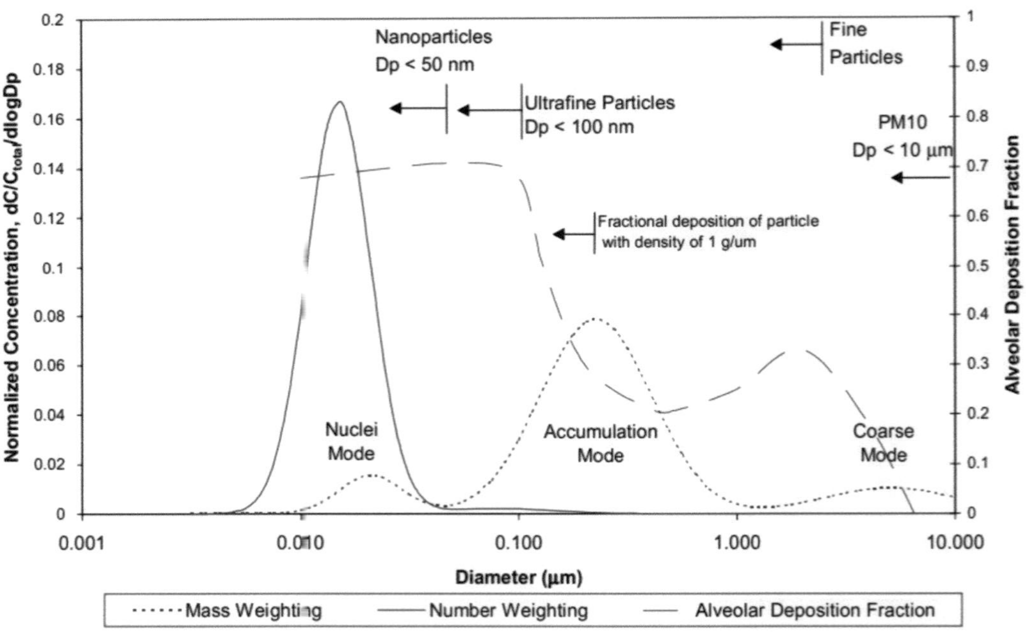

Figure 2. Typical diesel particle matter size distribution weighted by number, surface area, and particle mass [5].

The nuclei mode typically contains 1%-20% of the diesel particle matter mass and more than 90% of the particle number, and the coarse mode contains 5%-20% of the particle mass [5].

Nevertheless, particles are a complex, heterogeneous mixture that changes in time and space. It encompasses many different chemical components and physical characteristics, many of which have been cited as potential contributors to toxicity. Each component has multiple sources, and each source generates multiple components. Identifying and quantifying the influences of specific components or source-related mixtures on measures of health-related impacts, especially when particles interact with other co-pollutants, there-

http://dx.doi.org/10.5772/60110

fore represents one of the most challenging areas of environmental health research. Current knowledge does not allow precise quantification or definitive ranking of the health effects of PM emissions from different sources or of individual PM components and indeed, associations may be the result of multiple components acting on different physiological mechanisms [10]. In this universe, the impact of biofuel burning on particle emissions from vehicular exhaust is just as challenging.

However, chemical composition is an important propriety, generally, particles constitute biological materials, organic compounds, hydrocarbons, acid, metals adsorbed or attached on its carbonaceous structures. Chemical composition is directly related to the emitting source. Particles are compounded by a carbonaceous nuclei and a huge number of substances adsorbed on its surface, such as organic compounds (OC) – polycyclic aromatic hydrocarbons (PAH), PAH-derivatives (quinones, semi-quinones, nitro-PAH, carboxi-PAH) and inorganic compounds – metals, ions, inorganic acids, salts, among others [generated primarily by mechanical processes]. Much of the organic compounds are formed by complex secondary processes, through n-alkanes and hydroxyl radicals (OH ●) in the presence of NO_x (Figure 3). On these processes, the type of product formed depends on the conditions of the combustion process that gave rise to particle and atmospheric conditions [11,12].

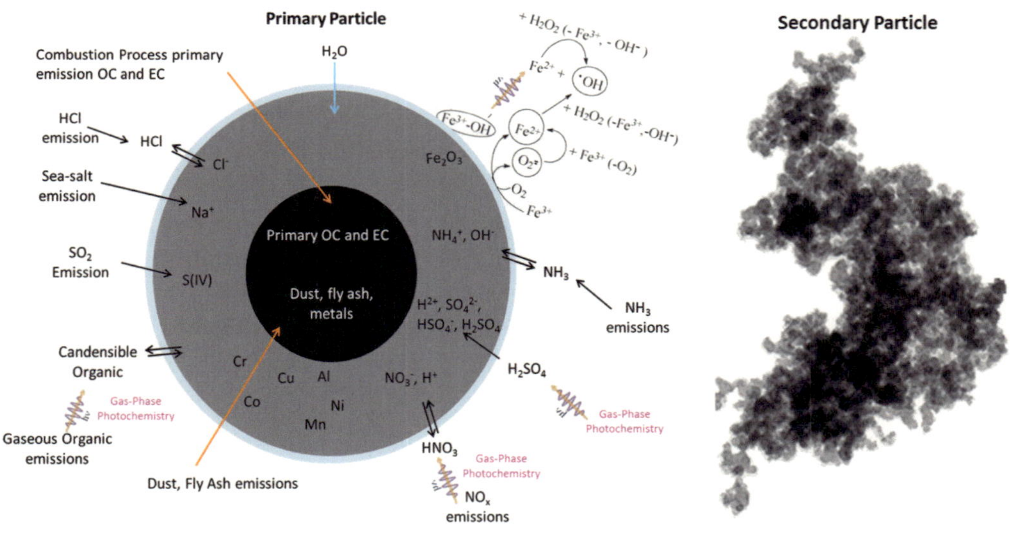

Figure 3. Primary and secondary particles, and chemical reactions and processes associated with the chemical composition.

Li et al. [13] demonstrated that in general, coarse particles of metal have a great contribution because fine metal particles and OC, and nanoparticles generally have OCs and PAHs. Due to the small size of the nanoparticles and large surface area of these particles, they may carry metals and a large number of organic compounds, which when inhaled can be absorbed into the respiratory tract. Many of these compounds are capable of generating reactive oxygen species (ROS) that promote toxicity cells [13].

Additionally, a study by Claxton et al. 2004 [14] reviewed the different classes of particle matter, including non-metallic organic, sulfur, and halogenated hydrocarbons, oxygenates, and nitrates. For hydrocarbons derived from combustion processes, there are various carcinogenic PAHs, such as benzo (a) anthracene, benzo (k) fluorene, Benzo (a) pyrene, benzo (b) fluoranthene, indeno (1,2,3-cd), pyrene, and dibenzo (ah) anthracene. Furthermore, many PAH are directly mutagenic as mono- and dinitro-HPA 1-nitropyrene and 3-nitrofluoranteno. Recent research has shown that quinones play a critical role in catalyzing the generation of ROS that promotes toxic effects on the human body [15,16]. Similarly, Kong et al. (2011) [17] demonstrated the ability of metals present in the MP as cobalt (Co), copper (Cu), iron (Fe), manganese (Mn), Nickel (Ni), vanadium (V) and titanium (Ti), to contribute to the increase of particle toxicity.

Thus, the purpose of this chapter is to describe the impact of biofuels on emissions of all particle size from vehicular exhaust. The particle emission profiles originating from both diesel and Otto cycle engines and the impact of the use of biofuels will be characterized.

3. Particle emissions from exhaust vehicles

The total particulate emission concentration from light-duty diesel engines is much smaller than that from heavy-duty diesel engines. In general, newer heavy-duty trucks emit diesel particulates at a rate 20 times that of catalyst-equipped gasoline-fueled vehicles [18]. The particle size distribution and chemical composition can vary greatly depending on the engine type, engine speed and load, and composition of fuel oil and lubricating emission control technology [19]. In addition, the reduction or increase in the emission of particles can be influenced by some factors as described below:

- The operating mode of the engine: operating in stratified condition, the total mass and the number of particulates are more than 20 times greater than in homogeneous operating conditions (air and fuel)
- The higher the engine speed, the shorter the time of vaporization of the fuel, and higher load regimes require a greater mass of fuel injected into the combustion chamber, which reduces the temperature within the chamber, thereby limiting the vaporization of fuel and generating larger amount of particles
- The type of fuel used in the engine.

3.1. Profile of particle emissions from diesel engines

Engines that use diesel as fuel have many applications, mainly due to its higher thermal efficiency and fuel economy. In general, the diesel emissions consist of a nonpolar fraction, a moderately polar fraction, and a polar fraction [20,21]; the remainder is unrecoverable (Figure 4).

http://dx.doi.org/10.5772/60110

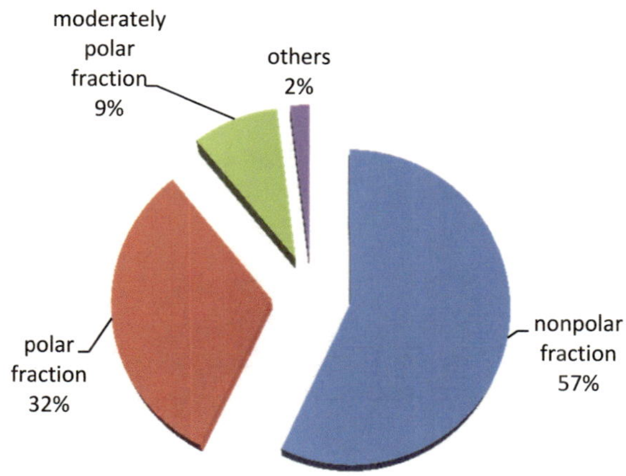

Figure 4. Diesel emission composition.

Particulate emission from diesel engines is receiving a great deal of attention due to its probable carcinogenic property. In the exhaust pipe of a diesel engine, the change of the exhaust gas temperature can result in nucleation and condensation of volatile materials and coagulation of particulates. These particles emitted from diesel engines are composed mainly of aggregates of spherical carbon particles coated with organic and inorganic substances, with the composition of the particles being predominantly 80%–90% organic and inorganic carbon (Figure 3). However, the particle composition may dramatically change depending on the engine type, engine speed and load, lubricating oil type, emission control technology, and fuel composition [19]. For this reason, it is not trivial to compare data from studies on carrier particles emitted in the exhaust using different parameters. Thus, it is necessary to study each parameter individually and setting the others to evaluate their effect on the concentration and distribution of particle size.

Sharma et al. (2005) [22] studied particle composition changing the engine load (Figure 5) and they observed the influence of the difference of engine load in the particle composition. The exhaust particulates from Mahindra direct injection transportation diesel engine (40 hp) were collected at four different engine operating conditions, namely idle, 40%, 70%, and full load. Figure 5 shows the diesel particle composition at 70% load in the study compared with the composition at 100% engine load. The broad composition of the particulates remains the same with the load and also when compared with the study of Volkswagen (1989). However, a closer examination suggested that the composition may dramatically change between OC and EC with a change in engine load. The authors observed that as the load increased from full load, the metal content in particulates, benzene soluble fraction (a marker for carcinogenicity), and OC gradually decreased. The trend for EC was quite the opposite, it increased with an increase in load.

The inorganic fraction of the particulate emissions consists primarily of small EC particles, ranging from 0.01 to 0.08 μm in diameter. Organic and elemental carbon account for approximately 80% of the total particulate matter mass [23].

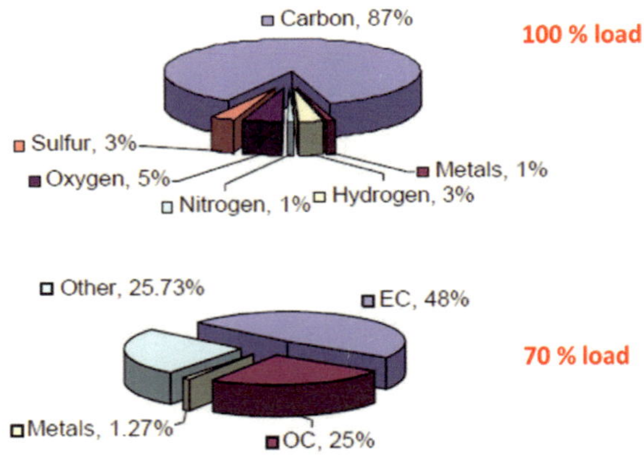

Figure 5. The influence of engine load in the particle composition [22].

In recent years, the emission of particles from vehicle exhaust is a phenomenon that has been much discussed because these are harmful to our health and the environment. Thus, in many countries, scientific research results were the basis for more restrictive legislation being implemented on emission of particulate matter from vehicle exhaust. This evolutionary process in search of a better quality of life for society by reducing the maximum allowable concentration of particulate emissions in vehicle exhaust, conditioned and demanded that the automotive industry and fuel producers seek innovative technologies to comply with the regulations. One of the ways that has been widely used to reduce emissions of some air pollutants, especially the emission of particulate matter are after-treatment devices such as urea-based selective catalytic reduction (SCR), diesel particulate filters (DPF), and diesel oxidation catalysts (DOC).

Nevertheless, the simultaneous reduction of particles and nitrogen oxides (NO_x) is a big challenge, because the strategies for reducing one component may lead to an increase in another. For this reason, a variety of exhaust after-treatment devices is essential. For NO_x reduction, SCR is commonly used in on- and off-road engines [24,25]. However, DPF and DOC have also become more standard in off-road engines and are already common in motor vehicles [26]. DPF significantly lowers the particle mass emissions, but its effect on particle number is two-sided. The mass is dominated by the soot accumulation mode, which is efficiently trapped in DPF, but the particulate number can be dominated by nuclei mode particles formed down-stream of the DPF [25], although the DPF seems to be capable of also removing ultrafine particles and nanoparticles effectively from the engine exhaust [27]. After-treatment of exhaust gas does not just lower emissions, but it also alters the chemical composition of vehicle exhaust [25].

Other technologies such as fuel injection pressure (FIP), the start of injection (SOI), and the application of exhaust gas recirculation (EGR) can also affect the particle emission profile. Li et al. (2014) [28] examined the effect of these technologies (FID, SOI, and EGR) on particle number size distributions (PNDs) and OC and EC emissions from a common rail diesel engine. In general, it was observed that increasing FIP and advancing SOI can improve combustion, soot and accumulation mode particle (AM) emissions decrease with increasing FIP and advancing SOI, the application of EGR increases soot and AM emissions, and soot-EC emission increases with the application of EGR at high load (Figure 6).

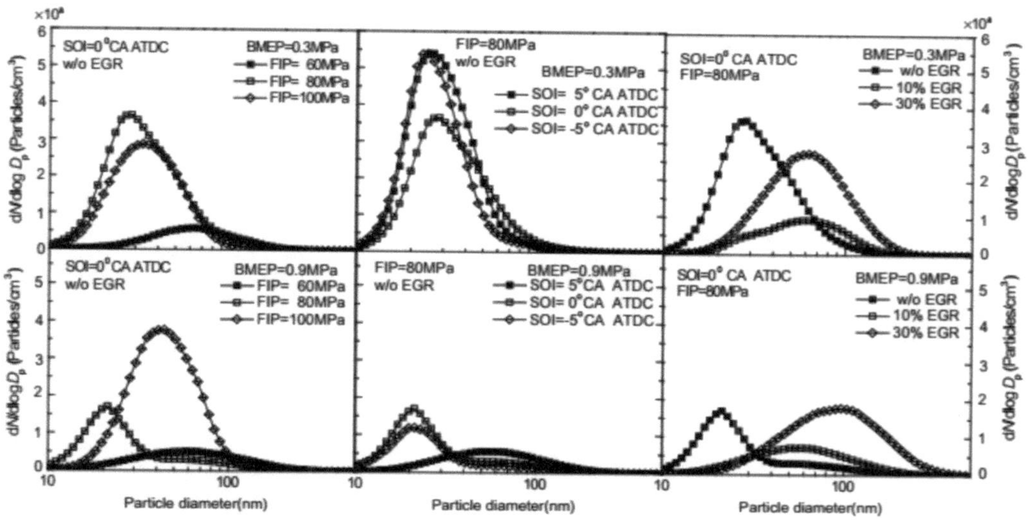

Figure 6. Effects of FIP, SOI, and EGR on particle number distributions [28].

In 2013, Agarwal et al. [29] developed an electrically heated diesel vaporizer to study the effect of use of different relative air fuel ratios and EGR levels on particle emission. They compared the emissions from conventional CI engines with an advanced combustion technology named as homogeneous charge compression ignition (HCCI). Figures 7a and 7b,show the results obtained for both mass and particle number concentrations. PM emissions were simultaneously reduced in HCCI combustion mode. However, particulate emissions from the HCCI engine largely depend on the EGR rate and relative air-fuel ratio. When the air-fuel mixture becomes leaner (increasing l), the PM mass emission decreases from diesel HCCI engine. With increasing EGR, the PM mass emission increases. The particle number concentration tends to increase also with an increase in EGR rate. Most of the diesel HCCI exhaust particles were ultrafine particles.

Ninga et al. (2004) [30] experimentally investigated the transformation of diesel particulates within the exhaust pipe when the exhaust gas is being cooled. The results showed that the transformation of the diesel particulates in the exhaust pipe depended mainly on the level of cooling, the concentration of the volatile materials, the initial concentration of the particulates in the exhaust, and the residence time of the exhaust gas within the exhaust pipe. The mass concentration and the soluble organic fraction of the particulates increased, while the gaseous hydrocarbon concentration decreased upon cooling the exhaust.

At high load conditions, although there is less volatile material in the exhaust, the original particulates in the exhaust can promote the condensation of the volatile materials and the coagulation between particulates upon cooling, so the particulate mass may also increase even

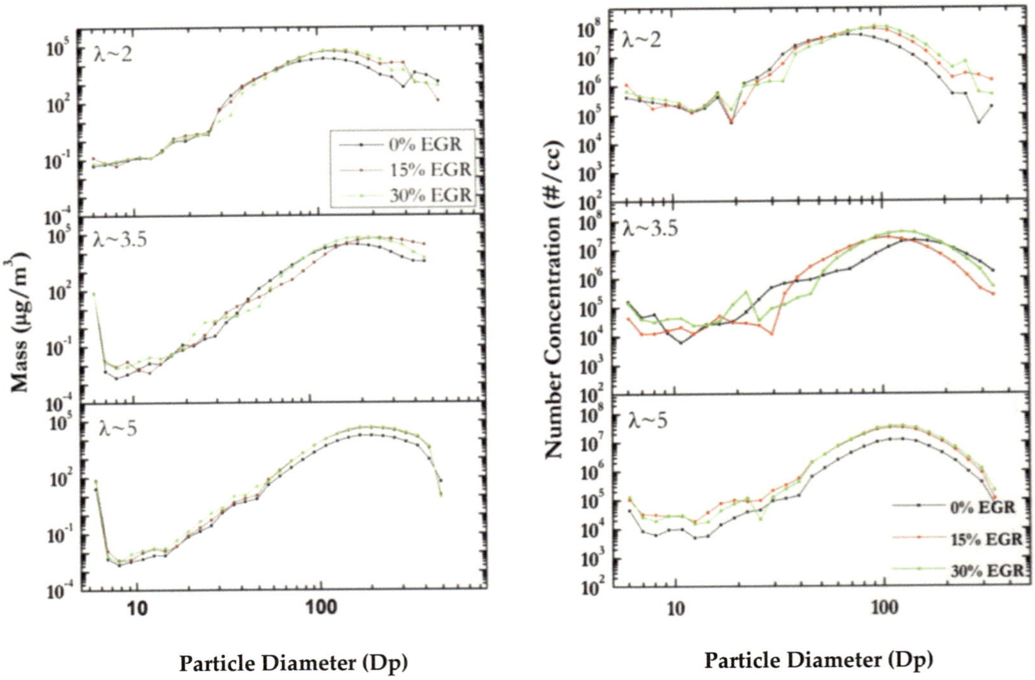

Figure 7. a) Particle mass and size distributions for various air-fuel ratios and EGR rates in diesel HCCI engine. (b) Number concentration of particles for various air-fuel ratios and EGR rate in diesel HCCI engine [29].

if the cooled temperature exceeds 200°C. Condensation of volatile materials and coagulation of particulates will be dominant in determining the number of small particulates when the engine load is higher than 30%, but when the cooled exhaust gas temperature is over 200°C, coagulation may be the primary mechanism leading to an increase in the number-average diameter of particulates [30].

3.2. Profile of particle emissions from Otto engines

Gasoline exhausts can be divided into three major components: gaseous phase, soot particles, and semi-volatile organics, which are distributed between the particulate and the gaseous phase. Correspondingly, its extracts include condensate (CD), particulate matter, and semi-volatile organic compounds (SVOC). Previous studies on gasoline exhausts focused primarily on the single component such as PM, CD, and SVOC. The studies on combination of these components are limited. In addition, efforts to reduce the total emission rate have led to modifications in fuel, engine, and after-treatment technology.

Currently, the automotive industry has been developing and applying the technology of direct injection engines in the Otto cycle to meet the challenges imposed by environmental legislation and to achieve energy efficiency. Despite the known advantages, such technology has negative factors, especially the formation of particulate.

The formation of PM in diesel engines is a phenomenon already known, however, in Otto engines, remains an issue to be further investigated, especially on the concentration of PM less than 2.5 microns ($PM_{2.5}$) in nominal average diameter gas exhaust. The topic is important given the recent revision and imposition of the European legislation on emission limits for PM on direct injection engines.

In the automotive industry, with the evolution of the control of pollutant emission programs and strong demand for optimizing the motor's fuel consumption, new vehicle technologies continue to be introduced. Thus, seeking to combine the specific power of a gasoline engine with the efficiency of diesel engines, direct fuel injection has been developed and applied in Otto engines.

The development of four-stroke, spark-ignition engines that are designed to inject gasoline directly into the combustion chamber is an important worldwide initiative of the automotive industry. The thermodynamic potential of such engines for significantly enhanced fuel economy, transient response, and cold-start hydrocarbon emission levels has led to a large number of research and development projects that have the goal of understanding, developing, and optimizing gasoline direct-injection (GDI) combustion systems (Figure 8) [31].

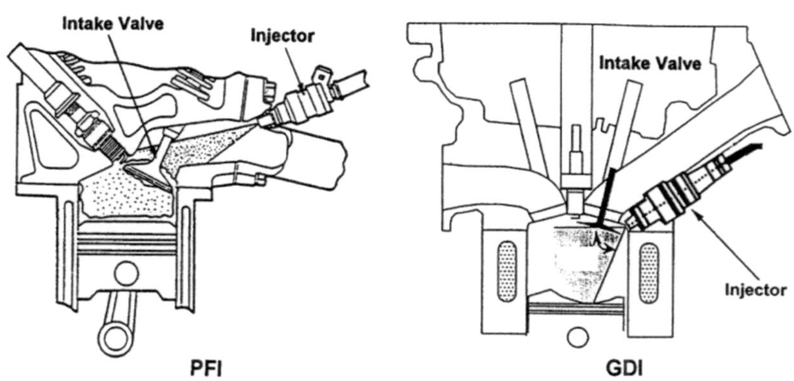

Figure 8. Differents injection fuel systems: port-fuel-injected (PFI) and gasoline direct injection (GDI) [31].

The break specific fuel consumption, and hence, the fuel economy, of compression-ignition, direct-injection (CIDI), diesel engine is superior to that of the port-fuel-injected (PFI) spark-ignition engine, mainly due to the use of a significantly higher compression ratio, coupled with unthrottled operation. The diesel engine, however, generally exhibits a higher noise level, a more limited speed range, and higher particulate and NO*x* emissions than the spark ignition (SI) engine.

In a study of the particle emission characteristics using modern GDI passenger cars with the focus on exhaust particle number emissions and size distributions, the results indicate that both particle size below 30 nm and the other with mean particle size approximately 70 nm consisted of soot but with different morphologies (Figures 9 and 10) [32]. Significant emissions of exhaust particles were observed also during decelerations conducted by engine braking and the particles most likely originated from lubricant oil ash components. The semi-volatile nucleation particles were observed at high engine load conditions. Thus, in general, the study indicates that a modern gasoline vehicle can emit four distinctive types of exhaust particles (Figure 10). Both during acceleration and steady conditions, the number size distribution of nonvolatile exhaust particles consisted of two modes, one with mean. In general, a major share of solid particles in the modern gasoline vehicle exhaust can be below this particle size limit, and during high engine load, vehicles can also emit small semi-volatile particles.

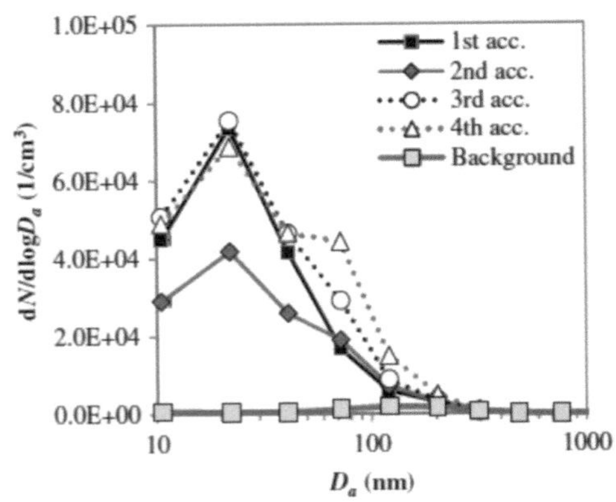

Figure 9. Particle number size distributions during the repetitions of acceleration tests from 30 km/h to 90 km/h [32].

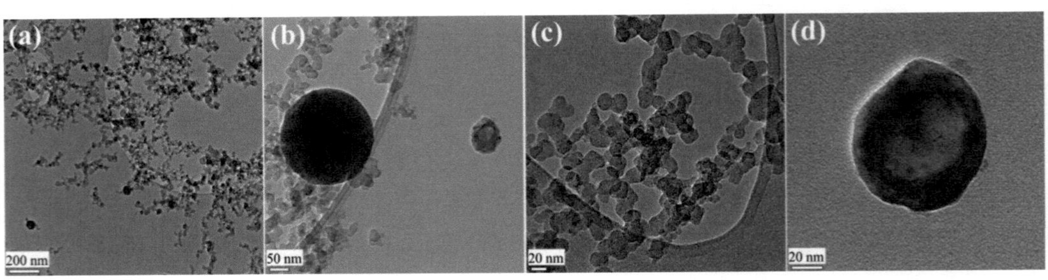

Figure 10. Transmission electron microscopy (TEM) images of collected exhaust particles during the New European Driving Cycle (NEDC) with various magnifications [32].

Two clearly distinct particle types were observed from samples collected over the whole NEDC. Firstly, around 10% to 20% of collected particles were nearly spherical (Fig. 10 a,b,d), often containing internal structure of lighter and darker areas. The size of those particles varied from 10 nm to even larger than 200 nm. These particles were composed of at least oxygen, zinc, phosphorous, and calcium where the metals are compounds of engine oil but not of fuel. The second particle type was agglomerated soot consisting of elemental carbon but also oxygen, zinc, phosphorous, and calcium. Note that also very small and nearly spherical soot-like particles were observed (Fig. 10b), possibly giving explanation for the bi-modal size distributions during acceleration and steady-state driving. However, the accumulated particles can agglomerate also on the grid which prevents the direct comparison of the number of collected particles with particle size distributions [32].

In another study using a chassis dynamometer, Maricq et al. 1999 [33] compared the mass of particulate matter emitted by a vehicle direct injection with a premix vehicle and a diesel vehicle. As shown in Figure 11, the direct injection engine emitted approximately 10 times as much particulate material as the premix engine in driving US FTP-75 cycle. However, the emission of 10 mg/mi of particulate matter from direct injection engine was far from the North American boundary current at the time diesel, 80 mg/mi.

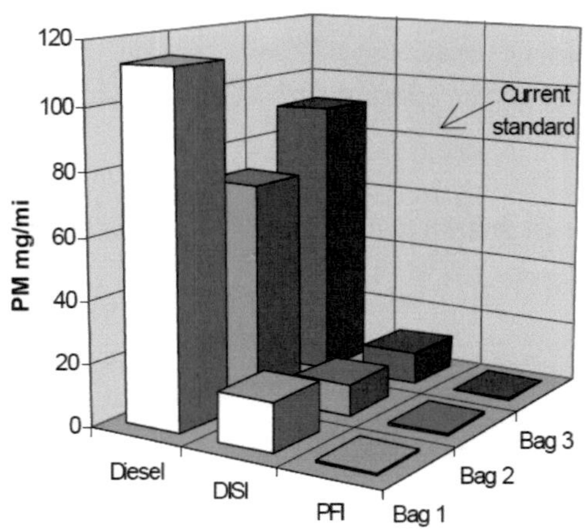

Figure 11. Comparison of the particulate matter emitted in the FTP-75 cycle [33], where DISI= direct-injection spark-ignited; and PFI= port fuel injected.

Another factor is that the exhaust gas fuel reforming has been identified as a thermochemical energy recovery technology with potential to improve gasoline engine efficiency, and thereby, reduce CO_2 in addition to other gaseous and PM emissions. The principle relies on achieving energy recovery from the hot exhaust stream by endothermic catalytic reforming of gasoline and a fraction of the engine exhaust gas. The hydrogen-rich reformate has higher enthalpy than the gasoline fed to the reformer and is recirculated to the intake manifold, that is, the reformed exhaust gas recirculation (REGR).

The REGR system was simulated by supplying hydrogen and carbon monoxide (CO) into a conventional EGR system. The hydrogen and CO concentrations in the REGR stream were selected to be achievable in practice at typical gasoline exhaust temperatures. Emphasis was placed on comparing REGR to the baseline gasoline engine, and also to conventional EGR. The results demonstrate the potential of REGR to simultaneously increase thermal efficiency, reduce gaseous emissions, and decrease PM formation [34].

In general, Kittelson & Kraft conclude that two mechanisms of precursor formation of particulate matter in the Otto engines' direct injection were identified. The first relates to the stratified operating condition, that is, when the fuel injection occurs in the compression phase. Due to the short time between injection and the spark, the fuel vaporization is not complete, so the air/fuel mixture presents heterogeneous characteristics, that is, fuel-rich regions with great potential for the formation of particulate matter. The second is mainly related to the homogeneous condition, that is, even when the fuel is injected in the admission phase, thus creating an accumulation of fuel on the cylinder walls, a potential source for the formation of particulate material [4].

In this context, the particle emissions of vehicles are restricted by emission standards which have significant variations depending on the country. In the US, since 2004, the same standards have been applied to vehicles regardless of the fuel and thus, the limits for the particulate mass emission have also covered the Otto vehicles. In the European Union, a particulate mass emission limit for direct injection Otto engines took effect in 2009 (Euro 5), and the first restrictions for particle number emissions will come into effect in 2014 (Euro 6). Thus, globally, the particle emission limitations for gasoline vehicles are under strong development [32,35].

4. Impact of the biofuel burning on particle emissions from the vehicular exhaust

Biofuels are obtained from biomass, the name given to the organic material in an ecosystem or a vegetable or animal population. As plants and animals may be continuously reproduced, it can be considered renewable energy sources. There are several types of biofuels that can be produced from biomass, such as alcohol (methanol and ethanol), biodiesel, bio-kerosene and others, and sources for this production can be both of animal origin (for example tallow or chicken fat) and vegetable (e.g. vegetable oils and cane sugar) [36].

In chemical terms, biodiesel is a mixture of alkyl esters from fatty acids and can be produced from plant-derived oils, waste oils and fats (resulting from domestic, commercial, and industrial processes such as, for example, frying) or animal fats. Dozens of plant species can be used for the production of this biofuel, such as soybean, palm, sunflower, babassu oil, peanut, castor, jatropha, and others [36].

We can highlight biodiesel and ethanol as among these fuels that can be used in internal combustion engines without requiring major modifications. The use of these biofuels can bring great changes in the emission of particulate matter profile, which will be discussed below. PM emissions have become a major concern due to their environmental impact [3]. In recent years, the law which governs the issuance of pollutants has forced manufacturers and automakers to develop engines and cleaner cars. In this scenario, fuels from renewable sources received considerable prominence and emerged as alternatives to fossil fuels. Several tests have been conducted with biodiesel and ethanol to ascertain the impacts on engine performance, fuel consumption, and exhaust emission, mainly in relation to diesel [3,37].

The differences on performance, combustion, and emission of biodiesel are caused by the difference existing between this and the diesel from fossil oil and chemical thermophysical properties such as density, cetane number, and oxygen content, being higher in biodiesel than in diesel [3].

http://dx.doi.org/10.5772/60110

Ethanol already has low cetane number which may lead to insufficient self-ignition quality for direct use of these alcohols in unmodified diesel engines. The key property of ethanol is its high octane number. The addition of ethanol to gasoline raises the octane value of gasoline and reduces engine knock, without affecting the efficiency of the catalytic converter [38]. Indeed, when Henry Ford designed his first automobile (Model T), it was built to run on both gasoline and pure ethanol [39].

Experimental studies have claimed that ethanol-blended fuels reduce exhaust emissions compared with gasoline-fueled engine. Generally, in these studies, the reductions in the exhaust emissions have been associated with the oxygen content of ethanol. It is well-known that the physical and chemical properties of ethanol are completely different from those of gasoline. In particular, their energy contents are lower than that of gasoline, both on mass and volume basis. This property shows that the engine will need more amount of fuel when it is fueled with ethanol blends to produce the same power output in a gasoline-fueled engine. This case will change air/fuel ratio in the cylinder and exhaust emission levels. One of the most important properties of both ethanol and biodiesel, compared with gasoline and diesel, is the oxygenated atoms in their molecular compounds which provide significant reduction in the CO and HC emissions, but it may adversely affect NO_x emissions [3,40,41].

Unlike biodiesel, which is applied only in diesel cycle engines, ethanol can be used in both diesel and Otto cycle engines. Due to the advantages of biodegradability, low toxicity as well as high miscibility with diesel fuel relative to ethanol, as an oxygenous biomass fuel, ethanol has also received considerable attention. In particular, its regenerative capability and cleaner burning characteristics make ethanol so attractive that it may also be considered as a predominant alternative fuel for diesel engines. Researches indicated that the ethanol–diesel blended fuels were technically acceptable for existing diesel engines. At present, there is a widespread interest in ethanol–diesel blended fuels for their potential to help reduce harmful exhaust emissions from current and future diesel engines. The first studies on the use of ethanol in diesel engines were conducted in South Africa in the 1970s, and continued in Germany and the United States during the 1980s through the work of Caro et al., 2001 [42].

Numerous experimental results indicate that ethanol/diesel blends could significantly reduce PM and smoke emissions. Table 1 shows some research results about PM emissions using biodiesel/diesel blends.

In a very comprehensive study on the impact of using biofuel (biodiesel) in the emission of particulate matter, the size, concentration, and number of particles are observed to have been directly influenced by the concentration of biodiesel added to diesel fuel, the engine load conditions, and also after-treatment technology adopted. Younga et al. [47] observed that the size distributions at 0% load were very different from other test modes and were bimodal, showing a predominant core mode of 15 nm and a substantially minor soot mode of ~ 68 nm. At above 25% load, the core mode disappeared. Instead, the size distributions were unimodal with a soot mode that increased in concentration and size with increasing load from 25%, 50% to 75% (Figure 12a) [47].

Ref.	Fuel	Embedded Technology	Emissions PM
[43]	10% and 15% ethanol to diesel	-	Reduce by 20%–27%
[44]	E0 (base diesel fuel), E5 (5%), E10 (10%), E15 (15%) and E20 (20%), base diesel fuel	-	Reduce by 30%–40%; the more ethanol was added, the less smoke emitted
[45]	blends containing 83%–94% diesel fuel, 5%–15% ethanol and 1%–3% additive cetane improver	-	15% ethanol–diesel blends could produce a drop of 33.3% in smoke and 32.5% in the soot mass concentration
[46]	blends containing ethanol fumigation is 20% improver	-	reduction of 51% in soot mass concentration
[47]	waste cooking oil biodiesel blends (B2, B10 and B20), engine loads (0%, 25%, 50% and 75%)	diesel oxidation catalyst plus diesel particulate filter (DOC + DPF)	Particle number concentration at a given load, reduce with increasing of biodiesel blend; DOC + DPF removed >99.84% nonvolatile particle

Table 1. Research results about PM emissions using biofuel/fossil fuel blends.

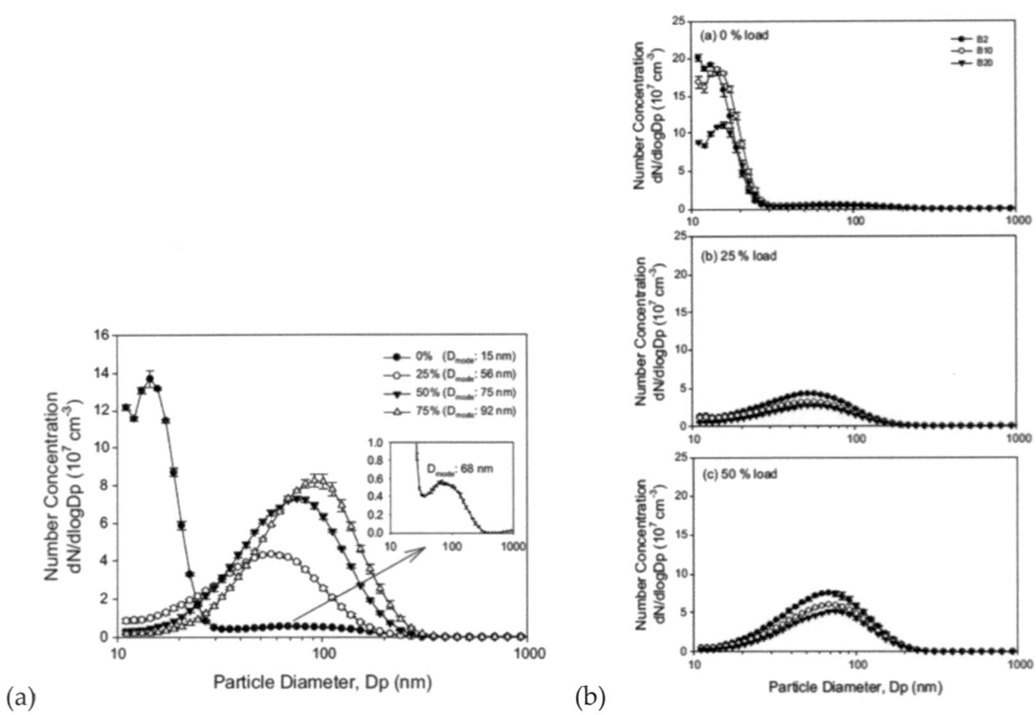

Figure 12. a) Average nonvolatile particle size distributions in pre-DOC + DPF exhaust at different engine loads using B2; (b) average nonvolatile particle size distributions in pre-DOC + DPF exhaust at (a) 0%, (b) 25%, and (c) 50% load using B2, B10 and B20 [47].

http://dx.doi.org/10.5772/60110

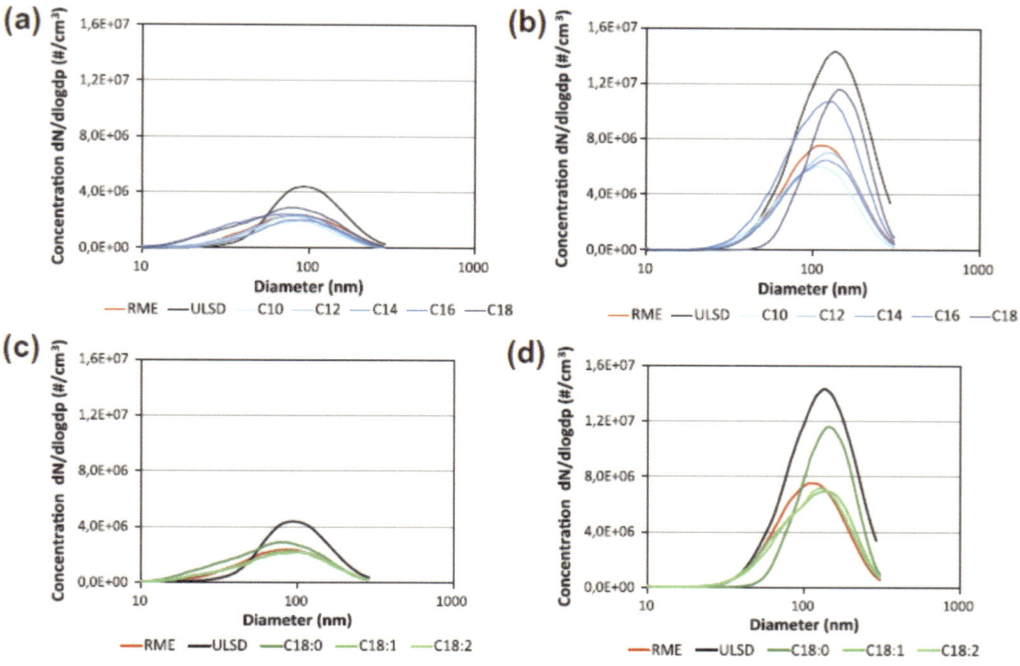

Figure 13. Particle number distribution: (a) effect of chain length, 0% EGR; (b) effect of chain length, 30% EGR; (c) effect of unsaturation, 0% EGR, and (d) effect of unsaturation, 30% EGR [48].

Figure 12b shows the effect of biodiesel concentration added to diesel in the particle size number concentration (Younga et al. research). At 0% load, the number of core particles decreased with increasing biodiesel blend. At 25% and 50% load, the number of soot particles decreased with increasing biodiesel blend. Therefore, the number reduction with increasing biodiesel blend was not limited to soot particles but also included the core particles. This is likely due to the increased oxygen content, lower aromatic content, prolonged soot oxidation time, and lower final boiling point with increasing biodiesel blend [47].

Besides the concentration of biodiesel present in the diesel fuel, it can be said that the properties and source of biodiesel used distribution can impact the size and number of particles emitted in the exhaust profile. This impact was observed by Pinzi and colleagues [48] in a study which showed varied effects of fatty acid methyl esters on the molecular structure (saturation degree and chain length) present in rapeseed oil methyl esters (RME - biodiesel). Furthermore, the effect of the use of EGR in the particle emission profile was also evaluated (Figures 13 and 14) [48]. The results were compared with those obtained for ultra low sulphur diesel (ULSD) burning.

Through Figures 13 and 14, one can observe that the results obtained by Pinzi et al. [4] showed that the ULSD fuel particle size distributions are greater and are predominantly at larger diameters than in the case of the fat acid methyl esters (FAME). All the methyl esters (including

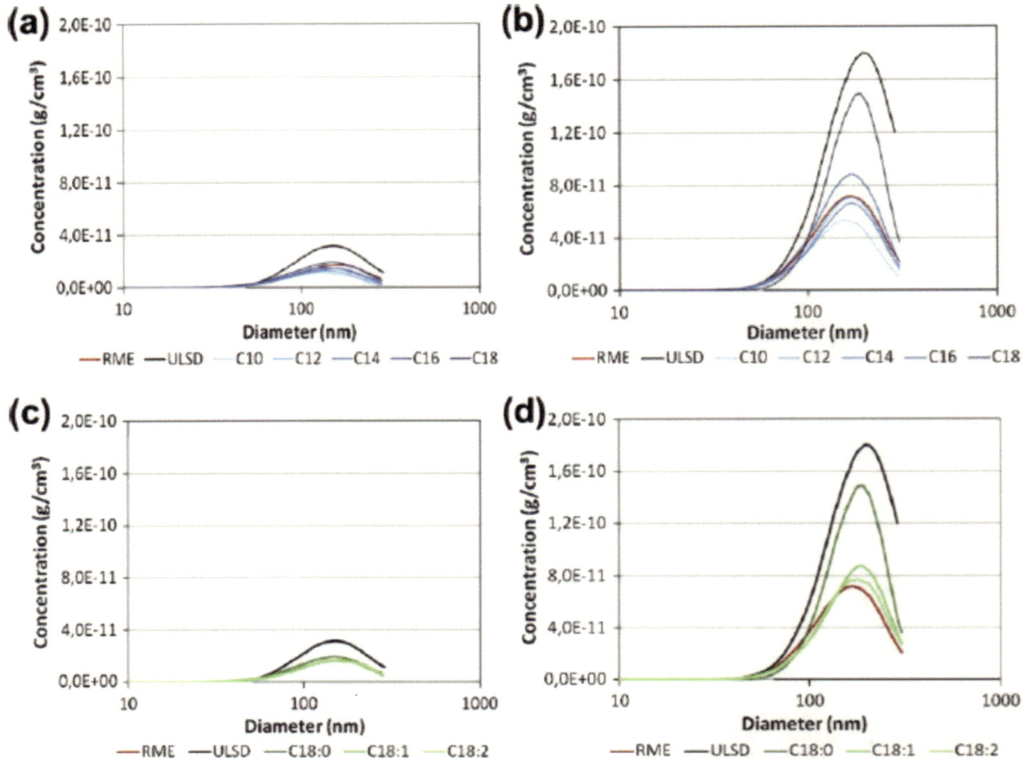

Figure 14. Mass particle size distribution: (a) effect of chain length, 0% EGR; (b) effect of chain length, 30% EGR; (c) effect of unsaturation, 0% EGR, and (d) effect of unsaturation, 30% EGR [48].

RME) gave lower total number (Figure 13) and mass emissions (Figure 14) of the particles than ULSD fuel. Moreover, during EGR conditions the total mass of particles of methyl esters was around 60% of the ULSD PM emission.

Guarieiro et al. [49] investigated the influence of the use of ethanol as an additive to biodiesel/ diesel blend in the size and number distribution of particles and the results obtained are shown in Figure 14. The fuels evaluated were B5 (diesel with 5% of biodiesel); B5E6 (ternary composition containing 89% diesel, 5% of biodiesel and 6% of ethanol); and B100 (100% of biodiesel). The burning of fuels showed concentrations of particles trendy accumulation of 50 > Da > 200 nm (Figure 15). In general, particles emitted from diesel engines are in the size range 20–130 nm. The geometric mean obtained for both fuels B5 and B5E6 was δ=86.6±3.7 nm, with a total number of particles of 9.6 × 106 particles/cm^3 for the B5 and 1.1 ×107 particles/cm^3 to the B5E6. The B100 showed geometric mean of δ=78.1±3.1 nm with total number of particles of 1.4 × 107 particles/cm^3. It was observed by the authors that there was an increase in the number of smaller particulate emissions when biodiesel is used instead of blended diesel with alcohol fuel.

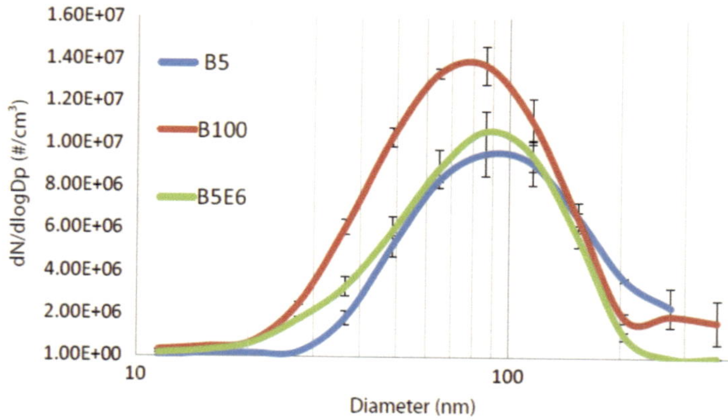

Figure 15. Distribution of number and size particles for fuels: B5, B5E6, and B100 [49].

The engine used in Guarieiro's research [49] has a mechanical injection, an anticipated injection can happen due to high modulus of volumetric compressibility of the B100, and this makes it longer to mix with air. Thus, there is an increase in premixed combustion fraction due to the ignition delay that can generate a lesser incomplete burning, reducing the size of the particles and consequently, increasing their concentration. However, the nucleation, condensation, and coagulation of the HC in the engine exhaust will generate some particles, leading to more particulates, both in number and in mass, than the B5 and B5E6.

Besides the physical characterization of particles emitted from burning diesel/biodiesel fuel blends, there are some studies in the literature that also evaluate the impact of biodiesel use in the chemical composition of the particles. The effects of diesel/biodiesel blends on the physical, chemical, and toxicological properties of diesel engine exhaust at low condition were investigated through the study of the changes in size-distribution and emission factors of PAH associated to PM [50]. For that, particle emissions from commercial petroleum-based diesel with 4% of soy biodiesel (B4), a biodiesel blend of 25% and 50% (B25 and B50), and also pure biodiesel (B100) were measured using a diesel engine at low load. PM was distributed in all sizes, while PAH size distribution was found in higher levels in the accumulation mode (30 nm < Dp < 2.5 lm). Total PAH emission factors (ng kg_1 fuel) for B4, B25, B50, and B100 were 237,111,182, and 319 ng.kg^{-1} fuel, respectively. Individual PAH emission factors showed that PAH containing four or more rings (MW > 202) such as BBF, BAA, PYR, and BGP were the main PAH emitted by the four studied fuels. The percentage reductions of individual PAH emission factors for the blended fuels in comparison with B4 were 37% and 22% for B25 and B50, respectively, and an increase of around 31% for B100. On the other hand, an increase in redox activity was observed for B25, B50, and B100 when compared with B4. In general, the results from our study suggest that emissions from pure waste cooking biodiesel may not be the better fuel choice in terms of PM, PAH, and BaPE particle size distribution and emission factors as well as redox activity (Figure 16). However, B25 and B50 blends presented some improvements in terms of PM, HPA, and BaPE size distribution and redox activity of engine exhaust in comparison with B4. This suggests that the addition of low percentages of biodiesel to diesel promotes benefits in both environmental and human health concerns [50].

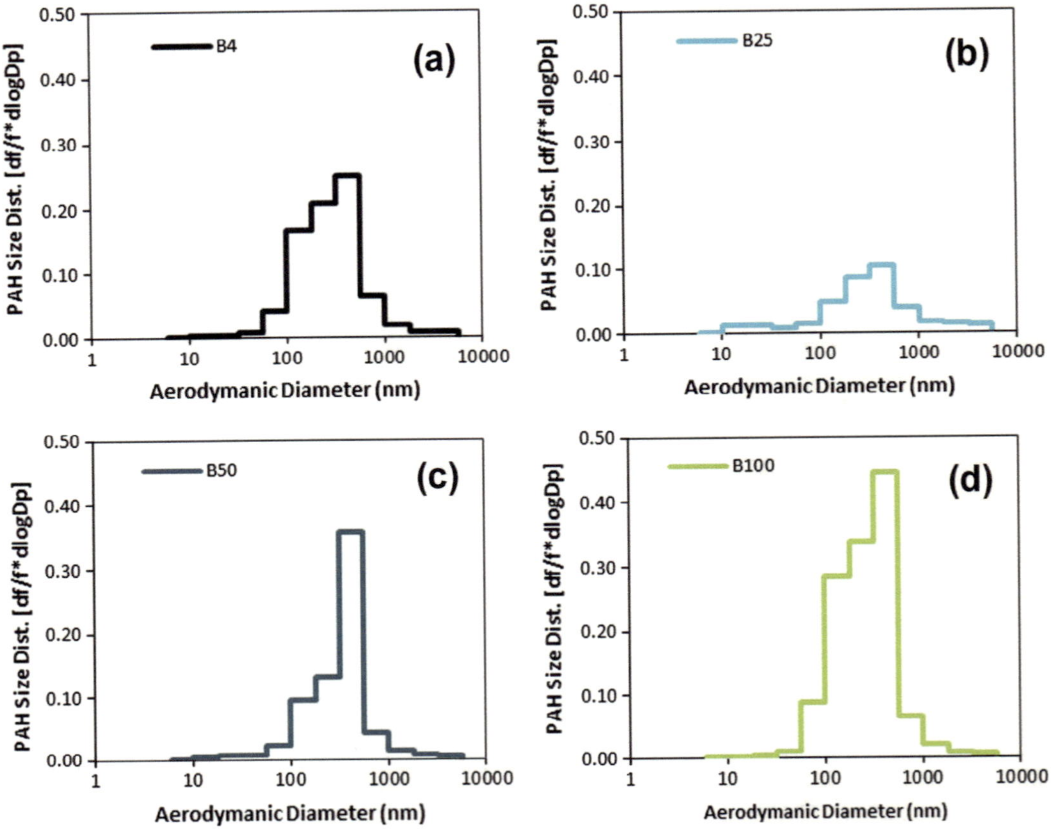

Figure 16. PAH size-distributed emission factors for the studied diesel/biodiesel fuels [50].

The study of PAH size distribution (Figure 16) shows a unimodal size distribution for all fuels peaking at 320–560 nm. Approximately 90%–99% of PAH were present in particles smaller than 1.8 μm for all fuels, and about 80% of them are found in particles between 56 nm and 1.8 μm. This broad size-distribution of PAHs starting in the range of nanoparticles and ending in small particle sizes is important for assessing the possible health effects associated with exposure to biodiesel emissions, as particle size will determine deposition and its chemical composition for possible adverse outcomes [50].

There are few studies on the impact of the use of ethanol/gasoline in emissions of particulate matter. However, some research points to the fact that the addition of ethanol in gasoline cannot have an effect on the number of emitted particles and their diameter by difference in the four-stroke moped engine [51] and particle number and diameter are reduced by 60% and 90%, respectively, when blends of ethanol/gasoline are applied over pure gasoline fuels, at all engine cycles [52].

Thus, the differences in particle characteristics and formation should be taken into account in the development of emission control strategies and of technologies for the assessment of the impact of particle emissions on the environment and human health. Nowadays, for PM, the regulated value is the total mass. Nevertheless, particle number and particle size distribution give more information than mass alone, because it is known that small particles have longer residence time in the atmosphere, and are more reactive and are more difficult to trap. Moreover, these small particles can reach the pulmonary alveoli, while larger particles deposited in the upper airways are easier to eliminate. Thus, small particles, especially ultrafine particles under 100 nm, are considered critical to human health and research about their impact with the use of biofuels should be developed more and more to understand what the real impact of biofuel in particle emissions is.

Acknowledgements

The authors wish to thank the SENAI CIMATEC and Brazilian agencies: CNPq, FAPESB, CAPES, FINEP, PRH55 ANP/MCTI and INCT for Energy & Environment.

Author details

Lílian Lefol Nani Guarieiro[1,2] and Aline Lefol Nani Guarieiro[3]

1 SENAI-CIMATEC, Salvador, BA, Brazil

2 INCT de Energia e Ambiente, UFBA, Salvador, BA, Brazil

3 Cetrel Odebrecht Ambiental, Rio de Janeiro, RJ, Brazil

References

[1] Lima MMC. Estimativa de concentração de material particulado em suspensão na atmosfera por meio da modelagem de redes neurais artificiais. Tese de mestrado defendida na Escola de Engenharia 2006, UFMG.

[2] EPA 2014 – Particle Matter disponible http://www.epa.gov/pm/ (accessed in Jully 15, 2015),

[3] Guarieiro LLN, Guarieiro ALN. Vehicle Emissions: What Will Change with Use of Biofuel? INTECH, Biofuels - Economy, Environment and Sustainability 2012; Chapter 14: 357-386.

[4] Kittelson D, Kraft M. Particle Formation and Models in Internal Combustion Engines. United Kingdom: University of Cambridge, 2014.

[5] Kittelson DB. Recent Measurements of nanoparticles emissions from engines. Current research on diesel exhaust particles japan association of aerosol science and technology 2001.

[6] Jung H, Kittelson DB. Measurement of electrical change on diesel particles. Aerosol Sci Technol 2005; 39: 1129-1135.

[7] EPA 2005; Chapter 4 - Air Quality: Characterization and Implications', United States Environmental Protection Agency, Office of Air Quality Planning and Standards, Research Triangle Park, NC 27711, USA.

[8] Morrow PE, Bates DV, Fisher BR, et al., Deposition and Retention Models for Internal Dosimetry of the Human Respiratory Tract (Report of the international Commission on Radiological Protection: ICRP: Task Group on Lung Dynamics). Health Phys. 1964; 12:173-207.

[9] Raabe OG. Deposition and Clearance of Inhaled Aerosols. Mechanism in Respiratory Toxicology 1982; 1: 27-76. H. Witschi and P. Nettesheim, Eds. Boca Raton, FL: CRC Press, 1982.

[10] Kelly FJ, Fussell JC. Size, source and chemical composition as determinants of toxicity attributable to ambient particulate matter. Atmos Environ 2012; 60: 504-526.

[11] Zhang Q, Stanier CO, Canagaratna MR, Jayne JT, Worsnop DR, Pandis SN, Jimenez JL. Insights into the chemistry of new particle formation and growth events in Pittsburgh based on aerosol mass spectrometry, Environ. Sci. Technol. 2004; 38: 4797–4809.

[12] Lim YB, Ziemann PJ. Products and mechanism of secondary organic aerosol formation from reactions of n-alkanes with OH radicals in the presence of NOx. Environ. Sci. Technol. 2005; 39: 9229–9236.

[13] Li N, Xia T, Nel AE. The role of oxidative stress in ambient particulate matter-induced lung diseases and its implications in the toxicity of engineered nanoparticles. Free Radical Biol Med 2008; 44: 1689-1699.

[14] Claxton LD, Matthews PP, Warren SH. The genotoxicity of ambient outdoor air, a review: Salmonella mutagenicity, Mutat. Res. 2004; 567: 347–399.

[15] Jakober CA, Riddle SG, Robert MA, Destaillats H, Charles MJ, Green PG, Kleeman MJ. Quinone emissions from gasoline and diesel motor vehicles. Environ. Sci. Technol. 2007; 41: 4548–4554. T otlandsdal AL, Herseth JL, Bolling AK, Kubátová A, Braun A, Cochran RE, Refsnes M, Ovrevik J, Låg M. Differential effects of the particle core and organic extract of diesel exhaust particles. Toxicol Lett. 2012; 208: 262-268.

[16] Kong S, LU B, JI Y, ZHAO X, CHEN LI, LI Z, HAN B, BAI Z. Levels, risk assessment and sources of PM10 fraction heavy metals in four types dust from a coal-based city. Microchem J 2011; 98: 280-290.

[17] IPCS. 1996. Environmental Health Criteria No. 171. Diesel Fuel and Exhaust Emissions. International

[18] Programme on Chemical Safety. http://www.inchem.org/documents/ehc/ehc/ehc171.htm.

[19] CAL EPA (California Environmental Protection Agency), 1998. Part B: Health Risk Assessment for Diesel Exhaust, Proposed Identification of Diesel Exhaust as a Toxic Air Contaminant. Office of Environmental Health Hazard Assessment (OHEEA).

[20] Schuetzle D, Perez JM. Factors influencing the emissions of nitrated-polynuclear aromatic hydrocarbons (nitro-PAH) from diesel engines. J Air Pollut Control Assoc. 1983; 33(8): 751-755.

[21] Schuetzle D, Jensen TE, Ball JC. Polar polynuclear aromatic hydrocarbon derivatives in extracts of particulates: Biological characterization and techniques for chemical analysis. Environ Int. 1985; 11(2-4): 169-181.

[22] Sharma M, Agarwal AK, Bharathi KVL. Characterization of exhaust particulates from diesel engine. Atmospheric Environment 2005; 39: 3023–3028.

[23] Pierson WR, Brachaczek WW. Particulate matter associated with vehicles on the road--II. Aerosol Sci. Technol. 1983; 2: 1-40.

[24] López JM, Jiménez F, Aparicio F, Flores N. On-road emissions from urban buses with SCR þurea and EGRþ DPF systems using diesel and biodiesel. Transport Res 2009; Part D, 14: 1–5.

[25] Maricq M.M. Chemical characterization of particulate emissions from diesel engines: a review. J Aerosol Sci 2007; 38: 1079–1118.

[26] Mooney JJ. Toxic solid nanoparticles, the importance of retrofitting diesel engine particle emission control systems to older in-use diesel engines, and available methods. Österreichische Ingenieur- und Architekten Zeitschrift (ÖIAZ) 2007; 152: 1–3.

[27] Niemi S, Lundin K, Karhu T, Laurén M, Ekman K, Nousiainen P, Paanu T. Exhaust particle number in off-road engines of different generations. SAE Paper 2009; 01-1869: 1–11.

[28] Li X, Xu Z, Guan C, Huang Z. Particle size distributions and OC, EC emissions from a diesel engine with the application of in-cylinder emission control strategies. Fuel 2014; 121: 20–26.

[29] Agarwal AK, Singh AP, Lukosea J, Guptab T. Characterization of exhaust particulates from diesel fueled homogenous charge compression ignition combustion engine. J Aerosol Sci 2013; 58: 71–85.

[30] Ninga Z, Cheung CS, Liu SX. Experimental investigation of the effect of exhaust gas cooling on diesel particulate. Aerosol Sci 2004; 35: 333–345.

[31] Zhao F, Lai MC, Harrington DL. Automotive spark-ignited direct-injection gasoline engines. Progr Energy Combust Sci 1999; 25: 437-562.

[32] Karjalainen P, Pirjola L, Heikkila, Lahde T, Tzamkiozis T, Ntziachristos L, Keskinen J, Ronkk I. Exhaust particles of modern gasoline vehicles: A laboratory and an on-road study. Atmos Environ 2014; 97: 262-270.

[33] Maricq MM, Podsiadlik DH, Brehob DD, Haghgooie M. Particulate Emissions from a Direct-Injection Spark-Ignition (DISI) Engine. SAE Technical Paper 1999; 1999-01-1530.

[34] Fennell D, Herreros J, Tsolakis A. Improving gasoline direct injection (GDI) engine efficiency and emissions with hydrogen from exhaust gas fuel reforming. Int J Hydrogen Energy 2014; 39: 5153-5162.

[35] Dieselnet, www.dieselnet.com, information retrieved: 14.1.2014.

[36] Guarieiro LLN, Torres EA, De Andrade JB. Energia verde. Ciência Hoje 2011; 48 (285): 36-41.

[37] Imtenan S, Masjuki HH, Varman M, Kalam MA, Arbab MI, Sajjad H, Ashrafur Rahman SM. Impact of oxygenated additives to palm and jatropha biodiesel blends in the context of performance and emissions characteristics of a light-duty diesel engine. Energy Convers Manage 2014; 83: 149-158.

[38] Agarwal AK. Biofuels (alcohols and biodiesel) applications as fuels for internal combustion engines. Prog Energ Combust 2007; 33: 233-271.

[39] Sward K. The legend of Henry Ford. 1st ed. New York: Rinehart; 1948.

[40] Canakci M, Ozsezen AN, Alptekin E, Eyidogan M. Impact of alcohol–gasoline fuel blends on the exhaust emission of an SI engine. Renew Energy 2013; 52: 111-117.

[41] Guarieiro LLN, Vasconcellos PC, Solci MC. Poluentes Atmosféricos Provenientes da Queima de Combustíveis Fósseis e Biocombustíveis: Uma Breve Revisão. Revista Virtual de Química 2011; 3 (5): 434-445.

[42] Caro PS, Mouloungui Z, Vaitilingom G. Interest of combining an additive with diesel–ethanol blends for use in diesel engines. Fuel 2001; 80: 565–574.

[43] Spreen K. Evaluation of oxygenated diesel fuels. Final Report for Pure Energy Corporation Prepared at Southwest Research Institute, San Antonio, TX, 1999.

[44] Kass MD, Thomas JF, Storey JM, Domingo N, Wade J, Kenreck G. Emissions from a 5.9 L diesel engine fueled with ethanol–diesel blends. SAE Technical Paper Series No. 2001-01-2018.

[45] Ahmed. Oxygenated diesel: emissions and performance characteristics of ethanol–diesel blends in CI engines. SAE Technical Paper Series No. 2001-01-2475.

http://dx.doi.org/10.5772/60110

[46] Abu-Qudais M, Haddad O, Qudaisat M. The effect of alcohol fumigation on diesel engine performance and emissions, Energy Convers Manage 2000; 41: 389–399.

[47] Younga L, Liou Y, Cheng M, Lu J, Yang H, Tsai YI, Wang L, Chen C, La J. Effects of biodiesel, engine load and diesel particulate filter on nonvolatile particle number size distributions in heavy-duty diesel engine exhaust. J Hazard Mater 2012; 199– 200.

[48] Pinzi S, Rounce P, Herreros JM, Tsolakis A, Dorado MP. The effect of biodiesel fatty acid composition on combustion and diesel engine exhaust emissions. Fuel 2013; 104: 170–182.

[49] Guarieiro LLN, Guerreiro ETA, Amparo KKS, Manera VB, Regis ACD, Santos AG, Ferreira VP, Leão DJ, Torres EA, de Andrade JB. Assessment of the use of oxygenated fuels on emissions and performance of a diesel engine. Microchem J 2014; 117: 94–99.

[50] Guarieiro ALN, Santos JVS, Eiguren-Fernandez A, Torres EA, da Rocha GO, de Andrade JB. Redox activity and PAH content in size-classified nanoparticles emitted by a diesel engine fuelled with biodiesel and diesel blends. Fuel 2014; 116: 490–497.

[51] Seggiani M, Prati MV, Costagliola MA, Puccini M, Vitolo S. Bioethanol– gasoline fuel blends: exhaust emissions and morphological characterization of particulate from a moped engine. J Air Waste Manage Assoc 2012; 62: 888–97.

[52] Costagliola MA, De Simio L, Iannaccone S, Prati MV. Combustion efficiency and engine out emissions of a S.I. engine fueled with alcohol/gasoline blends. Appl Energy 2013 ;111: 1162–71.

Chapter 12

Thermoanalytical Methods in Verifying the Quality of Biodiesel

Marcelo Y. Misutsu, Leandro F. Cavalheiro, Thiago G. Ricci, Luiz H. Viana, Silvio C. de Oliveira, Amilcar Machulek Junior and Lincoln C. S. de Oliveira

Additional information is available at the end of the chapter

http://dx.doi.org/10.5772/59479

1. Introduction

Fossil fuels (petrol, coal and natural gas) are nowadays the main used energy source to move most part of overland, aquatic and air means of transportation, likewise several industries and even public lighting [1]. Petrol is the most used raw material among these energetic resources, as it is a limited energetic source there is a great concern toward substituting it, petrol shows unstable price and its unbalanced distribution has even caused wars. The global petrol reserves had reached 1,668.9 billion gallons in the end of 2012, enough to guarantee 53 years of the worldwide energy production. The dada is in the annual statistic report, 2013 edition, by BP (multinational enterprise headquartered in United Kingdom, which operates in the energetic sector, mainly petrol and natural gas) thus a reference document to the sector. According to the study, throughout the last decade, global petrol reserves have increased 26%. Other Opep countries are still top ranking, controlling worldwide petrol reserves [2].

There also is the environmental concern to restraining the current climatic changing process caused mainly by CO_2, methane and nitrogen oxides. The current scientific and technological development context aims to emerge new technologies and the global goal of transiting to a sustainable economy [3].

However, Brazil, as well as other countries, is facing a severe energetic crisis due to the fuel consumption increase, either for production and for fuel consumption as for petrochemical industry. Likewise the climatic changes have caused long drought periods in months that supposedly had great quantity of rain. Each of these factors contribute to focusing efforts to establish alternative sources effectively applicable in energy production.

One way is the use of resources produced by agribusiness [3]. This transition to sustainability was mainly determining the need or strategic interest to replace oil with other materials. Vegetable oils as alternative fuels have begun being studied in the late nineteenth by Rudolph Diesel, inventor of the internal combustion engine century, and these were used in natura. But the direct use in engines came up with many problems, for example, oily material accumulation in the injection nozzles, the incomplete oil burnt, coal deposit formation in the combustion chamber, low power efficiency and, as a result of firing, the release of acrolein (propenal) which is toxic. Studies of alternatives have been considered for best performance of vegetable oils in diesel engines, for example, dilution, micro-emulsion with methanol and ethanol, catalytic cracking and trans-esterification reaction with short alcohols chains. Among these alternatives, the trans esterification reaction has been the most used, since the process is relatively simple and the product (biodiesel) has properties similar to those of petro diesel, such as viscosity and density [4].

Biodiesel can be produced from a variety of materials. These raw materials include most vegetable oils (soybean oil, cottonseed, palm, peanut, rapeseed/canola, sunflower, safflower, coconut oil) and animal fats (tallow) and discard oil (frying oils). The raw material choice for biodiesel production depends largely on geographical factors. Depending on the origin and quality of the raw material, changes in the production process may be required [5].

One of the great advantages of using biodiesel as fuel is linked to the environmental benefits; it has some features that represent advantages over petroleum-based fuels, such as absence of aromatics and sulfur; high cetane amount; average oxygen content; higher flash point; lower pollutants emission and good lubricity. Another advantage of biodiesel is the substitution of those in diesel engines without the need for any modifications to the engine [6].

The petroleum oils are stable at distillation temperature, even in the presence of excess oxygen. Unlike vegetable oils containing unsaturated compounds, the oxidation reaction can be observed up to room temperature and the temperature close to 250 °C because of the additional thermal decomposition reactions, leading to the formation of polymeric compounds by condensation reaction [7].

Toward such facts, the current context of scientific and technological development aimed at the emergence of new technologies and feasibility of application throughout the shortest time, as well as innovations that allow the search for global goal of transition to a sustainable economy. For this purpose, a possible alternative would be to replace the fossil diesel for biodiesel production or diesel/biodiesel mixture, which allows increasing the volumes of the mixtures reservations, what is highly interesting toward biodiesel production from a variety of renewable raw materials. Regarding to such purpose the use of analysis tools physico-chemical behavior of biodiesel or biodiesel blends containing, enabling the acquisition of information and the establishment of values that ensure the optimal operation of engines, are desirable. Therefore, the use of thermal analysis techniques, such as Thermogravimetry/ Derivative Thermogravimetry (TG/DTG) and Differential Scanning Calorimetry (DSC) has produced excellent results, contributing to a historical development of scientific applications. Thermal analysis is an important tool for the determination of thermodynamic properties, heat capacity, phase transition temperatures, among other perfectly applicable for determining the thermal characteristics of biodiesel, as decomposition temperature/combustion degree of oxidation, temperature crystallization, a polymerization potential and others.

Along the chapter, it will be discussed data about of methyl biodiesel obtainment from soybean oil at laboratory scale and its physicochemical characterization compared to two samples produced on an industrial scale: one obtained from recycled soybean oil and other from new soybean oil and diesel/biodiesel blends in varying proportions. The physical chemical parameters established by ANP were used as reference density, kinematic viscosity, acid number and iodine, flashpoint, accelerated oxidation test (RANCIMAT), cold filter plugging point, carbon residue and water content (Karl Fischer). In order to obtaining data from thermal behavior profile of the samples of biodiesel and diesel/biodiesel blends, such as thermal stability, thermal decomposition process, crystallization temperature and residue content were applied thermoanalytical TG/DTG and DSC techniques.

2. Traditional methods used in biodiesel characterization

The quality biodiesel must be produced following strict industry specifications, according to ASTM D6751 international level. In Brazil, the National Agency of Petroleum, Natural Gas and Biofuels (ANP) by ANP Resolution Nº 45, dated 08.25.2014 specifies the characteristics of the product. According to the resolution Physicochemical characteristics of biodiesel must be made through the use of the guidelines of the Brazilian Association of Technical Standards (ABNT) standards "American Society for Testing and Materials" (ASTM), "International Organization for Standardization" (ISO) and the "Comité Européen de Normalisation" (CEN). Some of the characteristics of the specification of biodiesel can be seen in Table 1.

In Brazil, through an aggressive government policy, thousands of Biodiesel production industries have been installed in all regions of the country, and this material is already mixed with diesel consumed at the pump stations. This fact makes the ANP hold a national monitoring program of the fuel sold in the country, with the association of accredited laboratories for this purpose. Testing methodologies validated by ASTM protocols are employed routinely evaluating samples collected in various parts of the country supply.

Biodiesel, when subjected to low temperatures is likely to present problems in the performance of fuel systems of cars. In winter, the crystallization of the methyl esters of saturated fatty acids can cause clogging of filters and pipes [8]. The cold flow properties of diesel are usually characterized by three temperature measurements following: Point cloud point (CP) connecting the cold (CFPP) filter and pour point (PP). Initially, the cooling temperatures causes the formation of seed crystals of wax which are solid at submicron invisible to the human eye level. Further decreases in temperature cause the nuclei to grow crystal [9]. Temperature at which crystals become visible is defined as the cloud point (CP), since the crystals form a generally hazy and cloudy suspension. The temperature at which the agglomeration of the crystal is large enough to prevent leakage of fluid free is defined as the pour point (PP). The filter plugging point at low temperature (CFPP) is then defined as the lowest temperature at which

40 mL of oil passes through the filter securely within 60 s. Crystallization of these compounds may lead to filters and pipes connected.

Previous studies concerning the cold flow properties of biodiesel have determined that the lengths of the hydrocarbon chains and the presence of unsaturated structures significantly affect its low-temperature properties [10-12].

In addition to these cold temperature tests, DSC should also be used to monitor the crystallization of multicomponent mixtures like biodiesel because CP results represent only the beginning of the crystallization process, and PP and CFPP cover the beginning of the operative problems; but none of these tests monitor the full range of crystallization. DSC is a well-established method for determining latent heat changes in a material upon either cooling (exothermic crystallization peaks) or heating (endothermic melting peaks). This method has been applied to monitor biodiesel crystallization and is generally considered to be more repeatable and more accurate than classical methods, such as PP or CP measurements, commonly used in industrial settings [13].

3. Biodiesel

Biodiesel is defined as alkyl esters (methyl, ethyl, propyl), commonly obtained from a chemical reaction of a vegetable oil or animal fat [14-17] with an alcohol (usually methanol) in the presence of a catalyst (usually sodium hydroxide or potassium hydroxide) in a call type for the trans-esterification [18-19] (Figure 1). This reaction produces glycerin as a byproduct, which ought to be widely applied in the chemical industry [5].This glycerin vegetable oil removed leaving the less viscous oil [20]. With its break, glycerin joins caustic soda (sodium hydroxide or potassium hydroxide) [21]. The remainder of the molecule (fatty acid) binds to the alcohol, forming biodiesel. Besides the glycerin, the biodiesel production chain generates a number of other co-products (pie, bran etc.) that can add value and constitute other important sources of income for farmers. Biodiesel is formed by different types of fatty acid esters, particularly the acids: oleic, linoleic and palmitic acids [22-25]; that can also be obtained by esterification and cracking process [26].

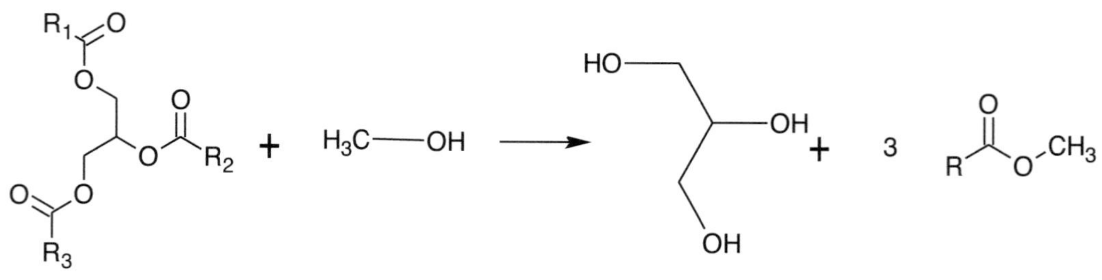

Figure 1. Transesterification Reaction

http://dx.doi.org/10.5772/59479

3.1. Biodiesel production

The transesterification reaction ought to be carried out in a reactor glass, jacketed for temperature control by circulating water, and mechanical agitation. The system temperature maintained around 50 °C. By adding 230 mL of anhydrous methanol and about 8.2 g of catalyst (NaOH), then completed dissolution of the catalyst by adding to 900 mL of soybean oil, establishing this moment as time zero of the reaction. The reaction time lasted about 45 min, in the first minutes were observed the conversion of esters by the sudden darkening of the mixture. Once the reaction is complete, the reactional mean is transferred a separator funnel and the formation of an upper layer corresponding to the methyl esters, a bottom phase containing glycerol formed from the reaction and add to excess methanol hydroxide sodium which does not react. Gathered to the reaction were formed soaps, and some traces of methyl esters and partial glycerides. Then the separated the two phases by decantation.

After the phases' separation, the obtained esters must be purified by washing them with a solution containing distilled water at 90 °C and 0,5 % concentrated HCl in order to neutralize the remaining catalyst of the reaction. The total neutralization is confirmed by adding phenolphthalein indicator 1 % (w/v), to the washing water. The aqueous phase, separated by decantation, followed ester and traces of moisture removed by subsequent filtration over anhydrous sodium sulfate. For the recovery of the excess methanol added in, there shall be the glycerin residue distillation at 80 °C under moderate vacuum [6].

3.2. Biodiesel physical-chemical characterization

Physicochemical parameters of biodiesel samples (methyl route) obtained by following the rules set by the Technical Regulation of ANP contained in Resolution Nº45/2014 of the National Agency of Petroleum, Natural Gas and Biofuels-ANP. Three samples of soybean biodiesel were analyzed, one that was produced in the laboratory (BL) and two industrial, one trans esterified with used oil (BI-01) and another with new oil (BI-02). All traits and methods used are in table 1.

Biodiesel developed in the laboratory (BL) has filed all within the parameters established by the Resolution of the ANP, except oxidation stability. Biodiesel oxidation occurs because the oils used as its raw materials containing unsaturated compounds, which are subject to oxidation reactions that take place at ambient temperature. Oxidation products cause corrosion in engine parts and deposit formation causing obstruction in the filters and injection system [5]. Therefore, the less subject to oxidation, the better the quality of biodiesel during its useful cycle. However, the value obtained for the oxidation stability out of specification is significant not to cause corrosion in the short term, long term only [5]. This parameter out of specification can be explained by the fact that after the transesterification process has not any added antioxidant (slow oxidative process biodiesel) unlike the industrial biodiesels.

The industrial biodiesel 01 (BI-01) showed three parameters that are out of specification the ester content, carbon residue, and total glycerol. The raw material used by the industry in producing biodiesel is the soybean oil used according to [5], oils used contain water, typically

CHARACTERISTICS	UNIty	LIMIT	MÉTHOD	BL	BI-01	BI-02
Aspect	-	*CIA		*CIA	*CIA	*CIA
Density at 20 ºC	kg/m³	850-900	ASTM D 4052	881,6	880,3	881,7
CKinematic Viscosity at 40°C	mm²/s	3,0-6,0	ASTM D 445	4,275	4,708	4,19
Water Teor, max.	mg/kg	500	ASTM D 6304	347	405	276
Flash Point, min.	°C	100	ASTM D 93	187,5	175,5	177,5
Esther Teor, min.	% mass	96,5	EN 14103	96,6	91,0**	97,4
Carbon Residue	% mass	0,05	ASTM 4530	0,0004	0,057**	0,01
Cold Filter Plugging Point, max.	°C	19	ASTM D 6371	-3,5	2,0	-4,0
Acid Value, max.	mg KOH/g	0,5	ASTM D 664	0,0081	0,2580	0,7391**
Free Glycerol, max.	% mass	0,02	ASTM D 6584	0	0,01	0,01
Total Glycerol, max	% mass	0,25	ASTM D 6584	0,175	0,27**	0,19
Mono, Di, Triacylglycerol	% mass	take note	ASTM D 6584	Mono - 0,3855; Di - 0,3266; Tri - 0,2814	Mono – 0,75; Di – 0,46; Tri – 0,08	Mono – 0,68; Di – 0,06; Tri – 0,00
Methanol, max.	% mass	0,2	EN 14110	<0,01	0,01	0,01
Iodine Index	g/100g	take note	EN 14111	119,02	88,997	122,97
Oxidation Stability at 110 ° C, min.	Hours	6	EN 14112	5,52**	6,9	11,34

*CIA – clear and impurities absent

**Out range parameters

Table 1. Biodiesel Samples Physical-chemistry Characteristics

from 2 to 7% FFA, Free Fat Acid, when an alkali catalyst (e.g. KOH) is added to these raw materials FFA react with the catalyst to form soap and water. The soaps formed during the reaction are removed with the glycerol in the aqueous washing step. When the concentration of FFA is too large soaps inhibit the phase separation between methyl ester and glycerol and contribute to emulsion formation during the aqueous rinse. With all this explains why the ester content, carbon residue and total glycerol present parameters out of specification. It is likely that the raw material should not have had a proper treatment for the transesterification reaction having a high content of FFA, with that, there was a decrease in the yield of the reaction, which may have been the reason for the ester content display a value specified below and the content of total glycerol be above specified. This above total content of glycerol may explain the residue carbon, as these fuels with high amounts of free glycerol present problems with deposition in

http://dx.doi.org/10.5772/59479

storage tanks glycerol creating a viscous mixture that can clog fuel filters and in the case being responsible for above the value specified for the carbon residue. To be able to prove it would be necessary to make analysis with the raw material.

The industrial biodiesel 02 (BI-02) showed only the acid out of the specified. In this case biodiesel, one possible explanation microbial growth in the fuel owing to the presence of water in its shell. The presence of microbes cause increased acidity and formation of sludge that can clog fuel filters. Another cause of high acidity could be related to the process of preparation of biodiesel, because if it is not washed with acid and the withdrawal is made correctly, the product may present acidity nonstandard.

4. Thermoanalytical methods

The International Confederation of Thermal Analysis and Calorimetry (ICTAC) defines thermal analysis as a group of techniques in which a physical property of a substance and / or its reaction products is measured as a function of temperature while the substance is subjected to a controlled program Temperature [27].

Thermal analysis provides information regarding: variation of density, thermal stability; and free water; bound water; purity, melting point, boiling point, heats of transition, specific heats, phase diagrams, reaction kinetics studies of catalysts, glass transitions, etc.

The advantages are many thermal analysis (requires little sample for tests, variety of results on a single graph), and its applicability occurs in several areas: food, catalysis, ceramics, civil engineering, pharmaceuticals, inorganic, organic, petrochemical, polymers, glass, among others. But there are some disadvantages in the use of thermal analysis such as the relatively high cost of equipment [28].

4.1. Thermogravimetry and Derivate Thermogravimetry (TG/DTG)

Thermogravimetry or thermo gravimetric analysis is based on studying the variation in weight of a sample, resulting in a physical change (sublimation, evaporation, condensation) or chemical (degradation, decomposition, oxidation) versus time or temperature. It is a technique with wide application area in determining the thermal behavior of materials. The principle is based on obtaining the thermogravimetric curve (TG), by plotting mass (mg) or percentage of weight loss (y-axis) versus temperature or time (X axis). The sample container is wrapped and placed on an analytical balance in a controlled environment, where it performs the continuous and programmed heating. Upon degradation, the sample loses mass, in the form of volatiles and the sensor registers the corresponding mass loss [29-30]. The method allows to obtain the first derivative of the TG curve, called DTG curve, which allows you to view the start and end of each event of mass loss, indicating the temperature range where a particular decomposition reaction occurs.

Typically TGA curve and its derivative DTG curve are presented as in Figure 2:

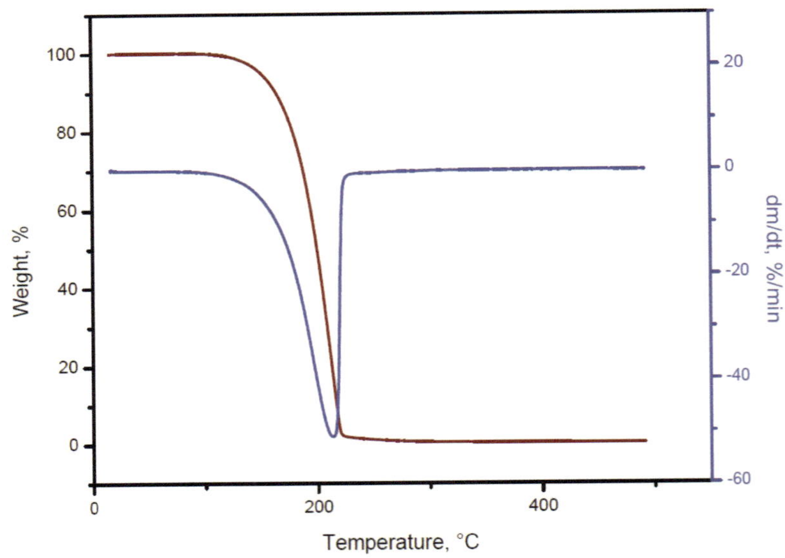

Figure 2. TGA curve (dark-red) and its derivative, DTG (blue)

4.2. Differential Scanning Calorimetry (DSC)

Differential Scanning Calorimetry (DSC) is defined as a technique which measures the temperatures and energy change associated with transitions in materials as a function of temperature or time.

DSC measures the difference in energy required for the substance and a reference material, thermally inert, while both are subjected to a controlled temperature variation so that the sample and reference are maintained under isothermal conditions, in relation the other, regardless the thermal event that is occurring in the sample.

The DSC curve provides qualitative and quantitative information about physical and chemical changes that involve endothermic processes (heat absorption), exothermic (releasing heat) or change in heat capacity [28].

There are two types of equipment that perform the Differential Scanning Calorimetry, the former is called a power compensation DSC and second DSC heat flow.

In power compensation DSC for the sample and the reference are placed in separate compartments of individual heat sources where the temperature and energy are monitored and generated by filaments of the same platinum, acting as resistive heaters and thermometers. This technique keeps constant the supplied heat. However, instead of measuring the difference in temperature between sample and reference during the reaction, a control system immediately increases the power supplied to the sample when the process is endothermic and increases the energy supplied to the process when reference is exothermic, thus preserving the sample and reference at the same temperature.

http://dx.doi.org/10.5772/59479

The second type of instrument is called by DSC "heat flow". In the oven the crucibles are arranged on a base of a highly conductive metal, usually platinum. The sample and the reference are then heated by the same power supply system. Each time the sample reacts energy flow is established between the crucibles through the base of platinum. The data in the form of electric potential (microvolts) corresponding to the temperature increase of both crucibles inside the oven should increase linearly and symmetrically. Thus, a uV versus time curve can be computed.

The flow is then measured by means of temperature sensors placed in each crucible, thereby obtaining a proportional to the thermal capacity difference between the sample and the reference signal.

Typically DSC curve with glass transition, crystallization and fusion events, respectively, are presented as in Figure 3:

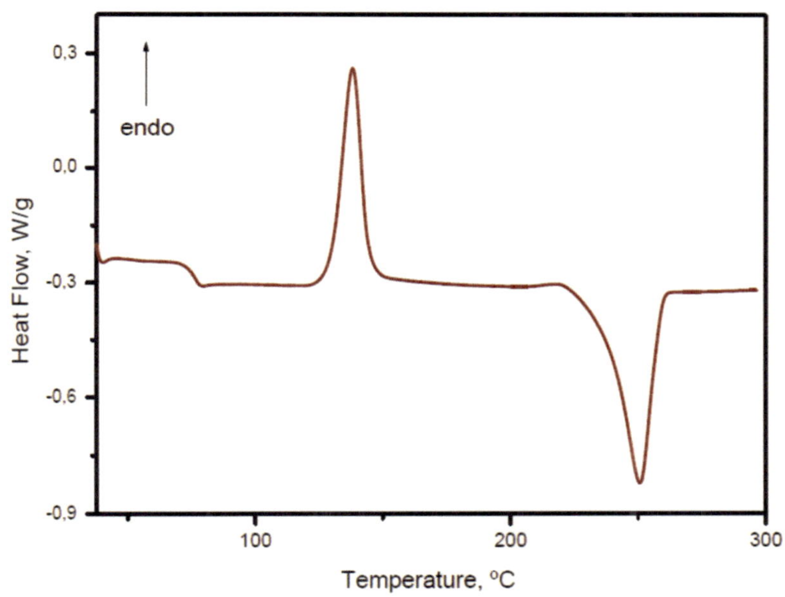

Figure 3. DSC curve with endothermic and exothermic events

5. TG profile of biodiesel samples

The TG curves were obtained in TGA equipment-Q50 from TA Instruments in an atmosphere of nitrogen at a flow rate of 100 mL min^{-1}, heating rate 20 °C min^{-1} in crucibles of platinum (Pt) as a support, in temperature range 30-650 °C.

In the TG curves of the methyl esters was only one event, which consumed more than 99 % of the sample mass. This step of weight loss is associated to vaporization of the methyl esters and the TG curve show the process end around 300 °C. Figure 4 and 5, show the comparison of the profile of the TG and DTG curves of the three samples, where it ought to be observed the similarity between them.

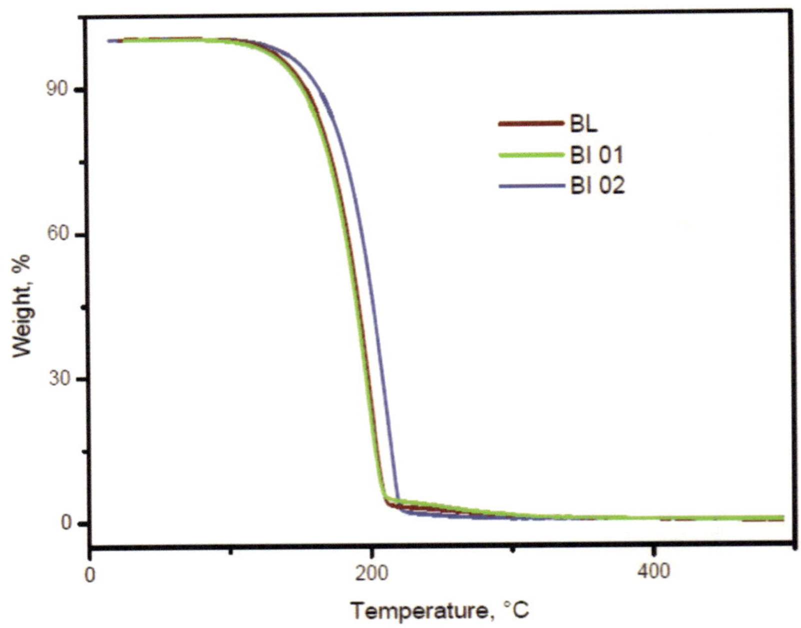

Figure 4. TG Profile of Biodiesel Samples

These curves showed very similar profile among them, the BI 02 showed a slightly higher thermal stability than the BI and BL-01, this fact can be related to oxidative stability, BI-02 showed higher oxidative stability than the other samples. The profile of the TG curves of the samples of biodiesel was similar to the soybean oil and fossil diesel, with only one step weight loss. The difference is in the temperature range in which vaporization of the methyl ester, soybean oil and fossil diesel occurs, begin around 155 °C, 357 °C and 30 °C respectively, as shown in Figure 6.

The large difference in thermal stability between the soybean oil and biodiesel is due to the fact that the oil molecules are larger than the samples of biodiesel oil is composed primarily of triacylglycerols while the biodiesel is composed of the methyl esters of fatty. Despite Soybean Oil satisfactory performance as a fuel or as a fuel source its spread is compromised, because its cost is high when compared to petroleum. Apart from being highly viscous, causing poor fuel atomization in the combustion chamber of the engine, causing serious operational problems such as the occurrence of deposits in their inner parts [5].

The biodiesel commercialization is made by mixtures with fossil diesel, the energy matrix was introduced in 2005 with the addition of 2 % of this fuel and currently the law determines to be

http://dx.doi.org/10.5772/59479

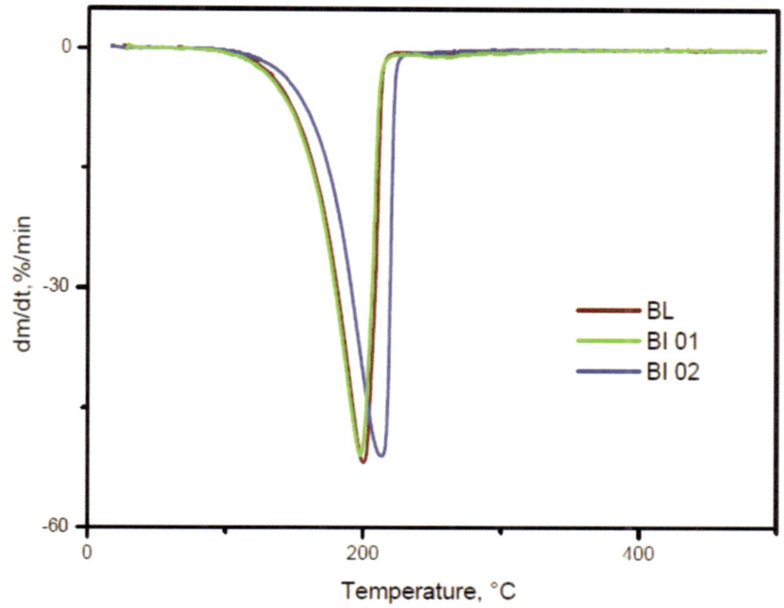

Figure 5. DTG Profile of Biodiesel samples

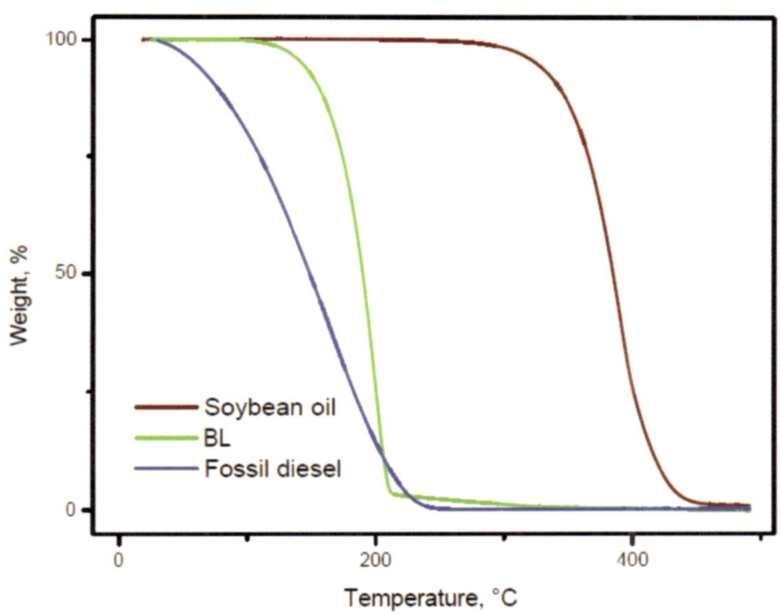

Figure 6. Soybean Oil, Biodiesel and Diesel Fossil TG Profile

plus 5 % biodiesel to fossil diesel. We observed the TG/DTG profile of the mixtures as shown in figure 7 and 8.

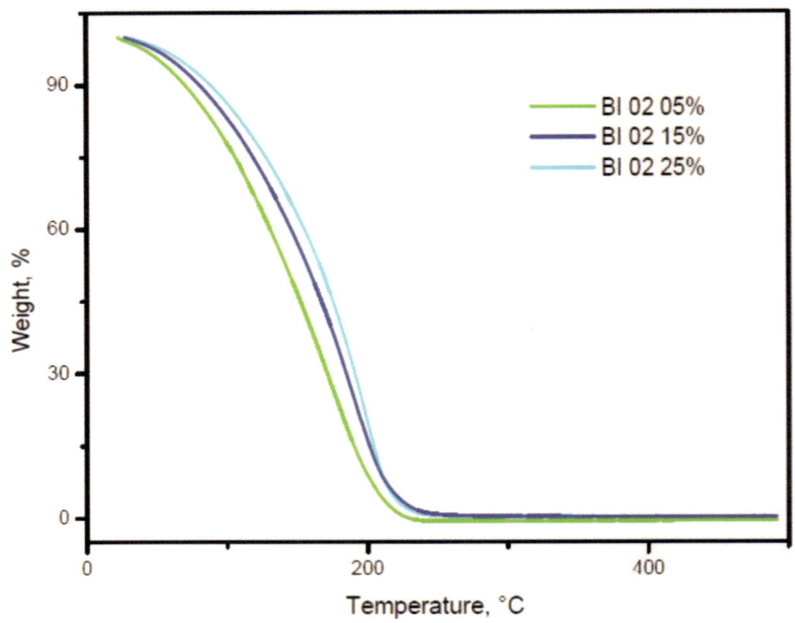

Figure 7. Profile of TG Fossil Diesel and biodiesel (BI 02) blends

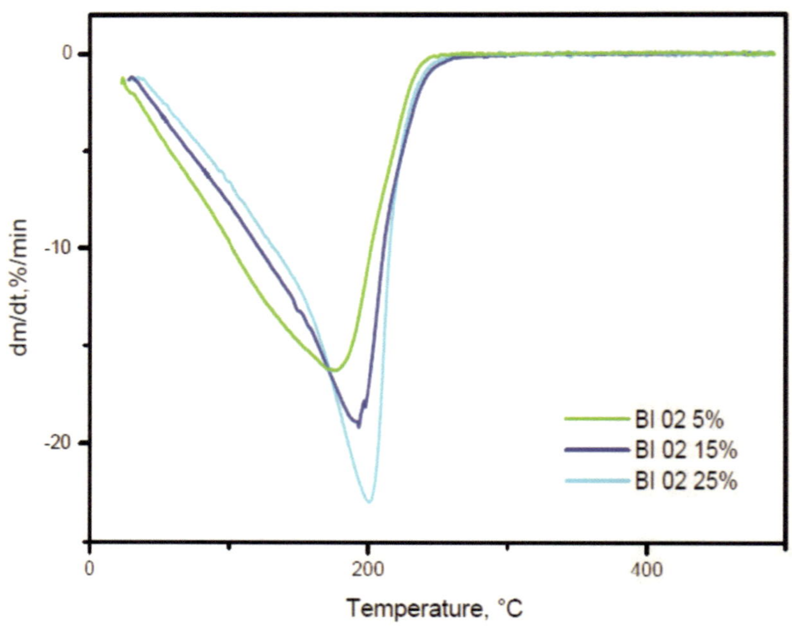

Figure 8. Profile of DTG Fossil Diesel and Biodiesel (BI 02) blend

The TG curves of mixtures of samples added to Biodiesel Diesel Fossil showed similar profiles with only one stage of mass loss (for mixtures with BL and BI 01 curves were similar to BI 02 profile, so only presented a figure 7. The curves profile of mixtures in all ratios, was similar to that of vaporization of fossil diesel profile, it was found that an increase of the concentration of biodiesel there was a displacement of the curve to higher temperatures, indicating a higher relative stability of the mixture to pure fossil diesel.

6. DSC biodiesel samples

The DSC curves were obtained in the device DSC-Q20 with cooling system RCS-90 from TA Instruments, using as a sample holder, aluminum crucibles with cover and stuck to both the crystallization and to volatilization. As a reference, a similar empty crucible was used. Nitrogen atmospheres with flow rate of 50 mL min^{-1} and heating rate of 20 °C min^{-1} (vaporization) and 5 °C min-^{1} (crystallization and melting) at temperatures from -80 to 550 °C.

6.1. DSC profile

For the three biodiesels, there was only one endothermic event with a temperature around 250 °C. Thus it can confirm that the event was indeed observed in the TG vaporization of biodiesel. The differential scanning calorimetry was used to check the physical and chemical transitions occurring in the vaporization of soybean biodiesel process. According to the DSC curve, there is an endothermic transition for each biodiesel: BL showed a ΔH=329.1 J g^{-1}, BI 01 showed a ΔH=252.5 J g^{-1} and BI 02 showed a ΔH=272.7 J g^{-1}. These transitions can be attributed to the processes of vaporization of methyl esters [31]. Such events can be seen in Figure 9:

6.2. Biodiesel crystallization

Comparing the curves of biodiesels (BL, BI-01 and BI-02) can be seen that all curves have two state transitions, the first liquid-liquid crystal formation, which is the solidification of a fraction rich in unsaturated compounds and the second, which is a freeze, which corresponds to the fraction rich in unsaturated compounds [32]. Crystallization of saturated compounds is represented by an exothermic peak at around-5 °C. In this range crystallization BL has the lowest crystallization temperature, -4 °C, followed by BI-02 and BI-01 which crystallize to -2.8 °C and 1.9 °C respectively. The crystallization points of unsaturated compounds are below -50 °C, as observed for the BL, BI-02 and BI-01 have the second crystallization temperature of -61.2 °C, -59.8 °C and -57.6 °C, respectively, as observed in the curves of cooling to -80 °C in Figure 10.

Because the greatest amount of unsaturated compounds in soybean biodiesel be unsaturated compounds that peak is observable below – 50 °C is larger in area, requiring more energy to crystallize than the saturated compounds. The peak around-50 °C needed for the phase transition of 68 and 58 J g^{-1} to BL and BI-02 respectively compounds. The peak at around-5 °C needed 16 and 15 J g^{-1} to BL and BI 02 respectively compounds. BI-01 for the peak around -50

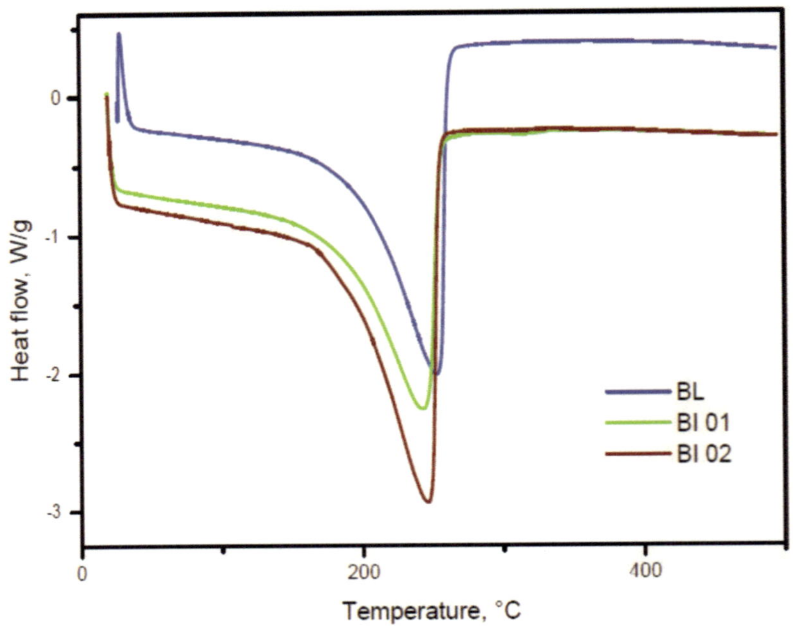

Figure 9. biodiesel sample DSC profile

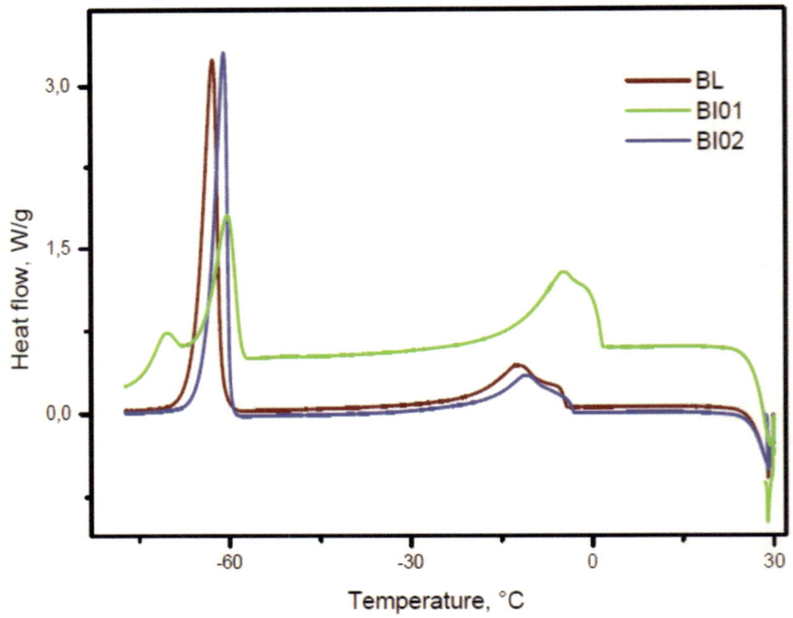

Figure 10. Crystallization of samples of biodiesel – Cooling curve

°C took 32 J g^{-1} for the largest peak and 7 J g^{-1} for the smaller peak. For the peak at around 0 °C took 30 J g^{-1} for the phase transition.

In the heating curves from – 80 °C to 30 °C may be seen two peaks of the curves BL and BI-02, referring to melting of the unsaturated (-46 °C) saturated (0 °C) compounds. To BI-01 may be seen three peaks related to the two unsaturated compounds (-53 to -40 °C) and references to unsaturated compounds (5 °C). BI-01 showed different behavior in crystallization of unsaturated compounds, there were two peaks of the onset of crystallization while BL and BI-02, showed only a peak of crystallization.

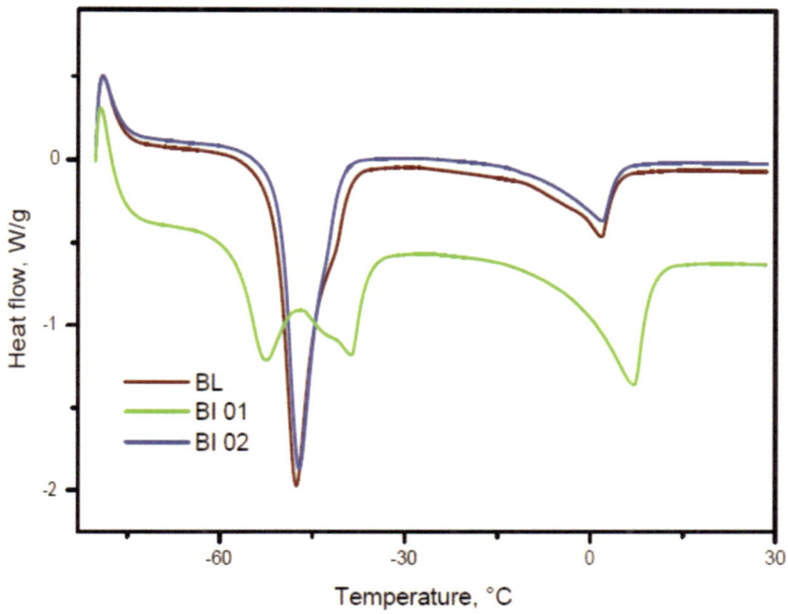

Figure 11. Crystallization of samples of biodiesel – heating curve

This fact may be related to the type of oil used in biodiesel production. As is reused oil underwent oxidation and hydrolysis of their triglycerides, causing increase in conjugated *dienes* and *trienes* unsaturated bonds [5], causing another peak appears in the unsaturated region.

7. Future development areas

Various types of vegetable oils have been tested in the preparation of biodiesel (canola, soybean, corn, sunflower, castor beans, cotton, etc.). The type of oil to be used in the production of biodiesel depends on geographical factors, as each region produces a certain kind of oil, according to their fitness. In some European countries, for example rapeseed oil, Brazil is already in use, depending on the region, can be produced from soy, castor oil or babassu [6] oil. Another highlight is the use of oils used in frying whose nobler purpose was hitherto the production of soap or disposed of in a sanitary [33] landfill. The high price of vegetable oils has become non-competitive biodiesel economically forward to petroleum diesel, programs and government incentives are needed. The search for new raw materials for the production of biodiesel and are more environmentally friendly and economically viable.

Biodiesel is a cleaner-burning alternative fuel, produced from renewable resources, does not contain oil, but it can be added to form a mixture. Can be used in compression-ignition engines (diesel) without modification. Non-toxic, essentially free of sulfur and aromatics [34] compounds. Its main disadvantages include low pour point (on cold) and maintain fuel quality during long-term storage [35]. A good source for development would address this disadvantage that has biodiesel.

8. Conclusions

Biodiesel production in the laboratory of alkaline transesterification reaction and the treatment of methyl esters were successful. All parameters except the oxidative stability were within the established in ANP Nº 45/2014, indicating a good quality biodiesel. The oxidative stability below specification may have been due to lack of the addition of antioxidant.

Industrial biodiesels showed some parameters out of specification. BI-01 showed three parameters out of specification (ester content, carbon residue content of glycerol), which is probably associated with the raw material. The BI-02 showed only one parameter out of spec (Acid Value) which can be associated to microbial growth in storage tanks or the fact that after neutralization of excess acid base with the same has not been washed properly.

By TG curves, soybean oil showed higher thermal stability than Biodiesel and Fossil Diesel. In oil, the onset of decomposition was 358 °C, while biodiesels began to vaporization around 160 °C. For fossil diesel mass loss started around 30 °C, whereas the increase in the binary mixtures Biodiesel slightly increased thermal stability.

The biodiesels were tested at a heating rate of 5, 10, 15 and 20 °C and all were obtained in the same profile.

The DSC curves of biodiesels calorimetric showed a similar profile. The analysis of biodiesels had only one endothermic peak on the vaporization of the esters. Comparing ΔH of biodiesels BL> BI-02> BI-01 with the respective ΔH 329.1> 272.7> 252.5 J g^{-1}.

The TG curves of the binary mixtures have very similar profile to that of fossil diesel, biodiesel from the addition made to increase the thermal stability of the mixture. When compared with fossil diesel, it begins to loss mass around 30 °C and the Biodiesel that begins to vaporization at 160 °C there has been a small increase in the temperature at which it begins to lose mass, and that the More mixture containing biodiesel B25, takes a longer time to fully decompose when compared with the mixture of 5 %, indicating greater thermal stability.

http://dx.doi.org/10.5772/59479

In the curves of crystallization of biodiesels (BL, BI-01, BI-02) shows that all the curves have two transitions of liquid-solid state, one that represents the solidification of a fraction rich in saturated compounds and the other corresponding to rich fraction unsaturated compounds. Crystallization of saturated compounds is represented by an endothermic peak at around 0 °C, and the unsaturated compounds endothermic peak around – 60 °C.

Acknowledgements

The authors thank the Brazilian funding agencies FUNDECT, CAPES and CNPq for financial support and UFMS by assigned space.

Author details

Marcelo Y. Misutsu, Leandro F. Cavalheiro, Thiago G. Ricci, Luiz H. Viana, Silvio C. de Oliveira, Amilcar Machulek Junior and Lincoln C. S. de Oliveira*

*Address all correspondence to: lincoln.cso@gmail.com

Institute of Chemistry, Federal University of Mato Grosso do Sul – INQUI/UFMS, Brazil

References

[1] Goldemberg J. Biomassa e Energia. Química Nova. 2009, 32,582

[2] Exame. As 20 maiores reservas de petróleo do mundo. http:/exame.abril.com.br (accessed 21 September 2013)

[3] Galembeck F, Barbosa CAS, Sousa RA. Aproveitamento Sustentável de Biomassa e de Recursos Naturais na Inovação Química. Quím. Nova. 2009, 32, 571.

[4] Lima JRO, Da Silva BR, Da Silva CCM, Dos Santos LSS, Santos JR, Ribeiro J, Moura EM, Moura VCR. Biodiesel de babaçu (ORBIGNYA SP) obtido por via Etanólica. Química Nova, v. 30, n. 3 p. 600-603, 2007.

[5] Knothe, G.; Gerpen, J. V.; Krahl, J.; Ramos, L. P. Manual de Biodiesel, 1ª ed. São Paulo: Editora Blucher, 2006

[6] Ferrari RA, Oliveira VS, Scabio A. Biodiesel de Soja – Taxa de conversão em ésteres etílicos, caracterização Físico-Química e consumo em gerador de energia. Química Nova, 2005, 28 (n.1), 19.

[7] Galvão LPFC. Avaliação termoanalitica da estabilidade oxidativa do biodiesel de Mamona. Dissertação de Mestrado. Universidade Federal do Rio Grande do Norte; 2006.

[8] Kerschbaum S, Rinke G, Schubert K. Winterization of biodiesel by micro process engineering. Fuel 87, 2590–2597, 2008

[9] Kaya C, Hamamci C, Baysal A, Akba O, Erdogan A, Saydut A. Methyl ester of peanut (Arachis hypogea L.) seed oil as a potential feedstock for biodiesel production. Renewable Energy 34, 1257–1260, 2009.

[10] Soriano NU, Migo VP, Matsumura M. Ozonized vegetable oil as pour point depressant for neat biodiesel. Fuel 85, 25–31, 2006.

[11] Ramos MJ, Fernández CM, Casas A, Rodríguez L, Pérez A. Influence of fatty acid composition of raw materials on biodiesel properties. Bioresource Technology 100, 261–368, 2009.

[12] Papadopoulos CE, Dazaridou A, Koutsoumba A, Kokkinos N, Christoforidis A, Nikolaou N. Optimization of cotton seed biodiesel quality (critical properties) through modification of its FAME composition by highly selective homogeneous hydrogenation. Bioresource Technology 101, 1812–1819, 2010.

[13] Knothe G, Dunn RO. A comprehensive evaluation of the melting points of fatty acids and esters determined by differential scanning calorimetry. Journal of the American Oil Chemists' Society 86, 843–856, 2009.

[14] Soares, C.M., Itavo, L.C.V., Dias, A.M., Arruda, E.J., Delben, A.A.S.T., Oliveira, S.L., de Oliveira, L.C.S. Forage turnip, sunflower, and soybean biodiesel obtained by ethanol synthesis: Production protocols and thermal behavior. Fuel, 89, 12, 3725-3729, 2010

[15] Canakci M, Van Gerpen J. Biodiesel Production From Oils and Fats with High Free Fatty Acids. Trans. ASAE, 2001, 44, 1429

[16] Monyem A, Van Gerpen JH. The effect of biodiesel oxidation on engine performance and emissions. Biomass & Bioenergy, 2001, 20, 317-325.

[17] Monyem A, Van Gerpen JH, Canakci M. The Effect of Timing and Oxidation on Emissions from Biodiesel-Fueled Engines. Transactions of the ASABE. Vol. 44(1), 35-42.

[18] Encinar JM, González JF, Rodríguez JJ, Tejedor A. Biodiesel Fuels from Vegetable Oils: Transesterification of Cynaracardunculus L. Oils with Ethanol. Energy Fuels, 2002, 16, 443.

[19] Noureddini H, Harkey D, Medikonduru VA. A continuous process for the conversion of vegetable oils into methyl esters of fatty acids. Journal American Oil Chemical Society, 1998, 75, 1775.

http://dx.doi.org/10.5772/59479

[20] Freddman B, Butterfield RO, Pryde EH. Transesterification kinetics of soybean oil 1. Journal of the American Oil Chemists Society 1986, 63, 1375-1380.

[21] Kusdiana D, Saka S. Methyl Esterification of Free Fatty Acids of Rapeseed Oil as Treated in Supercritical Methanol. Journal of Chemical Engineering of Japan 2001, 34, 383.

[22] Shuchrdt, U.; Sercheli, R.; Vargas, M. Transesterification of Vegetable Oils: a Review. Jounal of the Brazilian Chemical. Society, 1998, 9, 190.

[23] Noureddini H, Medikonduru V. Glycerolysis of fats and methyl esters. J. Am. Oil Chem. Soc., 1997, 74, 419.

[24] Dorado MP, Arnal JM, Gómez J, Gil A, Lopez FJ. The effect of a waste vegetable oil blend with diesel fuel on engine performance. Trans. ASAE, 2002, 45, 519.

[25] Lago RCA, Szpiz RR, Hartman L. Transformação de óleos vegetais de alta acidez em ésteres etílicos na presença de catalisadores ácidos para fins combustíveis. Química Industrial, 1988, 666, 8-11.

[26] SEBRAE. Biodiesel. http://www.sebrae.com.br/sites/PortalSebrae/bis/Conhe%C3%A7a-a-cartilha-do-biodiesel (acessed July 2012).

[27] Ionashiro M, Giolito I. Fundamentos da termogravimetria, Análise térmica diferencial, Calorimetria exploratória diferencial. São Paulo. 2004

[28] Mothé CG, Azevedo AD. Análise térmica de materiais. São Paulo: iEditora, 2002.

[29] Lucas, E. F.; Soares, B. G.; Monteiro, E. Caracterização de Polímeros.Determinação de Peso Molecular e Análise Térmica. Rio de Janeiro, LTC, 2001.

[30] Haines, P. J. Thermal Methods of Analysis, Blackie Academic & Professional, 1995.

[31] Gondim AD. Avaliação da estabilidade térmica e oxidativa do biodiesel de algodão e do efeito da adição de antioxidantes (α – tocoferol e BHT). Tese de doutorado. Universidade Federal do Rio Grande do Norte; 2009.

[32] Pérez, A. et. al. Winterization of peanut biodiesel to improve to cold flow properties. Bioresource Technology, 101, 2010, 7375-7381.

[33] Zaher F. Utilization of used frying oil as diesel engine fuel. Energy Sources 2003, 25, 819-826.

[34] Parente EJS. Biodiesel: uma aventura tecnológica num país engraçado. Fortaleza: Tecbio, 2003.

[35] Dunn RO. Effect of Antioxidants on The Oxidative Stability of Methyl Soyate (Biodiesel). Fuel Processing Technology. 2005, 86, 1071-1085.

Chapter 13

Qualitative Characteristics of Biodiesel Obtained from Sunflower Oil

Estelvina Rodríguez Portillo, Araceli Amaya Chávez,
Arturo Colín Cruz and Rubí Romero Romero

Additional information is available at the end of the chapter

http://dx.doi.org/10.5772/59673

1. Introduction

This chapter presents the results of analysis parameters of biodiesel produced from sunflower oil in Paraguay. The analysis was prepared according to the requirements of quality demanded by the Paraguayan standard methods for this purpose, based on parameters of the American Society for Testing and Materials (ASTM) and European Norms (EN). It was found that the biodiesel produced in this country comply with international standards.

The ASTM defines biodiesel as a monoalkyl ester of long chain fatty acids derived from vegetable oils, seeds, or animal fats. Biodiesel is obtained from vegetable oil or animal fat, has significant environmental benefits such as being nontoxic, having less emission, and being biodegradable [1, 2]. Biodiesel is currently defined in the European Union in the technical regulation (European Norms) EN 14214 or in the USA in ASTM 6751-02 [3]. The most common is that the esters, which are part of the composition of biodiesel, were methyl, so they are called fatty acid methyl ester (FAME).

Biodiesel characteristics as a fuel vary depending on its composition, and the fuel used to be stringently monitored to avoid adverse impacts on the environment and engines. A very important consideration in the efficiency of combustion is the ignition delay in the combustion, which is influenced by the ratio of compression adopted, which in turn is related to the kind and quality of fuel used. The physicochemical properties of the biodiesel may vary depending on the source from which the mixture of fatty acids was obtained, from the transesterification and separation efficiency process. Factors such as hydrocarbon chain length, branching, and degree of saturation influence their composition; hence, quality control is important to guaranty the combustion efficiency and to lower atmospheric emissions [4, 5].

In the specific case of biofuel, some physicochemical properties have great influence in the ignition, combustion, and contaminant formation when a diesel engine is used.

As a result, the evaluation of these kinds of properties has great relevance because it practically defines the fuel's usefulness. Therefore, these papers focus on the chemical properties, reviews according to the standardized quality requirements, and test methods for biodiesel that are mentioned in the Paraguayan and International regulation.

1.1. Biodiesel production processes and quality

The world production of biodiesel between 1993 and 2003 increased to an impressive annual rate of 28.5%, from 38 to 467 million gallons, while bioethanol production increased to an annual rate of 6.7% in the same period, reaching 5.770 million gallons in the year 2003 [6]. In the USA, the annual production of biodiesel was approximately 570 million liters in 2004 (80% from soy, 19% from animal fat, and 1% from other crops). From 1999 to 2001, the car fleet that used biofuel increased to a 100%. Due to the increasing environmental concerns related to the emissions of fuel-derived atmospheric pollutants, alternative sources of energy have been receiving greater attention.

One of the things that have permitted the growth of biodiesel market is the establishment of quality regulations, trying to homogenize the production as well as the soy subsidy.

The European Union is the world leader in biodiesel production and consumption. In the year 2004, it produced around 2.2 billion liters, and the three main countries that produce this energy-giving fuel are Germany (1.4 million tons), France (1.35 million tons), and Italy (0.32 million tons). Two factors have permitted the European Union to become the production leaders; the first has to do with a Common Agricultural Policy (CAP), which is oriented to a policy of crop of the European Union members, where the subsidy to grain, oil seed, and protein crop producers was promoted so they could dedicate 10% of their lands to produce input to obtain biodiesel.

The second factor would be the high fuel taxes that have established direct subsidy for biofuel from a partial or total tax exempt. In the year 2004, 90% of fuel taxes ware exempt by the European Parliament; in Germany, 100% of taxes were exempt.

The biodiesel in Paraguay is manufactured by the transesterification of vegetable oils, in this case, sunflower oil (*Helianthus annuus*) in presence of a catalyst. The main component of vegetable oils and animal fats are triacylglycerols (TAGs). TAGs react with long three chain fatty acids and alcohol (mostly methanol), with a 6:1 ratio, to produce a mix of fatty acid methyl esters (biodiesel), and glycerol is the bioproduct. Thus, biodiesel production depends on the origin of oil used, the transesterification process, and the distribution and storage (Figure 1).

The replacement of gasoline with biodiesel does not require any engine modification, except for the combined change of natural rubber gasket (in old models) and fuel filters after using biodiesel (used diesel cars). This will apply if it is good-quality biodiesel.

The quality of biodiesel can be described in two groups: (1) the general physicochemical properties, for example, density, viscosity, flash point, % sulfur, Conradson carbon residue, %

http://dx.doi.org/10.5772/59673

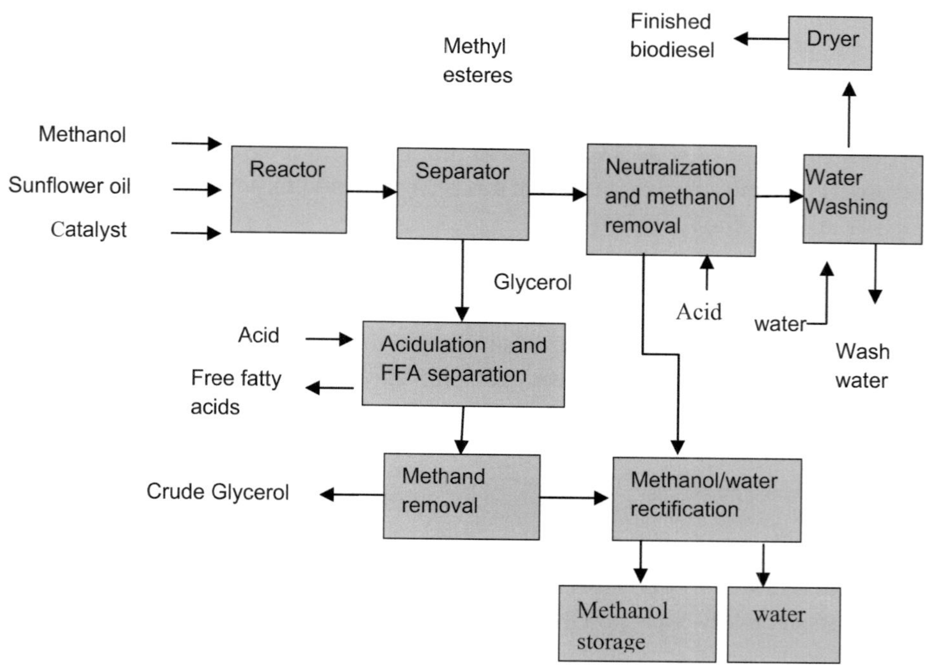

Figure 1. Schematic process flow for biodiesel production.

sulfate, ketonic number, and acid number, and (2) the composition and purity of fatty esters such as methanol, free glycerol, total glycerol, water, and esters contents, among others (Table 1) [7]. The evaluation of biodiesel quality is achieved through the determination of the chemical composition and physical properties of fuel [7]. In fact, some contaminants and other minor components are the major issues in the quality of biodiesel.

Quality factors	Indicator
Oil source	Acidity Phosphorus content Oxidation stability Iodine index
Production process	Water content, impurities, waxes Esterification Stages of separation Distillation
Distribution and storage	Oxidation stability Water content

Table 1. Quality factors of biodiesel.

As to normative reference, the International Standard Organization (ISO), the American Society for Testing and Materials (ASTM), and the European Norms (EN) have been applied in several countries for biodiesel quality control. The International Standard Organization (ISO) has developed quality standards for oil and products derived from them. Most of the other specifications for biodiesel are based on these standards. While the reference norm is comprehensive, the most representative of it is being adapted by several countries, especially the EU and the set by the ASTM.

The American Society for Testing and Materials (ASTM) describes different tests to ensure the correct function of the fuel. These specifications define the properties and the main checkpoints for the feasible use of biodiesel in the market, cited in the ASTM D 6751 guide, where most of the requirements demanded by the producer countries are based. This international organization of standardization has recently announced the publication of four rules regarding biodiesel, including the mentioned D 6751, which are as follows [8]:

a. The norm ASTM D975-08a (specification for diesel fuel oil) is applied to diesel engines for transportation, with an inspection that allows mixing 5%.

b. The norm ASTM D396-08b (specification for fuel oil) is referred to use of domestic heat and boilers. It has an inspection that also allows mixing 5% of biodiesel for the mentioned effects.

c. The norm ASTM D7467-08 (specification for diesel fuel oil with a mix of biodiesel B6 to 20) is a completely new specification because it includes mixtures of fuel finished between 6% (B6) and 20% (20%) for engines used for transportation.

d. The norm ASTM D 6751-08 (specification of biodiesel for mixture B100 destined to medium distilled fuel) is used to control the quality of pure biodiesel (B100) before mixing it with regular diesel, and it has an inspection that includes a requirement to control minor compounds through a new cold filtration test.

Nowadays, the use of 100% biodiesel is unknown since the mixture of 5–20% has been applied gradually. This is a way to ensure the correct operation of automotive machines.

According to ASTM, engine and car manufacturers, pipeline companies, biodiesel producers, and oil companies will use this group of specifications to prepare fuel for quality control, engine design, bidding processes, and acquisition contract.

The rule UNE EN 14214:2003 of the European Community specifies the requirements and the assay methods of the FAMES that are marketed and provided for use, such as the following:

a. Automotive fuel in diesel engines (100% biodiesel)

b. Mixture with diesel (rule EN 590)

This norm has the special feature of including the iodine value, which is not included in the ASTM since they generally use colza oil in biodiesel manufacturing in that continent. The maximum acceptable value is IV = 115, which would exclude other oils like soy and its esters because these exceed the limits [9].

http://dx.doi.org/10.5772/59673

Obviously pointing out once more that the properties are dependent from the raw material that is used; that way, it roughly had parameter deviation. With regard to the current state of biodiesel production technology, it can be said that is already tested, relatively mature, in dissemination period, capable of taking advantage of different raw material, and that it has reached market level in several countries. Currently, most of the biodiesel is produced by methanolysis in basic medium [10].

The challenge for any country or region consists in the implementation of processes based on native raw material, which should be optimized to obtain a low production cost biodiesel that would make it competitive but fulfilling the international specifications of quality in order to be used as a diesel engine. These parameters are of great importance since the characteristics have a major impact on the diesel engine behavior, as well as on the contaminant emission when it is used, and in securing storage and transporting conditions among others. Therefore, to know the mechanic and environmental efficiency, some indicators such as water presence, acid index, methanol content, triglycerides, etc., are of great importance. Impurities such as glycerides, glycerol, free fatty acids, and catalyzed waste bring adverse consequences to the engine performance, for example, soot deposits in the injectors. The mass calorific power of biodiesel is 13% lower than diesel and around 8% per volume unit; however, it is not exactly revealed in the loss of power because biodiesel has a slightly higher density than diesel [11]. In Paraguay (study field of this paper), the legislation NP 16 018 05 is the one that specifies the quality requirements of the produced biodiesel, and it corresponds with the American Union ASTM norm and the European Union EN (Table 2). In this legislation, the requirements and trial methods for pure biodiesel (B100) applied in diesel engines are established.

The values that overpass the limits marked by the norm would lead to engine problems as mentioned before. However, those are highly dependent on the raw material. Therefore, the main challenge when evaluating the quality is to improve the regularity and homogeneity of the raw material supply and to optimize the production process in order to standardize the final product. Despite its many advantages, it also has many problems. One of them comes from its better solvent capacity than the regular diesel, so the existent residues are dissolved and sent through the fuel line being able to clog the filters [12]. Another item is a less energetic capacity, approximately 5% less, although it is not that notorious in practice because it is compensated with the higher ketone index, which produces a more complete combustion with less compression. Certain hypotheses suggest that more combustion deposits are produced and that the cold starting of engines is degraded, but there are no records of it. Another difficulty is referred to the storage logistic area since it is degradable in a relatively short time. Therefore, an exact planning of its production and expedition is necessary, for which the quality parameters are essential.

Some potential problems of the net biodiesel-operated engine or with high-level mixtures as well as bad quality are as follows [12]:

a. Clogging and filter obstruction

b. Nozzle and injection hole block, ducts and passages, and draining of the fuel feeding system blockade

Requirement	Unit	Limits		Trial method[1]
Minimum	Maximum			
Ester content[2]	% (m/m)	96.5		EN 14103
Density at 20°C	g/mL	Report		ASTM D 1298//ASTM D 7042 ISO 3675 // ISO 12185
Viscosity at 40°C[3]	mm²/s	3	6.5	ASTM D 445 ISO 3104 IRAM-IAP A 6597
Flash point	°C	100		ASTM D 93//ISO/CD 3679
Sulfur content	mg/kg		10	ASTM D 5453
Carbon residue	g/100 g		0.3	ASTM D 189 ASTM D 4530 ISO 10370
Cetane number		Report[4]		ASTM D 613 ISO 5165
Sulfated ash	% (m/m)	0.05		ASTM D 874 ISO 3987
Water content	% (m/m)		0.080	ISO 12937
Silt content	ppm		24	ASTM D 2276
Copper foil corrosion (3 h at 50°C)			1	ASTM D 130 ISO 2160
Oxidation stability at 110°Chours		6		En 14112
Acid index	mg KOH/g		0.8	ASTM D 664 IRAM 6558 EN 14104
Total glycerin content [5]	% (m/m)		0.25	EN 14105//ASTM D 6584 (B)
Free glycerin[5]	% (m/m)		0.02	EN 14105//EN 14106 (C) ASTM D 6584
Free methanol or ethanol content[6]	% (m/m)		0.2	EN 14110
Alkaline metal (Na + K)	mg/kg		5	EN 14538
Phosphorus content	mg/kg		10	ASTM D 4951//EN 14107
Cloud point[7]	°C	Report		ASTM D 2500

Source: INTN (2008).

[1] The one contained in the first term four each essay is the discrepancy method.

[2] Includes carbon 17.

[3] The determination of biodiesel viscosity can be done through the Saybolt viscosimeter to be converted later to mm²/s according to what is indicated in the table (ASTM D 2161).

[4] The used raw material must be specified.

[5] The methodology indicated in B and C attachments is informative, and it can be applied in the industries as an alternative method but not as a reference until the publication of it as a Paraguayan norm.

[6] If the flash point is equal or more than 130°, it will not be necessary to analyze this item.

[7] The cloud point must be reported.

Table 2. Quality requirements in the Paraguayan norm and trial methods.

c. Increase of the injection pressure with an excessive pressure drop

d. Stuck and broken piston rings

e. Formation of deposits on the injectors, pistons, and piston slot

f. Clogging of the fuel pump because of high viscosities

g. Lack of fuel feeding to the engine (power drop) because of high viscosities

h. Accelerated wear of valves, needles, and injection puma pistons and of the injectors

Between the causes attributable to the biodiesel properties, we can find the deposit production due to excess metal that causes ash formation and ash abrasion, sediment formation through polymerization, or crystallization of heavy molecules or crystallization and jellification at low temperatures. In addition, it also causes acid, aldehyde and ketone oxidation, polymerization and degradation, ester hydrolysis with free acid formation, water accumulation, microbial growth and associated iodine formation, and low fuel volatility [13].

1.2. Raw material and quality of biodiesel

The most important biodiesel production to ensure trouble-free operation of diesel engines aspects is complete reaction, removal of glycerin, removal of catalyst, removal of alcohol, and absence of free fatty acids. If any of these aspects are not adequately able to meet specifications, different types of problems arise in the motor, such as excessive formation of soaps, injection deposit formation, corrosion, etc. Other aspects, such as the removal of methanol, are important from the standpoint of safe operation of the fuel. Biodiesel production from various raw materials is technically and economically feasible, including projects in small scale. This opens an opportunity for a large number of small and medium enterprises who wish to produce their own fuel by providing a wide range of possible raw materials.

The vegetable oil used in the production of biodiesel can be obtained from various oil seeds. These species differ in their agronomic characteristics, and relative to the oil content in the composition of grain and fatty acid profile.

The main characteristics of the raw materials used in the manufacture of biodiesel that have the greatest influence on its quality are as follows:

a. Oil content: The oil content is an important feature that can influence the choice and use of a raw material for the production of biodiesel from vegetable oil. Sunflower, rapeseed, jatropha, castor, and groundnut are the sources that have higher oil content in grain, a variation of 40–64% oil. Soybean, palm, and cotton have low oil level at around 15–25%. Associated with this feature, the oil production per hectare and the production cycle of each commodity must also be considered [14]. For crops of sunflower, rapeseed, and peanut oil, content in grain and oil yield per hectare are similar. The less output of oil per hectare is presented for growing cotton.

The inclusion of nonfood raw materials in the production of biodiesel is seen as an important ally, but that does not compete with the raw materials used in food.

b. Fatty acid composition: The variable composition profile and fatty acid content of the raw materials presented interferes with features and quality of biodiesel. The content of saturated and unsaturated fatty acids is highly variable among commodities. Palm oil has the highest content of saturated fatty acids (51.5%), while castor oil has a lower content (1.6%). Rapeseed oil has 6.5% saturated fatty acids, while soybean oil, sunflower, and groundnut have values between 11.7% and 17.8% [14].

c. Influence of the composition and acid ester content fatty raw materials on the properties of biodiesel: The composition and the content of fatty acid esters directly influence the properties of biodiesel, and consequently the quality and performance as fuel. Ester formed during transesterification with the alcohol has the same profile of fatty acids of vegetable oil source as the process transesterification does not affect the fatty acid composition [15].

d. Oxidative stability: This refers to the relative resistance to physical changes and produced by chemical interaction with the environment during storage of a liquid fuel. The composition of esters of fatty acids, the size of chain, and the presence of unsaturation directly affect the oxidative stability. Vegetable oils contain natural antioxidants, tocopherols being the most common. Antioxidants are highly reactive with free radicals to form stable compounds and, therefore, do not contribute to the oxidative process [16]. The biodiesel produced from raw materials rich in saturated and monounsaturated fatty acids has better oxidation resistance.

e. Low temperature properties: A major problem associated with the use of biodiesel is their low resistance to low temperatures and can be indicated by parameters such as cloud point, pour point, and filter plugging point at low temperature. At lower temperatures, the formation of crystal nucleation wax solids and the temperature decrease cause the crystals to increase in size. Saturated fatty compounds have higher melting points than the unsaturated. Therefore, the biodiesel produced from vegetable fats and oils with significant amounts of unsaturated fatty compounds should be lower for the cloud and pour points values, contrary have lower oxidative stability [16].The biodiesel from soybean, sunflower, rapeseed, jatropha, peanut, and cotton are the best low temperature properties. While palm biodiesel has a high point of obstruction cold filter. The castor biodiesel shows good oxidative stability and good low temperature properties, which contradicts what was observed for the other oilseeds.

With so many differences between raw materials and fatty acid profile, what we see is that each raw material has one or more desirable properties for biodiesel quality. Then it is necessary to examine what is the best raw material for biodiesel production. Choosing any of the raw materials must meet the standards and needs of each country, without competing with food availability. In countries with sufficient availability of grains, oilseeds may represent an alternative for diversification and promotion of agricultural industries. The high oil content of sunflower seeds produces a high-protein cake, and the crop shows good adaptability with respect to soil and temperature, making this oilseed a good alternative for biodiesel. The

http://dx.doi.org/10.5772/59673

production of biodiesel from sunflower involves simple procedures resulting in high-quality fuel, which offsets the higher market values

For sustainable development, the promotion of biofuel previous study should be done on food demand in each country. Currently, poverty is an important factor to consider owing the food security, much more than the use of land for the production of sunflower or the labor employed. The increase in biofuel production may cause problems on agricultural priority, that is, food crops, which could obstruct the availability of raw materials. In addition, the negative environmental impacts change native vegetation for farmland. It is necessary to perform the environmental impact assessment of the biodiesel production, but there are not sufficient data available [17]. For countries like Paraguay that do not have oil, it may represent an important alternative economic development. The level of competition between energy crops and food and fodder production would depend on the progress with regard to yields, efficiency of livestock feed, and conversion technologies for bioenergy.

The status of biofuel production and the possible implications for food production and security should be analyzed since the choice of feedstock depends on local availability, cultivation, relative prices, and government incentives for specific production.

A study of local realities and skills at the time of the creation of regulatory frameworks and the use of techniques such as crop planting interspersed would enable the production of both biofuel feedstock and food production.

2. Materials and methods

The biodiesel used for the study come from three national companies that use the sunflower oil species *Eliantus annus* and transesterification process with methanol, which are unloaded at the PETROPAR S.A. (Petróleos Paraguayos S. A, Villa Elisa plant), in a common reservoir of approximately 3000 L of capacity. Then the samples are taken from these tanks corresponding to pure biodiesel (B100). The sample conditions such as irreplaceability, minimum quantity, instable, or other conditions as in situ essay are considered as eliminatory criteria in the technique and method selection.

The sampling procedure was made according to the ASTM D 4057-95 norm (Reapproved 2000), Standard Practice for Manual Sampling of Petroleum and Petroleum Products.

Ten weekly samples of 2.5 L were taken from the surface in full tank (bottom sample), 10 cm away from the bottom in half tank, or 10 cm from the output level in tanks with reservoirs that were not filled.

The determination of biodiesel quality was performed through analysis, and the protocols followed are standardized in an international level and set by norms that value some properties (Table 3). Ten analysis and ten different samples were performed using the ASTM and EN methods, within the requirements of quality established by the Paraguayan norm NP 16 018 05 (Table 2).

Method	Determination	Available equipment
EN 14103-2003	Fatty and oily derivatives. Fatty acid metallic ester (FAME)—determination of linoleic acid ester content	Gas chromatographer Headspace Shimadzu 2010, with AOC 5000 (autoinjector)
ASTM D 1298 ASTM D 4052	1. Standard practice for density, relative density (specific gravity), by hydrometer method (ASTM D1928). 2. Standard test method for density and relative density meter	1. Hydrometer 2. Digital densimeter DE 40
ASTM D 455	Standard test method for cinematic viscosity of transparent and opaque liquids (calculation of dynamic viscosity)	Precision kinematic viscosity bath—Model "S"
ASTM D 189	Standard test method for Conradson carbon residue	Device for determination of Conradson carbon
ASTM D 93	Standard test methods for flash point by Pensky–Martens closed cup tester	Pensky–Martens closed cup flash tester
ASTM D 4928	Standard test method for water by coulometric Karl Fisher titration (water only)	Karl Fisher moisture titrator MKC-501
ASTM D 130	Standard test method for the detection of copper corrosion (3 h at 50°C)	Copper corrosion bath, copper corrosion test bomb
UNE EN 141104	Acid index—volumetric method	TAN titration and burette. Brand: Metrohm, Model 664
EN 14105:2003	Total glycerin content	Gas chromatographer Headspace Shimadzu 2010, with AOC 5000 (autoinjector)
EN 14110:2003	Free methanol and ethanol content	Gas chromatographer Headspace Shimadzu 2010, with AOC 5000 (autoinjector)
ASTM D2500	Cloud point	GB/T3535 GB/T6986

Table 3. Results obtained, trial methods, and equipment used.

3. Results and discussion

The obtained results in this investigation are presented, described, and discussed in this section.

The obtained results were averaged and grouped in a table comparing the established limits for each characteristic, with their previous statistic analysis based on the Statistical Package for the Social Sciences (SPSS) version 16.0.

In Figures 2 and 3, the observation clouds and the lineal adjustment obtained in those variables that present more lineal dependence are exposed. A correlation between viscosity and density, acid, and glycerin was found.

http://dx.doi.org/10.5772/59673

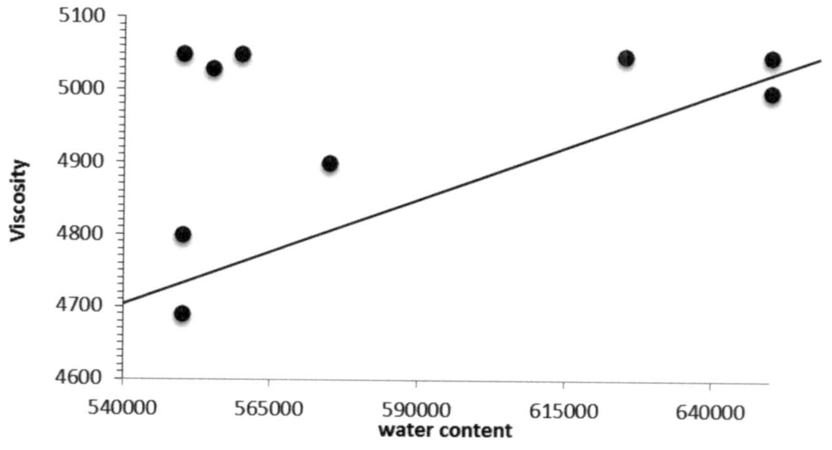

Figure 2. Water content–viscosity correlation in the biodiesel of sunflower oil.

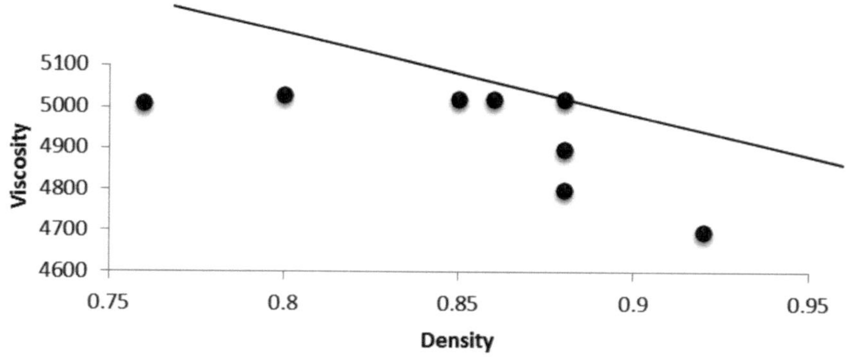

Figure 3. Density–viscosity correlation in the biodiesel of sunflower oil.

It can roughly be said that a functional dependence between the observations exists, but not the line type; therefore, the correlation is too small.

Excess water in the biodiesel causes the following:

a. Hydrolysis reactions (appearance of free fatty acids)

b. In the engine, the undissolved water can give corrosion problems

c. Bacteriological growths (clogged filters)

d. Trace of H_2SO_4

As for the viscosity, the parameter determines a minimum viscosity in which no power was lost through the injection pump and the injector outlet. The maximum value allowed is limited by considerations related to engine design, size, and characteristics of injection system.

Viscosity is a very important parameter in the fuel, as it directly affects the atomization process. Fuels with high viscosity tend to form larger droplets on injection and cause poor fuel atomization, encourage the formation of engine deposits, and cause the need for more energy to pump the fuel.

The density of the fuel defines the mass of fuel injected into the combustion chamber. This is because the meters of fuel injection pumps are designed to measure the fuel volume and not mass; denser fuel contains more mass for same amount of volume.

To protect the performance and durability of combustion equipment, the standard sets limits for impurities remaining in the biodiesel production process. These standards include methanol, glycerin (a by-product), and unconverted or partially converted raw materials, and sodium and potassium may be used with caustic soda to catalyze the transesterification reaction.

Methanol is one of the reactants used to produce biodiesel. To bring the reaction to completion, usually it uses a 4:1 excess methanol so that the methanol can be removed from the product. Methanol over 0.2 wt% is incompatible with some elastomers and metals in automotive fuel systems (Figure 4).

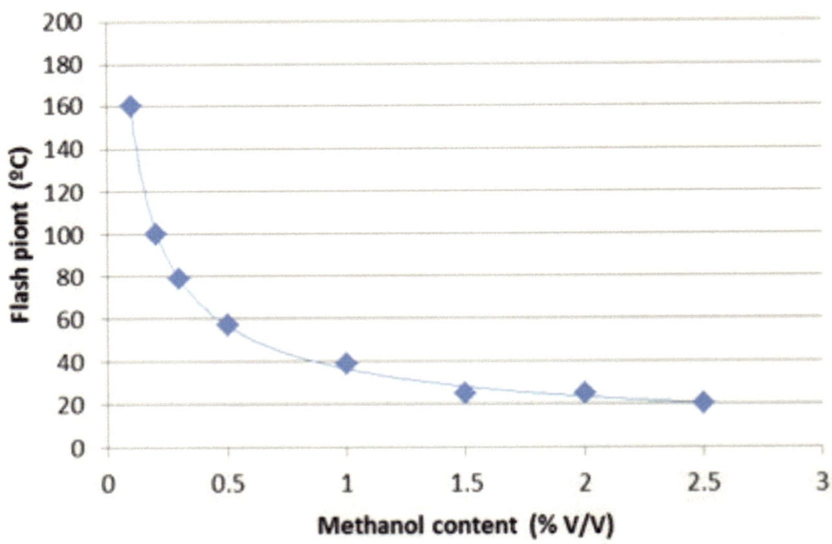

Figure 4. Relationship between flash point and methanol content.

Table 4 presents the biodiesel characteristics, and it is compared to the limits established by the norm. The values of the measured properties respond to parameters of international quality. Basically, the use of this biodiesel does not require any modification in the vehicle engine and ensures less contaminant emission when it is used as a result of the characteristics obtained in the results (Table 4).

http://dx.doi.org/10.5772/59673

Characteristics/unit	Average value obtained	Limits
Content of linoleic acid methyl ester (%)	97.5	Min. 96.5%
Density (g/mL)	0.8700 at 20°C	Report
Kinetic viscosity (mm²/S)	5.03	Maximum 6.5 mm²/S
Carbonaceous wastes (g/100 g)	0.06	0.3 g/100 g
Flash point (°C)	176	Min. 100°C
Water content (%m/m)	0.0553	Maximum 0.080%m/m
Copper corrosion (A, B, C)	1 A (3 h at 50°C)	1 A
Acid—volumetric method (mg KOH/g)	0.7	Max 0.8 mg KOH/g
Total glycerin (%)	0.22	Max 0.25%
Content of methanol or methanol free (%)	0.004	Max 0.2%
Cloud point (°C)	19	Report

Source: Compilation based on results obtained in the laboratory of the Department of Quality Control.

Table 4. Comparison of the average values obtained from the fuel quality parameters determination with the limits established in the norm.

About the presence of linoleic acid as a quality indicator of the oil used in the process of transesterification, the presence of this acid was found, with a minimum of 96.5% as C18:2. This is considered "ideal" by the established standards for this purpose.

The relative density is useful for the load data in the vehicle where the fuel is going to be used and also for the combustion efficiency [18]. Taking into account that the specific gravity range of the oily products used as fuel goes from 0.700 to 1.050 g/mL [18] in relation to the reported density in this case, it can be considered as optimum. The kinetic viscosity expressed in 5.03 mm^2/s indicates the property to resist changes in fluency with the temperature changes, so it is very interesting to consider an effective lubrication since it is closely related to temperature and defines the utility of fuel under different temperature ranges [12].

On the other hand, the carbon residue is presented between 0.06 g per each 100 g of biodiesel; therefore, it is less unlikely to form carbon residue deposits. The flash point at 176°C decreases the possibility of storage accidents as well as the tendency to vaporize in room temperature. This is also related to the quantity of emissions to the atmosphere [13].

The water in 553 ppm is inside the considered normal limits, so it does not stop being considerable in the biodiesel use since it could present chemical processes such as hydrolysis, corrosion problems, and bacterial growth [19].

Moreover, considering that corrosion is a very important factor in the utility of the whole fuel, in relation to the quality of biodiesel produced in Paraguay, the results indicate that they corrode steel to 1A (classification number of assigned 1–4 based on a comparison with the ASTM Copper Strip Corrosion Standards), which is considered the maximum allowed limit. This will probably reveal some sulfur presence, a consequence of an inadequate refining, as well as the emission of this type of contaminants in the use of biodiesel, although not in a considerable scale.

The presence of acid components could be the result of some additive components such as inhibitor agents and detergents or as degradation product formed during its use and storage [20]. Mainly, the degradation products contribute to a raise in the acidity number, which leads to the tendency to fuel corrosion on the metallic surface and increase of the degradation velocity. The acid number of crude sunflower oil was 2.8 mg KOH/g, and the moisture content and volatile matter were each 0.04%. The triacyl glycerides (TAGs) found in this oil contained 3.6% palmitic acid, 3.2% stearic acid, 78.8% oleic acid (18:1), 12.5% linoleic acid (18:2), 0.1% linoleic acid (18:3), 0.3% arachidic acid, 0.9% behenic acid, and 0.4% lignoceric acid as their main fatty acid components. The sunflower seeds used were of the high oleic acid type, as can be seen from the fatty acid composition.

The total glycerin quantity indicates a probable incomplete reaction because these are oil remains without reacting, and this could produce crystallization and deposits. However, this parameter was observed still within the limits.

On the other hand, the low presence of free methanol indicates that it was successfully removed during its process. This also influences in the increase of the flashpoint and the low possibility of corroding elements of the engine because of this factor.

As to the cloud point, fog or turbidity is seen in the sample at 19°C, indicating the starting of its own crystallization when it is under continuous cooling. This parameter is also within the allowed ranges.

4. Conclusion and recommendations

From the obtained results,can be specifically concluded that biodiesel fulfills the parameters of quality demanded by the international market, presenting itself as available alternative to replace oil. Ergo, elaborated with material from sunflower with a methanolysis process, it presents some characteristics that are suitable for its use in transportation with diesel vehicles as well as safe storage conditions. However, the manufacturing and reaction conditions to determine a more complete evaluation as well as the technology behavior against changes the raw material in order to ensure a stable product is still missing.

Therefore, it is presented as a valid potion to diverse the energetic matrix and to replace the fossil originated fuel contributing to the reduction of carbon emission to the atmosphere coming from the transportation.

http://dx.doi.org/10.5772/59673

It would be interesting to complement the investigation checking the environmental raw material production, as well as transportation and industrialization to ensure the sustainable production. A diversification of biodiesel feedstock, a detailed revision of the environmental and social impact, is necessary. Although major changes are not yet visible on the land with the increased production of biofuels, it is recommended to be implemented in public policies to support family farming and crop diversification in the production areas to avoid the risk of extensive monocultures.

Author details

Estelvina Rodríguez Portillo[1*], Araceli Amaya Chávez[2*], Arturo Colín Cruz[2] and Rubí Romero Romero[2]

*Address all correspondence to: aamayac@uaemex.mx

1 Research Department, Faculty of Science and Technology, National University of Itapúa, Encarnación, Paraguay

2 Facultad de Química, Universidad Autónoma del Estado de México, Paseo Colon Esq. Tollocan, Toluca, Mexico

References

[1] Encinar JM, Gonzalez JF, Rodríguez JJ, Tejedor A. Biodiesel fuels from vegetable oils: transesterification of *Cynara cardunculus* L. oils with ethanol. Energy Fuels 2002;16:443–50.

[2] Ma F, Hanna M.A. Biodiesel production: a review. Bioresour Technol 1999;70:1–15.

[3] Narayanan D, Zhang Y, Mannan MS. Engineering for sustainable development (ESD) in bio-diesel production. Process Saf Environ Protect 2007;85:349–359.

[4] García J, García J. Biocarburantes líquidos: Biodiesel y bioetanol. Informe de vigilancia tecnológica—CITME. Editorial Elecé Industria Gráfica, Madrid, España. 2006.

[5] Saraf S, Thomas B. Influence of feedstock and process chemistry on biodiesel quality. Process Saf Environ Protect 2007;85:360–365.

[6] Vargas M. Biodiesel. CONAE (Consejo Nacional de Ahorro Energético, Argentina). Buenos Aires. 2007.

[7] Monteiro M, Pepe AA, Morais L, Ferreira AG. Critical review on analytical methods for biodiesel characterization. Talanta 07/2008;77(2):593–605.

[8] American Society for Testing and Materials (ASTM). Web page www.astm.org.consulted in November 2009.

[9] Centro Multidisciplinar de Innovación y Tecnología. Caracterización de Biodiesel. Biodiesel en Sustitución del Gasoil-Biodiesel de Calidad. Publicaciones Técnicas. CEMITEC-Navarra- España. 2007.

[10] Komers K, Stloukal R, Machek J, Skopal F. Biodiesel from rapeseed oil, methanol and KOH. 3. Analysis of composition of actual mixture. Eur J Lipid Sci Technol 2001;103:363–71.

[11] Agudelo SJR, Benjumea R, Gómez ME, Pérez BJF. Biodiesel. Una revisión del desempeño mecánico y ambiental. Ingeniería & Desarrollo. Universidad del Norte. Medellín-Colombia 2003;13:1–14.

[12] Canakci M, Sanli H. Biodiesel production from various feedstocks and their effects on the fuel properties. J Ind Microbiol Biotechnol 2008;35(5):431–441. doi: 10.1007/s10295-008-0337-6.

[13] Landa, Itziar. Caracterización de biodiesel principales parámetros de calidad. CEMITEC, Navarra, España. 2006.

[14] Park JY, Kim DK, Lee JP, Park SC, Kim YJ, Lee JS. Blending effects of biodiesels on oxidation stability and low temperature flow properties. Bioresour Technol 2008;99(5):1196–1203. doi: 10.1016/j.biortech.2007.02.017.

[15] Sharma Y, Singh B, Upadhyay S. Advancements in development and characterization of biodiesel: a review. Fuel 2008;87(12):2355–2373. doi: 10.1016/j. fuel.2008.01.014.

[16] Ramos M. J., Fernández C. M, Casas A., Rodríguez L., & Pérez Á. Influence of fatty acid composition of raw materials on biodiesel properties. Bioresour Technol 2009;100(1):261–268. doi: 10.1016/j.biortech. 2008.06.039.

[17] Universiteit Utrecht. Copernicus Institute of Sustainable Development. Energy and Resource. Potential indirect land use change (ILUC). Available online: http://www.uu.nl/en/research/copernicus-institute-of-sustainabledevelopment/ research/energy-and-resources/potential-indirect-land-use-change-iluc

[18] INTN (Instituto Nacional de Tecnología y Normalización). Normas Paraguayas. Legislación alimentaria y Resoluciones MERCOSUR internalizadas. INTN Asunción, Paraguay. 2008.

[19] Proc Kenneth RL, McCormick KK, Chandler B, Buchhler B. Operating Experience and Teardown Analysis for Engines Operated on Biodiesel Blends (B20), NREL/CP-540-38509, 2005-01-3641.

[20] IDIADA Automotive Technology. Technical Report: Comparison of vehicle emissions at European Union annual average temperatures from E0 and E5 petrol. PEPSOL, YPF, Abengoa Bioenergía, CEPSA, Ebro Puleva. 2003.

Chapter 14

Valorization of Waste Cooking Oil into Biodiesel over Heteropolyacids Immobilized on Mesoporous Silica — A Kinetic Study

M. Caiado, A. Tropecêlo and J.E. Castanheiro

Additional information is available at the end of the chapter

http://dx.doi.org/10.5772/59584

1. Introduction

Biodiesel is a source of energy derived from renewable sources, which can be a substitute for diesel fuel. It is biodegradable, and it has limited greenhouse gas emissions because of the closed CO_2 cycle and a lower combustion emission profile (especially SOx). Biodiesel consists of fatty acid alkyl esters (usually methyl esters, FAME) derived from either the transesterification of triglycerides and/or the esterification of free fatty acids (FFAs) with low molecular weight alcohols [1-3].

Traditionally, the production of biodiesel is carried out in the presence of a homogeneous base catalyst; however, effluent disposal leads to environmental problems and economical inconveniences. These problems can be overcome by the use of heterogeneous catalysts. Another problem with commercial production of biodiesel is the high cost of raw materials. In order to overcome this problem, waste oils and fats can be used as feedstocks [4-7]. However, waste cooking oils contain a high amount of free fatty acids, which is a problem for biodiesel production by the traditional process (homogeneous alkali-catalyzed transesterification) [8,9]. Different solid acids such as resins with sulfonic acid groups [10-14], zeolites [15-17], solid super acid catalysts [18] and carbonaceous catalysts [19] have been used as catalysts in the conversion of waste cooking oil containing a high amount of free fatty acids into biodiesel.

Heteropolyacids (HPAs) have several advantages as catalysts, which make them economically and environmentally attractive. On the one hand, HPAs have a very strong Bronsted acidity, approaching the superacid region; on the other, they are efficient oxidants, exhibiting fast reversible multielectron redox transformations under rather mild conditions. Their acid-base and redox properties can be varied over a wide range by changing the chemical composition.

Solid HPAs possess a discrete ionic structure, comprising fairly mobile basic structural units, e.g., heteropolyanions and counter cations (H^+, H_3O^+, $H_5O_2{}^+$, etc.) unlike the network structure of, e.g., zeolites and metal oxides [20-22]. HPAs have low specific surface areas (1–10 m^2/g). In order to increase the specific area of HPAs or, even better, to increase the number of accessible acid sites of the HPAs, a variety of supports like activated carbon [23-26], silica [27-32], zeolite [33-38] and polymeric matrix [39-43] have been used as support to immobilize HPAs.

HPAs and HPAs supported on different supports have been used as acid catalysts for biodiesel production [44-48].

Transesterification of waste cooking oil with high acid value and high water content was performed using heteropolyacid $H_3PW_{12}O_{40}.6H_2O$ (PW) as homogeneous catalyst. PW was found to be the most promising catalyst, which exhibited the highest ester yield (87%) for transesterification of waste cooking oil and an ester yield of 97% for esterification of long-chain palmitic acid. The PW acid catalyst showed higher activity under the optimized reaction conditions compared with the conventional homogeneous catalyst sulfuric acid, and it can easily be separated from the products by distillation of the excess methanol and can be reused several times [45].

Biodiesel production in the presence of 20 wt% myristic acid from soybean oil was carried out over heteropolyacid immobilized on mesoporous Ta_2O_5 materials. Different catalysts were prepared. The network structures of the hybrid materials and the functions of the incorporated alkyl groups on the catalytic activity of the materials have been put forward [46].

Biodiesel synthesis from waste cooking oil was carried out over 12-tungstosilicic acid on SBA-15. The heteropolyacid was prepared using impregnation method. The effect of different reaction parameters like percentage loading, catalyst amount, mole ratio, time and temperature were studied for the supreme conversion. The catalyst was recycled up to four times after simple workup without notable loss in the activity [48].

In this work, we report the transesterification of waste cooking oil with different alcohols over heteropolyacids immobilized on SBA-15. We also studied the effect of free fatty acid addition into waste cooking oil. A kinetic model is proposed.

2. Waste cooking oil raw converted to biofuel

The McDonald's Corporation is the world's largest chain of fast-food restaurants, serving around 68 million customers daily in 119 countries across 35,000 outlets [49]. The waste cooking oil in about 90% of McDonald's restaurants is used for after-market uses, including biodiesel. The used cooking oil from the restaurants is collected, recycled into biodiesel, and put back into distributors' trucks to fuel their deliveries to McDonald's restaurants. In a few countries, McDonald's waste cooking oil are a closed-loop system. Since 2013, McDonald's developmental licensee in the United Arab Emirates operates a 100% closed-loop biodiesel system. More than 25,000 liters of waste cooking oil are being collected from McDonald's approximately 108 restaurants each month and converted into biodiesel. Waste cooking oil is stored at the

http://dx.doi.org/10.5772/59584

restaurants in a specially designed container. 1.75 million liters of biodiesel have been generated from waste cooking oil and used to power the distribution fleet. Excess biodiesel can be sold, helping other companies reduce their transportation impacts. In the United Kingdom (in 2013), McDonald's replaced more than 3 million liters of diesel fuel with biofuel made from used cooking oil collected at approximately 1,200 restaurants. Across markets in Europe, the waste cooking oil from more than 80% of the McDonald's Europe restaurants is being converted into biodiesel. In Brazil 19 restaurants are participating in a closed-loop system that has recycled more than 187,000 liters of used cooking oil into 65,000 liters of biodiesel since 2009, for use in five trucks that deliver to the restaurants [50].

3. Experimental

3.1. Catalyst preparation

Mesoporous silica SBA-15 was prepared according to the literature [51] An ethylene oxide (EO)/propylene oxide (PO) triblock copolymer (P123) with composition EO20PO70EO20 and with an average molecular weight of 5800 was used as the template. The synthesis consisted of 2.0 g of triblock P123 dissolved in 60 cm^3 of 2 $mol.dm^{-3}$ aqueous HCl and 15 cm^3 of distilled water under stirring. Then, 4.4 g of tetraethyl orthosilicate (TEOS) was added dropwise to the solution at room temperature. The mixture was stirred for 24 h at 313 K, and then the temperature was raised to 373 K and kept at 373 K for another 24 h in a Teflon-lined autoclave. Finally, the resulting precipitate was filtered, washed carefully with distilled water, air-dried, and calcined at 773 K in air for 5 h to remove the template and to obtain the final product SBA-15.

The heteropolyacid was immobilized on SBA-15 by the grafting technique. The grafting of tungstophosphoric acid (PW) was carried out by mixing 2.0 g dried SBA-15 with 3-aminopropyltriethoxysilane (23.4 μL) containing freshly distilled toluene refluxing for 48 h. The obtained solid material was immersed in an aqueous solution of PW with stirring for 5 h. The solid was then dried in vacuum to obtain the heteropolyacid anchored mesoporous catalyst [52].

3.2. Catalyst characterization

The textural characterization of the catalysts was based on the nitrogen adsorption isotherm, determined at 77 K with a Micromeritics ASAP 2010 apparatus.

The FTIR spectra were recorded with a Bio-Rad FTS 155 instrument.

The amount of W in silica catalysts was measured by dissolving the catalyst in H_2SO_4/HF 1:1 (v/v) and analyzing the obtained solution using inductively coupled plasma analysis (ICP), which was carried out in a Jobin-Yvon ULTIMA instrument.

The X-ray diffraction (XRD) patterns of the heteropolyacid, silica and catalysts were obtained by using a Bruker powder diffractometer with built-in recorder, using Cu Kα radiation, nickel filter, 30 mA and 40 kV in the high voltage source, and scanning angle between 0.7° and 5° of 2θ at a scanning rate of 1°/min.

Transmission electron microscopy (TEM) analyses were performed on a Hitachi S-2400 scanning electron microscope, at a current voltage of 25 kV.

Catalyst acidity was measured by means of potentiometric titration [53].

3.3. Catalytic experiments

The catalytic experiments were carried out in a stirred batch reactor at 60°C. In a typical experiment, the reactor was loaded with 30 mL of methanol and 0.2 g of catalyst. Reactions were started by adding 2.5 ml waste cooking oil (WCO).

Stability tests of the catalyst were carried out by running four consecutive experiments, using the same reaction conditions. Between the catalytic experiments, the catalyst was separated from the reaction mixture by filtration, washed with acetone and methanol and dried at 70° C overnight.

In order to study the reusability, the PW-NH_2-SBA-15 catalyst was filtered from the reaction mixture. After this operation, it was soaked in hexane overnight and it was dried overnight.

Undecano was used as the internal standard. Samples were taken periodically and analyzed by GC, using a Hewlett Packard instrument equipped with a 30 m x 0.25 mm HP-5 column.

4. Results and discussion

4.1. Catalyst characterization

SBA-15 and PW-NH_2-SBA-15 catalysts showed a typical IV adsorption isotherm with H1 hysteresis loop as defined by IUPAC. It was observed that the specific surface area and total pore volume of SBA-15 decreased with the immobilization of PW on SBA-15 (Table 1).

	PW load[a] (wt%)	**Surface area[b] (m^2/g)**	**V_Tc (cm3/g)**
SBA-15	-	1050	1.38
PW-NH_2-SBA-15	5.6	735	0.62

(a) HPA load determined by ICP analysis

(b) BET

(c) $(p/p^{\circ}) = 0.98$

Table 1. Physicochemical characterization of SBA-15 and PW-NH_2-SBA-15

X-ray diffraction patterns of the SBA-15 and PW-NH_2-SBA-15 catalysts are shown in Figure 1. All three reflections are still detectable after PW immobilization, suggesting that hexagonal pore structure of the support is retained.

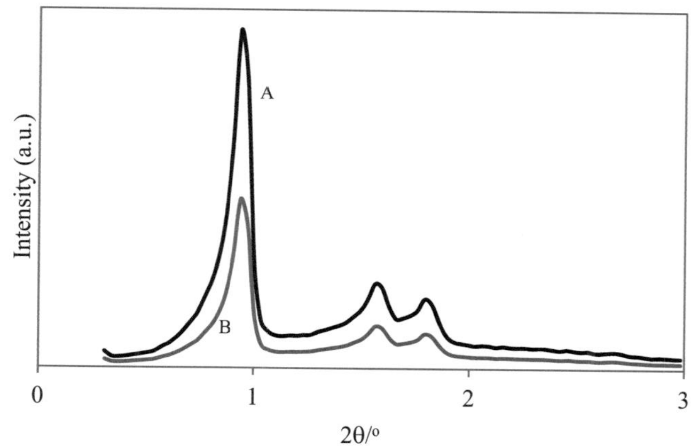

Figure 1. X-ray diffractograms of SBA-15 (A) and PW-NH_2-SBA-15 (B).

Figure 2 shows the FT-IR spectra of SBA-15 and PW-NH_2-SBA-15 samples. For SBA-15 bands were observed at 1084, 812 and 476 cm^{-1}, which are attributed to asymmetric stretching, symmetric stretching and bending modes of Si–O–Si, respectively. For PW-NH_2-SBA-15 catalyst the characteristic bands of SBA-15 are present along with the bands at 1083, 982, 890 and 808 cm^{-1}, which are the fingerprint of Keggin structure of HPW. However, some bands typical of the Keggin-type HPA structures are overlapped or partially overlapped with the bands of the SBA-15 matrix framework, probably due to the low loading of PW. Similar results were also observed by Liu et al. [52].

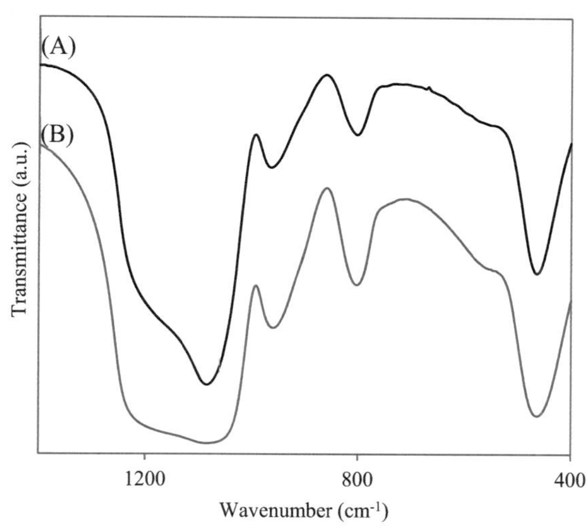

Figure 2. FT-IR spectra of catalysts: (A) SBA-15 and (B) PW-NH_2-SBA-15.

TEM image of PW-NH_2-SBA-15 catalyst was carried out. The morphology of the support remains after immobilization of PW on SBA-15 (Fig. 3).

Figure 3. Transmission electron microscopy (TEM) image of PW-NH_2-SBA-15.

Table 2 shows the initial electrode potential (Ei) of the materials. The Ei indicates the maximum acid strength of the surface sites [53]. It can be observed that Ei increased with the amount of PW immobilized in SBA-15, which can be due to the increase in the amount of protons with PW loading of the SBA-15 support.

Sample	E_i (mV)
SBA-15	+110
PW-NH_2-SBA-15	+408

Table 2. Initial electrode potential (mV) of materials

4.2. Catalytic experiments

The biodiesel production from WCO was carried out over PW-NH_2-SBA-15 catalyst at 60°C.

Different catalytic experiments were carried out at different stirring speeds to study the influence of external resistances to mass transfer. It was observed that experiments carried out with 700 rpm have got a good mix of the compounds and eliminate possible mass transfer problems.

4.3. Effect of the nature of alcohol

The alcohols most frequently used in biodiesel production are methanol and ethanol. Due to the low cost of methanol, this alcohol is the first choice for the esterification/transesterification reactions of WCO. However, for biodiesel production to be more environment friendly, ethanol is the ideal candidate for the synthesis of a fully biogenerated biodiesel, since ethanol is derived from agricultural products (renewable sources) [3].

Figure 4 shows the effect of the nature of alcohol (methanol, ethanol and 1-propanol) on the fatty acid ester concentrations obtained over PW-NH_2-SBA-15 catalyst, at 60°C. It was observed that the ester concentrations obtained with 1-propanol led to lower conversion, when compared with methanol and ethanol. Similar results were also observed by Sreeprasanth et al. [54], on the transesterification of rubber seed with ethanol over Fe-Zn-1 catalyst. This behavior can be explained not due to the reaction rate with ethanol is slower than with methanol, as ethyl nucleophile is less reactive than methyl nucleophile.

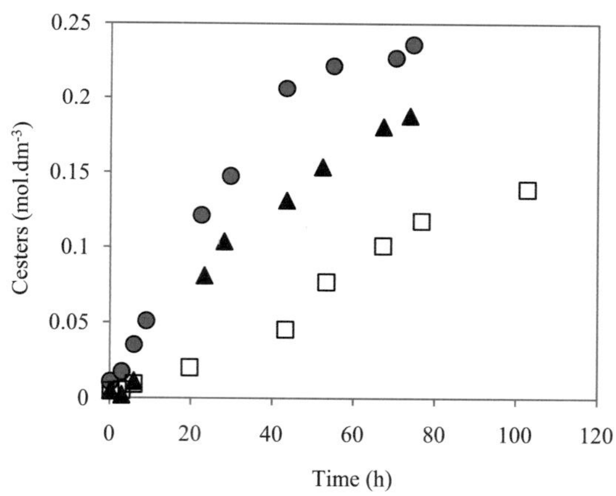

Figure 4. Biodiesel production from WCO with methanol over PW-NH_2-SBA-15 catalyst. Effect of alcohol nature. FAME concentration (mol.dm^{-3}) *versus* time (h): (○) Methanol; (▲) Ethanol; (□) 1-Propanol.

4.4. Effect of the initial amount of free fatty acid

In order to study the influence of the initial free fatty acid amount in the WCO, three different amounts of fatty acid (as a model) was added to the reaction mixture. The catalytic experiments were carried out at different amounts of initial palmitic acid in the WCO, over PW-NH2-SBA-15 catalyst, while the initial concentration of WCO (0.088mol.dm^{-3}) and the catalysts loading (m=0.2 g) were kept constant. Figure 5 shows the influence of the initial amount of FFA in WCO on the biodiesel production. It was observed that the initial reaction rate increases with the amount of FFA. When the initial amount of FFA increases, a slight increase on the conversion was observed. Similar results were also observed by Marchetti and Errazu [16]. Therefore, this effect could also be seen on the total FAME production since the final amount of biofuel will be produced from the triglycerides as well as from the fatty acids present in the reaction mixture.

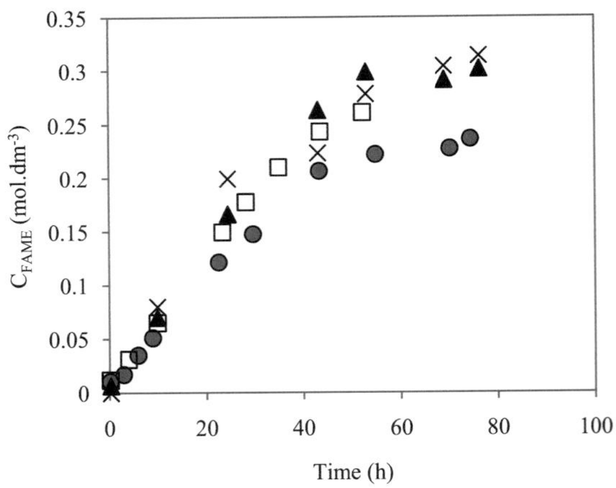

Figure 5. Biodiesel production from WCO with methanol over PW-NH_2-SBA-15 catalyst. Effect of the amount of palmitic acid. FAME concentration (mol.dm^{-3}) *versus* time (h): (○) 0%; (□) 4%; (▲) 12 % ; (×) 27 %.

4.5. Catalyst stability

In order to study the catalytic stability of the PW-NH_2-SBA-15, different batch runs with the same catalyst sample and under the same conditions were carried out. Figure 6 shows the catalytic activity of PW-NH_2-SBA-15 at different batch runs. It was observed that the catalytic activity decreases only about 10%, after the fourth use.

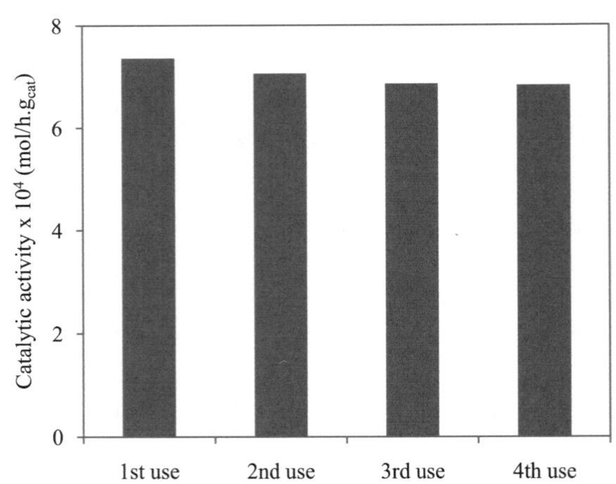

Figure 6. Catalytic stability of PW-NH_2-SBA-15 catalyst in biodiesel production with methanol. The initial activities are taken as the maximum observed reaction rate, which was calculated from the maximum slope of the methyl ester kinetic curve.

http://dx.doi.org/10.5772/59584

4.6. Kinetics modeling

A simple kinetic model can be established based on the following assumptions:

1. Isothermal and isobaric reaction conditions;
2. First order kinetics with respect to the reactants is assumed. The forward and reverse reactions follow second order overall kinetics;
3. Due to the excess of methanol used, the reverse reaction could be minimized and it was not considered in the reaction rate.
4. Triglycerides are consumed according to the consecutive reaction network:

$$T + A \xrightarrow{k_1} D + E$$

$$D + A \xrightarrow{k_2} M + E$$

$$M + A \xrightarrow{k_3} G + E$$

where T represents triglycerides, D represents diglycerides, M represents monoglycerides, A represents alcohol, G represents glycerol and E represents esters of fatty acids;

The reaction rate of these three pseudo elementary reaction are expressed as:

$$r_1 = k_1 C_T . C_A \tag{1}$$

$$r_2 = k_2 C_D . C_A \tag{2}$$

$$r_3 = k_3 C_M . C_A \tag{3}$$

For batch reactor the mole balance equations may be written as

$$\frac{dC_T}{dt} = -\frac{W}{V}(r_1) \tag{4}$$

$$\frac{dC_A}{dt} = -\frac{W}{V}(r_1 + r_2 + r_3) \tag{5}$$

$$\frac{dC_D}{dt} = \frac{W}{V}.(r_1 - r_2) \tag{6}$$

$$\frac{dC_M}{dt} = \frac{W}{V}.(r_2 - r_3) \tag{7}$$

$$\frac{dC_G}{dt} = \frac{W}{V}(r_3) \tag{8}$$

$$\frac{dC_E}{dt} = \frac{W}{V}(r_1 + r_2 + r_3) \tag{9}$$

The differential equations system was integrated using the Euler Method. The optimization was carried out by the *SOLVER* routine in a *Microsoft Excel* spreadsheet.

Figure 7 shows the concentration of fatty acid methyl ester (FAME) *versus* time (h) on the transesterification of WCO with methanol. The solid line represents the model fitted to the data points. It was observed that the kinetic model fits experimental concentration data quite well. The model parameters, k_1, k_2 and k_3, have got the value of 0.00979, 0.01348 and 0.01956 $dm^6.mol^{-1}.h^{-1}.g_{cat}^{-1}$, respectively. It was observed that $k_1<k_2<k_3$, which can be explained due to the molecular size of monoglycerides, diglycerides and triglycerides and due to the textural characteristics of PW-NH_2-SBA-15. The size of monoglycerides is smaller than that of diglycerides and triglycerides. Consequently, it is expected that, near the active sites of catalyst, the amount of monoglycerides is higher than the amount of diglycerides and triglycerides. As the reaction rate is dependent on reactant concentration, a high concentration of monoglycerides leads to high reaction rates.

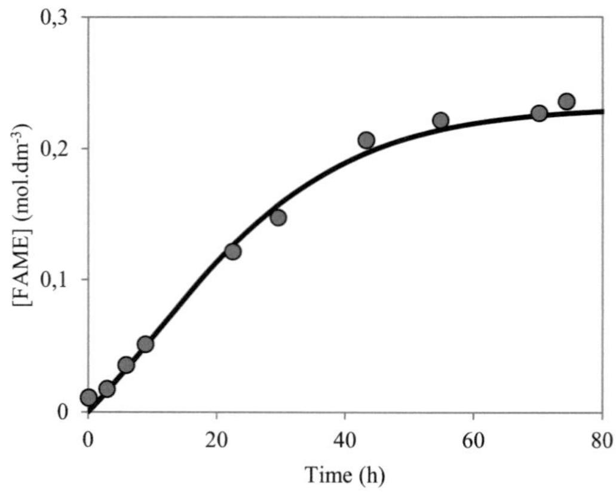

Figure 7. Concentration of FAME ($mol.dm^{-3}$) versus time (h).

5. Conclusions

Biodiesel production from WCO with methanol was carried out over tungstophosphoric acid immobilized on SBA-15 by grafting technique, at 60°C. After PW immobilization, the morphology of the support remained.

In order to optimize the conditions, the influence of various reaction parameters, the nature of alcohol and the amount of initial free fatty acid on the transesterification of WCO in the presence of PW-NH_2-SBA-15 catalyst were carried out. The catalytic stability of the material was also studied.

The esterification/transesterification of WCO with propanol and ethanol led to lower concentration of fatty acid esters than with methanol.

When different amounts of free fatty acids (palmitic acid) were added to the WCO, a slight increase on the concentration of FAME was observed.

In order to study the catalytic stability of PW-NH_2-SBA-15 catalyst, four consecutive batch runs were carried out. It was observed that the catalytic activity of PW-NH_2-SBA-15 tends to stabilize.

Acknowledgements

This work was carried out with support from the Project PTDC/CTM-POL/114579/2009.

Author details

M. Caiado, A. Tropecêlo and J.E. Castanheiro*

*Address all correspondence to: jefc@uevora.pt

CQE, Departamento de Química, Universidade de Évora, Évora, Portugal

References

[1] Ma F., Hanna M.A. Biodiesel production: a review. Bioresource Technology 1999; 70(1) 1–15.

[2] Szybist J.P., Song J., Alam M., Boehman A.L. Biodiesel combustion, emissions and emission control. Fuel Processing Technology 2007; 88 (7) 679–691.

[3] Lotero J. E., Liu Y., Lopez D. E., Suwannakarn K., Bruce D. A., Goodwin J. G. Synthesis of biodiesel via acid catalysis. Industrial and Engineering Chemical Research 2005; 44(14) 5353–5363.

[4] Jacobson K., Gopinath R., Meher L. C., Dalai A. K. Solid acid catalyzed biodiesel production from waste cooking oil. Applied Catalysis B: Environmental 2008; 85(1-2) 86–91.

[5] Hass M.J, McAloon A.J., Yee W.C., Foglia T.A. A process model to estimate biodiesel production cost. Bioresource Technology 2006; 97 (4) 671–678.

[6] Echim C., Verhe R., De Greyt W., Stevens C. Production of biodiesel from side-stream refining products. Energy & Environmental Science 2009; 2 (11)1131–1141.

[7] Atadashi I.M., Aroua M.K., Abdul Aziz A.R., Sulaiman N.M.N. Production of biodiesel using high free fatty acid feedstocks. Renewable and Sustainable Energy Reviews 2012; 16 (5) 3275– 3285.

[8] Kulkarni M.G., Dalai A.K. Waste cooking oil-An economical source for biodiesel:A review. Industrial and Engineering Chemical Research 2006; 45 (9) 2901–2913.

[9] Phan A.N., Phan T.M. Biodiesel production from waste cooking oils. Fuel 2008; 87 (17-18) 3490–3496.

[10] Talukder M.M.R., Wu J.C., Lau S.K., Cui L. C., Shimin G., Lim A. Energy Fuels 2009; 23 (1) 1-4.

[11] Shi W., He B., Li J. Esterification of acidified oil with methanol by SPES/PES catalytic membrane. Bioresource Technology 2011; 102 (9) 5389–5393.

[12] Park J.-Y., Wang Z.-M., Kim D.-K., Lee J.-S. Effects of water on the esterification of free fatty acids by acid catalysts, Renewable Energy 2010; 35 (3) 614–618.

[13] Tesser R., Casale L., Verde D., Di Serio M., Santacesaria E. Kinetics and modeling of fatty acids esterification on acid exchange resins. Chemical Engineering Journal 2010; 157 (2-3) 539-550.

[14] Tesser R., di Serio M., Guida M., Nastasi M., Santacesaria E. Kinetics of Oleic Acid Esterification with Methanol in the Presence of Triglycerides. Industrial and Engineering Chemical Research 2005;44 (21) 7978-7982.

[15] Chung K.-H., Chang D.-R., Park B.-G. Removal of free fatty acid in waste frying oil by esterification with methanol on zeolite catalysts. Bioresource Technology 2008; 99 (16) 7438–7443.

[16] Marchetti J. M., Errazu A. F. Comparison of different heterogeneous catalysts and different alcohols for the esterification reaction of oleic acid. Fuel 2008; 87(15-16) 3477-3480.

[17] Chung K.-H., Park B.-G. Esterification of oleic acid in soybean oil on zeolite catalysts with different acidity. Journal of Industrial and Engineering Chemistry 2009;15 (3) 388–392

[18] Li Y., Zhang X., Sun L. Fatty acid methyl esters from soapstocks with potential use as biodiesel. Energy Conversion and Management 2010; 51 (11) 2307–2311.

[19] Guo F., Xiu Z., Liang Z. Synthesis of biodiesel from acidified soybean soapstock using a lignin-derived carbonaceous catalyst. Applied Energy 2012; 98 (1) 47–52.

[20] Kozhevnikov I.V. Catalysis by heteropoly acids and multicomponent polyoxometalates in liquid-phase reactions. Chemical Reviews 1998; 98(1) 171–198.

[21] Kozhevnikov I.V. Heteropoly acids and related compounds as catalysts for fine chemical synthesis. Catalysis Reviews Science and Engineering 1995; 37 (2) 311–352.

[22] Okuhara T., Mizuno N., Misono M. Catalytic chemistry of heteropoly compounds. Adv. Catal. 1996; 41, 113–252.

[23] Dupont P., Védrine J.C., Paumard E., Hecquet G., Lefebvre F. Heteropolyacids supported on activated carbon as catalysts for the esterification of acrylic acid by butanol. Applied Catalysis A: General 1995;129 (2) 217-227.

[24] Pizzio L.R., Cáceres C.V., Blanco M.N. Adsorption of tungstophosphoric or tungstosilicic acids from ethanol–water solutions on carbon original. Journal of Colloid and Interface Science 1997; 190 (2) 318-326.

[25] Obali Z., Dog˘u T. Activated carbon–tungstophosphoric acid catalysts for the synthesis of tert-amyl ethyl ether (TAEE). Chemical Engineering Journal 2008; 138 (1–3) 548–555.

[26] Ferreira P., Fonseca I.M., Ramos A.M., Vital J., Castanheiro J.E. Acetylation of glycerol over heteropolyacids supported on activated carbon. Catalysis Communications 2011, 12 (7) 573–576.

[27] Chimienti M.E., Pizzio L.R., Cáceres C.V., Blanco M.N., Tungstophosphoric and tungstosilicic acids on carbon as acidic catalysts. Applied Catalysis A: General 2001;208 (1–2) 7–19.

[28] Caetano C.S., Fonseca I.M., Ramos A.M., Vital J., Castanheiro J.E. Esterification of free fatty acids with methanol using heteropolyacids immobilized on silica. Catalysis Communications 2008; 9 (10) 1996–1999.

[29] Schwegler M.A., Vinke P., van der Eijk M., van Bekkum H. Activated carbon as a support for heteropolyanion catalysts. Applied Catalysis A: General 1992; 80 (1) 41–57.

[30] Dias A.S., Pillinger M., Valente A.A., Mesoporous silica-supported 12-tungstophosphoric acid catalysts for the liquid phase dehydration of d-xylose. Microporous and Mesoporous Materials 2006;94 (1–3) 214–225.

[31] Caiado M., Machado A., Santos R.N., Matos I., Fonseca I.M., Ramos A.M., Vital J., Valente A.A., Castanheiro J.E. Alkoxylation of camphene over silica-occluded tungstophosphoric acid. Applied Catalysis A: General 2013;451 (1) 36–42.

[32] Kasztelan S., Payen E., Moffat J.B. The existence and stability of the silica-supported 12-molybdophosphoric acid keggin unit as shown by Raman, XPS, and 31P NMR spectroscopic studies. Journal of Catalysis 1990; 125 (1) 45–53.

[33] Haber J., Pamin K., Matachowski L., Much D. Catalytic performance of the dodecatungstophosphoric acid on different supports. Applied Catalysis A: General 2003;256 (1) 141–152.

[34] Sulikowski B., Haber J., Kubacka A., Pamin K., Olejniczak Z., Ptaszynski J. Novel "ship-in-the-bottle" type catalyst: evidence for encapsulation of 12-tungstophosphoric acid in the supercage of synthetic faujasite. Catalysis Letters 1996;39 (1–2) 27–31.

[35] Mukai S.R., Masuda T., Ogino I., Hashimoto K. Preparation of encaged heteropoly acid catalyst by synthesizing 12-molybdophosphoric acid in the supercages of Y-type zeolite. Applied Catalysis A: General 1997; 165 (1–2) 219–226.

[36] Mukai S.R., Lin L., Masuda T., Hashimoto K. Key factors for the encapsulation of Keggin-type heteropoly acids in the supercages of Y-type zeolite. Chemical Engineering Science 2001; 56 (3) 799–804.

[37] Tran M.H., Ohkita H., Mizushima T., Kakuta N. Hydrothermal synthesis of molybdenum oxide catalyst: Heteropoly acids encaged in US-Y. Applied Catalysis A: General 2005;287 (1) 129–134.

[38] Ferreira P., Fonseca I.M., Ramos A.M., Vital J., Castanheiro J.E. Esterification of glycerol with acetic acid over dodecamolybdophosphoric acid encaged in USY zeolite. Catalysis Communications 2009;10 (5) 481–484.

[39] Choi J.S., Song I.K., Lee W.Y. Simulation and experimental study on the polymer membrane reactors for the vapor-phase MTBE (methyl tert-butyl ether) decomposition. Catalysis Today 2000; 56 (1–3) 275–282.

[40] Choi J.S., Song I.K., Lee W.Y. Performance of shell and tube-type membrane reactors equipped with heteropolyacid-polymer composite catalytic membranes. Catalysis Today 2001; 67 (1–3) 237–245.

[41] Song I.K., Lee W.Y., Kim J.-J. Methyl t-butyl ether decomposition in an inert membrane reactor composed of 12-tungstophosphoric acid catalyst and polyphenylene oxide membrane. Applied Catalysis A: General 1993; 96 (1) 53—63.

[42] Castanheiro J.E., Ramos A.M., Fonseca I., Vital J. The acid-catalysed reaction of α-pinene over molybdophosphoric acid immobilised in dense polymeric membranes. Catalysis Today 2003; 82 (1–4) 187–193.

[43] Castanheiro J.E., Fonseca I.M., Ramos A.M., Oliveira R., Vital J. Hydration of α-pinene over molybdophosphoric acid immobilized in hydrophobically modified PVA membranes. Catalysis Today 2005; 104 (2–4) 296–304.

[44] Tropecêlo A.I., Casimiro M.H., Fonseca I.M., Ramos A.M., Vital J., Castanheiro J.E. Esterification of free fatty acids to biodiesel over heteropolyacids immobilized on mesoporous silica. Applied Catalysis A: General 2010;390(1–2) 183–189.

http://dx.doi.org/10.5772/59584

[45] Cao F., Chen Y., Zhai F., Li J., Wang J., Wang X., Wang S., Zhu W. Biodiesel production from high acid value waste frying oil catalyzed by superacid heteropolyacid. Biotechnology and Bioengineering 2008;101 (1) 93–100.

[46] Xu L., Li W., Hu J., Yang X., Guo Y. Biodiesel production from soybean oil catalyzed by multifunctionalized Ta2O5/SiO2-[H3PW12O40/R] (R = Me or Ph) hybrid catalyst, Applied Catalysis B: Environmental 2009; 90 (3–4) 587–594.

[47] Baig A., Ng T.T., A single-step solid acid-catalyzed process for the production of biodiesel from high free fatty acid feedstocks. Energy Fuel 2010; 24 (9) 4712– 4720.

[48] Nilesh N., Brahmkhatri V., Patel A. Efficient synthesis of biodiesel from waste cooking oil using solid acid catalyst comprising 12-tungstosilicic acid and SBA-15. Fuel 2014;135 (1) 253–261.

[49] http://en.wikipedia.org/wiki/McDonald's (accessed 14 October 2014).

[50] http://www.aboutmcdonalds.com/mcd/sustainability/planet/minimizing-waste/biodiesel.html (accessed 14 October 2014).

[51] Gao L., Wang Y., Wang J., Huang L., Shi L., Fan X., Zou Z., Yu T., Zhu M., Li Z. A novel ZnII-sensitive fluorescent chemosensor assembled within aminopropyl-functionalized mesoporous SBA-15. Inorganic Chemistry 2006; 45(17) 6844–6850.

[52] Liu H., Xue N., Peng L., Guo X., Ding W., Chen Y. The hydrophilic/hydrophobic effect of porous solid acid catalysts on mixed liquid phase reaction of esterification. Catalysis Communications 2009; 10 (13) 1734–1737.

[53] Pizzio L.R., Vásquez P.G., Cáceres C.V., Blanco M.N. Supported Keggin type heteropolycompounds for ecofriendly reactions. Applied Catalysis A: General 2003;256 (1–2) 125–139.

[54] Screeprasanth P.S., Srivastava R., Scrinivas D., Ratnasamy P. Hydrophobic, solid acid catalysts for production of biofuels and lubricants. Applied Catalysis A: General 2006;314(2) 148–159.

Chapter 15

Photoacoustic Spectroscopy in the Assessment of the Quantitative Composition of the Biomass — Barley Straw

Kingsley L. Iroba and TabilLope G.

Additional information is available at the end of the chapter

http://dx.doi.org/10.5772/59225

1. Introduction

Agricultural biomass residue is a naturally occurring polymeric compound comprised of cellulose, hemicellulose, lignin, and extractives [1]. Biomass has the potential to be used as raw material for the biofuel industry. Cellulose occurs as crystalline fibers with random amorphous portions. The crystallinity of cellulose fibers is a key obstacle for effective biomass enzymatic hydrolysis [2]. Hemicellulose functions as a matrix for the cellulose and acts as a linkage between the crystalline cellulose and amorphous lignin [1-2]. Lignin is a polyphenolic compound that acts as a cementing material for biomass [1]. Lignin binds the hemicellulose and cellulose within the matrix. It provides rigidity and mechanical strength, holding it together [3]. The complex structure of lignocellulosic biomass presents resistance and 'recalcitrance' to biological and chemical degradation during enzymatic hydrolysis/saccharification and the subsequent fermentation process. This leads to a very low conversion rate, which makes the process less economically feasible [4-7]. Pretreatment strategies are employed to overcome the aforementioned challenges by deconstructing/disrupting the lignocellulosic matrix [8], while making cellulose and hemicellulose accessible for the production of bioethanol.

Radio frequency (RF)-alkaline and steam explosion pretreatment of biomass barley straw have shown that pretreatment enhanced the modification of the natural binding characteristics of lignocellulosic biomass [8-9]. Subsequently, this improved the physical properties of the produced pellets from the pretreated biomass [10]. RF heating has been considered as an energy efficient heating method that is an alternative to conventional methods (such as hot air). In RF heating, there is a volumetric heat generation inside the product, which is a result of the interaction between the RF waves and the ions/molecules of the product [11]. Therefore, heat flows from inside the product to the outside, unlike conventional heating methods in which heat is transferred from the heating medium to the product via conduction or convection [8]. RF heating has the following advantages: 1) uniform electric field strength inside the application chambers, therefore preventing uncontrolled heating, overheating, local hot spots, and product degradation; 2) large penetration depth (10-30 m); and 3) higher energy efficiency [11-13]. RF has been successfully applied to leather drying [13], quantification of hydroxycin-namic acids and lignin in perennial forage and energy grasses [14], thermal therapy [15], and in other research fields such as food processing (blanching, tempering, pasteurisation, sterilisation) and medicine [11]. Recently, Izadifar et al. (2009) [12] demonstrated that RF can be used for the extraction of podophyllotoxin from rhizomes of *Podophyllum peltatum*.

Steam explosion is operated by introducing the feedstock into the reactor and heating under steam pressure (2000-5000 kPa; 200-260°C) for a few minutes [16]. Steam explosion induces chemical effects because "water itself acts as an acid at high temperatures" [17]. The sudden thermal expansion involved in the termination of the reaction causes the particulate structure of the biomass to open up [17]. Lam et al. 2011 [18] investigated the steam explosion of Douglas Fir (*Pseudotsuga menziesii*) at a reaction temperature of 200-220°C and a retention time of 5-10 min. Excoffier et al. (1991) [19] and Ferro et al. [20] (2004) have applied steam explosion on lignocellulosic biomass. These authors reported that steam explosion is effective in the pretreatment of biomass.

The traditional methods of determining the chemical compositions of biomass involve gravimetric and analytical procedures. These procedures are time-consuming, laborious and expensive to perform, with low sample throughput and often results in a corresponding degradation of natural polymers [1, 21-22]. In contrast, procedures involving infrared (IR) spectroscopy are useful tools in rapidly extracting information about the structure of biomass constituents and the associated chemical changes resulting from various biomass treatments [1]. Infrared spectroscopy offers researchers an alternative method that is easier, robust and rapid. Fourier transform infrared (FTIR) spectroscopy has been successfully applied in a variety of species in wood surface characterization, for estimating the carbohydrate and lignin components [1, 22]. The majority of the carbon based molecules in plants and animals alike are highly active in the IR. FTIR is associated with superior spectral resolution and provides information on the fundamental molecular vibrations [22]. This permits better discrimination of structural and compositional differences, and often better structural interpretation [22]. Furthermore, FTIR analysis requires only small amounts of biomass plant material, which helps when screening samples available in limited quantities [22]. Applications of FTIR have also been found in biomedical weld and medicine research such as cancer and bone [23-24].

Mid-infrared spectroscopy and near infrared reflectance spectroscopy (NIRS) are the two types of IR spectroscopy that have found application for the measurement of chemical composition in lignocellulosic biomass. NIRS has been used in the prediction of the chemical composition in bulk plant samples. While mid-IR spectroscopy in contrast to NIRS predicts the fundamental molecular vibrations, as such it gives a better information and insight into the molecular bonds present in the sample [25]. FTIR has been successfully applied for the comparison of Fourier transform infrared (FTIR) spectra of individual cells acquired using synchrotron and conventional sources [23]. The authors reported that the main beneficiaries of this rapid development (FTIR) on the instrumental front are the applications-based sciences, such as biomedical and cell-biological fields, for which results have been produced that are unavailable by standard methodologies. Stewart et al. (1995) [26] used FTIR spectroscopy to investigate the changes in composition and structure of oak wood and barley straw that has been subjected to chemical and biochemical treatments. The samples were also analyzed gravimetrically for residual neutral sugar composition, lignin and uronic acid contents. The spectroscopic techniques provided complementary information to the traditional approach. The FTIR fingerprint of wheat endosperm arabinoxylan (AX) has been investigated using a set of polysaccharides exhibiting variation of their degree of substitution and xylo-oligosaccharides comprising xylose units mono-or disubstituted by arabinose residues [27]. Yu (2005) [28] applied synchrotron FTIR microspectroscopy for the investigation of the molecular chemistry of various feed tissues. These images revealed spatial intensity and distribution of chemical functional groups in various feeds tissues within cellular dimensions. The researcher reported that such information can be used for plant breeding program for selecting superior variety of plant for targeted feed purposes and for prediction of feed quality and nutritive value. This proved that synchrotron FTIR microspectroscopy can be used for biological structure study.

Changes in plant cell wall composition following pretreatment largely influence the end use performance. The ability to easily monitor these compositional changes is, therefore, important. Hence, the objective of this investigation is to study and rapidly estimate the relative chemical composition of lignin, cellulosic and hemicellulosic polymers in pretreated and non-treated lignocellulosic barley straw using FTIR-PAS. This work focuses on the effects of radio frequency-alkaline and steam explosion pretreatments on the cell compositions of barley straw.

2. Material and methods

2.1. Material procurement and preparation

Barley straw of the "Xena" variety was grown in Maymont, SK (56.667°N, 107.794°W) and obtained from RAW Ag Ventures Limited (Maymont, SK) in October 2009. To increase the surface area of the biomass, the straw was ground using a hammer mill (Model No. GM13688, Glen Mills Inc., Maywood, NJ) with screen size of 1.6 mm. A dust collector (House of Tools, Model no. DC-202B, Saskatoon, SK) was connected to the outlet of the hammer mill to control dust during operation and to provide flow of the biomass in and out of the hammer mill. The initial moisture content of the straw was 8.09% (wet basis). The moisture content was measured based on ASABE standard method, ASAE S358.2 (2008). As a comparison between pretreatment methods, the 1.6 mm biomass grind was subjected to two different pretreatment methods: radio frequency (RF)-alkaline pretreatment using a RF machine (1.5 kW & 27.12 MHz laboratory dryer, Strayfield, Theale, Reading) in a blown glass reactor (volume 4.25 liters) stationed in University of Saskatchewan, Saskatoon, SK, and steam explosion (SE) pretreatment located at the Clean Energy Research Center, University of British of Columbia, Department of Chemical and Biological Engineering, Vancouver, BC. The material and operating variables considered in both methods of pretreatment are shown in Table 1 (a-b). Each pretreatment was performed in two replicates. For more details on the RF-alkaline and SE pretreatment, see [8-9].

Variables	Levels
Hammer screen size	1.6 mm
NaOH solution concentration	1% w/v
Biomass: NaOH solution ratio	1:4 - 110 g biomass and 440 g NaOH solution 1:5 - 110 g biomass and 550 g NaOH solution 1:6 - 100 g biomass and 600 g NaOH solution 1:7 - 100 g biomass and 700 g NaOH solution 1:8 - 90 g biomass and 720 g NaOH solution
Soaking time	1 h
Temperature	70, 80, and 90°C
Residence time	20 minutes

(a)

Variables	Levels
Temperature (°C)	140, 160, and 180
Moisture Content (% mass fraction of water)	8, 30, and 50
Retention Time (min.)	5 and 10

(b)

Table 1. a) RF-alkaline material and operating variables using blown glass reactor; (b) SE material and operating variables with corresponding levels

2.2. Chemical composition analysis of lignocellulosic biomass

The chemical composition analysis of the RF-alkaline and SE pretreated biomass grind was performed using the National Renewable Energy Laboratory standard (NREL) [30] at a laboratory facility at the Agriculture and Agri-Food Canada, Saskatoon, SK. Each sample was replicated twice. The NREL standard uses a two-step acid hydrolysis to fractionate the biomass into forms that are more easily quantified. The first step uses 72% H_2SO_4, while the second step uses 4% H_2SO_4. The lignin fractionates into acid insoluble and acid soluble material. The acid insoluble lignin is the residue (remaining solids) from the hydrolysis suspension. Acid-soluble lignin moieties were quantified using the Waters Acquity Ultra Performance Liquid Chromatography–MS system (Acquity 2004-2010, Waters Corp., Milford, MA), which has the capability of separating and quantifying the various lignin components. The complex

http://dx.doi.org/10.5772/59225

carbohydrates are hydrolyzed into monomeric forms (xylose, arabinose, mannose, glucose, and galactose) and subsequently quantified using UPLC-MS. The percentage hemicellulose was obtained by adding up the percentage xylose, arabinose, mannose, and galactose, while the percentage glucose was assigned to percentage cellulose. Further details on the material preparation, physical characteristics of the biomass grind, radio frequency alkaline technique, steam explosion process, and the chemical composition analysis using the NREL standard can be obtained in the research studies of [8-9]. The pretreated and non-treated samples were further ground to screen size of 0.354 mm using a precision grinder (Falling Number, Model No. 111739, Huddinge, Sweden).

2.3. Preparation of reference materials of known concentration

In order to develop a predictive model that will be rapidly used for the quantitative prediction of the chemical composition contained in the RF-alkaline and SE pretreated biomass, pure cellulose (microcrystalline powder), hemicellulose (xylan from birch wood), and lignin (hydrolytic) powders (Sigma-Aldrich Canada Ltd., St. Louis, MO) were mixed in different proportions (Table 2). These were used as reference spectra to determine the relationship between the respective quantity in the mixture and the representative sample FTIR spectra. The FTIR wavenumber-dependent instrumental effects were corrected by using carbon black reference spectrum.

Reference Mixtures	Cellulose (%)	Hemicellulose (%)	Lignin (%)
C1H0L0	100	0	0
C0H1L0	0	100	0
C0H0L1	0	0	100
C7H2L0	75	25	0
C2H7L0	25	75	0
C0H2L7	0	25	75
C0H7L2	0	75	25
C7H0L2	75	0	25
C2H0L7	25	0	75
C5H2L2	50	25	25
C2H5L2	25	50	25
C2H2L5	25	25	50
C3H3L3	33	33	33
Total	**13**	**13**	**13**

C, H, and L represent Cellulose, Hemicellulose, and Lignin, respectively

Table 2. Reference materials: Pure cellulose, hemicellulose, and lignin mixtures used to obtained the reference spectra (adapted from [21])

2.4. Fourier Transformed Infrared Photoacoustic Spectroscopy (FTIR-PAS)

Intensity of spectra generally increases as the particle size decreases [30]. To avoid moisture interference, the biomass samples were further dried using the forced-air convection dryer [31] (Shaw et al. 2007) set at 40°C for 48 h. Photoacoustic intensities are lower for samples with high moisture content. This might be due to lower efficiency of heat transfer between the moist cellulose surface and the carrier gas [32]. The Infrared data/spectra of the reference materials and biomass samples were collected using Mid-IR beamline (01B1-1) with energy range of 4000-400 cm^{-1}, at the Canadian Light Source Inc. (CLS, University of Saskatchewan, Saskatoon, SK). The beamline has a MTEC Model 300 photoacoustic cell (MTEC Photoacoustic Inc., Ames, IA) for FTIR-PAS of bulk samples. The sample cup was filled with reference biomass sample (52-75 mg, depending on the pretreatment type and combination) and purged with helium gas to remove water vapor and CO_2. Helium gas is also needed in the medium because of it sound propagating properties. The collected FTIR spectra of the reference materials and biomass samples were recorded using Globar source (silicon carbide rod). When the radiation is incident on the sample, the energy of the Infrared beam is being absorbed by the sample layer. The photoacoustic signal is generated by thermal expansion of the gas caused by heat associated with the thermal wave emanating from the sample. The photoacoustic signal is carried by a carrier gas (Helium) to a microphone which is transferred to the FTIR electronics (detector) for processing; this ultimately produces the needed spectrum [33-34]. The spectrum for each reference material and biomass samples were recorded separately averaging 64 interferograms (number of scans) collected from wavenumbers of 2000-400 cm^{-1} at a resolution of 4 cm^{-1}. The higher the number of scans the better the signal (lesser noise). Stuart (1997) [36] reported that the signal-to-noise ratio (SNR) is proportional to the square root of the number of scans, n (SNR α $n^{0.5}$). Therefore, the higher the number of scans, the higher is the SNR. Resolution of 4 cm^{-1} was used to be able to discriminate between too close overlapping peaks, help increase the SNR and subsequently obtain higher resolution. The OPUS 6.5 (Brucker Optics Inc. Billerica, MA) software was used for the collection of the FTIR-PAS spectra. Three replicates were performed for each reference and biomass samples.

2.5. Determination of concentration

Beer-Lambert discovered that the amount of light transmitted by a solid sample was dependent on the thickness of that sample [35]. The Beer-Lambert law which can be applied to all electromagnetic radiation, states that the absorbance of a material is directly proportional to the thickness and concentration of the sample as shown: $A=\varepsilon CL$. A=absorbance of the material, C=concentration, L=pathlength of the sample, and ε=constant of proportionality, which is referred to as the molar absorptivity [35]. Infrared spectra, particularly in the spectra of solid samples are often associated with the presence of asymmetric bands. As such, peak height cannot be used for the quantitative analysis of the spectra, because the baseline will vary from sample to sample. Instead, peak-area measurements should be used [35].

http://dx.doi.org/10.5772/59225

2.6. Quantitative and qualitative analysis of the FTIR-PAS spectra

The two quantities of greatest interest in virtually any type of spectroscopy are, of course, band positions (wavenumbers) and intensities, the former generally conveying qualitative information, the latter quantitative [30]. Therefore, these two variables were used for the FTIR data analysis.

2.6.1. Spectrum manipulation

There are techniques that assist in both qualitative and quantitative interpretation of spectra. OriginPro software (Data analysis and graphing Version 8.6, OriginLab Corporation Northampton, MA) was used for the spectrum manipulation, quantitative, and qualitative analysis of the FTIR-PAS Spectra:

2.6.2. Baseline correction, subtraction, and rescaling

The Baseline Mode is a tool for choosing a baseline mode and creates the baseline. A user-defined baseline treatment was applied in this analysis. A common flat baseline of 0.012, joining the points of lowest absorbance (via fitting Pro) on the peak was selected and applies across all the reference materials and biomass sample spectra. Thereafter, baseline subtraction was performed from the input data, such that the absorbance difference between the selected baseline and the top of the band is then used. This helps to improve the accuracy of the peak finding. Rescaling of the baseline to zero was subsequently performed, so that all the spectra will have a common origin of zero.

2.6.3. Smoothing

Smoothing is a signal processing technique typically used to remove or diminish noise from signals/spectrum. After a spectrum is smoothed, it becomes similar to the result of an experiment obtained at a lower resolution [35, OriginLab Corporation manual]. The features are blended into each other and the noise level decreased. A smoothing function is basically a convolution between the spectrum and a vector whose points are determined by the degree of smoothing one wish to apply [35]. There are multiple smoothing methods that work differently depending on the nature of the signal and the noise contained in the signal. Each method offers a different performance. In this present analysis, Savitzky-Golay was used. The Savitzky-Golay filter method performs a local polynomial (order of 2) regression around each point, and creates a new, smoothed value for each data point. This method is superior to other methods (such as adjacent averaging) because it tends to preserve features of the data, such as peak height and width, which can be "washed out" by adjacent averaging (OriginLab Corporation, Northampton, MA, 2012). To increase the smoothness of the result, one can increase the "window size," used in each local regression (OriginLab Corporation, Northampton, MA, 2012). As such, the window size in this analysis was increased from one to two. But for very large window size, adjacent averaging may depart from the input signal too much, whereas

Savitzky-Golay can still preserve the overall profile (OriginLab Corporation, Northampton, MA, 2012).

2.6.4. Peak finding settings

Manual peak editing was performed so as to effectively select the required respective peaks. The second (2nd) derivative was used to search for all the hidden peaks and heavily overlapped bands included in the spectrum data. Differentiation is used to resolve and locate peaks in an envelope. Sharp bands are enhanced at the expense of broad ones, and this may allow easier selection of a peak, even when there is a broad band beneath it [35].

2.6.5. Characteristic peak assignment

The characteristic wavenumbers for pure cellulose, hemicellulose, and lignin listed in Table 3, were used for the peak assignment. Five characteristic peaks were identified for the pure cellulose, six characteristic peaks for pure hemicellulose, and six characteristic peaks for pure lignin (Figure 1a-c). The number of the peaks identified for the respective reference materials depends on the mixture of the reference materials. All seventeen peaks were identified for the treated and non-treated biomass samples.

2.6.6. Peak integration

To obtain quantitative values from the area under the manipulated spectrum/peaks, the area under the respective peaks were integrated and output into excel file.

2.6.7. Normalization of photoacoustic infrared spectra

Prior to the spectrum manipulation, the FTIR-PAS biomass sample spectra were corrected for wavenumber-dependent instrumental effects by dividing the reference carbon black ("background") spectrum intensity. This strategy implicitly assumes that the stability of the instrumentation used is adequate to ensure reliable results, even though the sample and reference spectra were collected at different times [30]. Carbon black is featureless, in the sense that it does not show any major characteristics peak [30] Photoacoustic (PA) cell intensities varied with sample packing in the PA cell [36]. Stuart (1997) [35], also reported that absorbance varies linearly with the sample thickness. Therefore, the effect of bulk density of the reference materials and biomass samples was corrected by dividing the integrated areas with respective mass of the reference materials and biomass samples contained in the PA sample cup.

The model was further standardized by normalizing the corrected integrated area data from 0 to 1. This was performed by dividing individual reference materials and biomass samples corrected integrated area data by corresponding maximum corrected integrated area data value. The aforementioned steps were performed for the three major components of lignocellulosic biomass (cellulose, hemicellulose, and lignin). Therefore, this normalization approach ensures that the predictive model is adaptable for quantitative analysis of FTIR-PAS spectra obtained for any lignocellulosic biomass.

http://dx.doi.org/10.5772/59225

3. Regression analysis

This portion of the analysis is divided into two parts: the training and verification analysis:

3.1. Training analysis

The normalized data of cellulose, hemicellulose, and lignin components in the reference materials and the combined biomass samples (i.e. combination of data from RF and SE analysis) were correlated to their percentage compositions. The RF and SE normalized data were combined so as to develop a general model that can be applied on a barley straw sample, irrespective of the pretreatment method applied on the biomass. The percentage compositions of the reference materials (Table 2) and the percentage compositions (Tables 4 & 6-8) of the combined biomass samples obtained from the NREL Standard were used as the dependent variable, while two out of the three replicates from the combined normalized data (obtained from the PA spectra) were randomly selected and used as the independent variables. Five independent variables were used for the cellulose (because it has five characteristic wavenumbers), while six independent variables were used for both hemicellulose and lignin (they both have six characteristic wavenumbers each). Multiple linear regression analysis was conducted at 5% significance level using IBM SPSS Statistics (Superior Performing Statistical Software, version 20 for Windows, 2012; IBM, Armonk, New York, NY). Regression models (equations 1-3) for cellulose, hemicellulose, and lignin were developed based on the generated regression coefficients. The developed models have the capability to predict quantitatively, the percentage compositions of cellulose, hemicellulose, and lignin in any lignocellulosic biomass.

3.2. Verification analysis (Validation)

Subsequently, the normalized data (based on the respective characteristic peaks) from the third replicate of biomass sample was substituted into the developed predictive models to estimate/ predict the percentage compositions of lignocellulosic components present in the non-treated, RF-alkaline and SE pretreated biomass samples (Tables 4 & 6-8).

4. Results and discussion

4.1. Reference material spectra

Figure 1 (a-c) shows the normalized spectral results obtained from a typical reference cellulose (100%), hemicellulose (100%), and lignin (100%), respectively, after smoothening and baseline corrections. Generally, FTIR-PAS techniques permit to obtain spectra which present specific characteristic key bands of individual components. In principle, such band position allows the discrimination of different species and provides important information about the chemical compositions of the material.

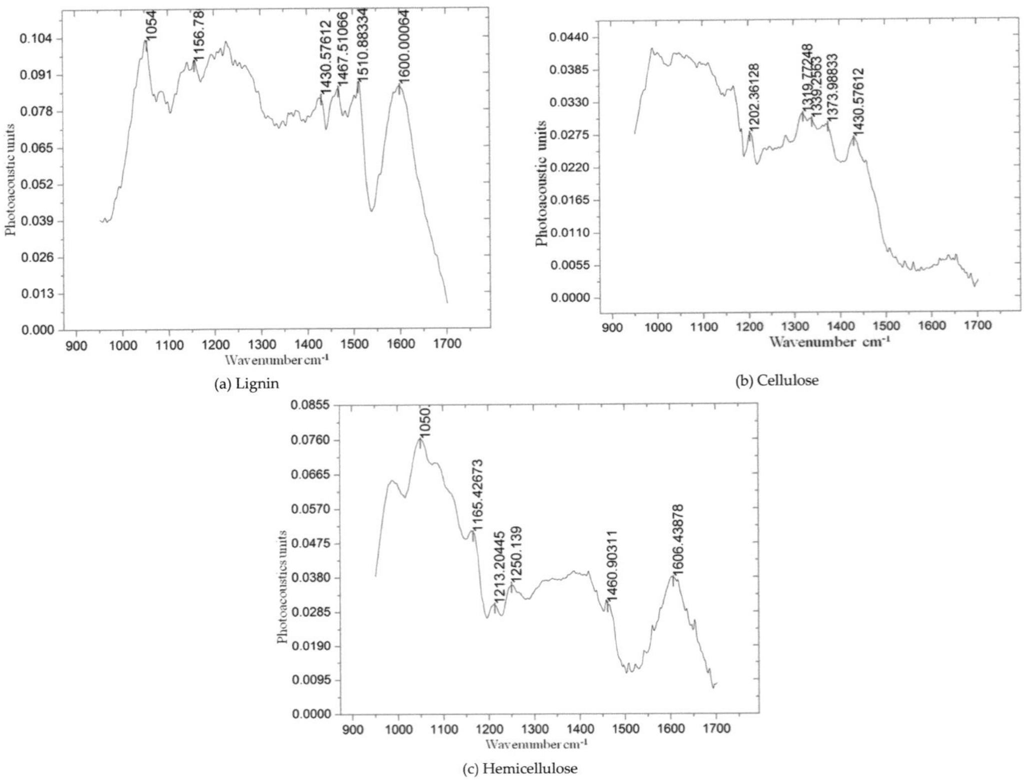

Figure 1. FTIR-PAS spectra of a) pure lignin (hydrolytic), b) cellulose (microcrystalline powder), and c) hemicellulose (xylan from birch wood) powders showing prominent peaks/band positions at the various characteristics wavenumbers.

4.2. Characteristic peak positions

The various prominent (characteristic) peak (band) positions and the corresponding peak assignments for cellulose, hemicellulose, and lignin are presented in Table 3. These are distinct peaks at different wavenumbers associated with these three biomass components. Lignin spectrum has six prominent peaks at wavenumbers of 1599 (X_1), 1511 (X_2), 1467 (X_3), 1429 (X4), 1157 (X_5), and 1054 cm^{-1} (X_6) (Figure 1a). Cellulose spectrum has five distinct peaks at wavenumbers of 1431 (X_1), 1373 (X_2), 1338 (X_3), 1319 (X_4), and 1203 cm^{-1} (X_5) (Figure 1b). Hemicellulose spectrum has six characteristic peaks at wavenumbers of 1606 (X_1), 1461 (X_2), 1251 (X_3), 1213 (X_4), 1166 (X_5), and 1050 cm^{-1} (X_6) (Figure 1c).

http://dx.doi.org/10.5772/59225

Wavenumbers (cm^{-1})	Cellulose 100%	Hemicellulose 100%	Lignin 100%	Peak Assignment
1650-1600	-	1606	-	1604-acetylated hemicellulosic polysaccharides [37].
1600-1550	-	-	1599	1595-aromatic ring stretch, strongly associated with aromatic C-O stretching mode [26]; 1595- phenylpropanoid polymer [37]; 1600-quadrant ring stretching (aromatic compound of lignin) [38-39]; 1600-aromatic skeletal vibration [1].
1550-1500	-	-	1511	1515-semicircle ring stretching (aromatic compound of lignin) [40-41]; 1510-lignin aromatic compound [28]; 1510-very strong aromatic ring stretch, aromatic C-O stretch [26, 40]; 1513-aromatic C=C stretch [41]; 1510-semi-circle stretch of para-substitute benzene rings [42]; 1550-protein [42]; 1502-aromatic skeletal vibration-softwood lignin [1]; 1505-aromatic skeletal vibration of benzene ring in lignin [1]
1500-1450	-	1461	1467	1462-C-H lignin deformation (methyl and methylene) [1].
1450-1400	1431	-	1429	1425- C-H in-plane lignin deformation with aromatic ring stretching [1]; 1430-CH_2 in-plane bending vibrations [43]; 1430- C-H cellulose deformation (asymmetric) [1]; 1433-aromatic C=C stretch [41].
1400-1350	1373	-	-	1370-weak C-O stretching [39, 44]; 1372-C-H cellulose deformation (asymmetric) [1]; 1380-C-H symmetric and asymmetric deformation [41]; 1382-C-O stretch [45]; 1378-cellulose (microcrystalline) [37].
1350-1300	1319, 1338	-	-	1335-weak C-O stretching [39, 44]; 1336-C-H ring in-plane bending vibrations [43]; 1336- O-H in-plane cellulose deformation [1]; 1318-CH_2 cellulose wagging [1].
1300-1250	-	1251	-	1258- acetylated glucomannan [37]; 1258-arabinoxylans (hemicellulose) [45].
1250-1200	1203	1213	-	1246 cellulosic material [39]; 1246-weak C-O stretching [44]; 1258- acetylated glucomannan [37]; 1250-acetylated hemicelluloses [43]; 1201- O-H cellulose deformation [1]
1200-1150	-	1166	1157	1160 glycosidic linkage [27]; 1162-C-O-C ring vibrational stretching [43]; 1165-arabinoxylans (hemicellulose) [45];
1100-1050	-	-	1054	1086- C-O of secondary alcohols [1].
1050-1000	-	1050	-	1044-arabinoxylans (hemicellulose) [45]; 1035-C-O, C = C and C-C-O vibrational stretching [43]; 1045-C-OH bending [44, 27]; 1025-non-structural CHO [39].

Table 3. Characteristics peaks/bands position and assignment of pure cellulose, hemicellulose, and lignin (adapted from [21])

4.3. Radio frequency-alkaline treatment on lignin

It was observed that NaOH concentration is a major factor in the pretreatment process. The swelling initiated by NaOH creates pores in the biomass matrix [46-48], which helps enhance the reactivity of the biomass matrix to any externally added material such as enzyme [8]. A similar finding was reported by [49]. The lignin values obtained from the traditional wet chemistry chemical composition analysis and through FTIR-PAS as shown in Table 4 depicts that in most cases, there is a decrease in the lignin in the pretreated samples as compared to the non-treated samples. This may be due to the unloosened lignified matrix in the non-treated sample which is tightly bound to the other constituents, unlike the RF alkaline pretreated samples with broken bonds which has been structurally separated, and disrupted. This disaggregation may be attributed to the interaction between the biomass and NaOH solution in the presence of the RF heating which is believed to be responsible for this solid loss [8]. It also seems that the disruption and deconstruction of the lignified matrix is associated with the dipole interaction, flip flop rotation, and friction generated between the electromagnetic charges from the RF radiation and the ions and molecules from the NaOH solution and the biomass [8]. Lignin removal is an important part of the pretreatment process, because lignin can effectively inhibit/prevent the cellulase enzymes from hydrolysing the cellulose. This lignin reduction resulting from the alkaline pretreatment had also been reported by [50-51].

Lignin estimated by FTIR-PAS was higher than those obtained with NREL procedures. The laborious and time consuming traditional method that uses 72% H_2SO_4 seems to create a more stringent condition that may lead to altering and further degrading the native cell wall compositions (such as lignin and complex carbohydrates), structure, and possibly generating artifacts. FTIR-PAS is a rapid, direct, non-invasive, and non-destructive chemical analytical technique. This rapid method can detect molecular chemical characteristics of biological materials at high spatial resolutions without altering the inherent biomass structure such as the tissue [40, 44].

The difference in the lignin values from the traditional approach and FTIR-PAS may also be attributed to the spectrum manipulations and the assumption that the reference lignin sample is 100% pure. The FTIR-PAS qualitative and quantitative analytical chemical information can be connected to the structural information within cellular dimension [42].

Equation 1 shows the predictive model for lignin generated from the training and verification analysis of the combined values from RF and SE using the methods described in section 6.3.8.

Lignin predictive model:

$$\%L = -33.92X_1 + 52.61X_2 + 32.16X_3 + 208.14X_4 + 98.46X_5 - 56.59X_6 + 17.33 + \varepsilon \qquad (1)$$

%L=%lignin (dry matter basis), X_i=regressors/explanatory variables (normalized data based on the respective characteristic peaks, as shown in Figure 1a), with the regressors representing the respective wavelengths as 1599 (X_1), 1511 (X_2), 1467 (X_3), 1429 (X4), 1157 (X_5), and 1054

http://dx.doi.org/10.5772/59225

Temperature (°C)	Biomass: NaOH solution ratio	Lignin (%)		
		Wet chemistry	FTIR-PAS	Standard error
Non-treated	-	20.12	22.62	1.25
70	1:4	19.56	15.80	1.88
80	1:4	19.22	15.78	1.72
90	1:4	20.66	20.40	0.13
70	1:5	17.81	16.77	0.52
80	1:5	18.22	20.63	1.21
80^{W}	1:5	15.62	18.15	1.27
90	1:5	17.64	19.38	0.87
90^{W}	1:5	15.37	15.90	0.27
24	1:6	21.57	23.65	1.04
70	1:6	17.91	15.98	0.97
90	1:6	18.38	21.63	1.62
24	1:7	17.99	22.34	2.17
70	1:7	17.82	17.55	0.13
80	1:7	18.10	25.22	3.56
80^{W}	1:7	12.39	20.57	4.09
80^{DW}	1:7	19.90	27.74	3.92
80^{TW}	1:7	20.38	26.49	3.06
90	1:7	18.05	20.19	1.07
90^{W}	1:7	10.98	18.78	3.90
70	1:8	18.53	22.20	1.83
80	1:8	18.86	21.60	1.37
90	1:8	19.27	17.45	0.91

Standard error=standard deviation between the wet chemistry and FTIR-PAS values divided by the square root of 2; W=washed after pretreatment; DW=pretreated with distilled water; TW=pretreated with tap water.

Table 4. Lignin composition of RF-alkaline pretreated and non-treated biomass grind obtained using the traditional wet chemistry and FTIR-PAS methods

cm^{-1} (X_6). 17.33=intercept, and ε=error term/stochastic variable which describes the noise (errors that could emanate from the equipment, environment, or the experimenter).

Table 5 shows the R^2 and mean square error values of the various biomass components from the regression analysis. However, the R^2 values are too low; this might be due to the associated stochastic variables and the spectrum manipulations.

Biomass composite	R^2 value	Mean square error
% Lignin	0.68	9.10
% Cellulose	0.34	13.86
% Hemicellulose	0.31	15.86

Table 5. R^2 and mean square error values from the regression analysis

Washing the pretreated samples reduces the lignin. This may be due to loss of solid lignin during the washing process [8]. This investigation shows that the concentration of NaOH solution and the ratio of biomass to the NaOH solution are the dominant contributing factors, while temperature plays a lesser role. The heat provided by the RF is needed to assist the alkaline solution in the deconstruction and disaggregation of lignocellulosic biomass matrix. It was also observed from this investigation that biomass can be alkaline pretreated even at room temperature if the required ratio of biomass and NaOH solution is applied.

4.4. Radio frequency-alkaline treatment on cellulose and hemicellulose

The difference in the cellulose and hemicellulose values from the traditional approach and FTIR-PAS may be attributed to the spectrum manipulations and the assumption that the reference cellulose and hemicellulose samples are 100% pure. Equations 2-3 show the cellulose and hemicellulose predictive models generated from the training and verification analysis of the combined values from RF and SE using the methods described in section 3.

Temperature (°C)	**Biomass: NaOH solution ratio**	**Cellulose (%)**			**Hemicellulose (%)**		
Wet chemistry	**FTIR-PAS**	**Standard error**	**Wet chemistry**	**FTIR-PAS**	**Standard error**		
Non-treated	-	42.51	50.37	3.93	29.98	23.82	3.08
70	1:4	22.25	28.65	3.20	23	18.10	2.45
80	1:4	22.37	27.69	2.66	22.14	28.47	3.16
90	1:4	26.93	26.09	0.42	26.24	20.73	2.75
70	1:5	24.21	4.98	9.61	21.63	30.37	4.37
80	1:5	21.07	29.42	4.18	21.38	24.56	1.59
80^{W}	1:5	27.69	26.55	0.57	21.6	15.81	2.90
90	1:5	24.65	37.03	6.19	21.05	8.44	6.31
90^{W}	1:5	33.44	34.50	0.53	26.08	22.58	1.75
24	1:6	30.93	14.95	7.99	29.12	21.00	4.06
70	1:6	28.25	29.02	0.39	23.18	22.88	0.15
90	1:6	30.37	20.85	4.76	22.36	39.31	8.47

Temperature (°C)	Biomass: NaOH solution ratio	Cellulose (%)			Hemicellulose (%)		
24	1:7	25.78	27.08	0.65	23.4	26.21	1.40
70	1:7	27.39	11.74	7.82	24.27	28.81	2.27
80	1:7	26.08	17.83	4.12	26.75	10.70	8.02
80^{W}	1:7	38.20	19.86	9.17	27.72	8.48	9.62
80^{DW}	1:7	21.88	2.06	9.91	19.12	5.76	6.68
80^{TW}	1:7	24.98	6.09	9.44	21.04	9.97	5.54
90	1:7	21.33	23.74	1.20	22.27	9.55	6.36
90^{W}	1:7	33.31	12.98	10.17	20.69	9.71	5.49
70	1:8	26.73	27.91	0.59	26.76	22.69	2.03
80	1:8	18.44	42.08	11.82	18.32	31.06	6.37
90	1:8	22.68	21.66	0.51	19	25.64	3.32

Standard error=standard deviation between the wet chemistry and FTIR-PAS values divided by the square root of 2; W=washed after pretreatment; DW=pretreated with distilled water; TW=pretreated with tap water.

Table 6. Cellulose and hemicellulose compositions of RF-alkaline pretreated and non-treated biomass grind obtained using the traditional wet chemistry and FTIR-PAS methods

Cellulose predictive model:

$$\%C = 28.63X_1 + 48.60X_2 + 35.83X_3 - 51.71X_4 - 29.24X_5 + 37.16 + \varepsilon \qquad (2)$$

Where %C=%cellulose value, X_i=regressors/explanatory variables (normalized data based on the respective characteristic peaks, as shown in Figure 1b), with the regressors representing the respective wavelengths as 1431 (X_1), 1373 (X_2), 1338 (X_3), 1319 (X_4), and 1203 cm^{-1} (X_5), 37.16=intercept, and ε=error term/stochastic variable.

Hemicellulose predictive model:

$$\%H = -14.25X_1 - 90.42X_2 + 34.14X_3 - 39.18X_4 + 71.17X_5 + 122.90X_6 + 30.42 + \varepsilon \qquad (3)$$

%H=%hemicellulose wet chemistry value, X_i=regressors/explanatory variables (normalized data based on the respective characteristic peaks, as shown in Figure 1c), with the regressors representing the respective wavelengths as 1606 (X_1), 1461 (X_2), 1251 (X_3), 1213 (X_4), 1166 (X_5), and 1050 cm^{-1} (X_6), 30.42=intercept, and ε=error term/stochastic variable.

Ramesh and Singh (1993) [2] reported that barley straw theoretically contains about 40% cellulose, 20% hemicellulose, and 15% lignin. Marsden and Gray (1985) [52] also reported that barley straw theoretically contains about 44% cellulose, 27% hemicellulose, and 7% lignin.

These values are comparable with the values obtained from the non-treated sample in this investigation. It should be noted that the variance in the chemical composition between the reported theoretical values and the values obtained from this investigation may be attributed to differences in locations where the crop was grown, weather conditions, the barley variety grown, and different methods of analysis.

4.5. Steam explosion treatment on lignin

The efficacy of FTIR-PAS techniques for studying changes in plant cell wall composition following steam explosion pretreatment has been evaluated. Table 7 shows that the SE pretreated samples have higher lignin content as compared to the non-treated. This may be due to the carbonization of the sample resulting from the direct contact of biomass with the walls of the reactor during the steam explosion pretreatment.

Temperature (°C)	Moisture content (%)*	Retention time (min)	Lignin (%)		
			Wet chemistry	FTIR-PAS	Standard error
Non-Treated	-	-	20.12	22.62	1.25
140	8	5	23.79	26.25	1.23
160	8	5	22.72	31.46	4.37
180	8	5	40.58	21.41	9.58
140	30	5	22.05	28.23	3.09
160	30	5	21.69	24.53	1.42
180	30	5	33.01	25.99	3.51
140	50	5	21.18	25.41	2.11
160	50	5	23.31	33.21	4.95
180	50	5	25.04	32.60	3.78
140	8	10	21.56	31.63	5.04
160	8	10	21.66	25.13	1.73
180	8	10	32.75	33.45	0.35
140	30	10	21.25	29.37	4.06
160	30	10	20.9	27.47	3.28
180	30	10	37.31	29.21	4.05
140	50	10	21	23.54	1.27
180	50	10	31.82	31.16	0.33

Standard error=standard deviation between the wet chemistry and FTIR-PAS values divided by the square root of 2; *= % mass fraction of water.

Table 7. Lignin composition of SE pretreated and non-treated biomass grind obtained using the traditional wet chemistry and FTIR-PAS methods

http://dx.doi.org/10.5772/59225

Hemicellulose degrades easily and some volatile organic compounds vaporize as volatile components, while cellulose behaves as a fixed carbon (solid combustible residue). This may account for the increase in the lignin content. Lam et al. (2011) [18] investigated the steam explosion of Douglas fir (*Pseudotsuga menziesii*) at a reaction temperature of 200-220°C and a retention time of 5-10 min. These researchers reported that there was increase in lignin content from 30 to 43% attributed to the thermal degradation of hemicellulose during the steam explosion treatment. Chen and Kuo (2011) [53] reported that cellulose and lignin are both locked in biomass from the mild carbonization process. This indicates that the degraded cellulose may appear as residue resulting to the increase in the lignin content.

The interaction between moisture content and temperature and also between retention time and moisture content had a statistically significant effect ($P<0.01$) on the lignin. The interaction among the three variables (moisture content, temperature, and retention time) also had a significant effect on the lignin. The difference in the lignin values obtained using the FTIR-PAS and the traditional approach may be attributed to the spectrum manipulations and the assumption that the reference lignin sample is 100% pure.

4.6. Steam explosion treatment on cellulose and hemicellulose

From Table 8, in comparison with the non-treated biomass, it is evident that retention time, moisture content, and temperature had a significant effect on the cellulose and hemicellulose content. The decrease in the sugars content increased at higher retention time and temperature. Wang et al. (2009) [54] also reported that the retention time and temperature are the process parameters required for the optimization of steam explosion process. In this present study, less degradation of the simple sugars was observed at higher moisture content. High feedstock moisture content acts as acid catalyst to hydrolyze biomass during steam explosion. However, the direct contact of biomass with the walls of the reactor will limit and affect the extent of the hydrolysis. Therefore, a combination of carbonization and acid catalyzed hydrolysis occurred which ultimately led to the degradation of the simple sugars and increase in the lignin. The obtained results from the wet chemistry demonstrated that the hemicellulose contained in the biomass was highly degraded (79% to 89%) compared to cellulose (58% to 77%) as reported in chapter 4. The high degradation of hemicellulose was due to its amorphous nature, which degrades easily and evaporates as volatile components during the carbonization process. Presumably, the crystallinity of cellulose was responsible for the less degradation of this component. These degradations can be explained by considering the monomers of hemicellulose and cellulose which consist primarily of sugars. Degradation of cellulose and hemicellulose during steam explosion/thermal pretreatment of biomass has been reported by [16, 19-20, 55-57]. It was reported that hemicellulose is very reactive and was nearly completely removed at 200°C, while both cellulose and lignin can be dissolved partially at higher temperatures. Shaw (2008) [57] performed steam explosion on poplar and wheat straw at 200-205°C, steam pressure of 1.66-1.73 MPa for 4-4.5 min. This author reported a decrease in cellulose and hemicellulose content with an increase in the lignin content after the steam explosion treatment of both biomass samples. Chen and Kuo (2011) [53, Yang et al. (2007) [59], and Khezami et al. (2007) [60] studied the effects of thermal process on biomass. These researchers showed that thermal pretreatment removes moisture and light volatiles from biomass. Bergman et al. (2005) [61] and Lipinsky et al. (2002) [62] reported that during the torrefaction process, biomass was partly decomposed giving off various condensable and non-condensable gases, with a carbon-rich solid as a final product. Mohammad and Karimi (2008) [7] reported a corresponding decrease in total sugar recovery with increasing temperature during steam explosion.

Furthermore, the difference between the cellulose and hemicellulose values estimated using the FTIR-PAS and the measured values using the traditional approach might also be associated with the spectrum manipulations and the assumption that the reference cellulose and hemicellulose samples are 100% pure. The main advantage of this correlation is that based on collection of FTIR-PAS spectra, it provides a rapid, easy, economical, non-destructive, and non-laborious estimation of the chemical composition of biomass. This may be of particular interest in the contexts where more sophisticated and expensive equipments for experimental measurement of biomass chemical compositions are not always available. However, care must be taken in selecting the steam explosion pretreatment conditions in order to prevent excessive degradation of the chemical properties of the complex carbohydrates, since the yields of hemicelluloses and cellulose were dependent on the pretreatment conditions of the steam explosion.

Tempt. (°C)	**M.C. (%)**	**R.T. (min.)**	**Cellulose (%)**			**Hemicellulose (%)**		
			Wet chemistry	**FTIR-PAS**	**Standard error**	**Wet chemistry**	**FTIR-PAS**	**Standard error**
Non-treated	-	-	42.51	50.37	3.93	29.98	23.82	3.08
140	8	5	11.19	20.53	4.67	3.42	12.39	4.48
160	8	5	11.17	17.93	3.38	3.29	13.10	4.90
180	8	5	9.58	18.54	4.48	3.74	6.60	1.43
140	30	5	14.82	19.87	2.53	3.8	17.79	7.00
160	30	5	14.63	17.53	1.45	4.13	12.54	4.21
180	30	5	12.07	17.30	2.62	3.88	11.44	3.78
140	50	5	15.81	22.06	3.12	6.58	15.95	4.69
160	50	5	14.51	19.74	2.62	5.11	15.18	5.04
180	50	5	15.77	20.24	2.23	5.23	14.02	4.40
140	8	10	10.02	16.91	3.45	3.9	16.01	6.05
160	8	10	9.11	19.49	5.19	4.47	9.44	2.49
180	8	10	8.74	16.08	3.67	3.23	12.03	4.40
140	30	10	15.38	15.66	0.14	5.03	14.91	4.94
160	30	10	13.57	16.95	1.69	6.59	6.50	0.05
180	30	10	12.7	16.24	1.77	4.54	12.47	3.96

http://dx.doi.org/10.5772/59225

Tempt. (°C)	M.C. (%)	R.T. (min.)	Cellulose (%)			Hemicellulose (%)		
			Wet chemistry	FTIR-PAS	Standard error	Wet chemistry	FTIR-PAS	Standard error
140	50	10	16.02	17.51	0.74	6.33	14.01	3.84
180	50	10	12.35	15.95	1.80	4.69	10.70	3.01

Standard error=standard deviation between the wet chemistry and FTIR-PAS values divided by the square root of 2; Tempt.=temperature; M.C.=moisture content; R.T.=retention time.

Table 8. Cellulose and hemicellulose compositions of SE pretreated and non-treated biomass grind obtained using the traditional wet chemistry and FTIR-PAS methods

5. Conclusion

Lignocellulosic biomass has been identified as a potential feedstock for the biofuel industry. Quantitation of lignocellulosic biomass components (lignin, cellulose, and hemicellulose) is often performed using the traditional acid hydrolysis followed by gravimetric determination. This approach is complicated and time consuming. FTIR-PAS was used in light of the need for rapid analysis of biomass materials and wood-based materials at large. The samples were initially pretreated using RF-alkaline and steam explosion techniques and analyzed gravimetrically using the traditional approach to elucidate compositional information. Thereafter, the effect of the pretreatment conditions on barley straw grind was further analyzed based on their FTIR-PAS spectra. In order to develop a predictive model that will be swiftly used for the quantitative prediction of the chemical composition of the biomass, reference materials: pure cellulose (microcrystalline powder), hemicellulose (xylan from birch wood), and lignin (hydrolytic) powders were mixed in different proportions with known concentrations. The reference materials were used to generate standard spectra to determine the relationship between the respective quantity of components in the mixture and the FTIR spectra of representative biomass sample. The FTIR wavenumber-dependent instrumental effects were corrected by using carbon black reference spectrum. Multiple linear regression models for cellulose, hemicellulose, and lignin were developed based on the generated regression parameters, with coefficient of determination 0.68, 0.34, and 0.31, respectively, and mean square error of 9.10, 13.86, and 15.86, respectively. This study reflects that pretreatment can also lead to the degradation of the energy potentials. However, non-treated biomass resulted in the least conversion yield during the enzymatic hydrolysis [63]. The FTIR-PAS technique has advantage because it is a quick, easier, and non-destructive method. The structure of the biomass is maintained when spectra are measured directly from the bulk of grind biomass surface. Consequently, this study has led to the following conclusions: i) Lignin matrix structurally disrupted and released during pretreatment process; ii) Pretreatment enhances the accessibility and digestibility of the cellulose and hemicellulose; iii) This increased the conversion rate and assisted in reducing the costs and amount of enzymes required for the next stage of process (enzymatic hydrolysis) by 64% and 33% for RF and SE pretreatment, respectively. This implies that the PA infrared spectra can be used for biofuel feedstock identification and analysis of the chemical composition of biomass before it is processed. This innovative approach could be easily adopted by the biofuel industry and extended to any form of lignocellulosic biomass feedstock.

Author details

Kingsley L. Iroba* and Tabil

*Address all correspondence to: kli931@mail.usask.ca

Department of Chemical and Biological Engineering, University of Saskatchewan, Saskatoon, Canada

References

[1] Pandey KK. A study of chemical structure of soft and hardwood and wood polymers by FTIR spectroscopy. Journal of Applied Polymer Science 1999;71: 1969-1975.

[2] Ramesh C.K, Singh A. Lignocellulose Biotechnology: Current and Future Prospects, Critical Review in Biotechnology 1993;13(2): 151-172.

[3] Iroba KL, Tabil LG. Lignocellulosic Biomass: Feedstock Characteristics, Pretreatment Methods and Pre-Processing for Biofuel and Bioproduct Applications, U.S. and Canadian Perspective. In: Zhang, B., Wang, Y. (Eds), Biomass Processing, Conversion and Biorefinery, New York: Nova Science Publishers, Inc.; 2013a pp. 61-98.

[4] Söderström J, Pilcher L, Galbe M, Zacchi G. Two-step steam pretreatment of softwood by dilute H_2SO_4 impregnation for ethanol production. Biomass and Bioenergy 2003;24(6): 475-486.

[5] Fan LT, Lee YH, Gharpuray MM. The nature of lignocellulosics and their pretreatments for enzymatic hydrolysis. Advances in Biochemical Engineering/Biotechnology 2006;23: 157-187.

[6] Chandra RP, Bura R, Mabee WE, Berlin A, Pan X, Saddler JN. Substrate pretreatment: The key to effective enzymatic hydrolysis of lignocellulosics? Advances Biochemical Engineering/Biotechnology 2007;108: 67-93.

[7] Mohammad JT, Karimi K. Pretreatment of Lignocellulosic Wastes to Improve Ethanol and Biogas Production: A Review in International Journal of Molecular Sciences 2008;9: 1621-1651.

[8] Iroba KL, Tabil LG, Dumonceaux T, Baik OD. Effect of alkaline pretreatment on chemical composition of lignocellulosic biomass using radio frequency heating. Biosystems Engineering 2013b;116: 385-398.

[9] Iroba KL, Tabil LG, Sokhansanj S, Dumonceaux T. Pretreatment and fractionation of barley straw using steam explosion at low severity factor. Biomass and Bioenergy 2014a; 66: 286-300.

[10] Iroba KL, Tabil LG, Sokhansanj S, Meda V. Pelleting characteristics of radio frequency and steam explosion pretreated barley straw. Int. J. Agric. Biol. Eng. 7(3): 68-82 lignocellulosic materials. Critical Reviews in Biotechnology 2014b;3(3): 235-276.

[11] Piyasena P, Dussault C, Koutchma T, Ramaswamy HS, Awuah GB. Radio frequency heating of foods: principles applications and related properties. Critical Reviews in Food Science and Nutrition 2003;43(6): 587-606.

[12] Izadifar M, Baik OD Mittal GS. Radio frequency-assisted extraction of *podophyllotoxin*: Prototyping of packed extraction reactors and experimental observations. Chemical Engineering and Processing 2009;48: 1437-1444.

[13] Balakrishnan PA, Vedaraman N, Sunder VJ, Muralidharan C, Saminathan G. Radio frequency heating-prospective leather drying system for future. Drying Technology 2004;22: 1969-1982.

[14] Allison GG, Thain SC, Morris P, Morris C, Hawkins S, Hauck B, Barraclough T, Yates N, Shield I, Bridgwater AV, Donnison IS. Quantification of hydroxycinnamic acids and lignin in perennial forage and energy grasses by Fourier-transform infrared spectroscopy and partial least squares regression. Bioresource Technology 2009; 100: 1252-1261

[15] Brausi M, Castagnetti G, Gavioli M, Peracchia G, de Luca G, Olmi R. Radio frequency ablation of renal tumours does not produce complete tumour destruction: results of a phase II study. European Urology Supplements 2004;3: 14-17.

[16] Toussaint B, Excoffier G, Vignon MR. Effect of steam explosion treatment on the physico-chemical characteristics and enzymic hydrolysis of poplar cell wall components. Animal Feed Science and Technology 1991;32: 235-242.

[17] Mosier N, Wyman C, Dale B, Elander R, Lee YY, Holtzapple M, Ladisch M. Features of promising technologies for pretreatment of lignocellulosic biomass. Bioresource Technology 2005;96: 673-686.

[18] Lam PS, Sokhansanj S, Bi X, Lim CJ, Melin S. Energy Input and Quality of Pellets Made from Steam-Exploded Douglas Fir (*Pseudotsuga menziesii*). Energy and Fuels 2011;25: 1521-1528.

[19] Excoffier G, Peguy A, Rinaudo M, Vignon MR. Evolution of lignocellulosic components during steam explosion. Potential applications. In Focher B, Marzetti A, and Crescenzi V (Eds), Proceedings of the international workshop on steam explosion

techniques: Oct 20-21, Milan Italy. Philadelphia (PA): Gordon and Breach;1991. Fundamentals and Industrial Application 1988;83-95.

[20] Ferro T, Torres A, Soler PB, Zanzi RD. Biomass Torrefaction. 2nd World Conference on Biomass for Energy, Industry and Climate Protection, 10-14 May 2004. Palazzo dei Congressi, Rome, Italy, Ed. by: Van Swaaij, Fjällström, Helm, Grassi, ISBN 88-89407-04-2, ISBN 3-936338-16-7, ETA-Florence and WIP-Munich, 2004; p 859-862.

[21] Adapa PK, Tabil LG, Schoenau GJ, Canam T, Dumoneaux T. Quantitative Analysis of Lignocellulosic Components of Non-Treated and Steam Exploded Barley, Canola, Oat and Wheat Straw Using Fourier Transform Infrared Spectroscopy. Journal Agricultural Science and Technology 2011;1: 177-188.

[22] Gordon GA. Application of fourier transform mid-infrared spectroscopy (FTIR) for research into biomass feed-stocks. In: Chapter 4, Fourier transforms-new analytical approaches and FTIR strategies. ed. Nikolic, G. ISBN 978-953-307-232-6. Intech 2011; p 71-88. http://www.intechopen.com/books/fourier-transforms-new-analytical-approaches-and-ftir-strategies (accessed 12 March 2014).

[23] Diem M, Romeo M, Matthaus C, Miljkovic M, Miller L, Lasch P. Comparison of Fourier transform infrared (FTIR) spectra of individual cells acquired using synchrotron and conventional sources. Infrared Physics and Technology 2004;45: 331-338.

[24] Huang RY, Miller LM, Carlson CS, Chance MR. In situ chemistry of osteoporosis revealed by synchrotron infrared microspectroscopy. Bone 2003;33: 514-521.

[25] Allison GG. Application of Fourier Transform Mid-Infrared Spectroscopy (FTIR) for Research into Biomass Feed-Stocks. Fourier Transforms-New Analytical Approaches and FTIR Strategies, chapter 4. Goran Nikolic, ISBN 978-953-307-232-6, Intech 2011; 71-88.

[26] Stewart D, Wilson HM, Hendra PJ, Morrison IM. Fourier-transform infrared and raman spectroscopic study of biochemical and chemical treatments of oak wood (*Quercus rubra*) and barley (*Hordeum vulgare*) straw. Journal of Agricultural and Food Chemistry 1995;43: 2219-2225.

[27] Robert p, Marquis M, Barron C, Guillon F, Saulnier L. FT-IR investigation of cell wall polysaccharides from cereal grains, arabinoxylan infrared assignment, Journal of Agricultural Food and Chemistry 2005;53: 7014-7018.

[28] Yu P. Molecular chemistry imaging to reveal structural features of various plant feed tissues. Journal of Structural Biological 2005;150: 81-89.

[29] Sluiter A, Hames B, Ruiz R, Scarlata C, Sluiter J, Templeton D, Crocker D. National Renewable Energy Laboratory, Laboratory Analytical Procedure (LAP). Determination of Structural Carbohydrates and Lignin in Biomass, Technical Report NREL/TP-510-42618 2007.

[30] Michaelian KH. Photoacoustic Infrared Spectroscopy: instrumentation, applications and data analysis. John Wiley & Sons publication, 2003; Vol 159.

[31] Shaw M, Meda V, Tabil LG, Opoku A. Drying and color characteristics of coriander foliage using convective thin layer and microwave drying. Journal of Microwave Power and Electromagnetic Energy 2007;41: 41-65.

[32] Gurnagul N, St-Germain FGT, Gray DGC. PA FTIR measurements on paper. Journal of pulp and Paper Science 1986;12: J156 – J159.

[33] McClelland JF, Jones RW, Bajic SJ. FTIR Photoacoustics spectroscopy in: J.M. Chalmers, P.R. Griffiths (Eds.), Handbook of Vibrational Spectroscopy, John Wiley & Sons, Ltd. 2002.;Vol. 2.

[34] McClelland JF, Jones RW, Luo S, Seaverson LM. A practical Guide to FTIR Photoacoustic Spectroscopy. MTEC Photoacoustics, Inc., CRC Press, Ames, IA, USA. 1992.

[35] Stuart B. Biological applications of Infrared Spectroscopy. John Wiley & Sons, Chichester, 1997;32-33.

[36] St Germain FGT, Gray DG. Photoacoustic Fourier Transform Infrared Spectroscopic Study of Mechanical Pulp Brightening. Journal of Wood Chemistry and Technology 1987;7: 33-50.

[37] Himmelsbach DS, Akin DE. Near-infrared–fourier-transform-raman spectroscopy of flax (*Linum usitatissimum* L.) stems. Journal of Agricultural and Food Chemistry 1998;46: 991-998.

[38] Colthup NB, Daly LH, Wiberley SE. Introduction to Infrared and Raman Spectroscopy, 3rd ed., Boston, Academic Press, MA. 1990.

[39] Yu P, Block H, Niu Z, Doiron K. Rapid characterization of molecular chemistry, nutrient make-up and microlocation of internal seed tissue. Journal of Synchrotron Radiation 2007;14: 382-390.

[40] Revol JF. On the cross-sectional shape of cellulose crystallites in valonia ventricosa. Carbohydrate Polymers 1982;2:123-134.

[41] Sun XF, Xu F, Sun RC, Fowler P, Baird MS. Characteristics of degraded cellulose obtained from steam-exploded wheat straw. Carbohydrate Research 2005;340: 97-106.

[42] Budevska BO. in: J.M. Chalmers, P.R. Griffiths (Eds.), Handbook of Vibrational Spectroscopy, Vol. 5, John Wiley and Sons, New York. 2002; pp. 3720-3732.

[43] Schulz H, Baranska M. Identification and quantification of valuable plant substances by ir and raman spectroscopy. Vibrational Spectroscopy 2007;43: 13-25.

[44] Wetzel DL, Eilert AJ, Pietrzak LN, Miller SS, Sweat JA. Ultraspatially-resolved synchrotron infrared microspectroscopy of plant tissue in situ. Cellular and Molecular Biology 1998;44: 145-167.

[45] Xu F, Sun JX, Geng ZC, Liu CF, Ren JL, Sun RC, Fowler P. Baird MS. Comparative study of water-soluble and alkali-soluble hemicelluloses from perennial ryegrass leaves (*Lolium peree*). Carbohydrate Polymers 2007;67: 56-65.

[46] Nlewem KC, Thrash ME. Jr. Comparison of different pretreatment methods based on residual lignin effect on the enzymatic hydrolysis of switchgrass. Bioresource Technology 2010;101: 5426-5430.

[47] Fan LT, Gharpuray MM, Lee YH. Cellulose Hydrolysis Biotechnology Monographs. Springer, Berlin, 1987; p. 57.

[48] Sun Y, Cheng JJ. Hydrolysis of lignocellulosic materials for ethanol production: A review. Bioresource Technology 2002;83(1): 1-11.

[49] Kumar P, Barrett DM, Delwiche MJ, Stroeve P. Methods for pretreatment of lignocellulosic biomass for efficient hydrolysis and biofuel production. Industrial and Engineering Chemistry Research 2009.48, 3713-3729.

[50] Xu J, Cheng JJ, Sharma-Shivappa RR, Burns JC. Sodium Hydroxide Pretreatment of Switchgrass for Enzymatic Saccharification Improvement. 2009 ASABE Annual International Meeting, Paper Number: 096062, June 21 – June 24, Nevada. 2009.

[51] Silverstein RA, Chen Y, Sharma-Shivappa RR, Boyette MD, Osborne J. A comparison of chemical pretreatment methods for improving saccharification of cotton stalks. Bioresource Technology 2007;98: 3000-3011.

[52] Marsden WL, Gray PP, Mandels M. Enzymatic hydrolysis of cellulose in lignocellulosic materials. CRC Critical Reviews in Biotechnology 1985;3(3): 235-276.

[53] Chen WH, Kuo PC. Torrefaction and co-torrefaction characterization of hemicellulose, cellulose and lignin as well as torrefaction of some basic constituents in biomass. Energy 2011;36: 803-811.

[54] Wang K, Jiang JX, Xu F, Sun RC. Influence of steaming explosion time on the physico-chemical properties of cellulose from *Lespedeza stalks* (*Lespedeza crytobotrya*). Bioresoure Technology 2009;100: 5288-5294.

[55] Yan W, Acharjee TC, Coronella CJ, Va´squez VR. Thermal Pretreatment of Lignocellulosic Biomass. Environmental Progress Sustainability Energy 2009;(28)3: 435-440.

[56] Chen WH, Kuo PC. A study on torrefaction of various biomass materials and its impact on lignocellulosic structure simulated by a thermogravimetry. Energy 2010; 35: 2580-2586.

[57] Heitz M, Capek-Menard E, Koeberle PG, Gagne J, Chornet E, Overend RP, et al. Fractionation of *Populus tremuloides* at the Pilot Plant Scale: Optimization of Steam Pretreatment Conditions using the STAKE II Technology. Bioresource Technology 1991;35: 23-32.

[58] Shaw MD. Feedstock and process variables influencing biomass densification. M.Sc. Unpublished dissertation submitted to Department of Agricultural and Bioresource Engineering Department, University of Saskatchewan, Canada; 2008.

[59] Yang H, Yan R, Chen H, Lee DH, Zheng C. Characteristics of hemicellulose, cellulose and lignin pyrolysis. Fuel 2007;86:1781-8.

[60] Khezami L, Chetouani A, Taouk B, Capart R. Production and characterization of activated carbon from wood components in powder: cellulose, lignin, xylan. Powder Technology 2005;157: 48-56.

[61] Bergman PCA, Boersma AR, Zwart RWH, Kiel JHA. Torrefaction for biomass co-firing in existing coal-fired power stations "biocoal"; Energy Research Centre of the Netherlands; 2005, Jul. 71 p. Report No.: ECN-C-05-013. 2005.

[62] Lipinsky ES, Arcate JR, Reed TB. Enhanced wood fuels via torrefaction. Fuel Chemist Division Preparation 2002;47(1): 408-410.

[63] Iroba KL, Tabil LG, Dumonceaux T, Sokhansanj S. Enzymatic saccharification of radio frequency alkaline pretreated and non-treated lignocellulosic biomass barley straw. Paper No. CSBE13-016. 2013 Conference for Interdisciplinary Engineering and CSBE/SCGAB AGM, University of Saskatchewan, Saskatoon; 2013.

Chapter 16

Biofuels in Aircraft Engines

Anna Maiorova, Aleksandr Vasil'ev and
Oganes Chelebyan

Additional information is available at the end of the chapter

http://dx.doi.org/10.5772/59871

1. Introduction

A large number of studies conducted in Russia and abroad have been devoted to the development of low-emission gas turbine engines for aircraft and power stations (see, e.g., [1]). However, the continual improvement of the environmental requirements of ICAO (International Civil Aviation Organization) forces new research to be carried out.

The use of renewable biofuels obtained from plants or fatty acids is very promising. At present, aviation accounts for about 2% of man-made emissions of CO_2 [2]. When using biofuel, the emission of smoke, solid carbon, carbon monoxide, sulfur and total carbon dioxide is decreased. The most economically feasible is a fuel that can be mixed in any proportion with conventional jet fuel and does not require the creation of an alternative ground fuel-supply infrastructure and ad hoc adjustment of aircraft engines. Thus, the use of bio-kerosene obtained from jatropha, instead of the traditional kerosene in aircraft would reduce "carbon trace" almost by 80%.

Fuel	Density, kg/m^3	Kinematic viscosity $\cdot 10^6$, m^2/s	Surface tension coefficient $\cdot 10^3$, N/m	Calorific value, M j /kg
Ethanol	788	1.550	22.3	27.2
Kerosene TS1	≥780	≥ 1.3	24.3	43.3
Summer diesel	≤ 860	3.0-6.0	28.9	42.5
Winter diesel	≤ 840	1.8-5.0	27.8	42. 5
FAME (biodiesel)	877-879	8.0	31.4	37.5
Vegetable oils	870-960	51-110	24.8 - 34.4	37 - 42.5

Table 1. Physical properties of fuels at 20°C

Foreign companies in recent years (2008-2014) have been intensively studying the possibility of using alternative fuels without the need for modification of aircrafts and engines. The first flight of the airplane on biofuel took place in 2008. The British Airline Virgin Atlantic Airways Ltd is the proprietor of that aircraft. Boeing and its international partners are already working hard to bring biofuels from the testing stage to the manufacturing stage. Boeing 747-8 Freighter and the 787 Dreamliner made the first demonstration of transatlantic and transpacific flights on biofuels in 2011 and 2012 [3]. In May 2014, KLM began weekly flights by an Airbus A330-200 between Queen Beatrix International Airport, in Oranjestad, Aruba, and Amsterdam's Schiphol Airport, Netherlands, using converted cooking oil as aircraft fuel [4]. So far, Russia has not done commercial-scale biofuel production. However, this trend has a great future because of the presence of large sown areas and water surfaces in our country [5]. Within the framework of the International Aviation and Space Salon MAKS-2013 Airbus and Rosteh State Corporation signed a partnership agreement in the field of aviation biofuels in Russia using only renewable resources.

One can find a large number of articles devoted to biofuels in the world literature (e.g. [6], [7]), a number of articles in [8]). An overview of current studies of the structure of such fuels as well as the characteristics of the processes of combustion and pollutant emissions in various types of engines is given in [9]. However, vast majority of the work are carried out in relation to internal-combustion engines or diesel engines. Studies on the atomization and combustion of biofuels compared with petroleum fuels in relation to gas turbine engines, as well as designing multi-fuel combustion chambers in the press are virtually absent. Nevertheless, one can see from Table 1 (the physical properties of various fuels corresponding to the Russian and international standards [10- 13]) that the spread in values of the fuel properties is rather wide, especially for viscosity. The present work is a continuation of researches [14] and [15]. In research [14] the design, manufacture and test of individual injectors and the burner as a whole for low-emission combustion chambers of gas-turbine engine or gas-turbine plant have been executed. Results showed that the designed spray unit can be used for different liquid fuels, both for fossil and for alternative fuels. The present work is devoted to the study of the influence of the physical properties of conventional fuels and biofuels on the characteristics of fuel–air aerosols and the combustion process. Fuel spraying was carried out by means of the developed burner.

2. The peculiarities of atomization of liquid fuels with different physical properties.

Experimental studies of the features of fuel–air sprays were performed at the Central Institute of Aviation Motors using laser diagnostics setup. The description of the test bench is given in reference [15]. The setup is equipped with instruments for laser measurements of the quality of spraying and the rate of droplets by the light scattering. In this work, the physical studies were carried out using the method of Phase-Doppler anemometry (PDPA TSI, United States). Digital photography was carried out using a Canon XL_H1 three-matrix color camera-recorder (Japan). As an object of study, a double-channel fuel burner with combined centrifugal-airblast

http://dx.doi.org/10.5772/59871

design was chosen [16]. The scheme of the spraying device is shown in Fig. 1. The channels of the nozzles are arranged concentrically. A pressure swirl pilot channel with a low rate of flow and cylindrical outlet nozzle is mounted on the burner axis. The main fuel feed channel is airblast with a ring nozzle. It is placed between the two air swirlers for better atomization of the liquid film and for stabilization of the fuel–air spray. The angles of the vane inclination of inner and peripheral swirlers relative to the axis of the device were 60° and 45°. The outer diameter of the fuel nozzle in a pneumatic atomizer is 22 mm and 1.1 mm in a centrifugal atomizer. A detailed description of the burner is given in [14], where it was tested on various petroleum and alternative fuels.

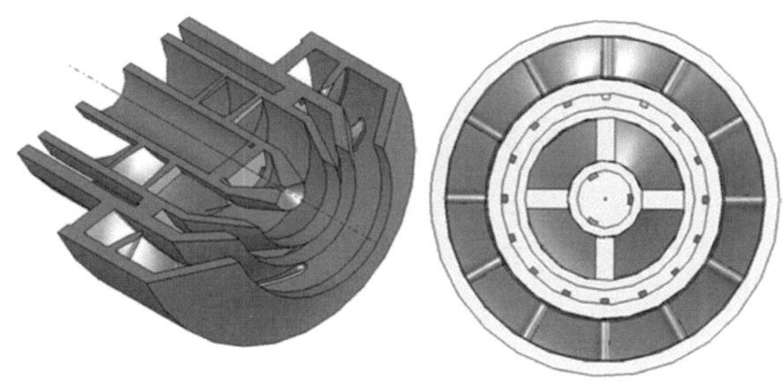

Figure 1. Test object

In this section, we studied the atomization at normal conditions for three types of liquid: (1) water (kinematic viscosity ν_F =1.05 × 10^{-6} m^2/s, surface tension σ_F = 73 × 10^{-3} N/m), (2) kerosene (ν_F = 1.9 × 10^{-6} m^2/s, σ_F = 25 × 10^{-3} N/m), and (3) a mixture of diesel fuel with rapeseed oil in the ratio of 50 : 50 (ν_F = 13.7 × 10^{-6} m2/s, σ_F = 30 × 10^{-3} N/m), which imitated liquid biofuel.

We used three methods of spraying: hydraulic (in which the energy of the liquid is used for spraying), pneumatic (spraying of liquid in the flows of air), and a combined centrifugal-pneumatic process, in which liquid spraying occurs due to the of the liquid state's own energy and the energy of air.

The first stage in the research on the atomization process is the investigation of the decomposition of liquid fuel films due to the loss of their own stability (hydraulic atomization). Figures 2–6 show the results of this investigation.

The results of this series of experiments allow us to make some assumptions about the mechanism of the liquid film decay into droplets. The film of the fuel is formed as a result of the interflow of swirled liquid streams into a single stream along the length of the swirl chamber and the nozzle of the injector. In this case, one can assert that when the fluid moves through the caves of the injector, its outer layer is decelerated due to the friction with the surface and the velocity components diminish in this layer. This gives rise to the shear stresses along the fuel film thickness. We can assume that with an increase in the velocity, depending on the properties of the fluid and geometrical parameters of the atomizer, the shear of the layers becomes so significant that the outer sublayer is swirled in the opposite direction relative to the velocity vector (the scheme is shown in Fig. 2). At the exit of the nozzle, after sudden

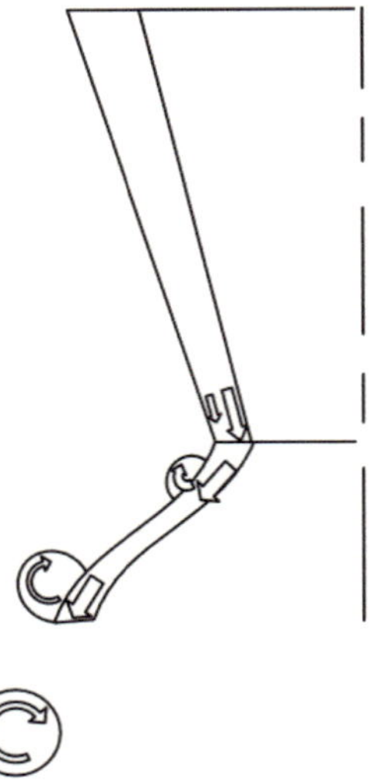

Figure 2. Scheme of the formation of waves–vortexes on the surface of the liquid film

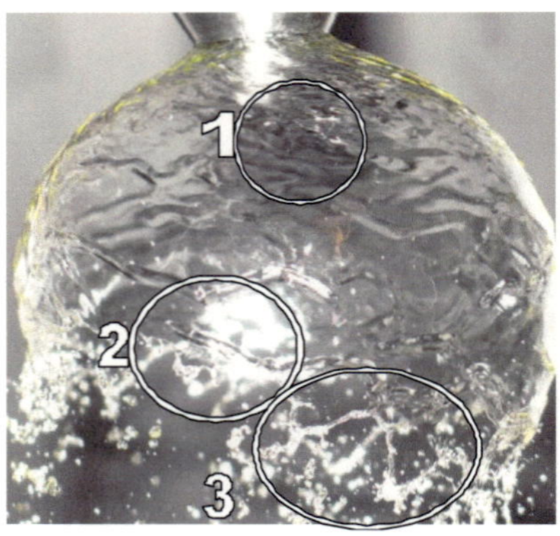

Figure 3. Development and decay of the waves upon hydraulic spraying of kerosene through an annular nozzle (photo) at G_F =17.5 g/s, G_A = 0

expansion and separation from the surface, the intensity of the vortex increases rapidly and waves are formed. When moving downstream, the wave height above the level of the film increases (see markers *1* and *2* in Fig. 3). The growth of these waves is caused by the fact that swirled formations move in the axial direction with lower velocity than the film and then disintegrate into bundles and individual droplets at a certain moment when capturing additional mass (marker 3 in Fig. 3). Such a vortex structure of the waves is confirmed by a series of images shown in Fig. 4, where we can clearly see that a sufficiently large number of droplets deviate from the direction of the stream outwards from the film at various expiration velocities and different nozzle designs. This can be explained by the disintegration of the swirling roller of the fluid as it moves in the direction backward to the main flow, and the roller (or its components) deviates to the periphery of the burner.

http://dx.doi.org/10.5772/59871

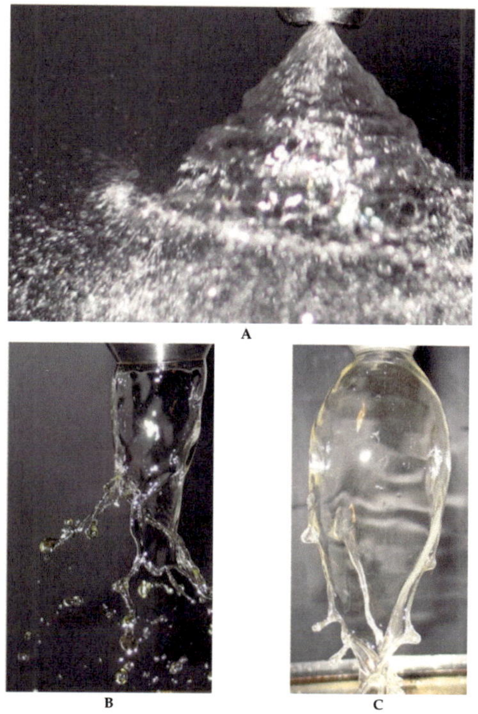

Figure 4. Liquid ejection across an expiration film upon hydraulic atomization of kerosene (a), (b) and biodiesel (c) at $G_A = 0$: (a) cylindrical nozzle, $G_F = 2.7$ g/s; (b), (c) annular nozzle, $G_F = 12.3$ g/s

Now we consider the dependence of the character of the fuel–air spray on the properties of the atomized liquids. Shown in Figs. 5 and 6 are pictures of the expiration of different fluids from cylindrical (Fig. 5) and annular (Fig. 6) nozzles with the same mass flow rate. The wave height above the level of the film depends apparently on the fluid viscosity because the relative shift of layers of the fuel becomes more difficult with increasing viscosity. Therefore, in Fig. 5a (water) one can easily see high wave formations. In Fig. 5b (kerosene) these formations have an appreciably lower height, and in Fig. 5c (a mixture of diesel and rapeseed oil), they are entirely absent.

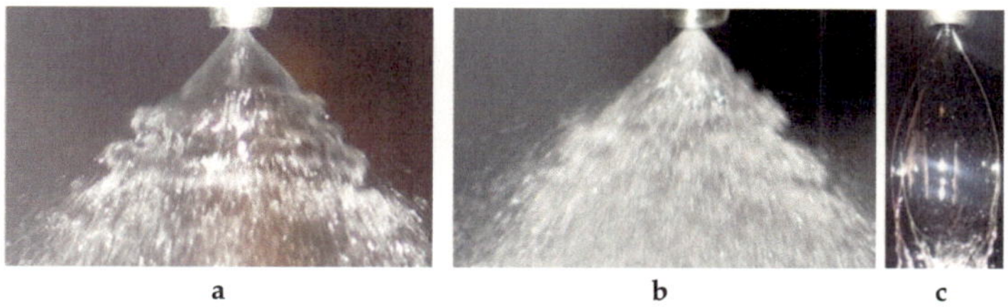

Figure 5. Comparison of the expiration of different liquids at the same mass flow through the cylindrical nozzle without air supply at $G_A = 0$, $G_F = 5$ g/s: (a) water, (b) kerosene, and (c) mixture of diesel with rapeseed oil (50%–50%)

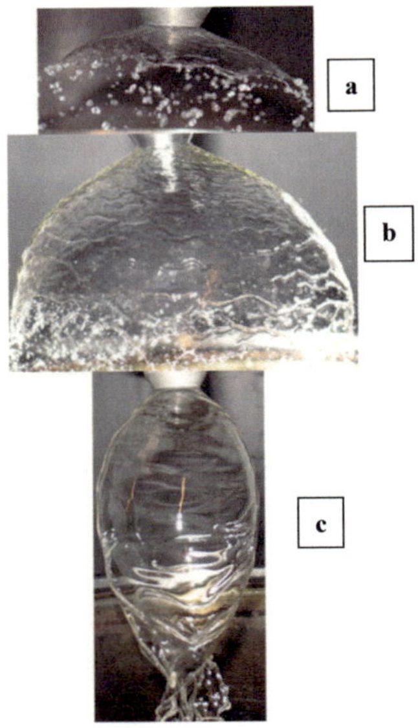

Figure 6. The same as in Fig. 5 in the case of an annular nozzle and $G_A = 0$, $G_F = 20$ g/s.

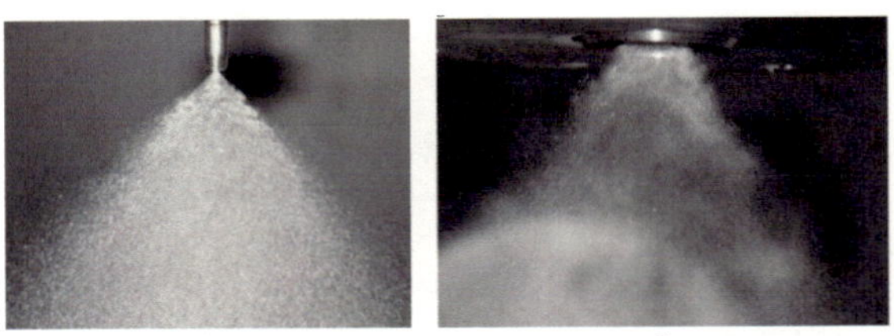

Figure 7. Expiration of kerosene through the centrifugal nozzle at $G_F = 5.5$ g/s: (a) without air supply ($G_A = 0$); (b) $G_A = 40$ g/s

Apart from the expiration velocity from the nozzle, the spray angle mainly depends on the viscosity of the fluid rather than on the surface tension (see Figs. 5, 6). The spray angle decreases with increasing viscosity. An analogous result stems from the experiments [17]. On the other hand, most likely, the remoteness of the point of the film disintegration from the output section of the nozzle depends mainly on the surface tension coefficient. As we can see from Fig. 6, the self-decay of kerosene and diesel-oil mix, with close surface tension coefficients occurs approximately on the same generator length of the film while the self-decay of water (having a considerably higher surface tension), occurs much earlier.

Having studied the waves formed on the surface of liquid films, one can assume the presence of similar effects in the case of mixed fuel–air flow. In reality, oscillations in the fuel concentration are usually observed upon pneumatic spraying. Their magnitude may change depending on the design of the sprayer unit, the injection velocity, and the properties of the ambient medium. In swirled flows in which the regions of the inverse fluid flows are formed, pulsations

http://dx.doi.org/10.5772/59871

are observed on both the outer and inner borders of the burner. The formation of these pulsations is similar to the wave formation on the surface of the film (Fig. 2), but at the same time these effects are not related physically. Apparently, these formations occur behind the exit of the air nozzle due to sudden expansion and braking of the layer on the boundary with the external medium. Figure 7 demonstrates the independence of these vortexes on the waves formed on the surface of the film. Figure 7a shows the fuel spraying using a pressure swirl nozzle without air supply. The vortexes caused by the film of the fuel are visible only near the nozzle, but we do not see any pulsations of concentration elsewhere. Figure 7b shows the same injector with an external air swirler. We can easily see large wave formations propagating down through the flow and weakly correlating with the waves in the film.

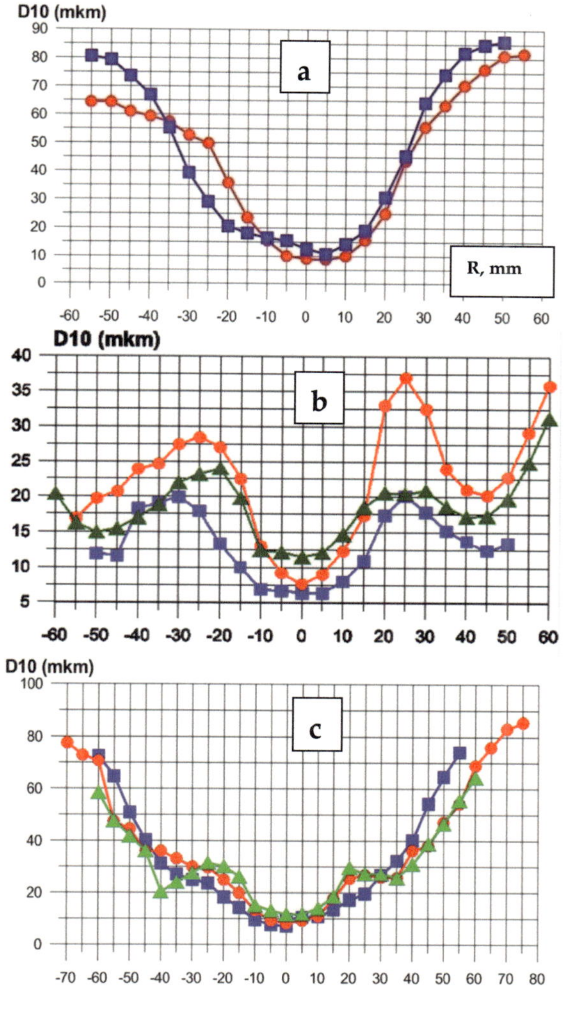

Figure 8. Linear distribution of the average droplet diameter over the burner diameter upon liquid atomization by three different methods: (a) centrifugal method, $G_F = 5$ g/s, $G_A = 0$; (b) centrifugal–pneumatic method, $G_F = 5$ g/s, $G_A = 40$ g/s; (c) pneumatic method, $G_F = 20$ g/s, $G_A = 40$ g/s; –● water, –■ kerosene, –▲ mixture of diesel with rapeseed oil (50%–50%)

Now we consider the impact of fluid properties and the related aforementioned phenomena on the dispersity of aerosol when using different methods of atomization (Figs. 8, 9). The first and most studied method of droplet atomization is that of centrifugation (hydraulic atomization). The liquid is fed through a near-axis pressure swirl nozzle without the external air flow (Figs. 8a, 9a). The average velocity of the fuel nozzle outlet of is 19–26 m/s depending on the type of liquid. The centrifugal–pneumatic method of atomization is shown in Figs. 8b and 9b. In this case, the average velocities of the fuel and air have the same order of magnitude. To implement centrifugal–pneumatic atomization, the fluid is fed through a pressure swirl nozzle with the same mass flow rate (5 g/s) as in the first case. Additionally, the air is fed through external swirlers with the total mass flow rate of 40 g/s. The average fuel velocity at the burner outlet is the same as in the first method (19–26 m/s) and on the order of 25 m/s for the velocity of air. In the third (pneumatic) atomization method (Figs. 8c, 9c), the injection velocity of the fluid is smaller than the velocity of air.

When implementing this method, a small part of liquid (3 g/s) is fed through the near-axis nozzle, while the main part (17 g/s) is fed through the annular airblast injector, and air velocity in this case is the same as in the second method. In this case, the average velocity is 3–4 m/s. The radial distribution of the diameter D_{10} of droplets is shown in Fig. 8, where D_{10} is the arithmetic mean of the size in an ensemble. This parameter determines the most probable size of droplets in the given region, and it can be used in predicting the engine wake-up mode: the greater the number of small droplets that enters the spark discharge zone, the simpler their evaporation and ignition.

Figure 9 shows the distribution of Sauter's mean diameter D_{32} of droplets - the ratio of the volume of a droplet to its surface area being averaged over the ensemble at a given point of space. This parameter is important when predicting the efficiency and homogeneity of the fuel burning.

As we can see from Figs. 8a and 9a, the viscosity of the liquid has a significant impact on the dispersity of droplets obtained by the centrifugal method of spraying. Thus, in the case of water having a surface tension three times higher, one can obtain droplets that are even smaller when compared with kerosene, which has a viscosity 1.9 times greater. In this case, a highly viscous mixture of diesel with rapeseed oil does not form waves on the surface of the film and is not sprayed at all (Fig. 5c). In the case of centrifugal–pneumatic spraying, as we can see from Fig. 8b, the most important role is played by the surface tension coefficient. The curves in Fig. 8b are arranged according to a consecutive increase in the surface tension and correspond to kerosene (bottom line), a mixture of diesel and rapeseed oil (middle line), and water (upper line). In Fig. 9b, we can note an insufficient secondary atomization of certain droplets with a large size in the flow of liquid with high viscosity. However, upon closer inspection, we can see that, at the locations of the maxima of concentration, the aforementioned dependence also takes place and it is violated only on the axis of the device near the zone of reverse flows due to insufficient intensity of the air flow. Further fragmentation of the drops of the viscous fluid near the separation zone can be done by aerating the root region of the fuel–air spray.

http://dx.doi.org/10.5772/59871

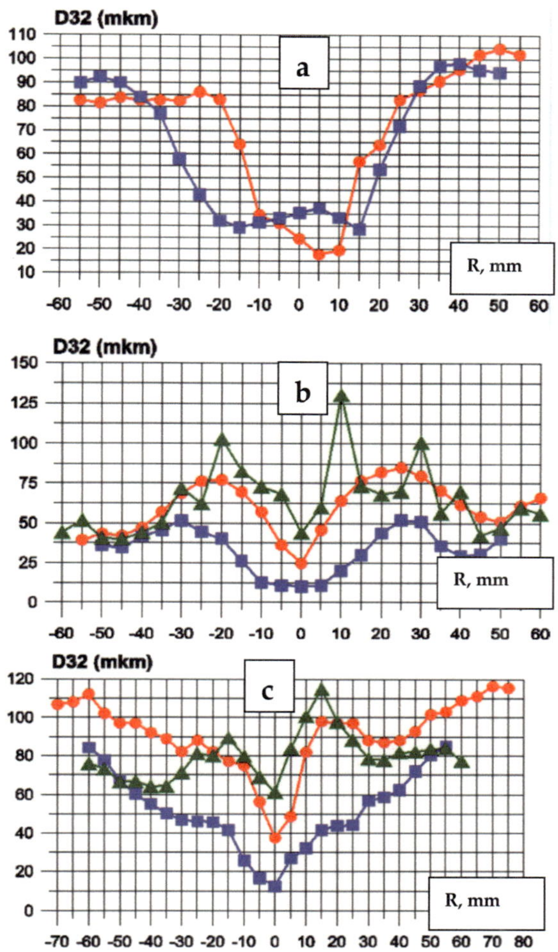

Figure 9. Distribution of the Sauter's drop diameters over the burner diameter upon liquid atomization by three different methods (designations are the same as in Fig. 8).

In the pneumatic method of spraying, as we can see from Fig. 8c, the influence of the properties of liquids on the linear size of droplets is nearly absent and all the curves merge into one. This parameter of aerosol is mainly determined by the air flow. Sauter's mean diameter of droplets (Fig. 9c) depends also on the surface tension coefficient. The influence of individual large drops is somewhat smoothened compared with the centrifugal–pneumatic method of spraying.

In the centrifugal–pneumatic method of spraying, the dispersity of droplets averaged over the whole section of the flambeau is the best. However, note that the ratio of the mass consumption of air to the mass flow rate of fuel (AAFR), which, in particular, determines the quality of spraying, equals 8 for this method and 2 in the case of the pneumatic method. Thus, in the case of low gas-turbine engine operating modes and high viscosity fuels like biodiesel or bio-kerosene, the centrifugal-pneumatic atomization method is optimal, while the pneumatic method is optimal for high operating modes.

3. The selection of mixed liquid fuel

For conducting the hot tests, ethanol and mixed biofuel on the basis of aviation kerosene (as most close relating to a turbine engine) have been chosen as alternative fuels. As one can see from Table 1, the combustion value of biofuels (especially ethanol) is significantly lower than that of fossil fuels. Furthermore, the viscosity of vegetable oils is ten times greater than the viscosity of the organic fuel. Therefore, for aircraft engines a blend of biofuels with conventional aviation fuels is more preferable then pure biofuels. At present, the use of industrially processed aviation biofuels in the Russian territory is not possible. Various versions of a percentage ratio of components of combustible mixtures on the basis of plant oil and ethanol (Table 2 and Fig. 10) have been investigated. Plant oil is necessary as surfactant for ethanol dissolution in the fuel. Aviation kerosene TS 1 or gasoline have been chosen as the main component of the mixture. The optimum ratio of components has been selected.

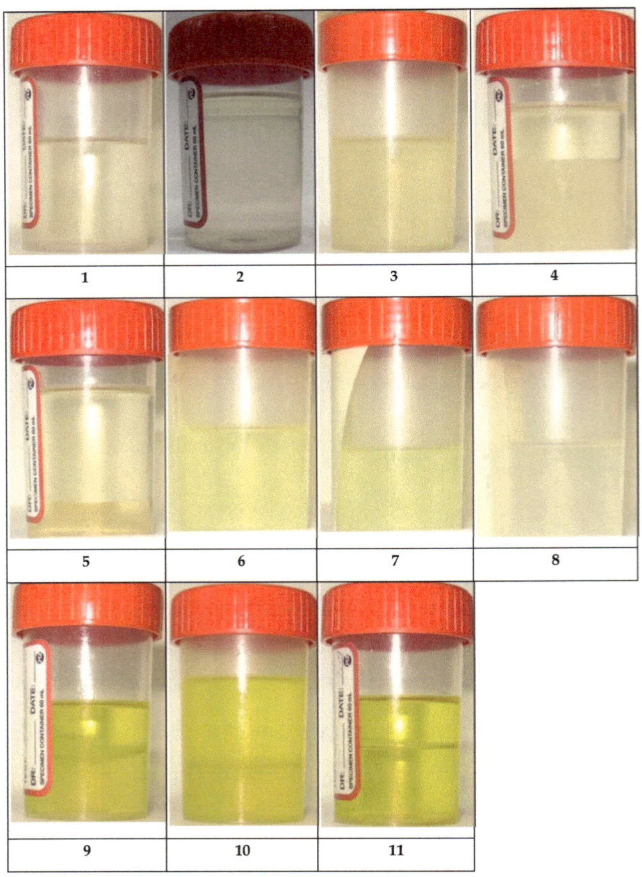

Figure 10. Photos of mixed fuels; signatures correspond to embodiments of the Table 2

http://dx.doi.org/10.5772/59871

	1	2	3	4	5
Kerosene TS-1	10%	30%	40%	50%	80%
Castor oil	10%	10%	20%	10%	10%
Ethanol	80%	60%	40%	40%	10%
	6	7	8		
Gasoline 95	80%	85%	10%		
Castor oil	10%				
Ethanol	10%	15%	90%		
	9	10	11		
Kerosene TS-1	40%	30%	50%		
Camelina sativa oil	20%	10%	10%		
Ethanol	40%	60%	40%		

Table 2. The embodiments of mixed liquid fuels

Analysis of the samples revealed that the use of different ratios of starting fuel, ethanol, and vegetable oil show results strikingly different from each other. It was possible to obtain well-blended homogeneous mixture only at certain narrow ranges of percentages of components. The embodiment 3 was chosen for further testing, as it showed the optimal ratio of components without settling on the bottom and without stratification. Variant 1, variants 6 and 7, and variant 8, which have also shown good mixing level, are notable for big maintenance of ethanol, kerosene or gasoline (up to 90 %). It is not beneficial in terms of the economic feasibility of introducing a new type of fuel.

For hot tests in the aviation combustion chamber, the mix in a ratio of 40% of kerosene TS-1, 20% of castor oil, and 40% of ethanol has been chosen as the most homogeneous and well mixed without any precipitations and stratifications. Its physical features are: $\varrho_F = 850$ kg/m^3, $\nu_F = 4.7 \times 10^{-6}$ m^2/s, and $\sigma_F = 27 \times 10^{-3}$ N/m.

4. Hot tests in aviation combustor.

Hot tests were performed at the Central Institute of Aviation Motors using a combusting chamber test rig. Fire tests of a burner with the low-emission aviation combustion chamber compartment have been conducted. Combustion chamber starting was conducted only on one pilot channel of a burner. Fuel mass flow rate ranged from 1 to 5.7 g/s. The part of the air arriving in the flame tube front passed through air swirlers of the burner. Thus, the centrifugal–pneumatic spraying was carried out, and as shown in the previous section, provided the best droplet dispersity. The kerosene TS-1, ethanol, and kerosene-ethanol-castor oil mixture were used as fuel. The operation mode corresponded to the altitude of an order of 2 km. The fixation of flame starting and blowout was carried out with the help of digital camera through a window at the liner outlet.

Test results are given in Figs. 11-20. The epures of the combustor's blowout characteristics at different excess air coefficients α_C and total air volume flow rates Q_C were obtained. Also the temperature fields behind an exit from the combustor in a pipe with a diameter of 110 mm have been taken out under various α_C.

Here α_C - the general excess air coefficient in the combustion chamber - the relation of total air mass flow rate passing through the chamber to the air flow rate was required theoretically for complete combustion of the fuel arriving at the same time in this chamber. Thus, $\alpha_C < 1$ means rich fuel-air mixture and $\alpha_C > 1$ means lean mixture.

The received blowout boundary line shows, that for conventional fuel (Figs. 11, 12), the combustor steadily works (the area within the curve) in the coefficient of air excess α_C range from 1 to 10 and till Q_C = 0,4 m³/c. The area boundary reaches satisfactory values on α_C, and comprehensible values on Q_C. The ignition domain (within the curve in Fig. 11) is sufficient on the square for assured firing of the combustion chamber.

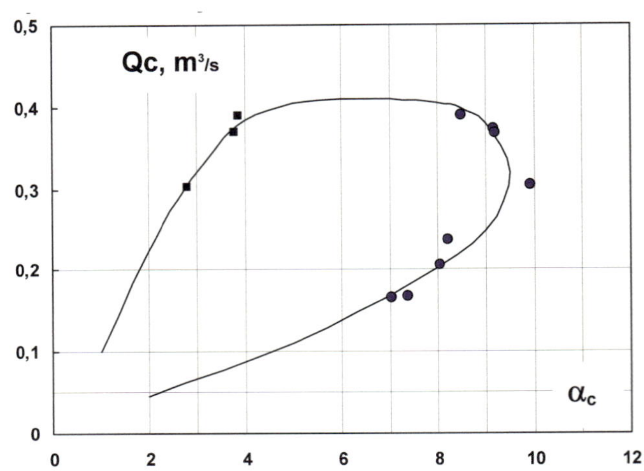

Figure 11.

Boundary lines of ignition and blowout in the combustion chamber compartment; fuel - kerosene TS-1; ● – lean blowout; ■ –rich blowout

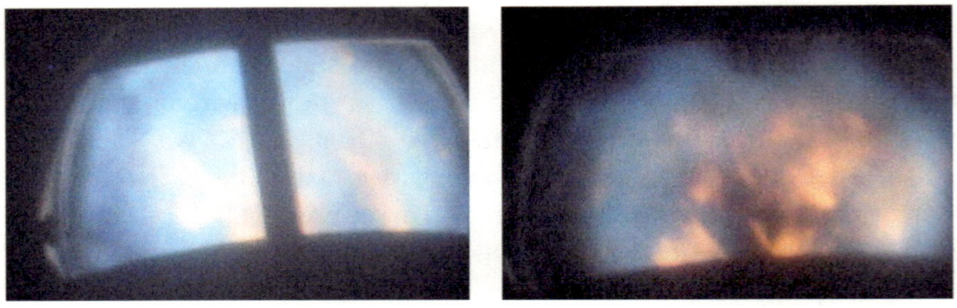

Figure 12. Flame photos at various α_C from wake-up to lean blowout (kerosene)

The radial temperature distribution at the combustor exit is shown in Fig. 13. The temperature field received has a symmetric appearance and a small non-uniformity on the value of the temperature - the minimum value differs from maximum on 50°C.

When using the ethanol (Figs. 14 - 16), lean blowout limit falls to $\alpha_C = 3$, and the combustor demonstrates stable operation only at major fuel flow rate (approximately $\alpha_C = 1.8$). This is due to the fact that alcohol is more volatile than the other liquid fuel, and thus, it will only burn before it can spread to a larger volume of flame front. The temperature reaches its maximum value with 300°C at $\alpha_C = 2.1$.

In view of the foregoing, the use of pure ethanol as an alternative type of aviation fuel is not possible, as a minimum, without the use of special fuel additives.

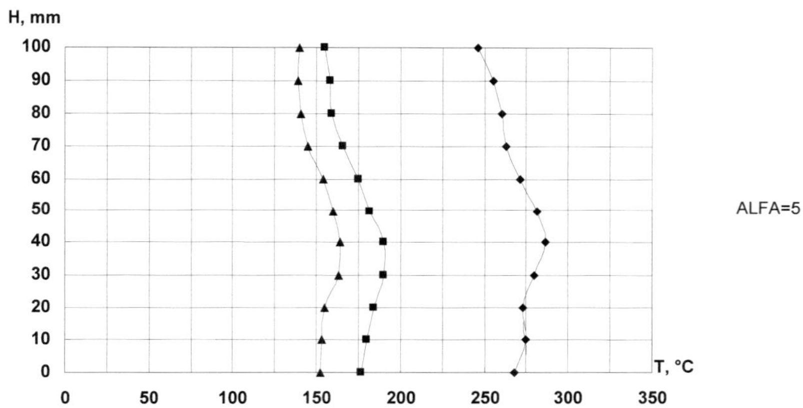

Figure 13. Temperature distributions along the height of the liner; ◆ - $\alpha_C = 3.5$, ■ - $\alpha_C = 5$, ▲ - $\alpha_C = 6.3$; fuel - kerosene TS-1; $Q_C = 0.3$ m³/s

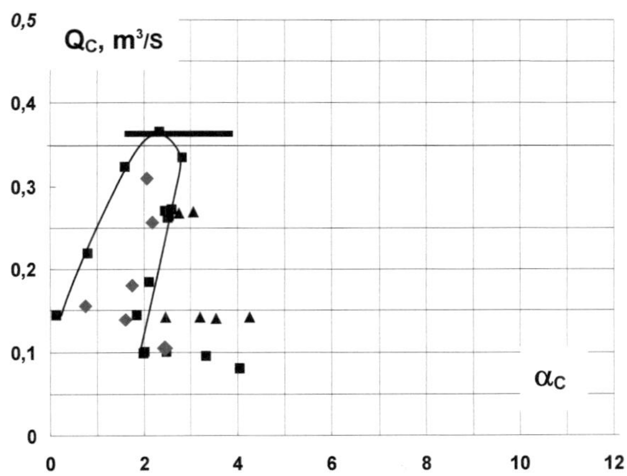

Figure 14. Boundary lines of ignition and blowout in the combustion chamber compartment; fuel - ethanol; ■ – blow-out, ◆ - combustor works, ▲ - combustor does not work

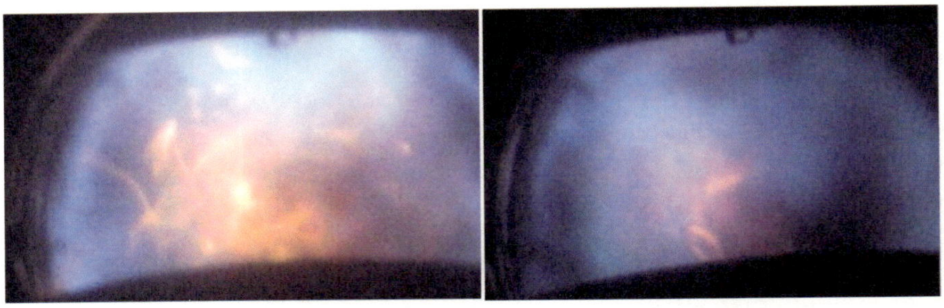

Figure 15. Flame photos at various α_C from wake-up to lean blowout (ethanol)

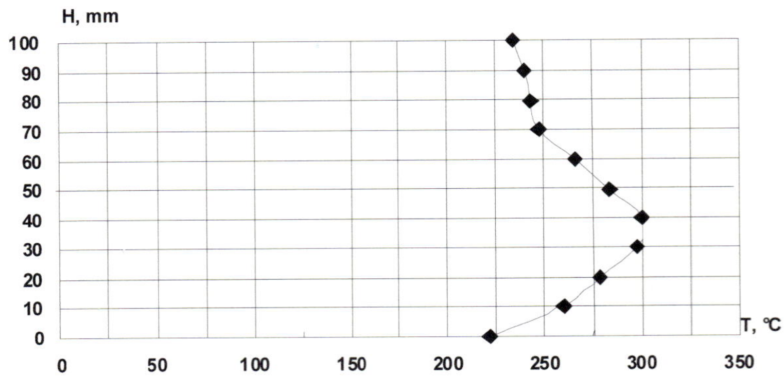

Figure 16. Temperature distribution along the height of the liner; fuel - ethanol; $\alpha_C = 2.1$; $Q_C = 0.21\ m^3/s$

When using blended fuel (kerosene-ethanol-castor oil mixture), the combustor works better than at pure ethanol. Nevertheless, the lean blowout boundary is reduced from 10 to 6.5 at maximum volume flow rate conservation in comparison with kerosene (Figs. 11 and 17). The flame color (Fig. 18) changes while maintaining its overall structure due to the reduction of combusting efficiency and flame temperature and the increasing soot production. One can see from the comparison of Figs. 13 and 19 to 20 that when using blended fuel, it is possible to reach a maximum outlet flame temperature of 290°C only by increasing the mass of the injected fuel (α_C changes from 3.5 to 2.6).

Thus, biofuel application results in poor combustion stability characteristics for aircraft engines when compared with kerosene. For biofuel use, it is necessary to provide a number of actions for the modernization of conventional aviation combustion chambers. Main activities include the incorporation of artificial flame stabilizers into the design to preserve the stability limits of the combustor and the optimization of fuel injection system for the purpose of reducing fuel-air aerosol dispersity to maintain combustion efficiency.

http://dx.doi.org/10.5772/59871

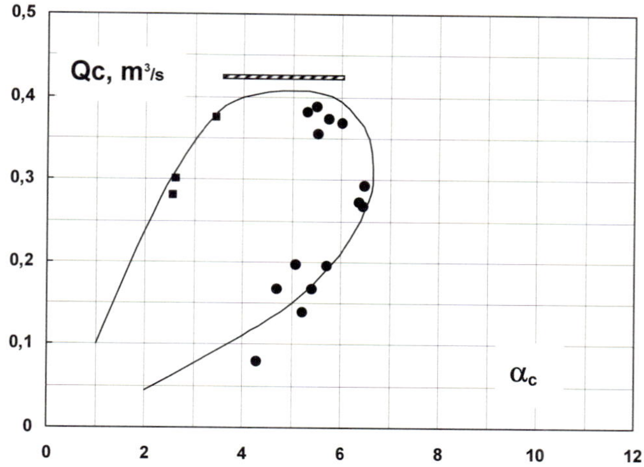

Figure 17. Boundary lines of ignition and blowout in the combustion chamber compartment; mixed fuel (the embodiment 3 from Table 2); ● – lean blowout; ■ –rich blowout

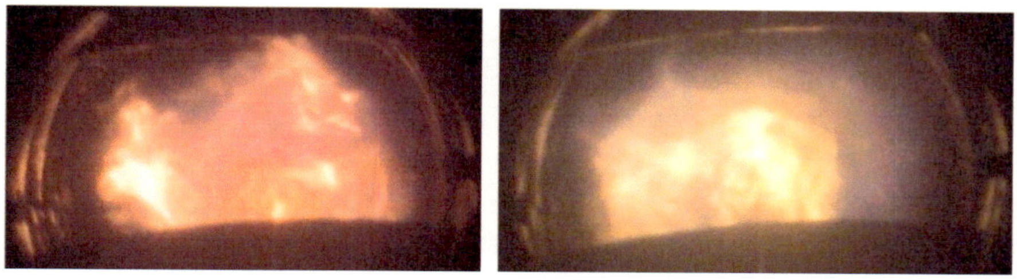

Figure 18. Flame photos at various α_C mixed fuel from wake-up to lean blowout; mixed fuel (the embodiment 3 from Table 2)

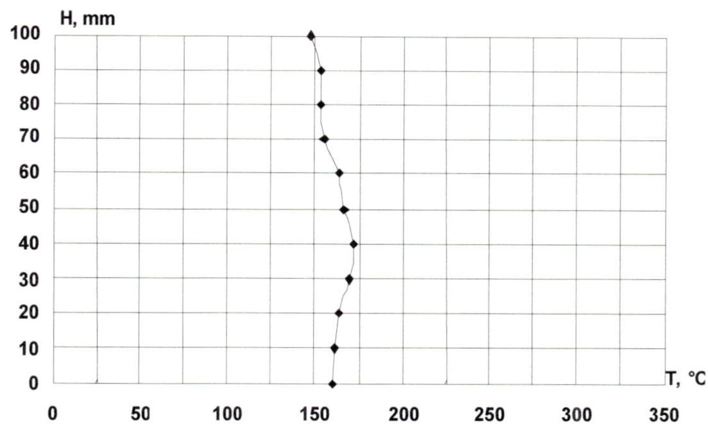

Figure 19. Temperature distribution along the height of the liner; α_C = 3.6; mixed fuel (the embodiment 3 from Table 2); Q_C = 0.29 m^3/s

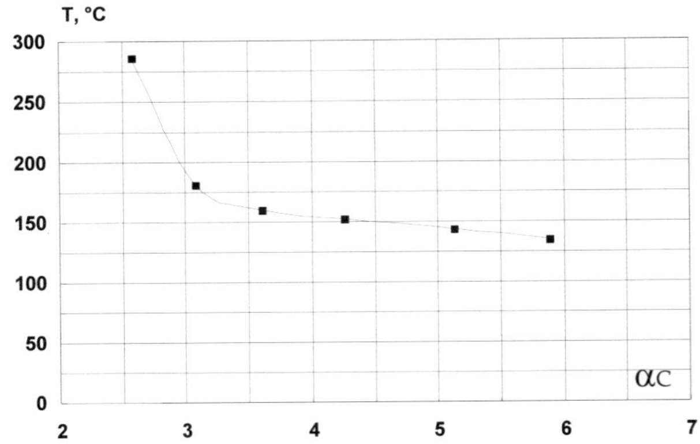

Figure 20. The dependence of axis temperature behind the combustor on excess air coefficient; mixed fuel (the embodiment 3 from Table 2)

5. Summary

An experimental study of the peculiarities of atomization of liquid fuels with different physical properties has been carried out. It has been shown that the spray angle upon hydraulic spraying is mainly determined by fluid viscosity, while remoteness of the point of the film decay from the exit section of the nozzle is determined by the surface tension coefficient. The effect of the properties of the liquid on the aerosol dispersity depends on the method of fluid crushing into droplets. In the case of the hydraulic atomization method without air supply, viscosity exerts the greatest impact on the dispersity of droplets. In the case of the centrifugal-pneumatic method (with the same order of magnitude of the velocity of liquid and air), the greatest impact is from the surface tension. In the pneumatic method of spraying, when the injection velocity of the fluid is lower than the velocity of air, the linear size of droplets is mainly determined by the air flow irrespective of the properties of the liquid, whereas Sauter's mean diameter depends also on the surface tension coefficient.

In the case of low gas-turbine engine operating modes and high viscosity fuels like biodiesel or biokerosene, the centrifugal-pneumatic atomization method is optimal, while the pneumatic method is optimal for high operating modes.

For conducting of hot tests in aviation combustor, 11 embodiments of mixed liquid fuels were proved. The mixture in a proportion of 40% of aviation kerosene, 20% of castor oil, 40% of ethanol had been chosen for the tests as the most uniform and well mixed, without deposition and stratification.

Fire tests of the compartment of aviation combustion chamber with fossil fuel (kerosene TS-1) have shown comprehensible characteristics. In particular, wide side-altars of the stable combustion, assured firing of the combustion chamber, with uniform enough field of gas temperature on exit.

http://dx.doi.org/10.5772/59871

The application of blended fuel (kerosene-ethanol-castor oil mixture) results in worse combustion stability characteristics for aircraft engines when compared with kerosene. For biofuel use, it is necessary to provide a number of actions for the modernization of conventional aviation combustion chambers.

Nomenclature

ν - kinematic viscosity, m^2/s

ϱ - density, kg/m^3

σ - surface tension coefficient, N/m

α- excess air coefficient

T – temperature, K

Q - air volume flow rate, m^3/s

G - mass flow rate, kg/s

D_{10} - droplet mean diameter, m

D_{32} - droplet mean Sauter diameter, m

Subscripts

A – air

C – combustion chamber

F - liquid fuel

Acknowledgements

This work was supported by the Russian Foundation for Basic Research, project No. 15-08-06293.

Author details

Anna Maiorova*, Aleksandr Vasil'ev and Oganes Chelebyan

*Address all correspondence to: majorova@ciam.ru

Central Institute of Aviation Motors named after P.I. Baranov, Russia

References

[1] Bernini E. Editor. Progress in Gas Turbine Performance. Rijeka: Intech. 2013. http://www.intechopen.com/books/progress- in-gas-turbine- performance (accessed 19 June 2013).

[2] Beginner's Guide to Aviation Biofuels, Red. 2. 2011.

[3] Boeing. 2013_environment report. http://www.boeing.com/aboutus/environment/environment_report_13

[4] https://www.klmtakescare.com/en/content/aruba-and-bonaire-on-biofuel

[5] http://www.biogas-rcb.ru/files/helpful/Biofuels-Market-Development-in-Russia-and-Worldwide.pdf.

[6] Marco Aurelio dos Santos Bernardes, Editor. Economic Effects of Biofuel Production. Rijeka: Intech. 2011.

[7] Marco Aurelio dos Santos Bernardes, Editor. Biofuel's Engineering Process technology. Rijeka: Intech. 2011.

[8] Manolis Gavaises (ed.). ILASS 2013. Proceedings of 25th European Conference on Liquid Atomization and Spray Systems, Chania, Greece, 1-4 September 2013.

[9] Charles K. Westbrook. Biofuels Combustion. Annual Review of Physical Chemistry. 2013. Vol. 64. pp 201-219.

[10] Russian State Standard 10227-86.

[11] Russian State Standard 305-82.

[12] European Standard EN 14214.

[13] J. Blin, C. Brunschwiga, A. Chapuisa, O. Changotadea, S. Sidibea, E. Noumia and P. Girard. Characteristics of vegetable oils for use as fuel in stationary diesel engines - towards specifications for a standard in West Africa. Renewable and Sustainable Energy Reviews. 2013; Vol. 22. pp. 580-597.

[14] Anna Maiorova, Aleksandr Sviridenkov, Valentin Tretyakov, Aleksandr Vasil'ev and Victor Yagodkin. THE DEVELOPMENT OF THE MULTI-FUEL BURNER In: Marco Aurelio dos Santos Bernardes (ed.). Economic Effects of Biofuel Production. Rijeka: Intech. 2011. pp. 281 - 298.

[15] A. Yu. Vasil'ev and A. I. Mayorova. Physical features of liquid atomization when using different methods of spraying. High Temperature, 2014, Vol. 52, No. 2, pp. 252–261. ISSN 0018_151X.

[16] A. Yu. Vasil'ev, A. I. Maiorova, A. A. Sviridenkov and V. I. Yagodkin. 2009. Patent of Russian Federation No. 86279.

[17] Soumik Mahapatra, Souvick Chatterjee, Swagata Shannigrahi, Achintya Mukhopadhyay and Swarnendu Sen. Experimental Investigation and Spray Characterization of Liquid Jet Atomization of Conventional Fuels and Liquid Bio-Fuels. ICLASS 2012, 12th Triennial International Conference on Liquid Atomization and Spray Systems, Heidelberg, Germany, September 2-6, 2012, proceedings. Contribution 1223b. ISBN 978-88-903712-1-9.

Section 5

Impact on the Environment and Policy

Chapter 17

Changes in Greenhouse Gas Emissions from Displacing Cattle for Biodiesel Feedstock

J.A. Dyer, X.P.C. Vergé, R.L. Desjardins and D.E. Worth

Additional information is available at the end of the chapter

http://dx.doi.org/10.5772/60047

1. Introduction

The adoption of biodiesel as a fossil fuel reduction strategy in Europe has created a demand for vegetable oil as a biodiesel feedstock and a new market for the canola oil grown in the Canadian Prairies. Canola is a variety of rapeseed developed in Canada which contains reduced levels of erucic acid, making its oil palatable for human consumption, and reduced levels of a toxic glucosin, which makes its meal a potential livestock feed. Because of the extracted canola oil being suitable for use as cooking oil, canola is considered to be a new and distinct crop from rape seed in Canada.

The market acceptability of canola oil as a biodiesel feedstock in Europe depends on the Carbon Footprint (CF) of its production being significantly lower than that of conventional diesel fuel [1]. In October 2012, The European Commission amended its directive relating to the quality of petrol and diesel fuel and the promotion of the use of energy from renewable sources. These amendments set mandatory margins for the amount of fossil CO_2 required to be offset by biofuels. For biofuel installations in operation on or before 1 July 2014, biofuels must achieve a greenhouse gas emission savings of at least 35% until 31 December 2017 and at least 50% thereafter. If the physical production of biofuels starts after 1 July 2014, the greenhouse gas emission saving from the use of biofuels must be at least 60% of the fossil CO_2 emissions of the equivalent petro-fuel energy.

These Directives also proposed a provision to address changes in the indirect land use given that current biofuels are mainly produced from crops grown on existing agricultural land. Therefore, if the demand for canola biodiesel continues to increase, questions arise about where the additional feedstock would be grown. The first goal of this chapter was to assess the impact of expanding canola production on beef production in the Prairie Provinces of Canada. The scope of this assessment excludes any substitution of beef with non-ruminant livestock in response to canola expansion. The second goal was to determine the feedback effect from this impact on the Greenhouse Gas (GHG) emission cost of canola biodiesel.

2. Background

The primary impact of canola expansion was expected to be the displacement of other grains and oilseeds from the land most capable of growing annual crops. The assessment in this chapter did not consider the small areas of canola that might be seeded into lower quality land because the expectation is that most of that land would not support the cultivation of canola. Since the expansion of canola between 1986 and 2006 happened concurrently with, and possibly as a consequence of, the shrinkage of summerfallow [2], there is little indication that canola expansion to date has caused the direct conversion of land under permanent year-round cover into annual crops.

Without this conversion of perennial forage areas to canola, a potential decrease in soil carbon could be ignored. However, the converse cannot be ruled out if beef production is forced to shift to a more forage (roughage) based diet. Because ruminants have the option of feeding on roughages, the land base that supports these livestock is likely to shift to more permanent cover if their feed grain supply is displaced by canola. Although beef cattle are the dominant ruminants in Canada, some consideration has been given to the potential expansion of sheep production [3]. The impacts of canola expansion on ruminant livestock production can be treated as secondary effects. An environmental effect is considered secondary when one environmental component is affected by another environmental component when the second component has been affected by a human activity [4, 5]. The activity being assessed in this chapter is the continued expansion of canola in Western Canada at the expense of livestock feed grains.

The Western Canadian beef industry is an intensive system that relies on finishing animals destined for slaughter in feedlots with a diet that is high in feed grains [6, 7]. Canadian lamb production is similarly intensive in this regard [3]. The conversion of these systems to extensive systems that are mainly based on grazing and hay consumption could be one of the indirect effects of canola expansion [8]. The main impact on the CF of beef production will be greater enteric methane emissions due to a higher share of roughage in the diet [7]. While Dyer et al. [8] qualitatively assessed the impacts on biodiversity from this potential land use change, a quantitative assessment of the GHG emissions from beef cattle displaced from a highly grain based diet into improved pasture or rangeland, and greater dependence on hay has not yet been carried out.

The GHG emission budgets of biofuel feedstock and livestock production have already been shown to strongly interact [9]. Instead of converting beef production from intensive to extensive production (as proposed in this chapter), Dyer et al. [9] replaced part of the beef population with hogs which, being non-ruminants, reduced enteric methane emissions. That beef-pork displacement scenario was based on the assumption that total protein supply must be maintained, and that hog and beef populations can be equated on the basis of their contribution to a constant supply of edible protein [10]. Unlike the assessment in this chapter, no change in the area for growing grain was allowed for in the replacement process in that analysis [9].

3. Methodology

While the expansion of canola can also displace baking quality grains or other food crops for humans, this assessment will only deal with the canola that displaces livestock feed grains, which would lead to changes in the livestock diet. The first set of impacts from canola expansion into ruminant livestock production will be on land use. The expected output variables from the analysis of land use changes included the weight and area of feed grain that will be displaced by the expanded canola crop, the areas of roughage crops, including, hay (for winter feed), pasture (improved) and rangeland (unimproved pasture), and the number of displaced grazing animals computed from the roughage crop yields and stocking rates [11, 7]. This chapter also considers the net changes in the GHG emissions budget for canola and the implications for protein supply.

3.1. Rangeland forage availability

If part of the increased forage in the ruminant diets is to come from more grazing, then one of the pools of available land would likely be rangelands. Therefore, the first land use change question addressed in this chapter will be how much rangeland could be allocated to grazing the livestock that are taken off feed grain due to the canola expansion. In addition to the impact on biodiversity [8], overgrazing would make the forage digestibility on rangeland less than the forage digestibility on tame pasture [12], which effectively lowers forage yields. Therefore, it is essential to set stocking rakes at a population density that is sustainable.

The Ecological Sustainable Stocking Rate (ESSR) was an essential indicator in quantifying the rangeland grazing resources. ESSR values have been quantified for rangelands in most of the agro-ecological sub-regions of the Prairie Provinces [11, 13-15]. The fraction of each agro-ecological sub-region in each of the three Prairie Provinces was extrapolated by Dyer et al. [16]. Integrating the ESSR fractions for these regions in each province gave the approximate provincial ESSR values shown in Table 1.

Each ESSR represents the sustainably grazed forage by one Animal Unit Month (AUM) from a given area of rangeland. One Animal Unit (AU) was defined as being equal to one 454 kg cow with calf, or five breeding sheep (ewes and their lambs), based on equivalent forage consumption [11, 17, 18]. One AUM is, therefore, a measure of forage production. Provincial ESSR and rangeland areas were combined to approximate the rangeland forage yields and the Total Sustainable Animal Units (TSAU) in each province shown in Table 2. To determine yields, an AUM must be converted to the required quantity of feed for each AU. The ESSR from Table 1 were converted to the required areas of rangeland per AU over six months (Column 3). The

ESSR (Column 1) were used to derive forage yield estimates for the three Prairie Provinces (Column 4). The hay needed to over-winter one AU (one breeding cow and her calf or five ewes and their lambs) must approximately equal the forage that a typical cow would have grazed from the rangeland during the six-month summer period [19].

Ecoregions		**DMG**	**MG**		**FF**		**PNF**	
			ESSR (AUM/ha)					
		0.37	**0.84**		**1.68**		**0.88**	**ESSR**
Province		**Share of province in each Ecoregion**						**AUM[1]/ha**
Manitoba							100%	0.88
Saskatchewan		30%	35%				35%	0.71
Alberta		20%	30%		25%		25%	0.97
		Canadian Prairie Ecoregions:						
DMG	=	Dry Mixed Grass		FF	=	Foothills Fescue		
MG	=	Mixed Grass		PNF	=	Parkland-Northern Fescue		

[1], AUM = Animal Unit Months

Table 1. Provincial Ecological Sustainable Stocking Rate (ESSR) factors for Canadian Prairie rangeland interpolated from ecoregion ESSR estimates

One AU has a daily requirement of 11.8 kg of dry matter forage [17]. Therefore, one AU-month (AUM) equals 355 kg of dry matter forage (30 days times the daily forage requirement). Six AUM (half year of feed) would equal 2.13 t of dry matter per AU. Forage yields in each province (Column 4 of Table 2), in t per ha, were the product of each provincial ESSR (Table 1) and 0.355 t dry matter. Table 2 also shows the rangeland area needed to support one AU for six months of summer grazing (Column 3), the total rangeland area (Column 2) and the TSAU (Column 5) in each province.

	ESSR	**Rangeland**	**Summer forage**	**Yield[1]**	**TSAU[2]**
Province	**AUM/ha**	**ha × 10^6**	**ha/AU {6 months}**	**t/ha**	**AU × 10^6**
Manitoba	0.88	0.72	6.9	0.31	0.10
Saskatchewan	0.71	4.55	8.4	0.25	0.54
Alberta	0.97	5.29	6.2	0.34	0.85

[1], Yield = Dry matter yield of forage from rangeland

[2], TSAU = Total Sustainable Animal Units

Table 2. Areas, carrying capacities and sustainable forage yields of rangeland in the Prairie Provinces of Canada.

http://dx.doi.org/10.5772/60047

3.2. Changes in arable land use

The next phase of this chapter considers the impact of expansion of canola on the areas seeded to livestock feed crops. Since they account for roughly 90% of the grains in cattle diets in western Canada, a mix of barley and oats was taken as representing a typical ration of feed grain in the prairie region. The area currently used to grow feed grain (oats and barley) and canola is shown in Table 3 in each province for the two most recent census years (2006 and 2011). The total provincial production and yields for these crops are also shown in Table 3. The respective yields were used to determine how much feed grain area would be displaced by expanded canola. The dry matter weights of production in Table 3 were used to determine how much new area in perennial forage would be needed the replace the lost livestock feed.

This assessment was based on four scenarios of how expanded canola could impact beef production (described in Section 3.3). The land use changes that are the basis of these scenarios are shown in Table 4. These changes include the expansion of canola, the feed grain displaced by canola and the areas of additional hay needed to replace the displaced feed grain. This table represents a dynamic area balance calculation for testing the quantitative response to assumed expanded areas of canola to be of the crops being displaced by canola. The controlling parameter for this table was the total area of new canola across all three provinces. While this table is a dynamic tool that changes depending on what value for this parameter is selected, the version of this table shown in this chapter assumed a total area of 0.7Mha for both census years. This canola area total was then distributed among the three provinces so that the rangeland in any one province would not be exceeded. The 0.7Mha of expanded canola was the maximum new canola area that the rangeland could have sustainably replaced the required forage.

	Crop area	Yield	Production	Crop area	Yield	Production
	000,ha	t/ha	000,t	000,ha	t/ha	000,t
		Canola			Feed Grain	
		2006			2006	
Manitoba	1,002	1.80	1,803	680	3.13	2,105
Saskatchewan	2,558	1.50	3,836	1,963	2.54	4,975
Alberta	1,728	1.90	3,283	1,716	3.11	5,301
Prairies	5,287	1.71	8,922	4,358	2.88	12,381
		2011			2011	
Manitoba	1,064	1.60	1,703	277	2.46	682
Saskatchewan	3,885	1.80	6,993	1,323	3.04	4,020
Alberta	2,438	2.20	5,364	1,475	3.54	5,204
Prairies	7,388	1.93	14,060	3,076	3.26	9,906

Table 3. Crop production comparisons for canola (biofuel feedstock) and barley (livestock feed grain) in the Canadian Prairie Provinces during two census years.

The crop type and year-specific crop yields from Table 3 were used to convert areas to production quantities. The computation sequence in this balance was:

1. set the area to produce canola,
2. let the area of canola define the displaced area of feed grain (barley and oats),
3. convert displaced feed grain area to lost feed grain production,
4. define the required production of forage to replace feed grain, and
5. determine the new forage area from the required forage production.

Since this chapter allowed for the contribution of canola meal to the ruminant diet, the lost feed grain production was reduced by the weight of canola meal from the expanded area of canola. The weight of extractable oil from canola is 39% of the harvested crop weight, which means that 61% of the harvested canola dry matter weight is available as livestock feed supplement [18, 19].

As a general rule for sheep and cattle, 1.8 kg of average quality hay can replace approximately one kg of barley or oats [22-24]. This broadly accepted rule of thumb allowed land under feed grains and under perennial forage to be equated on the basis of nutrient energy for ruminant livestock. This approximation also allowed the land diverted away from the feed grains into canola production to be expressed in terms of the additional tame hay or grazing land that ruminant livestock would need to maintain their dietary energy intake.

	New canola		Displaced feed grain		Required	Tame hay/pasture		Rangeland	
	area	production	area	production	forage	yield	area	yield	area
	000,ha	000,t	000,ha	000,t	000,t	t/ha	000,ha	t/ha	000,ha
				2006					
Manitoba	109	197	70	218	393	4.22	93	0.31	720
Saskatchewan	315	473	201	511	919	3.43	268	0.25	4,184
Alberta	276	524	171	532	958	4.65	206	0.34	2,792
Prairies	700	1,193	442	1,261	2,269	4.10	567	0.29	7,695
				2011					
Manitoba	63	101	38	94	169	4.22	40	0.31	543
Saskatchewan	301	542	192	584	1,052	3.43	307	0.25	4,164
Alberta	336	739	208	734	1,321	4.65	284	0.34	3,852
Prairies	700	1,382	438	1,412	2,541	4.10	631	0.30	8,559

Table 4. Changes in areas and production resulting from displacement of barley by 700, 000 ha of canola needed for biodiesel feedstock, and the areas of tame hay, improved pasture or rangeland to grow enough forage to replace the lost feed grain for cattle (represented by barley and oats) during two census years in the Prairie Provinces.

Tame hay yields are less accurately reported (by survey) than the yields of annual field crops. A typical yield of about 4.1 t/ha across Canada, however, has been estimated [25]. Bootsma et al. [26] demonstrated that perennial forage yields on improved land vary with regional climate and soil types. For simplicity, it was assumed that the spatial variance among these tame hay yields (Column 6 of Table 4) would be the same as among the rangeland forage yields (Column 8) and that the provincial tame hay and improved pasture yields could be scaled to the rangeland yields (Column 8 of Table 4 or Column 4 of Table 2). The steps in the above computation sequence relate to the column numbers in Table 4 as follows: Step 1 is in Column 1, Step 2 (allowing for canola meal) is in Column 3, Step 3 is in Column 4, Step 4 is in Column 5, and Step 5 is in Column 7 for tame hay and Column 9 for rangeland. Table 4 also shows the canola production in Column 2.

3.3. Defining the canola expansion scenarios

The second goal of this chapter was to determine the change in the GHG emissions budget for the ruminants undergoing a diet. Prior to this determination for livestock, two preliminary scenarios were considered for the additional forage crop resulting from canola expansion. The difference between not including canola meal in the diet of displaced beef cattle (Scenario 1) and including canola meal (Scenario 2) served to demonstrate the feedback effect of canola meal in partially mitigating the secondary effects of canola expansion into the feed supply for beef cattle.

Two additional scenarios were used to assess the secondary impact of the canola expansion on livestock production. The first (Scenario 3) involved relocating the displaced feedlot cattle to pasture and rangeland, and a diet much richer in hay. The second (Scenario 4) assumed that the steers and heifers destined for finishing in feedlots would be butchered as veal at the calf or pre-yearling life stage, rather than being relocated to feedlots, or to pasture and hay. In order to avoid a major drop in protein supply in this scenario, these pre-yearlings would be replaced with sheep to be grazed and wintered on hay.

Six age-gender categories define the lifecycle of western Canadian beef cattle based on the feed intake and live weight differences among these categories [7, 27]. This grouping put breeding bulls and cows in one category. Figure 1 defines the age-gender categories and shows the ages and their intake of annual feed grains. This grouping ignores the newborn calves because at this age these animals do not consume grain. The bottom three categories in Figure 1, which include the animals destined for slaughter, consume proportionally more feed grain than do the replacement categories. This dietary difference was essential to the GHG emission assessment described below.

To help understand the two livestock scenarios, the structure of the beef cattle population in the three Prairie Provinces is shown in Figure 2. The breeding stock included 0.7, 1.6 and 2.1 million head of cattle in Manitoba, Saskatchewan and Alberta, respectively in 2006, making up 46% of the beef cattle population of the Prairie Provinces. Bulls account for 5% of these breeding cattle. The animals that are less than one year old are split almost equally between bull calves and young heifers. About 7% of the animals shown as steers and slaughter heifers category in Figure 2 are slaughter calves. Although the younger age-gender categories

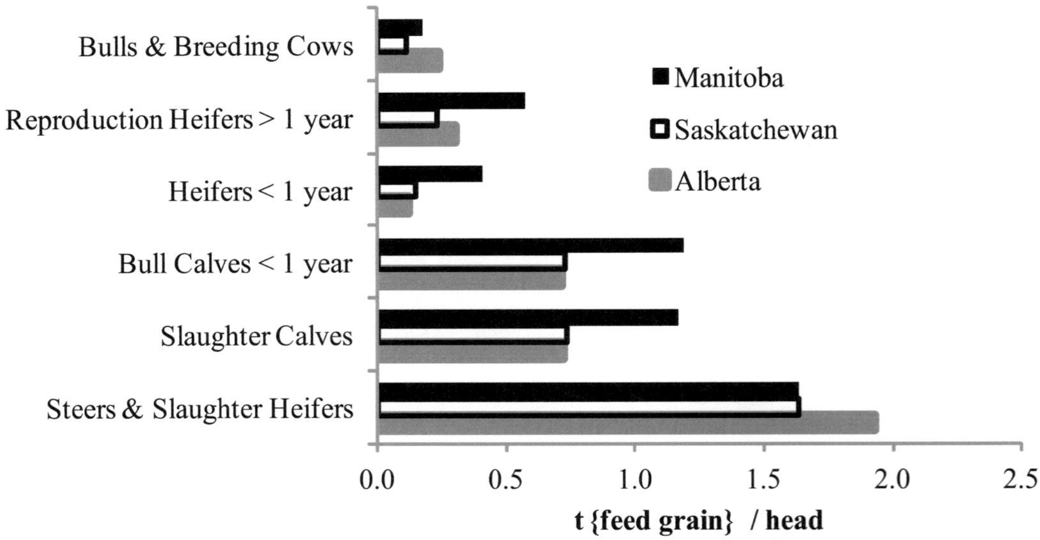

Figure 1. Annual feed grain consumption in each age-gender category of beef in the Prairie Provinces.

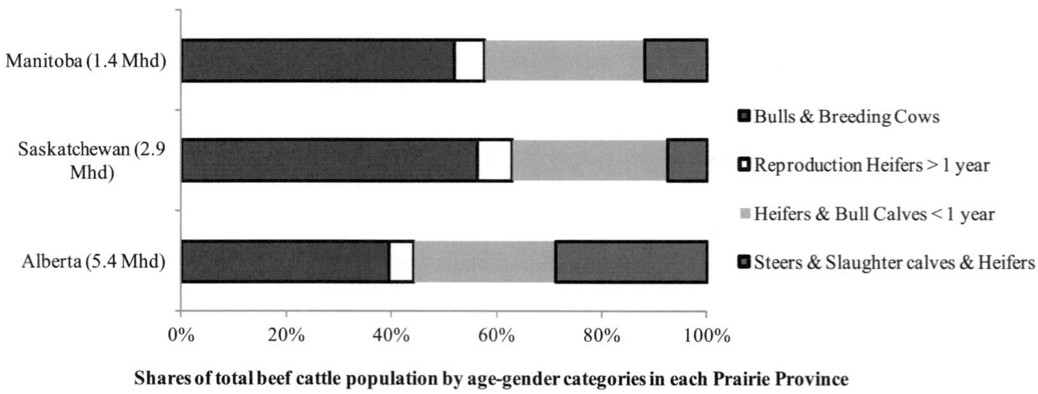

Figure 2. Age-gender based population distributions and total cattle populations (in millions of head (Mhd)) of the beef industry in the three Prairie Provinces of Canada during 2006.

fluctuate depending on market prices, the relative populations of the steers and slaughter heifers indicate that there is a net flow of these animals to the feedlots which are mostly located in southern Alberta.

In order to fully understand the full impact of livestock production on GHG emissions, the GHG emissions from the areas that provide livestock feed, not just emissions of enteric methane, must be included in this budget. The Livestock Crop Complex (LCC) defines the crop areas required to feed Canada's livestock populations [27, 28]. Five specific crop complexes have been defined in Canada, including the BCC, DCC, PCC, ACC and SCC, respectively, for beef, dairy, pork, poultry (avian) farms [7, 20, 29, 30] and, most recently, for sheep

http://dx.doi.org/10.5772/60047

[3]. The LCC concept has been used to quantify the cropland that was not used to support livestock in Canada [28]. This LCC-excluded land concept is similar to the LCC application in this chapter, since the land designated for canola expansion was removed from the BCC.

The crop complex area includes both the roughage and grain crops in the animal diet. Only the BCC, DCC and SCC in the Prairie Provinces include land in perennial forage. The grain area in each LCC is the product of population, diet and the yield of each feed grain, integrated over all grain crops in the livestock diet, although (as in Table 3) feed grain in this region is mostly a mix of barley and oats. In this chapter the potential changes in the BCC due to anticipated canola expansion were assessed for each of the three Prairie Provinces and for the Prairie Province region of Canada.

3.4. GHG emissions budget for ruminants

The GHG emissions for the two livestock scenarios were simulated for 2006 with the Unified Livestock Industry and Crop Emissions Estimation System (ULICEES) model [27]. ULICEES was created by assembling the five sets (discussed above) of livestock-specific GHG computations from the Canadian beef, dairy, pork and poultry industries [7, 20, 29, 30] in one spreadsheet model. Figure 3 shows the total GHG emissions for beef production in each prairie province calculated by ULICEES for 2006. The livestock GHG emission assessments include fossil CO_2, CH_4 and N_2O. Since these calculations provided a baseline for Scenarios 3 and 4, separate totals for the three GHGs are shown in Figure 3. These emissions are expressed as fossil CO_2 emission equivalent quantities.

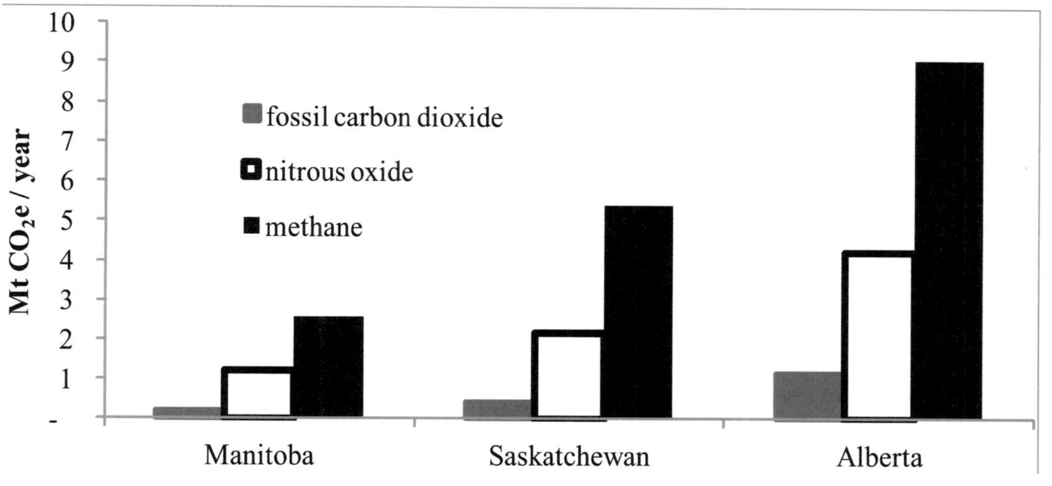

Figure 3. Total Greenhouse Gas (GHG) emissions from the beef industry in the Prairie Provinces of Canada during 2006.

ULICEES uses the Tier 2 methodology from IPCC [31], modified for Canadian conditions [32], to estimate nitrous oxide emissions for each age-gender livestock category. Methane emissions from enteric fermentation and manure storage were calculated separately by ULICEES [27].

Both methane source estimates also relied on IPCC Tier 2 methodology [31]. Both types of methane emissions were then calculated on a per-head basis for each age-gender category and multiplied by each respective category population. Using the six farm energy terms defined in [33], the provincial fossil CO_2 emission rates for 2006 were simulated by Dyer et al. [34]. These estimates were incorporated into ULICEES [27]. Unlike the CH_4 and N_2O mission estimates, the fossil CO_2 emission estimates were not distributed over age-gender categories within each livestock type.

Three ULICEES simulations for beef and lamb production in the three Prairie Provinces were required to describe the two livestock scenarios. The first ULICEES simulation was the baseline set of GHG emissions by the beef industry with no assumed changes in the population structure of the industry. To run the two additional ULICEES simulations, the changes in the age-gender livestock populations described above were implemented in the inputs to the ULICEES model. For ULICEES to implement the grass beef scenario (#3), the replacement heifers were used as an analog for grass beef because their diet is mostly forage [7, 27]. This meant that in the grass beef scenario (#3) the populations of steers, slaughter heifers and slaughter calves in ULICEES were transferred to the replacement heifer age-gender category. To apply ULICEES to the veal/lamb scenario (#4), the populations of steers, slaughter heifers and slaughter calves were transferred to the newly born calves' category, to which ULICEES attributes no GHG emissions [27]. In addition, the sheep populations had to be expanded to consume the forage no longer consumed by those animals that were converted to veal production. This was achieved by inflating the sheep populations by the ratios of meat animals in the beef industry to the sheep population expressed as protein in each province.

Before the reallocation of steers and slaughter heifers to the reproduction heifer category (Scenario 3), these populations were redistributed to match the distribution of breeding cows among the provinces. This was done to remove the influence of the concentration of feedlots in southern Alberta, to which the cattle destined for finishing for market before slaughter tend to gravitate. Before inflating the sheep populations in Scenario 4, the GHG emissions from sheep were redistributed to match the distribution of GHG emissions from beef cattle given by ULICEES. This was done to reduce instability caused by the populations of sheep in western Canada being very small relative to beef cattle.

3.5. Changes in the Carbon Footprint (CF) of canola

The CF of expanded canola must combine initial GHG emission costs of actually growing the canola crop with the secondary impact assessment of the crops being displaced by the canola. In addition, it must include potential benefits stemming from the shift from annual to perennial ground cover for both scenarios. The change in beef production (from feed grain to hay) would mean that the soil surface is never bare between crops which would cause atmospheric CO_2 to be sequestered as soil carbon. For the Prairie Provinces the average yearly carbon storage would be approximately 0.55 t{carbon}/ha [35], or 2.02 t/ha of sequestered CO_2. In this chapter when the CF determination takes all of these terms into account, it is then deemed to be the net CF of canola.

http://dx.doi.org/10.5772/60047

Table 5 shows the GHG emission rates used for all four scenarios normalized to areas of expanded canola so that all of these coefficients have the same area basis. The first two columns show the emission rates for canola and feed grain [36], while the last two columns show the changes in the GHG emission rates of the two livestock scenarios (#3 and #4). Columns 3, 4 and 5 of Table 5 all represent emission rates for the new areas of hay in the expansion scenarios, with Column 3 showing the rates as reported by [36]. Columns 4 and 5 have been normalized to the areas in expanded canola. The differences between these two columns demonstrate the importance of substituting canola meal for part of the displaced feed grain. Column 6 of Table 5 gives the sequestration rate for CO_2 by the conversion to perennial forage normalized from the forage area to the expanded canola area in each province. The negative signs on these values illustrate that sequestration flux direction is apposite that of GHG emissions. Columns 7 and 8 include the increase in GHG emissions compared to the baseline GHG simulations normalized to the area remaining in feed grains in each respective scenario. Columns 7 and 8 do not include the emission cost of growing canola (Column 1) or the benefit of sequestered CO_2 (Column 6).

	Expanded	**Displaced**	**Required forage**			**Sequestered**	**Livestock scenarios[5]**	
	canola	**feed grain**	**initial[1]**	**no meal[2]**	**with meal[3]**	**soil carbon[4]**	**grass beef**	**veal/lamb**
				t CO_2e/ha				
Manitoba	1.30	1.19	0.52	0.69	0.44	-1.72	1.91	-2.52
Saskatchewan	1.03	0.68	0.39	0.52	0.33	-1.71	2.65	4.85
Alberta	1.28	0.95	0.53	0.63	0.40	-1.51	0.35	4.44
Prairies	1.16	0.88	0.48	0.60	0.39	-1.63	1.48	3.30

[1], GHG emission intensity of hay per unit area of hay grown and harvested, not normalized to canola.

[2], GHG emission intensity of forage with no substitution by canola meal, normalized to canola area.

[3], GHG emission intensity of forage with substitution by canola meal, normalized to canola area.

[4], fossil CO_2 sequestered by new forage, normalized to canola area.

[5], livestock GHG emission intensities not including areas of expanded canola and not normalized to canola areas.

Table 5. GHG emission intensities per unit area for canola, feed grains (represented by oats and barley combined) and hay, rates of CO_2 sequestration under new forage areas and two ruminant production scenarios, (area basis of intensities shown as footnotes) for the Prairie Provinces during 2006.

Figure 4 shows the changes in GHG emissions that can be attributed to the land use changes induced by the proposed expansion of canola. These changes were measured by the differences between the two scenario simulations and the baseline simulations shown in Figure 3. These differences were expressed as emission rates per unit area of feed grains in the baseline ULICEES simulations. Unlike the emission differences shown in Columns 7 and 8 of Table 5, those in Figure 4 include the cost of growing the expanded canola crop and the new forage crop, and the CO_2 sequestered by the land use change from feed grains to perennial forage.

The rates shown in Figure 4 were also normalized to baseline feed grain areas so that they have the same area base.

The GHG emission cost of growing canola and additional forage, and the sequestration of CO_2 under the new forage area were added to the assessment after the ULICEES simulation process. This was necessary because ULICEES computes the forage component of ruminant diets in the BCC and SCC (as well as the DCC) by partitioning areas from fixed pools of land in hay and in improved pasture to the regional beef, sheep and dairy populations [27]. Unlike the grain components of those diets, ULICEES cannot, therefore, create new areas of forage to meet changes in the BCC. Adding both of these terms to the simulations from ULICEES, required them to be expressed on the basis of the expanded area of canola (as shown in Table 5). Of the 2.3 Mha of feed grain area used in ULICEES to support cattle and sheep in the Prairie Provinces, Scenario 3 converted 44% to expanded canola while Scenario 4 converted 77% to expanded canola.

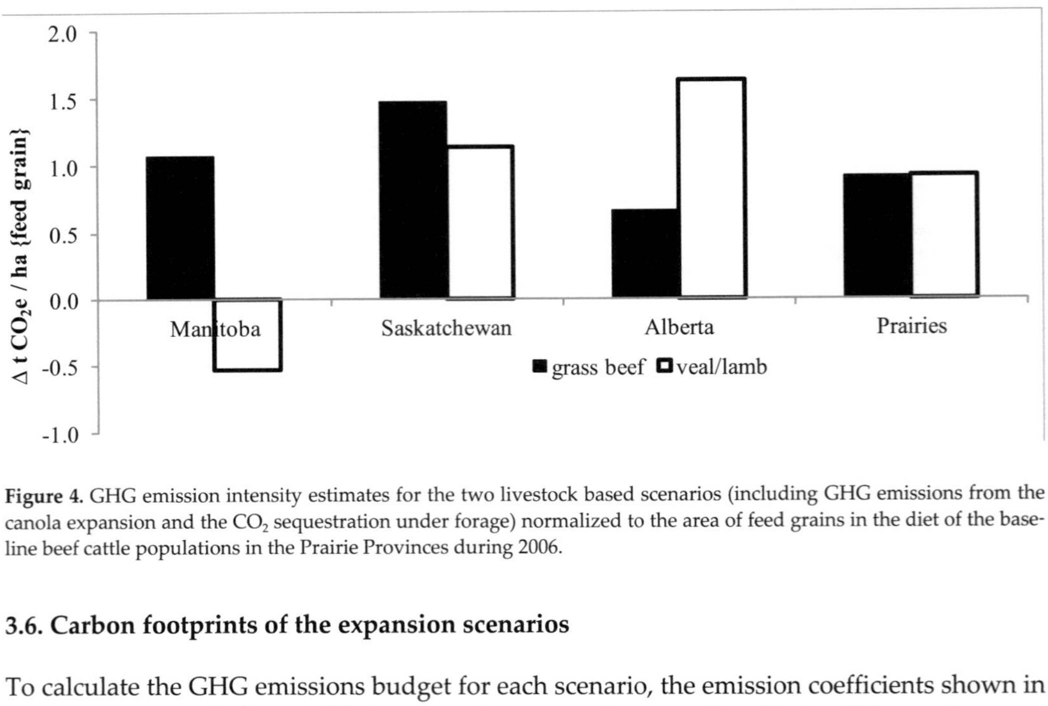

Figure 4. GHG emission intensity estimates for the two livestock based scenarios (including GHG emissions from the canola expansion and the CO_2 sequestration under forage) normalized to the area of feed grains in the diet of the baseline beef cattle populations in the Prairie Provinces during 2006.

3.6. Carbon footprints of the expansion scenarios

To calculate the GHG emissions budget for each scenario, the emission coefficients shown in Columns 4, 5, 7 and 8 from Table 5 were each integrated separately with the difference between the canola and feed grain emission coefficients, and the sequestration of CO_2 (Columns 1, 2 and 6 of Table 5). Table 6 shows the GHG emission rates from the four secondary impact scenarios for expanding canola in the three Prairie Provinces. Scenarios 2, 3 and 4 assume that canola meal can compensate for part of the displaced feed grain in the ruminant diet.

The GHG emissions intensities (EI_{1-2}) of the two scenarios based on crop differences (#1 and #2) were the result of straight forward addition of emission terms from feed grain (barley and oats combined) and perennial forages to the CF of canola. The inclusion of the CO_2 sequestration rates (SR) from Table 5 in Equation 1 reduced these GHG emissions intensity estimates. These terms were summarized as follows.

$$EI_{1-2',canola,net} = EI_{canola} - EI_{feed\ grain} + EI_{forage} - SR_{forage} \tag{1}$$

The crop-specific EI values in Equation 1 for each province and the region were taken from Columns 1 and 2, and either 4 or 5 of Table 5, depending on whether canola meal was assumed to be a feed supplement.

Using Columns 1 and 2 of Table 5, and either Column 7 for the grass beef (#3) scenario or Column 8 for the veal/lamb (#4) scenario, the GHG emissions intensities ($EI_{3\text{-}4}$) of the two livestock based scenarios (#3 and #4) can be represented symbolically by the following equation, which includes the same SR_{forage} term as Equation 1 and the emission cost of the additional hay (EI_{forage}) in each livestock scenario.

$$EI_{3-4,canola,net} = EI_{canola} - EI_{feed\ grain} + \Delta EI_{livestock} + EI_{forage} - SR_{forage} \tag{2}$$

However, the feed grain EI values in Equation 2 were generated as part of the ULICEES simulation of the livestock-specific Scenarios 3 and 4. The emission cost of the additional hay production appears as a separate term in Equation 2 because it had to be calculated externally from ULICEES. Since EI values generated by ULICEES were expressed on the basis of the scenario feed grain areas, it was necessary to convert the EI for canola and the SR for the new forage from Table 5 back to the feed grain area basis. This was done by multiplying the canola GHG emission rates from [36] and the SR terms from Table 5 by the areas freed from feed grain and adding the difference between these two GHG emission quantities to the respective GHG emission differences.

These new GHG emissions (for canola plus livestock and additional hay, minus the sequestered CO_2) were then divided by the respective new canola areas to give the emission rates shown in Table 6 to represent the net CF for canola under the two livestock scenarios. To convert from the feed grain to the new canola area basis, the area ratios of feed grain to canola were taken for the whole Prairie region for both livestock scenarios as a way of smoothing these normalized estimates over the three provinces.

Scenarios 1 and 2 were useful for demonstrating the role of canola meal in minimizing the crop displacement by the expanded canola area. As well, Column 4 of Table 6 (for Scenario 2) showed what the inclusion of carbon sequestration without accounting for GHG emissions from livestock would mean for the CF of canola. The difference between the sequestered soil carbon shown in Column 6 of Table 5 and Column 4 of Table 6 was that in Table 5 the sequestration rates did not include the GHG emission costs of growing the new forage. It was the CO_2 sequestration rates from Table 6 that were incorporated into Scenarios 3 and 4.

	Expanded	Required forage			Livestock		Fossil CO_2
	canola	no	with	with meal	scenarios		offset by
	only	meal[1]	meal[1]	and soil C[2]	grass beef[2]	veal/lamb[2]	canola
Scenario #		1	2	2	3	4	
				t CO_2e/ha{canola}			
Manitoba	1.30	0.80	0.55	-1.17	2.41	-0.69	-1.64
Saskatchewan	1.03	0.87	0.68	-1.03	3.32	1.48	-1.63
Alberta	1.28	0.96	0.72	-0.78	1.50	2.12	-2.00
Prairies	1.16	0.88	0.67	-0.97	2.05	1.20	-1.77

[1], meal = canola meal after oil extraction which is available as substitute livestock feed.

[2], includes CO_2 sequestered by the land use change from annual feed grain to perennial forage.

Table 6. Area based GHG emission intensity estimates for canola, and four canola expansion impact scenarios normalized to the area of the expanded canola crop, and potential the fossil CO_2 emissions offset by canola oil as a biodiesel feedstock in the Prairie Provinces during 2006.

Only Scenarios 3 and 4 represent the net CF of the expanded canola because the secondary impact on ruminant livestock production was incorporated by ULICEES into these two scenarios. The measure of this impact and the net CF of the new canola was a comparison with fossil CO_2 emissions that were expected to be offset by the expanded canola. The offset fossil CO_2 emission intensities (FI) are shown in each province and for the Prairies in the last Column of Table 6. They are also shown as negative values to reflect the opposite direction from the net CF. These fossil CO_2 emission offsets vary with provinces because their calculations accounted for the variations in the provincial canola yields (Y_{canola}). Each prairie yield was the production-weighted average from the 2006 and 2011 census years (Table 4). The offset fossil CO_2 emission intensities (FI) per ha of canola were calculated as follows.

$$FI = 2.8 \times 88\% \times 0.39 \times Y_{canola} \tag{3}$$

The diesel fuel to fossil CO_2 conversion factor is 2.8 kgCO_2/kg of fuel [36]. Equation 3 also took account of the 12% difference in energy content between petrodiesel and biodiesel [37] and the 39% by weight of canola yield (kg oil/kg canola seed) that is canola oil.

3.7. Protein based GHG emission intensities of scenario livestock

The assessment of canola expansion must also take into account the CF of the protein production from the proposed new distributions of age-gender categories of the ruminant livestock industries. The differences in GHG emission intensities between the two scenarios were assessed on the basis of protein supply and compared to baseline simulations for this indicator from ULICEES [3]. In this context, protein is taken to include only human edible protein (excluding blood meal, pet food, edible offal and leather). This comparison did not allow for potential nutritional differences between the protein derived from beef and lamb. Figure 5 shows the protein based GHG emission intensities for both livestock scenarios. As a reference baseline for this comparison, the 2006 protein based intensities of beef and lamb [3] are also shown in Figure 5. For this indicator, the actual GHG emission intensity simulations, rather than the differences from baseline GHG emissions, were used.

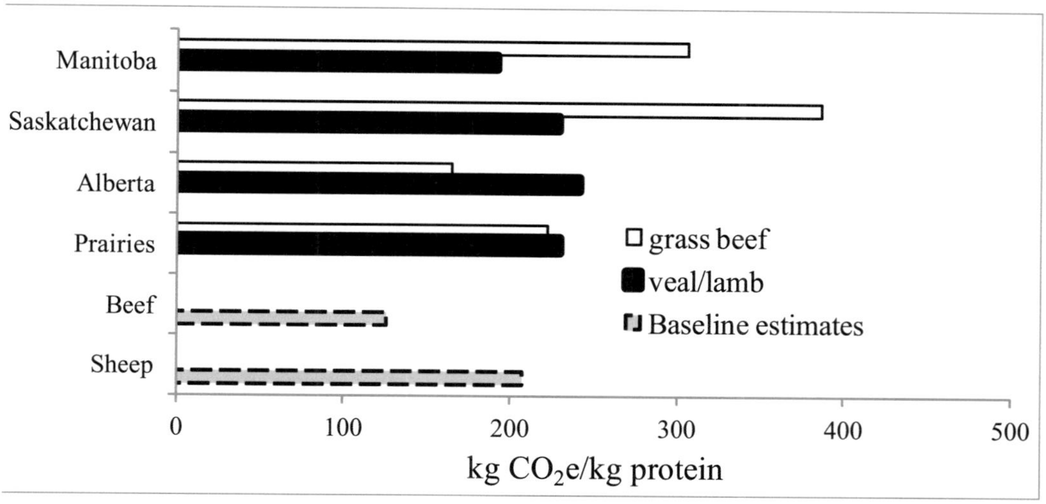

Figure 5. Protein based GHG emission intensities for the two livestock based canola expansion scenarios and for the baseline beef and sheep populations in the Prairie Provinces during 2006.

To calculate the protein based GHG emission intensities for Scenarios 3 and 4, the weight of animal Protein (P) was computed from the number of head (H) and the live weight (W) of the age-gender categories involved in simulating the two livestock scenarios that provide slaughter animals. The age-gender categories involved in the assumed population redistributions were steers and slaughter heifers (*s&sh*), replacement heifers (*rh*), culled cattle (*cc*), slaughter calves (*sc*), culled ewes (*ce*) and slaughter lambs (*sl*). The live weight conversions to protein were 6.4% for slaughter lambs [38, 39] and 8.3% for slaughter steers and heifers [10, 39]. Because breeding cows were culled every six years and ewes were culled every 4.5 years [3], reduction factors of 0.17 and 0.22 were applied to culled cow and ewe populations, respectively, in Equations 4 and 5.

$$P_{grass\ beef} = 0.083\left(H_{s\&sh} \times W_{rh} + 0.17 \times H_{cc} \times W_{cc}\right) \tag{4}$$

$$P_{veal/lamb} = 0.083\left(H_{s\&sh} \times W_{sc} + 0.17 \times H_{cc} \times W_{cc}\right) + 0.064\left(H_{sl} \times W_{sl} + 0.22 \times H_{ce} \times W_{ce}\right) \tag{5}$$

The Prairie average live weights (W) used in ULICEES were 495 for steers and slaughter heifers, 506 for replacement heifers, 616 for culled cattle, and 380 for slaughter calves, while the live weights were 57 for ewes and 48 kg and for lambs.

4. Results and discussion

Alberta had the highest rangelands ESSR by virtue of being the only province with the Foothills Fescue (FF) Ecoregion, which had the highest ESSR value (Table 1). Because it had the largest share of the Dry Mixed Grass (DMG) Ecoregion, with the lowest ESSR of the four ecoregions, Saskatchewan had the lowest provincial ESSR value. The Mixed Grass (MG) and the Parkland-Northern Fescue (PNF) ecoregions, with similar ESSR values, are more or less evenly divided between Saskatchewan and Alberta. Table 2 shows that Manitoba has only 7% of the prairie rangeland while Alberta has 50%. The yields of forage from Prairie rangelands (Table 2) were roughly one tenth of those of tame hay (Table 4). However, in both cases these forage yields were general approximations since these yield statistics are not regularly surveyed in the prairie region.

All of the statistics for annual crops (Table 3) showed considerable variability over the three provinces and the two census years, which helps to explain some of variability in the results of this assessment. The area in canola in the Prairies grew between 2006 and 2011 while the area in feed grains (barley and oats) shrank by a similar proportion (Table 3). The dry matter yield of two feed grains was 62% higher than the yield of canola. Even though the proposed expansion area for canola is only a little over 10% of the total canola crop land in the Prairies, Table 4 shows that the proposed expansion could lead to the conversion of roughly 80% of the rangeland to full time grazing by domestic ruminants in order to make up for the feed lost to the canola expansion.

Figure 3 demonstrated the importance of the province of Alberta to the CF of the beef industry in the Prairie Provinces. Alberta beef generated as much of the total GHG emissions as did Saskatchewan and Manitoba combined. Methane was the dominant type of GHG in the western Canadian beef industry. Not only was CH_4 the type of GHG with the highest quantity, it is the GHG that will increase the most if cattle are fed a more roughage based diet. This is because most of this gas is enteric methane which is the direct result of the ruminant digestion of roughage, the dominant component of the diet when cattle are displaced from a feed grain diet. This trend is partly counteracted, however, by the decreases in both N_2O and fossil CO_2 emissions when cattle are less intensively managed (as in Scenario 3), or when sheep are substituted for the feedlot finished cattle (Scenario 4). This feedback effect is accounted for in ULICEES.

Because the emission rates for Scenarios 3 and 4 in Table 5 were estimated from differences between the scenario simulations and the baseline, they were, not surprisingly, rather unstable. This instability carried over into Table 6. The wider spread among provinces for Scenario 4 in Table 5 indicates that they were a bit less stable than Scenario 3. For the Prairies, Scenario 3 was closer to the baseline ULICEES simulations (the baseline being zero in this regard) than

http://dx.doi.org/10.5772/60047

Scenario 4. Even with Manitoba having a negative difference, the Scenario 4 emissions rate for the Prairies was more than double the Scenario 3 emissions rate for the Prairies.

Being normalized to the same area basis, the prairie-wide GHG emission rates were almost equal for the two scenarios. In Figure 4 both scenarios show lower values for the prairie region than the respective rates in Table 5 because the rates in Figure 4 included the differences between GHG emission rates from the expanded canola and the respective CO_2 sequestration rates, which were negative quantities. For both Figure 4 and Table 5 (Columns 7 and 8), Scenario 3 in Manitoba was the only negative emission rate difference. In Figure 4, the only province where Scenario 4 was greater than Scenario 3 was Alberta.

The difference between Columns 4 and 5 in Table 5 (Scenarios 1 and 2) shows that the potential contribution by canola meal to ruminant diets could decrease the requirement for replacement perennial forage by over a third in the Prairie region. The inter-provincial variations in GHG emission intensities from the differences between the assumed livestock systems in Scenarios 3 and 4 were several times higher than the inter-provincial variations in just the direct emission intensities of the expanded canola crop (Column 1 in Tables 5 and 6). All of the GHG emission intensities from Table 5 showed considerable inter-provincial variations with Saskatchewan having the lowest GHG emission intensities for the crops of the three provinces, but the highest intensity differences from the baseline simulations for the two livestock scenarios. This inter-provincial variation was evident even before the livestock GHG emission rates were normalized to the canola areas.

The areas in feed grain production in the BCC that were freed to expand canola production were 1.75 Mha in Scenario 4 and 1.01 Mha in Scenario 3. The regional total feed grain area was 2.28 Mha.

Hence, the expanded canola areas were smaller than the feed grain areas in Scenario 3 but greater in Scenario 4. Thus, normalizing from feed grain to canola areas (for Table 6) would inflate the rates in Column 7 of Table 5 and deflate the rates in Column 8. The greater area changes associated with Scenario 4 helps to explain why Scenario 4 was more sensitive to the inclusion of GHG emission from the new canola and the sequestered CO_2 under the new forage area.

Without considering the livestock in Table 6, Scenarios 1 and 2 would actually appear to reduce the net CF of the expanded canola. Additionally, when CO_2 sequestration is considered, Scenario 2 suggests that just the growing of canola reduces GHG emissions without having to consider the fossil CO_2 emission offset potential. The ranking of the two livestock scenarios reversed in Table 6 compared to Table 5, with Scenario 3 having the greater net CF for expanded canola. Manitoba showed the biggest difference between the two livestock scenarios, whereas Alberta (with the largest beef industry) showed the least difference. The offset fossil CO_2 emission intensities (Table 6) were highest in Alberta because of the higher canola yields in that province. The net CF of the expanded canola in Table 6 was below the offset fossil CO_2 emission intensities (Column 7) for Scenario 3 in Alberta and for Scenario 4 in Saskatchewan. At the prairie region scale, however, the potential fossil CO_2 emissions offset by canola oil was less than the net CF of the expanded canola for Scenario 3. The potential fossil CO_2 emissions offset for the prairie region exceeded net CF of the expanded canola for Scenario 4, but not by a high enough to meet the EC directives [1].

To understand the role of CO_2 sequestration in the net CF of the expanded canola, a sensitivity test was run on the soil carbon storage rate [35] with a plus or minus 20% range. For Scenario 3 the range on the net CF was from 1.7 to 2.4 tCO_2e/ha, for a range about 2.1 tCO_2e/ha of ±16%. For Scenario 4 the range was from 0.9 to 1.5 tCO_2e/ha, for a range about 1.2 tCO_2e/ha of ±27%. Whereas a 20% increase in soil CO_2 sequestration rate would change Scenario 4 to 51% below the fossil CO_2 emission offset by canola, the result for Scenario 3 would only be 3% below that offset level. If the expanded canola described in this chapter were considered to be a continuation of the current operation of Canadian canola production, rather than a new installation, Scenario 4 might be deemed to just barely qualify for export to the EU [1] with soil carbon sequestration made 20% higher than reported by [35]. The increased sensitivity of Scenario 4 compared to Scenario 3 was due to the greater area of feed grain that was freed from the BCC for expanded canola in Scenario 4.

The protein based emission intensities in Figure 5 were close to equal for the two livestock scenarios in the Prairie region. Saskatchewan had the highest protein based GHG emission intensities in Scenario 3, while Alberta had the highest intensity for Scenario 4, but only slightly higher than Saskatchewan for Scenario 4. Scenario 3 exceeded Scenario 4 in Manitoba and Saskatchewan, while Scenario 4 was higher in Alberta. For the region, both scenario protein based emission intensities were higher than the baseline intensities for both beef and sheep, although only slightly higher than for sheep.

5. Conclusions

There are appreciable margins of error associated with a theoretical assessment such as described in this chapter that should be viewed with caution. Given the number of assumptions and approximations that this assessment relied on, the hope is that this chapter would set the scope for studying the potential canola-beef industry interactions more intensely and with more empirical data. Instead of being able to use individual simulations of GHG emission budgets from ULICEES, this assessment had to rely on incremental changes between scenario and baseline simulations. This meant that the incremental results from ULICEES were very sensitive to the random noise from the inputs to ULICEES. This noise means that the results are more meaningful on the basis of the prairie region than on the provincial scale. The need for terms external to ULICEES to be integrated with ULICEES output to account for the additional hay, as well as for the expanded canola, was the result of ULICEES not having the capacity to generate additional hay area in the LCC. The need for these external terms made Scenarios 1 and 2 important steps in this assessment.

On the other hand, the strength of this analysis stems from the use of the ULICEES model which has undergone both peer review in the scientific literature and a wide range of successful applications, also described in scientific form. It was reassuring also that, at least regionally, the two livestock scenarios provided comparable quantities of protein. Scenario 4 embodied an additional challenge. To use the feed resources freed by the early slaughter of so many cattle an increase in the Prairie sheep population by a factor of about 50 was called for, given the very small size of the current prairie sheep industry relative beef in the region. It was not surprising that such an exchange between two livestock types would result in the greater differences among the provinces seen in Table 6 for Scenario 4 compared to Scenario 3. It was also not surprising that 73% more land was made available for the expanded canola crop by Scenario 4 than Scenario 3.

Two issues regarding the methodology need clarification. First, in the CF stage of the assessment, grazing land, either tame pasture or rangeland, was mostly left out of the GHG emission budget calculations. This omission was mainly because ULICEES does not attribute any GHG emissions directly to these lands, electing instead to treat all enteric methane emissions as direct emissions from each animal, regardless of where that animal is located, and also because almost no inputs can be directly attributed to pasture. Manure voided directly onto pasture was also considered to have no methane emission cost in ULICEES.

The second issue was the ethical implications for the choices for scenarios. These scenario choices were made strictly for their value as boundary conditions in reallocating cattle to other categories in the assessment and forcing ULICEES to redistribute the resulting GHG emissions. Raising and slaughtering young calves for veal, is considered by many to be inhumane and, therefore, ethically unsustainable, regardless of the outcome of the CF assessment. Although this chapter does not advocate or condemn veal as a meat source, this assumption facilitated the expansion of sheep in Scenario 4. Also by assuming an all-roughage diet for the inflated sheep population for Scenario 4 the problem that the actual diet for Canadian sheep contained too much feed grain for this assessment was bypassed. However, removing all feed grain from the diet of the expanded sheep ignored the need to have a small share of grain in the diet of breeding ewes.

The third issue was the use of GHG emission intensities based on land areas in this assessment, rather than on measures of productivity. Land based emission intensities are generally not practical in describing the CF of ruminant livestock because these farming systems involve three different land uses, including annual crops, hay and pasture (both improved and unimproved), which are difficult and rather arbitrary to equate to a single indicator of land value. Land based emission intensities were the only way that terms external to ULICEES could be integrated with ULICEES output. The land basis for emission intensity was applicable in this assessment only because it was the incremental changes in these intensities, rather than the integral values, that were used. Otherwise, the land based GHG emission intensities in Tables 5 and 6, particularly for the two livestock scenarios, are not likely to be applicable outside of the context of this assessment.

This chapter explored three parameters of sustainability. The first was land use change in which it was revealed that the needed increase in forage production cannot be acquired from the use of rangeland. Given the very low yields of livestock feed that can be achieved within the limits set by the ESSR, small increases in canola area require too large portions of the remaining natural grasslands in Western Canada to be grazed. This deprives wild native ungulates of their feed sources in these areas and it could threaten the natural plant diversity in these lands as well, even with the co-grazing of cattle with sheep. In contrast to the greater net CF for canola, a shift to tame hay or improved pasture as a way of increasing forage, would protect both biodiversity and reduce soil erosion, because the soil surface is never bare.

The degree to which rangelands are already grazed by cattle is not known. Even if all of the rangeland shown in Table 2 were available for expanded livestock grazing, the 0.71, 1.60 and 2.14 million head of breeding cattle in Manitoba, Saskatchewan and Alberta, respectively (from Figure 2), in 2006 greatly exceeded the 0.10, 0.54 and 0.85 million AU that can be supported for six months on Prairie rangeland (Table 2). For the Prairies, the breeding cow population (the basis for defining the AU) was three times larger than the carrying capacity of rangeland defined on this basis. Also modern beef cattle are appreciably larger than the breeding cows at the time the AU indicator was devised.

The second sustainability parameter, and the main target of this assessment, was the extended scope of the CF of the new canola areas. The net CF of the expanded canola exceeded the fossil CO_2 emission offsets associated with petrodiesel by 16% in Scenario 3 and was exceeded by the fossil CO_2 emission offsets by 32% in Scenario 4, leaving little hope of this expansion option ever complying with the EC directives on biofuel feedstock production. In spite of the limitations of the modeling approach used for this assessment, the findings from both livestock scenarios send a message that expansion of canola for biodiesel feedstock is unlikely to be sustainable if ruminant livestock are displaced into a more forage dependant production system by the expansion.

Without CO_2 sequestration under the new hay area, the margin between the net CF of canola and the fossil CO_2 emission offsets would have been much greater. Because CO_2 sequestration declines to almost zero by about 40 years as the soil carbon sink is recharged [27] (a consideration in all GHG mitigation strategies), this term is not perpetual. The magnitude by which the fossil fuel GHG emissions to be offset were too low in relation to the change in scenario GHG emissions was further demonstrated by the sensitivity to the yearly soil carbon storage rate. The need for a 20% increase in the CO_2 sequestration to bring just Scenario 4 into complying with EC directives indicates that allowing canola to displace feed grains from the BCC is unsustainable. This suggests that a shift from ruminant to non-ruminant livestock farming [9] would be a better strategy for expanded canola feedstock to interact positively with Canadian livestock industries with respect to GHG emissions.

The failure of Scenarios 3 and 4 was in spite of not including several factors that would have made the net CF of the expanded canola even higher. The main factor was that no allowance was made for the processing side of the canola oil, or the fuel that was required to collect and transport the canola seed to processing plants. While the canola expansion described in this chapter called for more perennial forage to replace feed grain in the ruminant diet, it was not known if sufficient new land would be available to grow the required forage. Both of the livestock scenarios assumed that canola meal could be incorporated into the livestock diet. While this is possible in principle, the poor palatability of canola meal to livestock is a limitation. In order to minimize this limitation, that meal would have to be spread throughout the prairie beef population so that it appeared in smaller portions in individual diets.

The third sustainability parameter was the protein based GHG emission intensity. This protein based indicator for the livestock described in both Scenarios 3 and 4 was higher than the protein based GHG emission intensities for the current beef and sheep industries. So in addition to more GHG emissions, this canola expansion option contributed less protein to the human diet for the same level of GHG emissions. This provides yet more argument for not allowing canola expansion into the beef industry to make that industry more dependent on a high roughage diet.

This assessment does not condemn all options for expanding canola production. Canola is a valuable cash crop for Canadian farmers and, in the right circumstances, can be a viable GHG mitigation option as a biodiesel feedstock. However, as the conversion of land that was in summerfallow to other crops in western Canada nears completion, continued displacement by canola of any other land use in the Prairie Provinces of Canada needs close assessment, including attention to secondary land use changes.

From a policy perspective, this assessment has one more limitation, because it may not always be clear exactly what land is being displaced. For example, canola expansion was more the beneficiary than the cause of shrinking areas in summerfallow in the Prairie Provinces. Similarly, feed grains may be displaced by food quality crops that were displaced as the direct result of canola expansion. In this case the causative role of canola expansion in livestock displacement may not be recognized, even though it would be the main driver of this land use change in this situation. In spite of these potential policy implementation hurdles, the general lesson from this assessment may still give some valuable guidance for international pasture and rangeland managers, particularly given the close similarity between canola and rapeseed. This chapter may also provide insight into the CF of more extensive, forage based, beef production, regardless of whether or not biofuel feedstock is what is driving the shift away from intensive, feed grain dependant beef production.

Author details

J.A. Dyer[1*], X.P.C. Vergé[2], R.L. Desjardins[3] and D.E. Worth[3]

*Address all correspondence to: jamesdyer@sympatico.ca

1 Agro-environmental Consultant, Cambridge, Ontario, Canada

2 Agro-environmental Consultant to AAFC, Ottawa, Ontario, Canada

3 Agriculture & Agri-Food Canada, Ottawa, Canada

References

[1] Summaries of EU legislation. Directive 2009/28/EC of the European Parliament and of the Council of 23 April 2009 on the promotion of the use of energy from renewable

sources and amending and subsequently repealing Directives 2001/77/EC and 2003/30/EC. Available from http://europa.eu/legislation_summaries/energy/renewable_energy/en0009_en.htm (Accessed 19 November 2014)

[2] Shrestha BM, Desjardins R.L McConkey BG, Worth DE, Dyer JA, Cerkowniak DD. Change in Carbon Footprint of Canola Production in the Canadian Prairies from 1986 to 2006. Renewable Energy 2014;63: 634-641. Available from http://www.sciencedirect.com/science/article/pii/S0960148113005533 (Accessed 19 November 2014)

[3] Dyer JA, Vergé XPC, Desjardins RL, Worth DE. A comparison of the greenhouse gas emissions from the sheep industry with beef production in Canada. Sustainable Agriculture Research 2014;3(3):65-75. Doi: 10.5539/sar.v3n3p65.

[4] Bisset, R. Methods for environmental impact analysis: recent trends and future prospects. Journal of Environmental Management 1980;11: 27-43.

[5] Wright DS, Greene GD. An environmental impact assessment methodology for major resource development. Journal of Environmental Management 1986; 24:1-16.

[6] Elward M, McLaughlin B, Alain B. Livestock feed requirements study 1999-2001. Catalogue No. 23-501-XIE. Ottawa: Statistics Canada; 2003; p.84.

[7] Vergé XPC, Dyer JA, Desjardins R.L, Worth D. Greenhouse gas emissions from the Canadian beef industry. Agricultural Systems. 2008;98(2):126–134, doi:10.1016/j.agsy.2008.05.003

[8] Dyer JA, Hendrickson OQ, Desjardins RL, Andrachuk HL. An Environmental Impact Assessment of Biofuel Feedstock Production on Agro-Ecosystem Biodiversity in Canada. In: Contreras LM. (Ed.) Agricultural Policies: New Developments., Hauppauge, NY: Nova Science Publishers; 2011. Chapter 3:87-115. Available from https://www.novapublishers.com/catalog/product_info.php?products_id=29219&osCsid=d6ac9accf428e090851d95f620e23e31 (Accessed 19 November 2014)

[9] Dyer JA, Vergé XPC, Desjardins R.L, McConkey BG. Implications of biofuel feedstock crops for the livestock feed industry in Canada. In: Editor: Dos Santos Bernardes MA.Environmental Impact of Biofuels., Rijeka: InTech; 2011. Chapter 9, pages 161-178. Available from http://www.intechopen.com/books/references/environmental-impact-of-biofuels/implications-of-biofuel-feedstock-crops-for-the-livestock-feed-industry-in-canada (Accessed 19 November 2014)

[10] Dyer JA, Vergé XPC, Desjardins R.L, Worth DE. The protein-based GHG emission intensity for livestock products in Canada. Journal of Sustainable Agriculture 2010; 34(6):618-629. Doi: 10.1080/10440046.2010.493376.

[11] Bailey AW, McCartney D, Schellenberg MP. Management of Canadian Prairie Rangeland, Agriculture and Agri-Food Canada, 2010; AAFC No. 10144, Cat. No. A52-178/2010E-PDF, Available from http://publications.gc.ca/collections/collection_2010/agr/A52-178-2010-eng.pdf. (Accessed 10 May 2013)

[12] Kloppenburg PB, Kiesling HE, Kirksey RE, Donart GB. Forage quality, intake and digestibility of year-long pastures for steers. J. Range Management 1995;48:542-548. Available from https://journals.uair.arizona.edu/index.php/jrm/article/view/9070/8682 (Accessed May 17 2013)

[13] Adams BW, Poulin-Klein L, Moisey D, McNeil RL. Rangeland Plant Communities and Range Health Assessment Guidelines for the Dry Mixedgrass Natural Subregion of Alberta. Rangeland Management Branch, Public Lands Division, Alberta Sustainable Resource Development: Lethbridge, Alberta: 2005; Pub. No. T/040. 106 pp. Available from www3.gov.ab.ca/srd/land/publiclands/range.html (Accessed 10 May 2013)

[14] Burkinshaw AM, Willowby MG, France K, Loonen H, McNeil RL. Range plant communities and range health assessment guidelines for the Central Parkland Subregion of Alberta, First approximation. Sustainable Land Development, Lands Division: Lethbridge, Alberta; 2009; Pub.No.T/125. Available from http://srd.alberta.ca/LandsForests/GrazingRangeManagement/documents/FoothillsParklandGuide-Aug24-2012.pdf. (Accessed 10 May 2013)

[15] Willoughby MG, Stone C, Hincz C, Moisey D, Ehlert G, Lawrence D. Guide to range plant community types and carrying capacity for the Dry and Central Mixedwood subregions in Alberta. Public Lands and Forests Division, Edmonton, Alberta: 2006; Pub No. T/103. 254 pp. Available from http://esrd.alberta.ca/fish-wildlife/species-at-risk/albertas-species-at-risk-strategy/documents/GeneralStatusOfAlbertaWildSpecies-2000.pdf (Accessed 19 November 2014)

[16] Dyer JA, Vergé XPC, Desjardins R.L, Worth DE. An assessment of greenhouse gas emissions from co-grazing sheep and beef in western Canadian rangeland. In: (Eds) Lac S, Kulshreshtha S, McHenry M. Agriculture Management for Climate Change. Nova Publishers; 2014; (In press).

[17] Prairie Farm Rehabilitation Administration (PFRA). Animal Unit Months, Stocking Rate and Carrying Capacity, Regina: Agriculture and Agri-food Canada; 2003. Available from http://agr.gc.ca/pfra/land/fft1.htm (1 of 4)5/14/2007 Date modified: 17 March 2003. (Accessed 10 May 2013)

[18] Ruyle G, Ogden P. What is an A.U.M.? In: Gum R, Ruyle G, Rice R. (Eds) Arizona Ranchers' Management Guide. Arizona Cooperative Extension, College of Agriculture, University of Arizona; 1993. Available from http://www.ars.usda.gov/SP2UserFiles/Program/304/ActionPlan2008-2013/2a.pdf. (Accessed 22 April 2013)

[19] Sooksom R. Guide for Beginning Farmers in Nova Scotia. Nova Scotia Department of Agriculture: Truro, NS: 2010;48 pp. Available from http://www.gov.ns.ca/agri/thinkfarm/guide.pdf, Accessed 22 April 2013.

[20] Vergé XPC, Dyer JA, Desjardins R.L, Worth D. Greenhouse gas emissions from the Canadian dairy industry during 2001. Agricultural Systems 2007;94(3):683-693. doi: 10.1016/j.agsy.2007.02.008

[21] Yacentiuk M. 2001. Full fat soybeans in swine rations. Manitoba agriculture, food and rural initiatives. Available from http://gov.mb.ca/agriculture/livestock/pork/bab02s57.html2001. (Accessed 18 March 2010)

[22] Institute of Food and Agricultural Sciences. Replacing hay with grain. In: The Disaster Handbook - 1998 Edition. Cooperative Extension Service, University of Florida: 1998;Chapter 6. Available from http://disaster.ifas.ufl.edu/PDFS/CHAP06/D06-08.PDF (Accessed 19 November 2014)

[23] Neel JB. Rules of thumb for winter feeding., University of Tennessee, Agricultural Extension Service. Animal Science. 2012; Info Series: AS-B-260 Available from http://animalscience.ag.utk.edu/beef/pdf/RulesOfThumb.260.pdf. (Accessed 18 June 2012)

[24] Schoenian S. The truth about grain: Feeding grain to small ruminants. Small Ruminant Info Sheet, University of Maryland Extension: 2011. Available from http://www.sheepandgoat.com/articles/graintruth.html. (Accessed 10 May 2013)

[25] The Canadian Encyclopedia. Forage crops. http://www.thecanadianencyclopedia.ca/en/article/forage-crops/ (Accessed 14 October 2014)

[26] Bootsma A, Boisvert JB, Dumanski J. Climate-based estimates of forage yields in the prairie region of western Canada. Soil Use and Management 2007;11(2):55–62. DOI: 10.1111/j.1475-2743.1995.tb00497.x

[27] Vergé XPC, Dyer JA, Worth D, Smith WN, Desjardins RL, McConkey BG. A greenhouse gas and soil carbon model for estimating the carbon footprint of livestock production in Canada. Animals 2012;2:437-454. doi:10.3390/ani20x000x.

[28] Dyer JA, Vergé XPC, Desjardins RL, Kulshreshtha SN, MConkey BG. Areas and greenhouse gas emissions from feed crops not used in Canadian livestock production in 2001. Journal of Sustainable Agriculture 2011; 35(7):780-803. DOI: 10.1080/10440046.2011.606493.

[29] Vergé XPC, Dyer JA, Desjardins R.L, Worth D. Greenhouse gas emissions from the Canadian pork industry. Livestock Science 2009;121:92-101, doi:10.1016/j.livsci.2008.05.022.

[30] Vergé XPC, Dyer JA, Desjardins R.L, Worth D. Long Term trends in greenhouse gas emissions from the Canadian poultry industry. Journal of Applied Poultry Research 2009;18: 210–222. doi:10.3382/japr.2008-00091

[31] Intergovernmental Panel on Climate Change. IPCC. Guidelines for National Greenhouse Gas Inventories. 2006;Volume 4: Agriculture, Forestry and Other Land Use. Chapter 10: Emissions from Livestock and Manure Management. Available from http://www.ipcc-nggip.iges.or.jp/public/2006gl/index.htm. (Accessed 10 May 2013)

[32] Rochette P, Worth DE, Lemke RL, McConkey BG, Pennock DJ, Wagner-Riddle C, Desjardins RL. Estimation of N2O emissions from agricultural soils in Canada – De-

velopment of a country-specific methodology. Can. J. Soil Sci. (Special edition) 2008;88, 641-654.

[33] Dyer JA, Desjardins R.L. A review and evaluation of fossil energy and carbon dioxide emissions in Canadian agriculture. Journal of Sustainable Agriculture 2009; 33(2): 210-228. doi:10.1080/10440040802660137.

[34] Dyer JA, Desjardins RL, McConkey BG, Kulshreshtha S, Vergé XPC. Integration of farm fossil fuel use with local scale assessments of biofuel feedstock production in Canada. Chapter 4, In: Biofuel Economy, Environment and Sustainability, Editor: Zhen Fang. InTech Open Access Publisher. Rijeka, Croatia; 2013. ISBN 980-953-307-471-4. Pages 97-122. doi:10.5772/50478

[35] VandenBygaart AJ, McConkey BG, Angers DA, Smith W, de Gooijer H, Bentham M, Martin T. Soil carbon change factors for the Canadian agriculture national greenhouse gas inventory. Can. J. Soil Sci. 2008;88: 671-680. Doi: 10.4141/CJSS07015

[36] Dyer JA, Vergé XPC, Desjardins R.L, Worth DE, McConkey BG. The impact of increased biodiesel production on the greenhouse gas emissions from field crops in Canada. Energy for Sustainable Development 2010;14(2):73–82. doi: 10.1016/j.esd.2010.03.001.

[37] Tickell J, Tickell K. From the Fryer to the Fuel Tank—The Complete Guide to Using Vegetable Oil as an Alternative Fuel. 3rd edition. Ashland, OH: BookMasters; 2003. Available from http://www.pssurvival.com/ps/biodiesel/From_Fryer_To_Fuel_Tank_1999.pdf (Accessed 19 November 10 2014)

[38] Pouliot E, Garie C, Theriault M, Avezard C, Fortin J, Castonguay FW. Growth performance, carcass traits and meat quality of heavy lambs reared in a warm or cold environment during winter. Can. J. Anl. Sci. 2009;89(2): 229-239, doi: 10.4141/CJAS08101

[39] United States Department of Agriculture, National Nutrient Database for Standard Reference, Release 21 Protein (g) Content of Selected Foods per Common Measure, sorted alphabetically. 2009;25pp. Available from http://www.ars.usda.gov/Services/docs.htm?docid=8964 (Accessed 10 May 10 2013)

Chapter 18

Bioethanol and Biodiesel as Vehicular Fuels in Brazil — Assessment of Atmospheric Impacts from the Long Period of Biofuels Use

Thiago Nogueira, Denise de Sales Cordeiro,
Rodrigo Alejandro Abarza Muñoz,
Adalgiza Fornaro, Antonio H. Miguel and
Maria de Fatima Andrade

Additional information is available at the end of the chapter

http://dx.doi.org/10.5772/60944

1. Introduction

The research field on alternative fuels has been developed in order to reduce the current dependence on petroleum derivatives and to improve air quality. Since the transportation sector represents one of the largest contributions to greenhouse gas (GHG) and other gas pollutants inventories, increasing the use of renewable fuel potentially reduces GHG emissions, such as carbon dioxide (CO_2), which contribute to global warming and climate change [1, 2]. Brazil is an important scenario for the study of environmental impact from the large-scale use of biofuels because a significant fraction of the vehicular fuels is derived from biomass (ethanol and biodiesel). In Brazil, anhydrous ethanol (maximum water mass content 0.7%) is added as an anti-knock additive to regular gasoline at a concentration ranging from 20% to 25% (known as gasohol or Gasoline C, E20, or E25 in Europe). Hydrous ethanol (maximum water concentration of 7.4%) is also used as a fuel in light-duty vehicles (LDV). Additionally, Brazil has experienced the use of biodiesel as additive to regular diesel since 2008. The minimum biodiesel content was set to 2% in 2008 and this content has been increasing, reaching 10% in 2015.

1.1. Bioethanol and biodiesel: The main biofuels in Brazil

Ethanol biofuel has characteristics that favor cleaner combustion, therefore reducing the emission of air pollutants. Ethanol is produced from the fermentation of agricultural products such as sugar cane, corn, wheat, beet, and cassava. The vast majority of ethanol produced worldwide is from sugar cane, mainly in Brazil [3]. Biodiesel consists of long-chain fatty acids esters mixture, which is produced by transesterification reaction of triglycerides present in vegetable oils or animal fats with short-chain alcohols in the presence of a catalyst (normally a base such as sodium (NaOH) or potassium (KOH) hydroxide) [4]. Figure 1 shows transesterification reactions of triglycerides for biodiesel production.

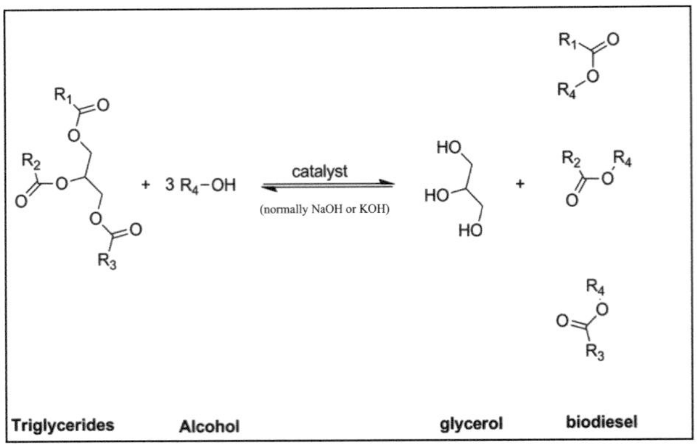

Figure 1. Transesterification reaction of triglycerides for biodiesel production.

Biodiesel use has been stimulated as an environmentally favorable alternative, since many studies have shown that biodiesel blends reduce particulate matter (PM), carbon monoxide (CO), and total unburned hydrocarbon (THC) emissions compared to diesel fuel [1, 5-8]. Biodiesel blends have been shown to reduce the overall life cycle emissions of CO_2, when evaluated using a total carbon life cycle analysis [1, 9], although this can depend on a variety of factors, such as land use change and transportation [10, 11]. A drawback in using biodiesel blends, however, is the potential to increase nitrogen oxide (NOx) emissions compared to ultra-low sulfur diesel fuel [5-7, 9, 12].

The Brazilian specification for biofuel is similar to the European and American, with some flexibility to meet the characteristics of domestic raw materials. The Brazilian National Agency of Petroleum, Natural Gas and Biofuels (ANP) is the governmental agency that regulates fuels sold in Brazil through resolutions and technical specifications. The ANP resolution Nº 42 [13] provides a biodiesel specification (B100) according to the provisions contained in the Technical Regulation No. 4/2004. ANP Resolution Nº7 provides a specification for bioethanol (anhydrous and hydrous ethanol) according to the provisions contained in the Technical Regulation Nº 3/2011. Acceptable values or ranges for physical and chemical properties of both Brazilian biofuels, the respective analytical methods, and the importance of each analysis can be found on specialized literature [4, 14-16].

Biofuel analysis should be performed on a representative sample obtained according to the same methods described on ABNT NBR 14883 - Petroleum and petroleum products - Manual sampling or ASTM D 4057 - Practice for Sampling Petroleum and Liquid Petroleum Products

http://dx.doi.org/10.5772/60944

(Practice for Manual Sampling of Petroleum and Petroleum Products) or ISO 5555 (Animal and vegetable fats and oils - Sampling). The monitoring of metals in bioethanol, such as iron and copper, is recommended by ANP as well as European and American norms, but not in biodiesel. Trace metals in biodiesel can catalyze oxidation processes which lead to the degradation of biodiesel and therefore this should be a parameter included in controlling parameters [15]. Inorganic anions such as chloride and sulfate in bioethanol can induce corrosion inside engine and storage tanks. Additionally, the monitoring of residual glycerol in biodiesel is related to the possible formation of acrolein, a pollutant known to cause severe adverse human health effects, from the burning of biodiesel containing glycerol. Glycerol is the main co-product formed during the production of biodiesel (Figure 1). Other impurities can be found in biodiesel such as residual alcohol, catalysts, and free fatty acids [17, 18].

Fuel quality, improvements in the technology applied to the engine, and the maintenance conditions are the main factors that influence the issuance of vehicle pollutants. To reduce emissions considerably, it is necessary to develop advanced combustion technologies and control devices, as well as environmentally clean fuels (with low pollution potential). Although many efforts have been done to develop environmentally friendly technologies and "clean fuels" to reduce environment pollution, vehicular emissions are still the main pollution source in many countries due to the increase in the urbanization and in the necessity of daily increase of displacements. Therefore, the development of public policies to reduce the contribution from this sector to air pollution becomes the priority for several urban areas. In the following, the historic overview of Brazilian biofuel uses and public programs of vehicular emissions control are addressed.

2. A historic overview of biofuels use in Brazil

In 1975, the Brazilian National Alcohol Program (Programa Nacional do Álcool - PROALCOOL) was created in Brazil to stimulate the use of domestic sources as a substitute for imported petroleum, and since then important changes occurred in the fuel composition used by motor vehicles in the country [19]. In 1979, 15% (v/v) of anhydrous ethanol started to be added to gasoline, reaching up to 22% (v/v) in the following years. The addition of 22% ethanol in gasoline was adopted by the Brazilian National Environment Council (CONAMA) [20] in 1990, on the recommendation of the energy sector. Production of ethanol-powered vehicles started in 1980.

The availability in the domestic market of hydrous ethanol and gasohol has brought benefits to the environment and to public health, highlighting the drastic reduction of toxic lead concentrations in the atmosphere. Ethanol acts as an anti-knock agent, replacing the lead-based additives, which have been completely removed from the national gasoline since 1991. In 1998, the Federal Government increased the content of anhydrous ethanol in gasoline to 24% through the Provisional Federal Law No. 1662-3 [21]. This increase did not result in significant changes in the vehicular emissions profile, since the vehicles manufactured in the country with more advanced technology, such as electronic fuel injection and oxygen sensors, were endowed with autocompensation system of the air/fuel ratio for variations of the ethanol content of that order. In 2006, the country's fuel blenders had to reduce ethanol content to 20% of its blended fuel because of ethanol shortages and returning to 25% soon after. In Brazil, gasoline volume sold was constant at around 2 × 10^7 m^3 between 1996 and 2009 and then increased to 4 × 10^7 m^3 in 2013 (Figure 2). During that same period, ethanol sold decreased from 1996 to 2003, restarted to increase, which peaked in 2009 (at 1.6 × 10^7 m^3) and decreased to 1.2 × 10^6 m^3 in 2013 [22].

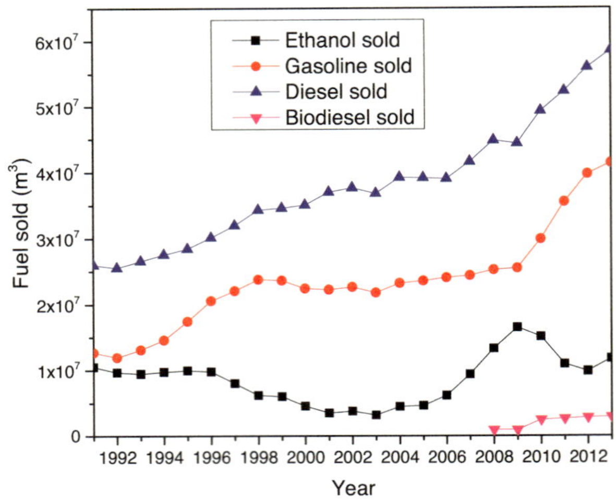

Figure 2. Fuel consumption in Brazil. Data from the Brazilian National Agency of Petroleum, Natural Gas and Biofuels [22]. The biodiesel consumption was calculated considering its minimum content in diesel: 2% in 2008-2009 and 5% 2010-2013.

During all periods shown in Figure 2, diesel sales increased 2.2 times, from 2.6 × 10^7 m^3 in 1991 to 5.8 × 10^7 m^3 in 2013. Biodiesel started to be sold in 2008, and since then sales increased from 0.9 × 10^6 m^3 to 2.9 × 10^6 m^3. In Brazil, the diesel composition has considerably changed in the last years. Diesel sold in Brazil, known as type B (containing biodiesel), the minimum biodiesel content was set to be 2% (v/v, B2) in 2008, 5% (v/v, B5) in 2010, 7% (v/v, B7) in 2014, and 10% (v/v, B10) after January, 2015. Details of the new regulations and for non-road applications are available at the ANP website [23].

The growth in the production of hydrous ethanol-fueled vehicles during the 1980s was very high due to the creation of the Brazilian National Alcohol Program, when about 94% of the new passenger vehicles were fueled by this biofuel, as can be seen in Figure 3. On the other hand, between the years 1989 and 1990, the surplus of gasoline associated with other factors resulted in a change from national alcohol overproduction to a deficit production situation [24]. In 1990, a combination of bad climatic conditions together with a rise in the international price of sugar forced the Brazilian Government to import alcohol (including methanol) for the first time to meet the national demand. Consumers had serious difficulties in getting alcohol to fill the tanks of their cars. In that year the production of ethanol-powered vehicles dropped from about 47% to about 11% of new vehicles produced. Due to shortage of anhydrous ethanol in Brazil, it was introduced, on an emergency basis, the mixture ethanol-gasoline-methanol

http://dx.doi.org/10.5772/60944

(7%-60%-33% by volume, respectively), for use in ethanol-powered vehicles. After 2003, flex fuel vehicles were introduced into the Brazilian market. These vehicles are designed to be fueled with gasohol, ethanol, or any blend of both. These vehicles had become attractive because their owners no longer had to be concerned with the price and market availability of ethanol. Increases in the pump price of ethanol lead to large-scale consumption of gasohol, and then of ethanol when ethanol prices subsequently fell to a competitive level. These vehicles types use special electronic sensors and an on-board computer to recognize the fuel composition and properly adjust the engine combustion parameters, without any interference from the driver. As shown in Figure 3, sales of these vehicles increased quickly in the Brazilian market. The vehicular industry in Brazil plays an important role on the economy, and has received governmental incentives. Currently, the government's role in ethanol fuels is much less and much different than in earlier times, and problems with the competition between the food and fuels sectors for sugarcane remain common.

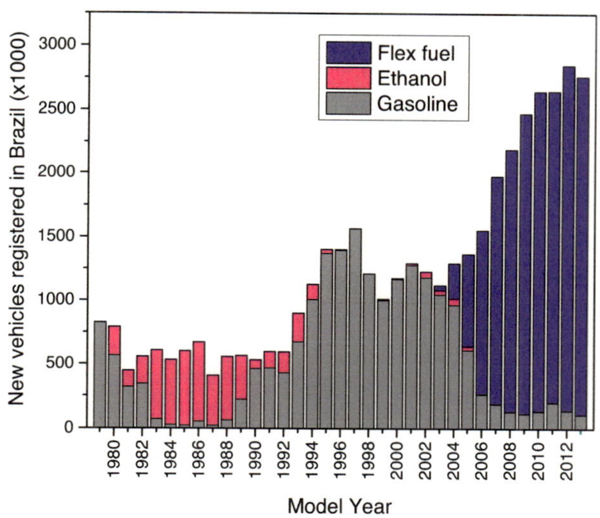

Figure 3. New light-duty vehicles registered in Brazil by fuel type. Data from the Associação Nacional dos Fabricantes de Veículos Automotores (ANFAVEA).

3. Brazilian vehicular emissions and control program of emissions

3.1. Light-duty vehicles emission controls

To control vehicular emissions, the Brazilian Government created in 1986 the Brazilian Motor Vehicle Air Pollution Control Program (PROCONVE) [20]. Its main goal was the reduction of atmospheric contamination by setting emission standards for vehicles, inducing technological improvements of manufacturing processes, enforcing those vehicles and engines to meet emission limits in standardized tests using a reference fuel. PROCONVE's target for controlling pollution from otto cycle LDV is based on the US Programs LEV (Low Emission Vehicle) and the California Air Resources Board (CARB) [25]. The compliance of requirements is assessed through protocols developed based on standardized dynamometer procedures and "reference fuels" trials. Furthermore, PROCONVE also imposes certification of prototypes and the statistical monitoring of vehicles in production (production trials). The PROCONVE program for LDV was performed at different phases that are described in Table 1.

Stage	Year of implementation	Description
Phase 1 (L-1)	1988-1991	Characterized by elimination of the most polluting vehicles and improvement of vehicles under production. The control of evaporative emissions also had begun at this stage. The main technological innovations that occurred in this phase were recycling of exhaust gases to control NOx emissions; secondary air injection in the exhaust manifold to control CO and HC; and optimization of the ignition stage.
Phase 2 (L-2)	1992-1996	Adaptation of catalysts and fuel injection systems for use with a mixture of ethanol was performed. The main innovations in vehicles were the electronic injection, electronically assisted carburetors, and catalytic converters. In 1994, the control of noise from vehicles started.
Phase 3 (L-3)	1997-2004	Given the requirement to meet the limits established from January 1, 1997 (CONAMA Resolution No. 15/1995), there were highly significant reductions compared to previous limits, and the manufacturer / importer employed the best available technology for the formation of mixing and electronic engine control, e.g., the oxygen sensor.
Phase 4 (L-4)	2005-2008	Referring to the CONAMA Resolution No. 315/2002, the priority at this phase that began in 2005 was the reduction of emissions of HCs and NOx (ozone precursor compounds). To attend this phase, engines with new technologies such as the optimized combustion chamber, increase in pressure of the injection pump, and the electronic fuel injection were developed.
Phase 5 (L-5)	2009-2013	The same emission limits of phase L-4 (CONAMA Resolution No. 315/2002) was applied at this phase. The priority in phase L-5 was to reduce emissions of HC and NOx. Analogous to the phase L-4, technological innovations were developed (optimization of combustion chamber, increase in pressure of the injection pump, and the improvement in the electronic fuel injection). At this phase there was a reduction of 31% in emissions of non-methane hydrocarbons (NMHC) for light vehicles. In addition, NOx emissions showed reduction of 48% and 42% for light vehicles and diesel-powered vehicles, respectively. Additionally, aldehyde emissions were reduced by approximately 67% for the light-duty vehicles.

Table 1. Phases during the implementation of Brazilian Motor Vehicle Air Pollution Control Program for light vehicles (PROCONVE-L) [20].

Figure 4 shows the evolution of the vehicular emission limits of CO, HC, NOx, and total aldehydes (RCHO) for LDV (Phases PROCONVE -L). As noted, the phases become increasingly restrictive. For instance, in PROCONVE (L-1) new vehicles had to emit no more than 24 g of CO per kilometer traveled; this value decreased to 12 g/km in phase 2 (L-2) and remained at 2 g/km between phases 3 and 5, a reduction of approximately 92% in the emission of this pollutant. Nitrogen oxides had a gradual reduction during PROCONVE phases. In phase 1 (L-1), vehicles could emit a maximum of 2 g/km. This value was reduced to 1.4 g/km at L-2 (reduction of 30%). In subsequent phases, the reductions ranged between 52% and 58%. In phase 5, the maximum emission allowed for NOx from new vehicles was 0.12 g/km. Volatile organic compounds such as HC and RCHO also had a very significant reduction in emissions. In phase 1, the maximum emission of HC was 2.1 g/km, whereas in phase 5 this value was 0.05 g/km, a reduction of 98%. Limits were imposed to RCHO only in phase 2, when the maximum emission allowed for RCHO was 0.15 g/km in phase 2 and decreased to 0.02 g/km in phase 5, a reduction of 87%.

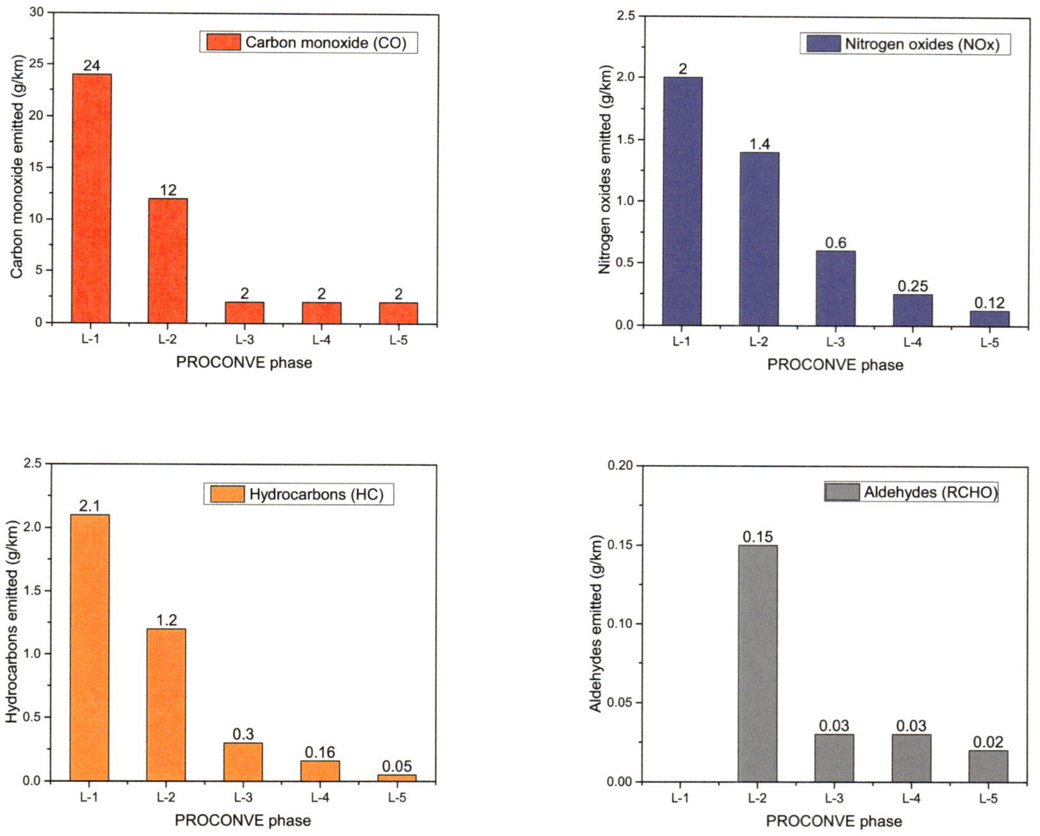

Figure 4. Evolution of the emission limits permit of CO, HC, NOx, and CHO for light vehicles during the phases of PROCONVE-L.

3.2. Heavy-duty vehicles emissions control

HDV have long been recognized as an important source of air pollution in urban areas [26, 27]. According to the São Paulo State Environmental Protection Agency [28], HDV are responsible for emitting into the atmosphere of Metropolitan Area of Sao Paulo 60.9% of all anthropogenic NOx, and 35.3% of all inhalable particulate matter (PM_{10}) during the 2012 year. HDV diesel-powered also contribute to emissions of fine particulate matter ($PM_{2.5}$), with black carbon (BC) as the main constituent. Emissions of CO and VOC, mainly aldehydes, also contribute to the air pollution from diesel exhaust [29, 30]. Efforts to reduce the emission of pollutants by HDV were delayed in Brazil. PROCONVE regulations for HDV began in 1990, but during phases 1 and 2 the limits for gaseous emission (P-1 phase) and PM (P-2 phase) were not legally required. The PROCONVE phases for controlling HDV emissions (Phases "P") are presented in Table 2 (CONAMA Resolution No. 18/86 [20]).

Stage	Year of implementation	Description
Phase P-1 and P-2	1990-1993	In 1990, engines with lower emission levels than that required in 1993 (the year that began the emission control for vehicles of this type with the introduction of phases P-1 and P-2) were produced. During this period, the limits for gaseous emission (P-1 phase) and particulate matter (P-2 phase) were not legally required.
Phase P-3	1994-1997	The development of new engine models aimed at the reduction of fuel consumption, to increase power and to reduce emissions of NOx by adopting the intercooler and turbo engines. This phase promoted a drastic reduction of CO (43%) and HC (50%).
Phase P-4	1998-2002	Emission limits have become even more restrictive than the P-3 phase.
Phase P-5	2003-2008	Aimed to reduce emissions of PM, NOx, and HC.
Phase P-6	2009-2011	In January 2009, the phase P-6 started, according to CONAMA Resolution No. 315/2002. The main objective of this phase, as well as the P-5 phase, was the reduction of emissions of particulate matter (PM), NOx, and HC.

Table 2. Phases for the implementation of Brazilian Motor Vehicle Air Pollution Control Program for heavy vehicles (PROCONVE-P) [20].

Figure 5 shows the evolution of the emission limits of CO, HC, NOx, and PM for HDV (Phases PROCONVE-P). As can be seen, the phases become increasingly restrictive. In phase P-1, HDV could emit a maximum of 14 g/km of CO, a value that decreased during the following phases, and in phase P-5 was 2.1 g/km, a reduction of approximately 85% in the emission of CO. Reduction of NOx started from 18 g/km in phase P-1 to 5 g/km in phase P-5 (reduction of 72%). The maximum emission of HC was 3.5 g/km in phase P-1, whereas in phase 5 this value decreased to 0.66 g/km, (reduction of 81%). Emissions of PM had regulatory limits, started only in phase P-2; in this phase, new HDV had to emit less than 0.6 g/km. This value decreased, and in the phase P-6, it was 0.02 g/km, which represents a reduction of 97% from the initial phase of PROCONVE for HDV.

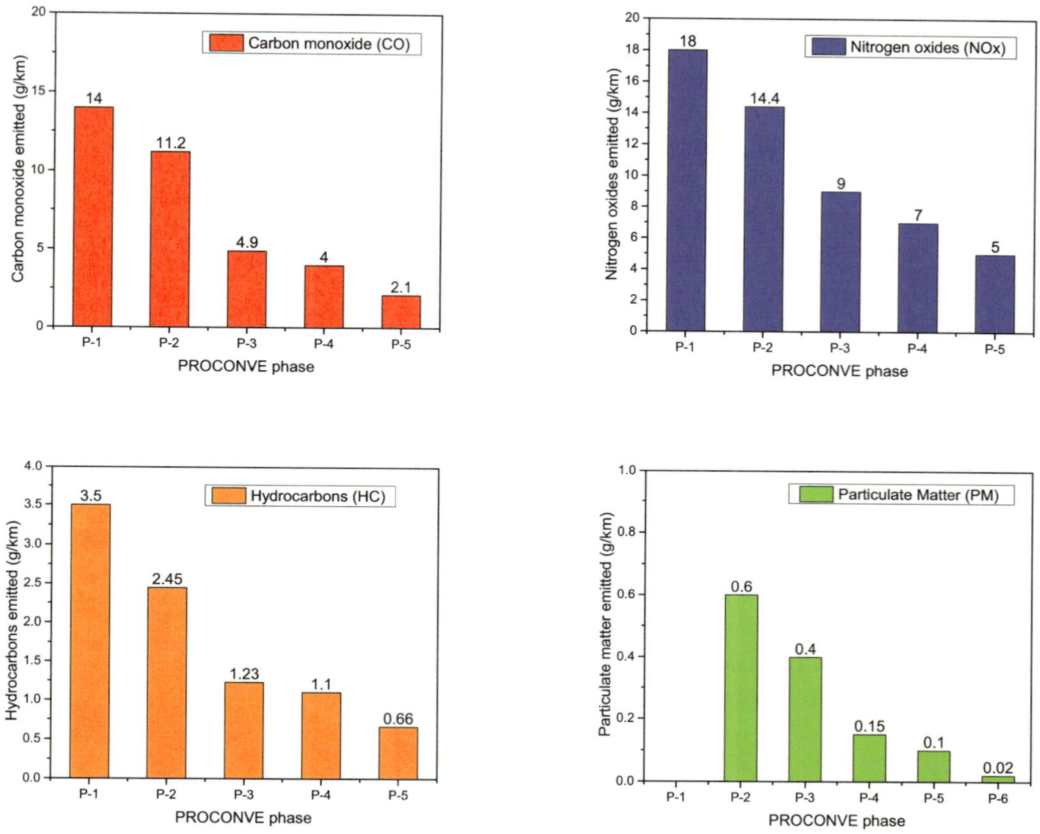

Figure 5. Evolution of the emission limits of CO, NOx, HC, and PM for HDV during the phases of PROCONVE-P.

Although established by the CONAMA Resolution 315/2002 [20], P-6 phase was not implemented on time due to delays in the specification of fuel (diesel) to be sold in Brazil. This fact caused delays in production of diesel with lower sulfur content and delays on the production of technological innovation engines of new vehicles. Reductions in the concentration of sulfur in diesel constituted a precondition for meeting the limits in P-6 phase, since the formation of sulfur compounds in combustion contributes to the so-called "poisoning" of the catalyst, not providing good operation, even in reducing emissions of NOx and HC.

In 2005, diesel S-2000 (2000 ppm of sulfur as a maximum limit) started to be sold in Brazil while S-500 (500 ppm of sulfur content in diesel) started to be sold only in the metropolitan areas. Reducing sulfur content in diesel from 13,000 ppm to 500 ppm provided a significant reduction of sulfur emissions in recent years. According to Resolution 315/2002 [20], which was not attended on schedule, after January 1, 2009, the P-6 phase should have started with S-500 diesel been distributed in Brazil and diesel S-50 in metropolitan areas. However, this phase was not implemented on schedule, so CONAMA established in November 2008, through Resolution No. 403 [20], a new phase (P-7) for HDV, which has stricter emission limits. Low sulfur content fuel legislation is being established in Brazil as a result of new legislation (P-7), which requires exhaust gas recirculation (EGR) or selective catalytic reduction (SCR) aftertreatment systems implementation in new HDV, starting in January 1, 2012, which is equivalent to meet the Euro V European emission standards for heavy-duty diesel engines. This phase (P-7), started on January 1, 2012, and enabled the availability and commercialization of a diesel fuel with content of 10 ppm sulfur (S-10). The automobile and fuel industries are required until 2016 to adapt to new technical standards, providing diesel engines and fuels to Brazilian markets following patterns already adopted in Europe, where diesel vehicles emit sulfur content up to 200 times smaller than is released by the Brazilian buses and trucks.

3.3. Emissions of gaseous pollutants from vehicles in Brazil

The effects of adding different blends of hydrous ethanol to a reference gasoline on flex fuel engine was evaluated in a study developed by de Melo et al. [31]. The results showed that, in general, CO emissions were reduced with hydrous ethanol addition due to the higher oxygen content of the ethanol contributing to oxidation into CO_2. Total hydrocarbons emissions were also reduced, while aldehydes and unburned ethanol increased with hydrous ethanol addition. Emissions of NOx presented a complex behavior, without a particular defined trend. At 3875 rpm, knocking occurrence limited spark timing advance leading to lower NOx emissions when using gasoline E25 (without hydrous addition) and fuel blend content of 30% hydrous ethanol. With hydrous ethanol addition, there was a trend of NOx reduction at lower speeds (1500 and 2250 rpm), while for high speed (4500 rpm), there was a trend of NOx increase.

A comparative study was carried out of pollutant emissions produced by a mid-size sedan powered by 1.4-L spark ignition engine on a chassis dynamometer operating with three different fuels. Commercial gasoline with 22% of ethanol (E22, gasohol), compressed natural gas (CNG), and hydrous ethanol showed that in the cold start tests the E22 produced the lowest CO and HC emissions, while CNG produced the lowest NOx emissions [32]. Considering the full test cycle, CNG emitted the lowest CO, NOx, and CO_2 concentrations, and the lowest fuel consumption. Gasohol (E22) presented the lowest emission levels of HC and CH_4. Hydrous ethanol showed the highest fuel consumption and higher pollutant emission levels than the other fuels, except for CO_2, which was higher than CNG and lower than gasohol.

Randazzo et al. [33] investigated the effects of use of diesel/biodiesel blends with concentrations of 3% (B3), 5% (B5), 10% (B10), and 20% (B20) on a passenger vehicle exhaust emissions, and the results showed that increasing biodiesel concentration in the fuel blend increases CO_2 and NOx emissions, while CO, HC, and PM emissions were reduced. Additionally, the work evaluated the effects of anhydrous ethanol as an additive to B20 fuel blend with concentrations of 2% (B20E2) and 5% (B20E5). The results showed that the addition of anhydrous ethanol to B20 fuel blend can be a strategy to control NOx exhaust. However, the authors concluded that it may require fuel injection modifications, since in this condition increases in CO, HC, and PM emissions were observed.

Pérez-Martinez et al. [34] showed that emissions factors for on-road LDV in the MASP in 2011 were 5.8 and 0.3 g/km, for CO and NOx, respectively. The values estimated in this study showed a significant reduction when comparing the values of emissions factors calculated in the experiment conducted in 2004 [35]. The values estimated in 2004 were 14.6 and 1.6 g/km, for CO and NOx, respectively. The reduction ratio was 2.5 times for CO and 3.2 for NOx. The authors attributed this fact to increased use of modern three-way catalysts using platinum and rhodium surfaces, which changes the nitrogen oxides back to nitrogen and elemental oxygen and complete the oxidation of CO to CO_2 [36]. In addition, there was the increased number of vehicles able to burn ethanol. In 2011, from the cars running in the MASP, about 55 % gasohol, 4 % burned hydrous ethanol, and 38 % were capable of burning both gasohol and hydrous ethanol (flex-fuel vehicles), while in 2004 about 69.5% of vehicles burned gasohol and 14.5% of the fleet burned hydrous ethanol. Regarding HDV, in 2011 the emissions factors estimated by Pérez-Martinez et al. [34] were 3.6 and 9.2 g/km, for CO and NOx, respectively, which represents a reduction of 5.7 and 2.4, for CO and NOx, in comparison with the data obtained for HDV running in 2004 in the MASP. In 2011, HDV burned a mixture containing 5% biodiesel, which may contributed to reduction in CO emissions [37], while minor decrease in emissions of NOx pollutant reported in the study may be associated with the use of biodiesel.

Guarieiro et al. [38] evaluated the exhaust emissions of a diesel engine running on biodiesel and operated by a steady-state dynamometer which provided a composition profile of the carbonyl compound emissions from exhaust of pure diesel (B0), pure biodiesel (B100), and biodiesel-diesel mixtures (B2, B5, B10, B20, B50, B75). This work showed that the mean concentration sum of total carbonyl compounds emission were 20.5 ppmv for B0 and 15.7 ppmv for B100, while for fuel blends the total concentration of carbonyl compounds were 21.4, 22.5, 20.4, 14.2, 11.4, and 14.7 ppmv, respectively, for B2, B5, B10, B20, B50, and B75. The study showed that major contributors to the total carbonyl compounds were formaldehyde and acetaldehyde, and that except for formaldehyde and acrolein, all carbonyl compounds showed a clear trend of reduction in the emissions from B2 to B100 (40% reduction, on average). The lowest total carbonyl emission factors were found when B50 was used, 2271 pg/g of fuel burned, while the individual emission factors determined (pg/g of fuel burned) were 539.7 (formaldehyde), 1411 (acetaldehyde), 30.83 (acrolein), and 310.7 (propionaldehyde).

Carbonyl compounds were measured in vapor-phase samples in a study developed in a Bus Station impacted by HDV fuelled with diesel/biodiesel fuel blend (B5) in Salvador, Bahia, Brazil, in 2012 [39]. Among them, formaldehyde, acetaldehyde, and propanone were the major quantified compounds, ranging from 28.4 to 287.3 ppbv (formaldehyde), 24.9 to 171.3 ppbv (acetaldehyde), and 5.8 to 72.3 (propanone). The data obtained in this site were compared to formaldehyde and acetaldehyde concentrations found in other sites impacted by HDV fuelled with pure diesel and diesel/biodiesel blends. The authors reported that the addition of concentrations higher than 3% of biodiesel to diesel showed an improvement in the carbonyl concentration profile at these places with high flow of HDV, and that the use of these biofuels revealed profiles similar to those found for sites less impacted by these vehicles. Higher concentrations of these carbonyls were observed in the current study compared to the results obtained in the same bus station in 1997 [40]. The average formaldehyde and acetaldehyde concentrations increased approximately 1.5 and 7 times, respectively, between the two observation periods. The authors suggested that these significant differences found between the two sampling periods can be explained by the increase of the vehicle fleet currently circulating in the Lapa Bus Station. While in 1997 the vehicular fleet at the Lapa bus station was 150 buses per hour, in 2010 it was over 330 buses per hour. In this manner, an increase of the vehicle fleet circulating in Lapa station can reflect a rise of both formaldehyde and acetaldehyde emission concentrations at that site. Other factors suggested by the authors was that the reason for the observed increase of carbonyl compounds concentration at this site was related to changes of the fuels used in vehicles. In 1997, buses were fuelled with pure diesel and in 2012 they were fuelled with B5 fuel blend. The results of an experimental campaign at a central bus station in Londrina, Brazil, where only pure diesel powered vehicles circulated showed that the formaldehyde concentrations were significantly higher than acetaldehyde [41]. The formaldehyde levels ranged from 6.17 to 10.43 ppbv with an average of 7.94 ppbv, while acetaldehyde presented levels in the interval of 0.49 to 2.12 ppbv with an average of 1.26 ppbv. The formaldehyde concentration was six times higher than the acetaldehyde concentration.

Nogueira et al. [42] determined on-road emissions of carbonyls from Brazil's current vehicle fleet based on two experimental campaigns conducted in traffic tunnels located in the MASP. Among carbonyl species, formaldehyde and acetaldehyde were the most abundant compounds found during all sampling time intervals. The higher carbonyl emissions were associated with HDV, which were fueled with a blend of regular diesel and 5% biodiesel from soy. LDV were responsible for high emissions of acetaldehyde, since in Brazil this type of vehicle burns a mixture of 75% (v/v) gasoline and 25% ethanol (gasohol), or hydrous ethanol. Brazilian LDVs reported emission of 5.7 mg/km and 7.4 mg/km for formaldehyde and acetaldehyde, respectively. When compared with data from LDV in the California, these values are 352% and 263% higher than that emitted by vehicles running on E10 (gasoline with 10% ethanol). The HDV average emission factor for formaldehyde and acetaldehyde was 28 and 20 mg/km, respectively, indicating that there was a reduction in the HDV emission of formaldehyde (42%) and acetaldehyde (58%) when comparing with the values obtained in an experiment conducted in the MASP in 2004 [35]. HDV running in 2011 in the MASP with diesel + 5% biodiesel showed formaldehyde emissions 33% higher than HDV running in California in 2010 with regular diesel [43].

Figure 6 shows the aldehyde emissions in milligrams per kilometer traveled by new vehicles sold in Brazil, together with the limits specified in each phase of the PROCONVE. As it can be seen, significant reductions in emissions occurred due to the evolution of pollution control public policies. Prior to 1992 (before and during phase 1 of the PROCONVE), no limits existed regarding aldehyde emissions. During that period, vehicles running on ethanol emitted about 0.13 g of total aldehyde (RCHO) per kilometer, about three times higher than vehicles running on gasohol. In phase 2 of the PROCONVE, which started in 1992, when the emission limit was initially 0.15 g of aldehyde per km traveled but later reduced to 0.03 g of aldehyde per km

http://dx.doi.org/10.5772/60944

traveled by 1997 (the beginning of Phase 3). As observed, vehicles running on ethanol emit more total aldehydes than do vehicles running on gasohol and the quantity of aldehydes emitted by ethanol-powered vehicles is similar to that emitted by flex-fuel vehicles running on ethanol. However, the production of vehicles powered by ethanol only was discontinued in 2006, and there are therefore no longer any records of emissions for new vehicles. It is also noteworthy that flex-fuel vehicles manufactured in 2010 emitted an average of 0.007 g of aldehyde per km when new and using ethanol, which was about 65% less than their 2003 counterparts under the same conditions. The PROCONVE is currently in phase 6, and the limit for aldehyde emissions is 0.02 g per km traveled.

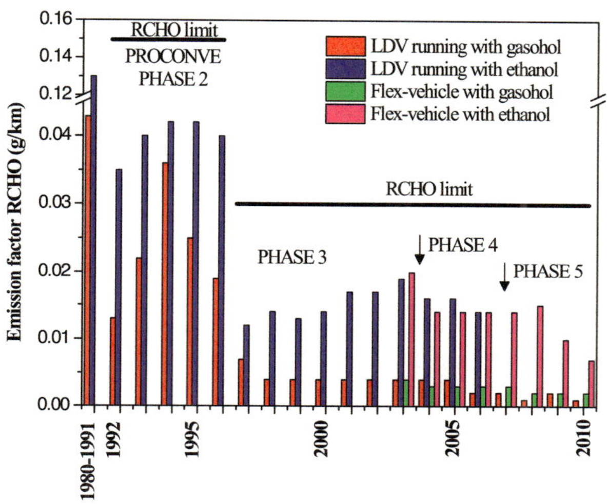

Figure 6. Total aldehydes (RCHO) emission factors from 1980 to 2010 for Brazilian LDV running with gasohol and ethanol fuels, and the phases of the PROCONVE.

3.4. Emissions of particulate matter from vehicles in Brazil

Atmospheric aerosol or particulate matter (PM) is a complex mixture of extremely small particles and liquid droplets, presenting also a complex chemical composition [44, 45]. The atmospheric PM is a stable suspension of liquid or solid particles with an aerodynamic diameter lower than 100 μm. Actually, the classification of PM size is very important due to its effects on human health and/or its direct and indirect effects in the environment and climate. The concentrations, size, and number distributions of particles are affected by physical and chemical processes or kinds of sources (natural or anthropogenic, and secondary atmospheric aerosol production). The inhalable particles have aerodynamic diameter equal or lower than 10 μm (PM_{10}), being divided into fine ($PM_{2.5}$) and coarse ($PM_{2.5-10}$) particles [44, 45]. It is important to emphasize that to control both the PM emissions by burning fuels and air quality, the legislation considers the mass concentration (μg/m^3) for a given particle size range ($PM_{2.5}$ and PM_{10}), without any relationship with the chemical composition. This parameter is valid for the national air quality standards, and also for the WHO guidelines [46]. For PM, even though the mass concentration, the most important parameter in legislation, the chemical composition measurements can contribute to better knowledge not only to the emission processes but also to the effects on health and environment. The PM is a complex mix of different elements and compounds both inorganic and organic/carbonaceous, which includes the ions (sulfate, nitrate, chloride, ammonium, sodium, potassium, calcium, and magnesium), the trace elements and/or metals (Pb, Zn, Ni, Cu, V, Cr, Cd, Al, P, S, Si, Ti, Ca, Fe, and others), elemental carbon (and black carbon) and organic compounds from simple hydrocarbons to oxygenated, aromatics, and polycyclic aromatic hydrocarbons. The fraction of these different compounds in the PM mass is due to sources and/or of the physical-chemical processes during the particles' life cycles [44, 45]. The heavy metals in the form of free elements are non-toxic. However, they are dangerous in their cationic form and when bound to short carbon chains. The metal ions form complexes with a large amount of binders and influence the biological functions, affecting the normal development of living beings' tissues and their adequate functioning [47]. Heavy metals, also referred to as trace elements, are an important fraction of the PM because they represent a risk to human health. Trace metals in airborne PM were considered to represent a health hazard since they can be absorbed into human lung tissues during breathing. Although many of these metals are constituents of tissues, their toxic effects are known even at low levels. Inorganic ions, as well as other PM constituents, result from fuel burning. In addition, these pollutants affect climate change, contribute to the acidification of aerosol, changing its conductivity [48]. Therefore, it is extremely important to have a qualitative and quantitative understanding of the PM composition and the contribution of fuel burning as well as the role of biofuel addition.

The speciation of organic compounds is still a major challenge, being an important advancement the differentiation between organic carbon and elemental carbon (or graphitic). Special attention is necessary for black carbon (BC) and elemental carbon (EC) measurements and reports [49]. BC is an important component of the PM, produced by incomplete combustion processes (fossil fuels, biofuels, and biomass), having a graphite-like microstructure, being strongly absorbent of visible light, refractory and insoluble in water and in organic solvents including methanol and acetone [49]. As BC is the most strongly light absorbing component of PM, it absorbs both incoming and outgoing radiation of all wavelengths, which contributes to warming of the atmosphere and darkening of the surface. It is, therefore, a pollutant that has important climatic influences [50].

One of the most important groups of organic compounds is the polycyclic aromatic hydrocarbons (PAH) that are present in the gas- and particle-phase in ambient air as well as other environmental compartments. These organic compounds have known mutagenic and carcinogenic properties. Due to this harmful effect of some PAH to humans, the US EPA shows them in its priority pollutants list, being some PAH structures shown in Figure 7. The lipophilicity, environmental persistence, and genotoxicity increase with heavier molecular weight from 4 to 6 aromatic rings [51-54]. PAHs are formed by the pyrolysis and incomplete combustion of organic compounds [55, 56]. A major source of this class of pollutants in large urbanized centers is the vehicular emission. Fuel burning produces, among other pollutants, measurable concentrations of PAH, depending on the characteristics of the fuel.

http://dx.doi.org/10.5772/60944

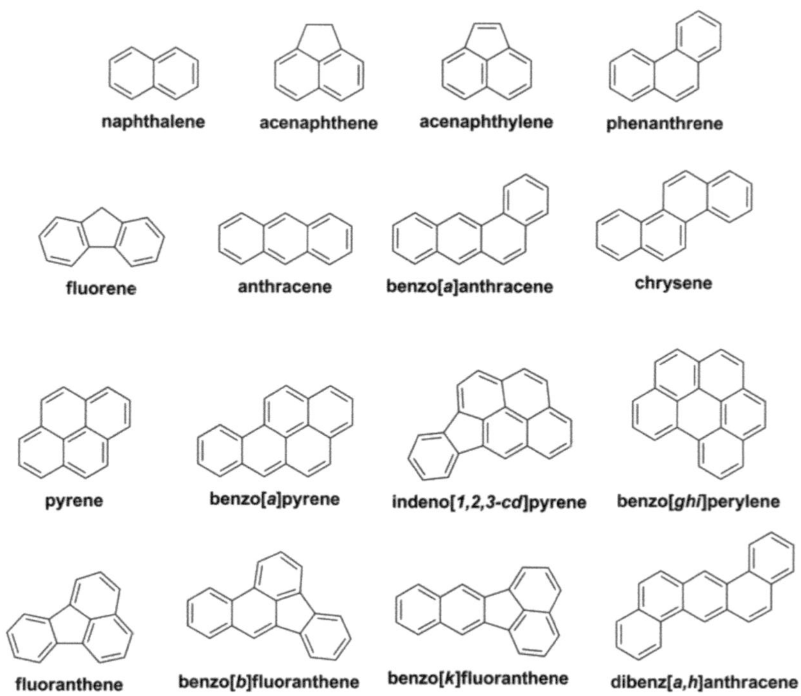

Figure 7. Structures and nomenclatures of the 16 PAHs on the EPA priority pollutant list.

The analytical techniques used for determination and quantification of the PM components, not only for emissions but also for atmospheric measurements, are shown in Table 3. These techniques include continuous aerosol sampling and analysis systems or analysis methods that are commonly applied to aerosol filter samples. The details of the analytical methods, advantages, and disadvantages are available in an important review about PM sampling and measurements [57]. Another important discussion involves the question of the appropriateness of sampling and analysis methods of PM for the purpose of adapting the requirements needed to establish emission and air quality standards. This discussion involves size-selective inlets (cyclonic flow, impactor, virtual impactor, and selective filtration), if the sampling is continuous or integrated in time (24 h standard), filters characteristics for only mass quantification or also for chemical composition (membranes of Teflon, polycarbonate, glass fiber, cellulose esters, or silver), plus the details regarding the flow measurement, flow control, and flow movers [58]. In the past decades, new techniques development enabled measurements of aerosol integral properties (total number concentration, cloud condensation nucleus concentration, optical coefficients, etc.), aerosol physical chemical properties (density, refractive index, equilibrium water content, etc.), measurements of aerosol size distributions, and measurements of size-resolved aerosol composition [59].

Emission quantifications (mass per kilometer traveled or mass per fuel mass burned) are calculated based on dynamometers (vehicles or motors) or tunnels experiments, being that the former corresponds an individual emission, while the latter represents the real fleet. Evaluation of influence of direct emissions on air pollutants concentrations from fuel burning by vehicles can be happened by experiments in special sites, like bus stations, close streets or avenues with heavy traffic, etc.

Constituents of PM	Usual analytical techniques
Mass	Gravimetric (filter weighing); beta attenuation monitor (BAM); tapered element oscillation microbalance (TEOM); quartz crystal microbalance (QCM)
Black carbon	Optical reflectometer (also called spotmeter and filter smokemeter); aethalometer; multi-angle absorption photometer (MAAP); photoacoustic soot sensor (PASS)
Total carbon: organic vs. elemental (graphitic)	Thermal/optical reflectance (TOR) or thermal/optical transmittance (TOT)
Elements and/or metals (Pb, Zn, Ni, Cu, V, Cr, Cd, Al, P, S, Si, Ti, Ca, Fe, Mg, Mn, K, As, Sb, Se, Co, In, Eu)	X-ray fluorescence analysis (XRF); particle induced X-ray emission (PIXE); atomic absorption spectrometry (AA); emission spectrometry (EE); inductively coupled plasma-mass spectroscopy (ICP-MS); instrumental neutron activation analysis (INAA) anodic stripping voltammetry (ASV)
Inorganic ions (Cl_-, NO_3^-, SO_42^-, Na^+, K^+, Ca^{2+}, Mg^{2+}, NH_4^+) and/or carboxylate anions	Ionic chromatography (IC)
Speciation of organics	Solvent extraction by Soxhlet techniques / cleanup by solid-phase extraction (SPE) follow analysis by high-performance liquid chromatography (HPLC) with various detectors (absorption, fluorescence, and mass spectrometry) / gas chromatography flame ionization detector (CGFID) / gas chromatography mass spectrometry (CGMS)

Table 3. Some analytical techniques for analysis of the main constituents of particle matter.

Tunnel experiments were performed in MASP in different times and even sampling in two tunnels, one with predominance of light-duty vehicles (LDV) that burn gasohol (75% gasoline + 25% anhydrous ethanol) and ethanol, and one that had LDV plus significant number of heavy-duty vehicles (HDV) that burn diesel and/or diesel + biodiesel [60]. Experiments during 2011 showed that LDV and HDV running in the Metropolitan Area of São Paulo has emitted less $PM_{2.5}$ in recent years than in the past [34]. In 2004 [60], LDV emitted 92 mg of particles ($PM_{2.5}$) per kilometer traveled, while in 2011 [34] this value decreased to 20 mg/km, a reduction of 4.6 times. Emissions from HDV decreased from 588 mg/km in 2004 to 277 mg/km in 2011, a reduction about 2 times. These data showed that LDV have a greater reduction in PM emissions, since in this period of study there was an improvement in the engine technology of LDV. Furthermore, in this period there was the introduction of flex fuel vehicles, and consequently the consumption of ethanol by vehicles has increase. On the other hand, HDV showed no major advances in engine technology, occurred only improves in the composition of diesel oil, for example, a reduction in sulfur content, and the addition of biodiesel to diesel (in 2011 vehicles circulated with 5% biodiesel).

Guarieiro et al. [61] evaluated the emission effects of biodiesel addition of 4% of soy biodiesel (B4), a biodiesel blend of 25% and 50% (B25 and B50) in diesel, and also pure biodiesel (B100). In this study, PM emissions were measured on a steady-state dynamometer in a test using a diesel engine at low load. Size-fractionated PM samples were collected using the NanoMOUDI impactor and analyzed for the 16 priority PAHs. In addition, $PM_{2.5}$ samples were collected and analyzed for redox activity by DTT assay. PM was distributed in all sizes, while PAH size distributions were found at higher levels in the accumulation mode (30 nm < Dp < 2.5 μm). Total PAH emission factors for B4, B25, B50, and B100 were 237, 111, 182, and 319 ng/kg fuel burned, respectively. Individual PAH emission factors showed that PAH containing four or more rings (MW > 202) such as benzo(b)fluoranthene, benzo(a)anthracene, pyrene, and benzo[ghi]perilene were the main PAH emitted by the four studied fuels. The study showed that percentage reductions of individual PAH emission factors for the blend fuels in comparison with B4 were 37% and 22% for B25 and B50, respectively, and an increase around 31% for B100. This work also showed an increase in redox activity for B25, B50, and B100 when compared to B4. The results from this study suggest that emissions from pure waste cooking biodiesel may not be the better fuel choice in terms of PM, PAH, BaPE (corresponding carcinogenicity index) particle size distribution and emission factors as well as redox activity. However, B25 and B50 blends presented some improvements in terms of PM, PAH, BaPE size distribution, and redox activity of engine exhaust in comparison to B4. The results suggest that addition of low percentages biodiesel to diesel promotes benefits in both environmental and human health concerns [61].

Exhaust emissions of 17 polychlorinated dibenzo-p-dioxins and polychlorinated dibenzofurans (PCDD/Fs) were compared from LDV fueled with gasohol and fueled with hydrous ethanol [62]. In this study were evaluated the influence of fuel type and quality, lubricant oil type, and use of fuel additives on the formation of these compounds. The results showed that emission factors of PCDD/Fs for the gasohol vehicle varied from undetected values to 0.068 pg international toxic equivalency/km traveled, whereas in the ethanol vehicle the observed variation ranged between 0.004 to 0.157 pg international toxic equivalency/km. The study also showed that use of fuel additive significantly diminished the emission of octachlorodibenzo-p-dioxin, in the gasohol-powered vehicle, whereas in the ethanol vehicle no significant associations were observed between the investigated variables and the emissions. The results of another study showed that vehicles running on gasohol emitted more of total PAH; the emission factors varied from 41.9 μg/km to 612 μg/km than vehicles running on ethanol from 11.7 μg/km to 27.4 μg/km [63]. The authors reported that in terms of benzo(a)pyrene toxicity equivalence, emission factors varied from 0.00984 μg TEQ/km to 4.61 μg TEQ/km for the gasohol vehicle and from 0.0117 μg TEQ/km to 0.0218 μg TEQ/km in the ethanol vehicle. The results also showed that the use of fuel additive for the gasohol vehicles cause a significant increase in the emission of naphthalene and phenanthrene, while the use of synthetic oil, instead of mineral oil, also contributed significantly to a decrease in the emission of naphthalene and fluorene. Regarding hydrous ethanol-powered vehicle, the same compounds were tested and showed no statistically significant influence on PAH emissions. Emissions of PAH from LDV running with ethanol or gasohol are lower than vehicles running with diesel, being that the diesel-powered vehicles showed emission about 200 times higher than vehicles running with ethanol [64]. Total PAH values from diesels ranged from 1.133 to 5.801 mg/km. Naphthalene, phenanthrene, fluoranthene, pyrene, and chrysene were detected in all test samples. Another study developed by da Silva et al. [65] evaluated the composition of inhalable particles and their trace metal content in LDV fueled with ethanol and gasohol. The results showed that the total emission factors ranged from 2.5 to 11.8 mg/km in the gasohol vehicle, and from 1.2 to 3 mg/km in the ethanol vehicle. The majority of particles emitted were in the fine fraction ($PM_{2.5}$), in which Al, Si, Ca, and Fe corresponded to 80% of the total weight. The results also showed that PM_{10} emissions from the ethanol vehicle were about threefold lower than those of gasohol.

Correa and Arbilla [66] evaluated the emissions of mono- and polycyclic aromatic hydrocarbons (MAHs and PAHs, respectively) from a six cylinder heavy-duty engine, typical of the urban buses of Brazilian fleet, fueled with pure diesel (D) and biodiesel blends (v/v) of 2% (B2), 5% (B5), and 20% (B20). The results showed the following average reduction of MAHs: 4.2% for B2 blend, 8.2% for B5 blend, and 21.1% for B20 blend. The average reductions for PAHs were 2.7% (B2), 6.3% (B5), and 17.2% (B20). However, some PAHs and MAHs emissions increased due to the biodiesel blends like phenanthrene, ethyl benzene, and trimethyl benzenes.

In order to characterize current concentrations of $PM_{2.5}$ from LDV and HDV in the MASP, Brazil, measurements of physical and chemical properties of aerosol were undertaken inside two tunnels located in the MASP in 2011 [67]. The two tunnels showed very distinct fleet profiles: in the Janio Quadros (JQ) tunnel, the vast majority of the circulating fleet are LDV, fuelled on average with the same amount of ethanol as gasoline. In the Rodoanel tunnel (RA), PM emission is dominated by HDV fuelled with diesel + 5% biodiesel blend. The study shows that in the JQ tunnel, $PM_{2.5}$ concentrations were on average, 52 μg/m^3, with the largest contribution from organic mass (42 %), followed by elemental carbon (17 %) and crustal elements (13 %). While in the RA tunnel, $PM_{2.5}$ was on average 233 μg/m^3, mostly composed of elemental carbon (52 %) and organic mass (39 %). The work showed that average organic mass:elemental carbon ratio in the JQ tunnel was 1.59, indicating an important contribution of elemental carbon despite the high ethanol fraction in the fuel composition. The study also shows that the sum of the PAHs concentration was 56 ± 5 ng/m^3 and 45 ± 9 ng/m^3 in the RA and JQ tunnel, respectively. In the JQ tunnel, benzo(a) pyrene (BaP) ranged from 0.9 to 6.7 ng/m^3 (0.02-0. 1 parts per thousand of $PM_{2.5}$) whereas in the RA tunnel BaP ranged from 0.9 to 4.9 ng/m^3 (0.004-0.02 parts per thousand of $PM_{2.5}$), indicating an important relative contribution of LDV emission to environmental BaP concentration.

The concentrations of $PM_{2.5}$ and PM_{10}, including their ionic composition was evaluable at an underground bus terminal, being that buses burning fuel blend of 95% diesel + 5% biodiesel (B5) in Salvador, Bahia, Brazil, in 2010 [68]. The results showed that the mean mass concentrations of $PM_{2.5}$ were 201 μg/m^3 during the daytime (8 am-7 pm), while the PM_{10} were 309 μg/m^3, during the same period of the day. Three times lower PM_{10} concentration (110 μg/m^3) was obtained from an experiment conducted at the same place in 2005, before addition of biodiesel to diesel [69]. The mean concentrations for the total carboxylate anions during the day were 139 and 180 ng/m^3 for $PM_{2.5}$ and PM_{10}, respectively. Results showed that monocarboxylate anions (propionate, acetate, and formate) were the most abundant, followed by ketocarboxylate anions (pyruvate), while the dicarboxylate anions (oxalate and malonate) were the least abundant. The ion Mg^{2+} (0.43 μg/m^3) was the most important cation and NO_3^- (0.083 μg/m^3) the main anionic species in $PM_{2.5}$, while Na^+ (0.60 μg/m^3) and SO_4^{2-} (0.62 μg/m^3) were the most abundant in [68].

The PM_{10} concentration (24 h) varied from 38.8 to 92.2 μg/m^3 on a bus station, for buses running with diesel/biodiesel fuel blend (B3) in the city of Londrina, Paraná, Brazil, during July 2008 [70]. Nitrate (8 μg/m^3), sulfate (1.4 μg/m^3), and ammonium (0.3 μg/m^3) presented the highest concentration levels, suggesting that biodiesel may also be a significant source for these ions, especially nitrate. The authors also demonstrate that a higher fraction of PAH particles was found in particles with diameters smaller than 0.25 μm in the Londrina bus station, and that the fine and ultrafine particles were dominant among the PM evaluated, suggesting that biodiesel decreases the total PAH emission. However, the authors suggested that use of biodiesel increased the fraction of fine and ultrafine particles when compared to diesel emissions by heavy-duty vehicles obtained in a 2002 campaign [71].

4. Vehicular emissions and the impacts on air quality

Vehicular emissions contain lead to a generation of various toxic substances such as carbon monoxide (CO), nitrogen oxides ($NOx = NO + NO_2$), hydrocarbons (HC), sulfur oxides (SOx), and particulate matter (PM) which, when in contact with the respiratory system, can produce several harmful health effects. The effects are related to respiratory diseases, increased incidence of cancer, cardiovascular diseases, neurological problems, and reproductive problems [72]. Biofuels have met a niche in the growing market as a consequence of economic policies (substitution of fossil fuels) or efforts to reduce air pollution caused by vehicular emissions. In Brazil, the CONAMA Resolution Nº 3 by 28/06/1990 determines the standards for air quality and suitable measurement methods. This resolution establishes standards for PM, smoke, inhalable PM, SO_2, CO, O_3, NO_2, and several sampling methods and analysis of pollutants. Table 4 shows the maximum and the desirable levels of standard concentration of air pollutants, which aims to protect people's health and the environment.

Despite the guidelines of the World Health Organization [46] for PM_{10} and $PM_{2.5}$, there is only national air quality standard for PM_{10} in Brazil (Table 4), being the Brazilian values higher than the WHO guidelines (50 μg/m^3 24 h mean and 20 μg/m^3 annual arithmetic mean for PM_{10}; and 25 μg/m^3 24 h mean and 10 μg/m^3 annual arithmetic mean for $PM_{2.5}$). The monitoring of air quality in Brazil is the responsibility of the States. However, few states have established air quality network, and only eight metropolitan areas (States' capitals) have been continuously monitored: Distrito Federal, Vitória, Belo Horizonte, Rio de Janeiro, São Paulo, Porto Alegre, Curitiba, and Salvador. In Rio de Janeiro and São Paulo States, other cities have also been monitored [73].

Pollutant	Averaging Time	Maximum Level (µg/m³)	Desirable Level (µg/m³)	Method of measuring
Total suspendend Particles	24 hours	240	150	High volume air sample / gravimetric
	Annual geometric	80	60	
Inhalable particles	24 hours	150	150	Gravimetric / inertial separation
	Annual arithmetic	50	50	
Smoke	24 hours	150	100	Reflectance
	Annual arithmetic	60	40	
SO_2	24 hours	365	100	Fluorescence pulse / pararosaniline method / ion chromatography
	Annual arithmetic	80	40	
NO_2	1 hour	320	190	Chemiluminescence
	Annual arithmetic	100	100	
CO	1 hour	40.000	40.000	Nondispersive infrared sensor
	8 hours	10.000	10.000	
O_3	1 hours	160	160	Ultraviolet

Table 4. National standards for air quality [99].

Thus, MASP quality monitoring data are more complete (in time and space), enabling the evaluation of air quality trends as a result of actions to control pollutants emissions. Almost 30 years after the creation of PROCONVE, results show that the implemented strategy is correct, and the adoption of increasingly restrictive phases was successful. As observed in Figure 8, atmospheric concentration data show a clear downward tendency for CO, a recognized pollutant emitted from incomplete fuel combustion, in the MASP over the past years. These data show that LDVs emitted 20 g/km on average in the 1980s and 0.3 g/km in 2010. Meanwhile, CO concentrations recorded in CETESB monitoring network reached values close to 30 ppmv. On the other hand, since 2008, the air-quality standards for 8-hour carbon monoxide (9 ppmv) have not been exceeded at any of the automatic monitoring stations in the MASP. In agreement, analysis about the trends of air quality in the MASP showed that although the fleet had increased at a substantial rate, annual mean values showed that all pollutants except for ozone have been decreasing in the latest years [74], as a result of the vehicular emission control program established by PROCONVE. The trend evaluation of all monitoring stations showed a decrease in the annual mean concentration levels of CO, NOx, PM_{10}, and SO_2, being on average -0.09 ppm/year, -1.82 µg/m³/year, -1.97 µg/m³/year, and -0.82 µg/m³/year, respectively [74].

An important analysis about air quality studies in MASP reported that atmospheric concentrations of acetaldehyde and ethanol were higher in Brazil than in other areas of the world,

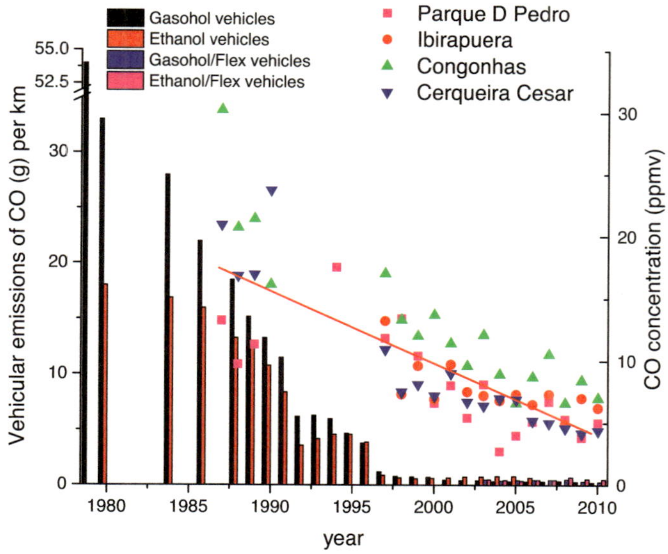

Figure 8. Carbon monoxide (CO) emissions from LDV by type of fuel and CO environmental concentrations (annual average) recorded by CETESB at automatic monitoring stations Parque Dom Pedro, Ibirapuera, Congonhas, and Cerqueira Cesar.

whereas the concentrations of aromatic compounds and carboxylic acids were lower [75]. More recent results analyzed atmospheric concentrations of ozone, nitrogen oxides (NO and NO_2), formaldehyde, and acetaldehyde measured in the MASP, over four seasons in 2012 and 2013 [76]. These results demonstrated that although there was a large increase in the number of vehicles in the MASP that use ethanol ("flex-fuel" vehicles), technological advances in vehicle emissions control have prevented any significant increase in the atmospheric concentrations of aldehydes. The average concentrations of formaldehyde and acetaldehyde were 8.6 ± 6.7 ppbv and 5.4 ± 5.2 ppbv, respectively. Both carbonyls are important constituents of the urban troposphere in the MASP (third and fourth in terms of concentration), representing an important fraction of the total VOCs.

In recent years, Brazilian LDVs have shown a reduction in aldehyde emissions. During the 1980s, most Brazilian vehicles did not use catalytic systems for the conversion of exhaust gases and the engines were less efficient. Since 2003, flex-fuel vehicles available on the Brazilian market have been equipped with modern three-way catalytic converters. Additionally, regulations in the PROCONVE program have been of great importance in reducing the emissions of pollutants (e.g., aldehydes) in the atmosphere. In the study developed by Martins et al. [77], the data also suggest a decrease of aldehydes concentration in the MASP between years 1986 and 2003. On the other hand, comparing to previous data, this study showed that the concentration of HC presented no decrease in the concentration. Among the HCs species analyzed, the highest concentrations observed were those of toluene (7.5 ± 3.4 ppbv), n-decane (3.2 ± 2.0 ppbv), benzene (2.7 ± 1.4 ppbv), and 1,3,5-trimethylbenzene (2.2 ± 1.5 ppbv) [77]. In some years, the highest formaldehyde values in Brazil have been recorded in the city of Rio de Janeiro, where those values increased considerably between 1998 and 2004, whereas formaldehyde and acetaldehyde levels both decreased sharply between 2004 and 2009. In the latter period, values ranged from 1.52 to 54.3 ppbv for formaldehyde and from 2.36 to 45.6 ppbv for acetaldehyde [78, 79]. The authors attributed high aldehyde concentrations to the increasing use of biofuel. Several researchers have worked out the reaction pathways in the combustion of different biofuels and have concluded that such processes can generally be expected to produce carbonyl compounds, particularly formaldehyde [80, 81]. In a study developed in 2000 showed ambient concentrations of up to 61 carbonyls measured in Rio de Janeiro, Brazil. The most abundant carbonyls were formaldehyde and acetaldehyde (8.8 pbbv and 5.8, respectively) followed by acetone, 2-butanone, and benzaldehyde. Ambient concentrations of other carbonyls (except acetophenone) correlated well with those of acetaldehyde and formaldehyde [82].

The concentration of atmospheric VOCs in the MASP, such as alcohols and aldehydes, were measured and compared with the data obtained in Osaka, Japan [83]. The results showed that concentrations of several pollutants found in Brazil were higher than in Japan. Ethanol concentrations found in Sao Paulo were significantly higher than those in Osaka, where the average concentrations of atmospheric methanol, ethanol, and isopropanol were 5.8 ± 3.8, 8.2 ± 4.6, and 7.2 ± 5.9 ppbv, respectively, while the average ambient levels of methanol, ethanol, and isopropanol measured in Sao Paulo were 34.1 ± 9.2, 176.3. ± 38.1, and 44.2 ± 13.7 ppbv, respectively. The levels of aldehydes, which were expected to be high due to the use of alcohol fuel during this period, were also measured at these sampling sites and the atmospheric formaldehyde average concentration measured in Osaka was 1.9 ± 0.9 ppbv; the average acetaldehyde concentration was 1.5 ± 0.8 ppbv. The atmospheric formaldehyde and acetaldehyde average concentrations measured in Sao Paulo were 5.0 ± 2.8 and 5.4 ± 2.8 ppbv, respectively. The ethanol/methanol and acetaldehyde/formaldehyde were compared between the two measurement sites and elsewhere in the world, which have already been reported in the literature. Due to the use of ethanol-fueled vehicles, these ratios, especially ethanol/methanol, are much higher in Brazil than those measured elsewhere in the world. Colon [84] compared environmental concentrations of some VOCs in the MASP with data obtained from EPA in Los Angeles. In their study, the overall MASP results demonstrated that the mean concentrations of single-ring aromatics are 2-3 times higher; volatile aldehydes are 5-10 times higher; and simple alcohols 10-100 times higher as compared to results of an EPA in the Los Angeles basin. In addition, n-alkanes containing between 4 and 11 carbons were only slightly elevated in Sao Paulo.

Particulate matter, main for PM_{10}, had more data about mass concentrations, due to States air quality monitoring stations, as well as studies by different research groups around the country. The first diagnosis of air quality monitoring network in Brazil showed the representation of differences both in space and in time. In the analysis for PM_{10} concentration, few stations had values above the national primary standard in 2012, but considering WHO guidelines the majority showed values above 20 reaching up to 100 $\mu g/m^3$, being that the higher values were observed in São Paulo and Rio de Janeiro, more populated urban areas in Brazil (IEMA, 2014).

In urban areas, such as the MASP, the fuel burning by vehicles is an important source of PM; when comparing current levels to those observed in the past, it appears that there has been an improvement in concentration levels of this pollutant, as result of the actions and emission control programs that took place over time. In recent years, the average concentrations tended to stabilize, indicating that even with diminishing vehicle emissions these levels are only sufficient to offset the increase in the fleet and the ensuing traffic conditions. Figure 9 shows the reduction on PM_{10} concentrations recorded in selected CETESB monitoring network. Since 2004, the value of Brazilian air quality standard (50 μg/m^3) for PM_{10} was not exceeded, but for all monitoring stations the annual mean values were above the WHO (20 μg/m^3) guidelines (Figure 9). Although there is no national air quality standard for $PM_{2.5}$, CETESB has monitored this pollutant; and since 1987 it corresponds to 60% of the PM_{10} mass in MASP atmosphere [28].

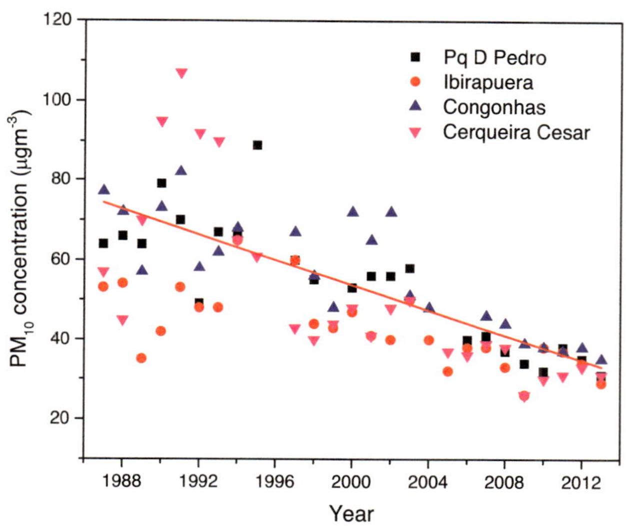

Figure 9. Annual arithmetic mean of MP_{10} concentrations recorded at CETESB automatic monitoring stations: Parque Dom Pedro, Ibirapuera, Congonhas, and Cerqueira Cesar, from 1987 to 2013.

Sampling and analysis of chemical composition of $PM_{2.5}$ were held for six Brazilian capitals (São Paulo, Rio de Janeiro, Belo Horizonte, Curitiba, Recife, and Porto Alegre) between 2007 and 2008 [85, 86], evaluating differences between summer and winter concentrations, meteorological influences, physicochemical profiles, and the effects on human health. The sources evaluation by receptor models identified the principal factors: soil and crustal material; vehicle emissions and biomass burning; and fuel oil combustion in industries (sulfur factor), being that vehicle emissions explained at least 40% of the $PM_{2.5}$ mass [86]. The concentrations of mass, black carbon (BC) and major ions in $PM_{2.5}$ are shown in Table 5, together with concentrations obtained in other places of Brazil, like as industrial site and region where sugar cane is routinely burned. The $PM_{2.5}$ mass concentrations were higher than the WHO guidelines, except in Recife. In urban areas, BC concentrations were higher than ions concentrations, being the sulfate concentrations higher than those of any other ion. The lowest sodium concentrations were observed in Belo Horizonte and Curitiba, which are farther from the ocean, on the contrary those observed in Seropédica and Santa Cruz (Rio de Janeiro) [87], which are near the ocean (Table 5). São José dos Campos and Araraquara had the highest potassium contents, which were associated with the occurrence of biomass burning events (for instance, sugar cane). Regarding trace elements and/or metals (Table 6), iron concentration was the highest followed by aluminum and zinc. The $MP_{2.5}$ PAH results are shown in Table 7, highlighting the high concentrations of total PAHs observed in São Paulo and Cubatão, more polluted areas among the cities evaluated.

		Average Concentrations / $\mu g\ m^{-3}$										References
City/State	Sampling data	Mass	BC	Cl^-	NO_3^-	SO_4^{2-}	NH_4^+	Na^+	K^+	Ca^{2+}	Mg^{2+}	
Seropédica/Rio de Janeiro	Aug. 2010-July 2011	12.1	-	1.07	1.70	3.30	-	1.59	0.23	0.24	0.07	[87]
Santa Cruz/Rio de Janeiro		12.4	-	1.00	1.64	3.81	-	1.56	0.23	0.24	0.05	
São Jose dos Campos/São Paulo	Feb. 2004-Feb. 2005	15.7	-	0.36	0.37	2.17	0.96	0.34	0.46	0.30	0.03	[100]
Araraquara/São Paulo	April 1999-Feb. 2001	-	-	0.18	0.38	1.58	0.26	0.07	0.38	0.16	0.03	[101]
São Paulo/São Paulo	13 May-19 July, 2002	11.9	-	0.11	0.43	1.20	0.30	0.13	0.14	0.10	0.02	[102]
São Paulo/São Paulo	June 2007-August 2008	28.1	10	0.24	1.22	3.09	1.25	0.24	0.25	0.08	-	[85, 86]
Rio de Janeiro/Rio de Janeiro		17.2	3.4	0.11	0.56	1.91	0.80	0.21	0.18	0.04	-	
Belo Horizonte/ Minas Gerais		14.7	4.5	0.04	0.19	1.15	0.34	0.08	0.19	0.10	-	
Porto Alegre/Rio Grande do Sul		13.4	3.9	0.15	0.43	1.17	0.35	0.19	0.27	0.04	-	
Curitiba/Paraná		14.4	4.4	0.07	0.16	1.08	0.37	0.10	0.28	0.04	-	
Recife/ Pernambuco		7.3	1.9	0.15	0.11	0.61	0.18	0.33	0.14	0.05	-	

Table 5. Particulate matter ($PM_{2.5}$ or fine fraction) mass, black carbon (BC) and ions average concentrations in some Brazilian regions.

http://dx.doi.org/10.5772/60944

		Trace elements Concentrations / ng m^{-3}								References
City/State	Sampling location	Mn	Fe	Zn	Cu	Cr	Al	V	Pb	
Rio de Janeiro/Rio de Janeiro	Urban area	5	307	20	35	<LOD	<LOD	2	<LOD	[103]
Campinas/São Paulo	Urban area	14.8	585	80.0	18.1	4.2	50.3	3.4	3.6	[104]
São Paulo/São Paulo	Urban + heavy traffic	6	181	75	10	2	55	2	16	[85, 86]
Rio de Janeiro/Rio de Janeiro		4	75	25	8	2	50	4	12	
Belo Horizonte/Minas Gerais		39	133	15	1	0.6	53	2	6	
Porto Alegre/RS		3	74	17	3	1	43	9	4	
Curitiba/Paraná		2	70	19	3	0.7	46	0.7	8	
Recife/Pernambuco		2	65	17	1	0.5	40	0.7	4	

Table 6. Trace elements concentrations in $PM_{2.5}$ (fine fractions) in some Brazilian cities.

City/State		Sampling location	PAH concentrations (ng m^{-3})	References
Canoas/Rio Grande do Sul		Roadside/traffic	1.32 (summer)	[105]
			2.02 (winter)	
Sapucaia /Rio Grande do Sul		Urban road/traffic	1.57 (summer)	
			3.05 (winter)	
Porto Alegre / Rio Grande do Sul		Metropolitan area	1.08 (summer)	[106]
			1.81 (winter)	
Salvador / Bahia	Lapa	Station bus	4.35 (winter)	[69]
	Aratu	Industrial center	2.94 (spring)	
	Bananeira	Non-vehicle influence	2.50 (spring)	
Cubatão / São Paulo		Industrial center	16.7 (spring)	[107]
São Paulo / São Paulo		Urban + heavy traffic	10.8 (winter)	[108]
Sapucaia do Sul, São Leopoldo and Novo Hamburgo / Rio Grande do Sul		Basin of the Sinos Region	1.52 (summer)	[109]
			1.68 (winter)	

Table 7. Atmospheric concentrations of total PAHs in some Brazilian cities.

5. Role of the biofuel use on secondary pollutants formation

Although the levels of primary air pollutants have decreased over the past 30 years in the MASP, secondary pollutants, like ozone and fine particles, frequently exceed established international and local air quality standards. Ozone and fine particles are secondary pollutants, products of VOCs, NO(x), and sunlight in the case of ozone and also sulfur oxides, ammonia, nitrogen oxides in the case of fine particles. Thus, it is very difficult to elaborate efficient strategies for their reduction. Since the frequent episodes of high concentrations of ozone occurring in the MASP are primarily associated with vehicular emissions, some reports in literature examined the impact that the implemented control program for mobile emissions (PROCONVE) had on tropospheric ozone concentrations and evaluate the impact of using reformulated gasoline-ethanol blend (gasohol) and of hydrous ethanol on the ozone formation [88-93]. Studies related to the formation of fine particles are more sparse, Albuquerque et al. [94] analyzed the impact of reducing the sulfur content in the diesel and the secondary formation of sulfate-nitrate-amonnium system, showing that an increase in the nitrate-ammonium formation may occur while the decrease in the sulfate-ammonium system occurs.

Sanchez-Ccoyllo et al. [92] evaluated the impact of the PROCONVE implementation based on an episode of high surface ozone concentrations that occurred in the MASP during March, 2000, employing the California Institute of Technology/Carnegie Mellon University three-dimensional photochemical model (CIT model [95-97]). In this work, different scenarios of emissions were considered according to the implementation of the PROCONVE. Scenario 1 assumed that all vehicles in the fleet were operating within PROCONVE guidelines. Scenarios 2 and 3 considered hypothetical situations in which PROCONVE had not been implemented. Scenario 2 established the premise that all vehicles were using pre-1989 technology, whereas scenario 3 allowed the existence of technological advances. That work showed that tropospheric ozone concentrations predicted for scenario 2 (vehicles pre-1989 technology) were higher than those predicted for scenarios 1, 3, and for the base case. The authors demonstrated that the PROCONVE had great influence on the improvement of air quality concerning ozone in MASP. In another studied developed by Sanchez-Ccoyllo et al. [93], the analysis was concentrated in the meteorological impacts on the ozone formation. Three meteorological variables: mixing height, wind speed, and air temperature, were considered in the study. The study also evaluated the role of having a reactive hydrocarbon (RHC) limitation or NOx limitation configuration on ozone formation in the MASP. In addition, NOx and RHC emission inventory reductions were used to evaluate their sensitivities with the CIT model. The results showed that changes in mixing height, wind speed, and air temperature input files have the greatest effect on peak ozone production, and the isolated effect of RHC emission reduction lead to 26% lower ozone levels compared to the base case. Based on the results of this study, the authors concluded that reductions on RHC emission could provide the best scenario for promoting lower ozone concentrations in the MASP. In agreement to these results, Orlando et al. [91] showed that the occurrence of high ozone production is more dependent on VOC in a study employing the trajectory model OZIPR together with the SAPRC chemical mechanism. In that work, five base-cases were created to verify the variation in maximum ozone concentration related to the ozone formation potential of each VOC that was injected in the airbase. The NO(x) and VOC emissions were independently and simultaneously reduced by 5%, 10%, 20%, and 30% to induce variations in ozone formation and the results showed that the most frequent compounds found among the ten main ozone precursors in Sao Paulo, using the reactivity scales created from the five base-cases, were formaldehyde, acetaldehyde, propene, isoprene, cis-2-butene, and trans-2-butene, with formaldehyde being always the main ozone precursor. In addition, the simulations data showed that an efficient strategy to decrease ozone concentrations in the MASP would be to reduce the total VOC emissions, while the same

http://dx.doi.org/10.5772/60944

strategy is not possible for NOx, since the reduction of these pollutants would increase ozone concentrations.

The impact of the use of reformulated gasoline/ethanol blend (gasohol) or hydrous ethanol on the ozone formation was evaluated in a study conducted by Martins and Andrade [89]. In this work, a three-dimensional photochemical model was employed to estimate the sensitivity of ozone formation and evaluate the implementation of emission scenarios, considering various fuel formulations in the MASP. Six scenarios were analyzed in the work; scenarios 1 to 5 involved different reduction scenarios in the compounds found in gasohol at different proportions related to the base case emission inventory, and the scenario 6 specified that the entire LDV running in the MASP would burn hydrous ethanol. The results showed that in scenario 3 (reductions in olefins, aromatics, and benzene) and scenario 5 (reductions in the five species that are associated with higher ozone sensitivity), ozone concentrations were below the national standard only at the air quality monitoring stations (not domain-wide). In addition, these results suggest that implementing scenario 6 (entire LDV fleet burn hydrous ethanol) would improve air quality in the MASP. In contrast, the work developed by Salvo and Geiger [98] suggest that the use of ethanol in flex-fuel vehicles has a negative effect on air quality as regards the formation of ozone. Their work reports the consequences of a shift in fuel use in the MASP, brought on by large-scale fluctuations in the price of ethanol relative to gasohol between 2009 and 2011. The work uses highly spatially and temporally resolved observations of road traffic levels, meteorology and pollutant concentrations, together with a consumer demand model, to show that ambient ozone concentrations felt by about 20% as the share of flex-fuel vehicles burning gasohol rose from 14% to 76%. On the other hand, in this work data suggest that NO and CO concentration increased. Thus, considering that tropospheric ozone production over MASP is hydrocarbon-limited, high NOx emissions (from gasohol burning) resulted in reductions in ambient ozone. These results are consistent with modeling studies previously, which concluded that the atmosphere in the MASP is VOC-limited [89, 91-93].

6. Concluding remarks

Almost 30 years after the creation of Brazilian Motor Vehicle Air Pollution Control Program (PROCONVE), results show that the adoption of increasingly restrictive phases was successful and environmental concentration data show a clear downward tendency for several pollutants over past years, except for ozone and fine particles. The reduction on concentrations of these atmospheric pollutants became the big challenge for the public policies to elaborate efficient control strategies in urban areas of Brazil.

The Brazilian experience demonstrates how the impact of vehicular emissions can be reduced with the use of biofuels associated with technological developments. Results addressed in studies conducted in Brazil demonstrate that, although a considerable increase occurred in the number of LDV able to run on hydrous ethanol, and the increased use of biodiesel by HDV, key technological improvements in the design of the vehicles mitigated increases in the emissions of pollutants such as aldehydes, CO, and NOx.

7. Abbreviations

BC, black carbon; CETESB, São Paulo State Environmental Protection Agency; GHG, greenhouse gas; HC, hydrocarbons; HDV, heavy-duty vehicles, LDV, light-duty vehicles; MASP, metropolitan area of São Paulo; NOx, nitrogen oxides; PM, particulate matter; PROCONVE, Brazilian Motor Vehicle Air Pollution Control Program; VOC, volatile organic compounds.

Acknowledgements

The authors gratefully acknowledge financial support from the Fundação de Amparo à Pesquisa do Estado de São Paulo (FAPESP, process 2008/58104-8), Fundação de Amparo à Pesquisa do Estado de Minas Gerais (FAPEMIG, BPD 00512-13 and APQ-01537-14), and the Conselho Nacional de Desenvolvimento Científico e Tecnológico (CNPq, 308174/2013-5). TN thanks FAPESP for fellowships supporting this research (Process 2011/18777-6). We would also like to thank CETESB (São Paulo State Environmental Agency) for the data they contributed to this report.

Author details

Thiago Nogueira[1*], Denise de Sales Cordeiro[2], Rodrigo Alejandro Abarza Muñoz[2], Adalgiza Fornaro[1], Antonio H. Miguel[3] and Maria de Fatima Andrade[1]

*Address all correspondence to: thiago.nogueira@iag.usp.br.

1 Department of Atmospheric Sciences, Institute of Astronomy, Geophysics, and Atmospheric Sciences, University of São Paulo (IAG/USP), Brazil

2 Institute of Chemistry, Federal University of Uberlandia, Brazil

3 A. H. Miguel & Associates, Pasadena, CA, USA

References

[1] Hoekman SK, Gertler A, Broch A, Robbins C. Investigation of biodistillates as potential blendstocks for transportation fuels, CRC AVFL-17, Final Report June. CRC Project No. AVFL-17; Final Report. 2009.

[2] Knothe G. Biodiesel and renewable diesel: a comparison. Prog Energ Combust 2010;36:364-73.

[3] Goldemberg J, Coelho ST, Guardabassi P. The sustainability of ethanol production from sugarcane. Energy Policy 2008;36:2086-97.

[4] Munoz RAA, Fernandes DM, Santos DQ, Barbosa TGG, Sousa RMF. Biodiesel: production, characterization, metallic corrosion and analytical methods for contaminants. In: Fang Z, ed. Biodiesel - Feedstocks, Production and Applications: Intech 2012;p.498.

[5] Hajbabaei M, Johnson KC, Okamoto RA, Mitchell A, Pullman M, Durbin TD. Evaluation of the impacts of biodiesel and second generation biofuels on NOx emissions for CARB diesel fuels. Environ Sci Technol 2012;46:9163-73.

[6] U.S Environmental Protection Agency. A comprehensive analysis of biodiesel impacts on exhaust emissions, EPA Draft Final Report. 2002.

[7] U.S Environmental Protection Agency. Draft regulatory impact analysis: changes to renewable fuel standard program, Draft Final Report. 2009.

[8] Graboski MS, McCormick RL. Combustion of fat and vegetable oil derived fuels in diesel engines. Prog Energ Combust 1998;24:125-64.

[9] Hoekman SK, Broch A, Robbins C, Ceniceros E, Natarajan M. Review of biodiesel composition, properties, and specifications. Renew Sust Energ Rev 2012;16:143-69.

[10] California Air Resources Board. California Air Resources Board LCFS [Internet], Available from: http://www.arb.ca.gov/fuels/lcfs/workgroups/workgroups.htm#pathways.

[11] Cronin K. Life-cycle assessment of biofuels in California's low carbon fuel standard, available on http://www.sage.wisc.edu/people/cronin/finaldraft_es539_kc.pdf. 2010.

[12] Hoekman SK, Robbins C. Review of the effects of biodiesel on NOx emissions. Fuel Process Technol 2012;96:237-49.

[13] Quiterio SL, Escaleira V, Sousa CRS, Maia LFPG, Arbilla G. Metals in airborne particulate matter in downtown Rio de Janeiro, Brazil. Bulletin of Environmental Contamination and Toxicology. 2004;72:916-22.

[14] Santos AL, Takeuchi RM, Fenga PG, Stradiotto NR. Electrochemical methods in analysis of biofuels, Applications and Experiences of Quality Control, Prof. Ognyan Ivanov (Ed.), ISBN: 978-953-307-236-4, InTech, DOI: 10.5772/16212. Available from: http://www.intechopen.com/books/applications-and-experiences-of-quality-control/electrochemical-methods-in-analysis-of-biofuels2011.

[15] de Oliveira MF, Saczk AA, Okumura LL, Stradiotto NR. Analytical methods employed at quality control of fuel ethanol. Energ Fuel 2009;23:4852-9.

[16] Korn MDA, dos Santos DSS, Welz B, Vale MGR, Teixeira AP, Lima DD, et al. Atomic spectrometric methods for the determination of metals and metalloids in automotive fuels - a review. Talanta 2007;73:1-11.

[17] Tormin TF, Gimenes DT, Silva LG, Ruggiero R, Richter EM, Ferreira VS, et al. Direct amperometric determination of tert-butylhydroquinone in biodiesel. Talanta 2010;82:1599-603.

[18] Lôbo IP, Ferreira SLC, Cruz RSd. Biodiesel: parâmetros de qualidade e métodos analíticos. Quimica Nova 2009;32:1596-608.

[19] Rosillo-Calle F, Cortez LAB. Towards proalcool II - a review of the Brazilian bioethanol programme. Biomass Bioenerg 1998;14:115-24.

[20] MMA. Current Conama Resolutions published between September 1984 and January 2012. / Ministry of the Environment. Brasilia: MMA 2012. http://www.mma.gov.br/port/conama/. Accessed 04.10.2014.

[21] CETESB. Relatório Anual de Qualidade do Ar no Estado de São Paulo 2009. CETESB-Companhia de Tecnologia de Saneamento Ambiental, São Paulo, Brazil. Relatório Anual de Qualidade do Ar no Estado de São Paulo 2009 CETESB-Companhia de Tecnologia de Saneamento Ambiental, São Paulo, Brazil2010.

[22] ANP. Agência Nacional de Petróleo, Gás Natural e Biocombustíveis (National Agency for Oil, Natural Gas, and Biofuels) (accessed 05.02.2014). http://www.anp.gov.br/.

[23] ANP. http://www.anp.gov.br/?pg=68406&m=diesel&t1=&t2=diesel&t3=&t4=&ar=0&ps=11&cachebust=1386157882674, accessed 14.01.2014. 2014.

[24] Brilhante OM. Brazil's alcohol programme: from an attempt to reduce oil dependence in the seventies to the green arguments of the nineties. J Environ Plann Man. 1997;40:435-49.

[25] CARB. California Air Resources Board (ARB). http://www.arb.ca.gov/msprog/levprog/levprog.htm, accessed on January, 15 2015.

[26] Lloyd AC, Cackette TA. Diesel engines: environmental impact and control. J Air Waste Manage 2001;51:809-47.

[27] Sawyer RF, Harley RA, Cadle SH, Norbeck JM, Slott R, Bravo HA. Mobile sources critical review: 1998 NARSTO assessment. Atmos Environ 2000;34:2161-81.

[28] CETESB. Relatório Anual de Qualidade do Ar no Estado de São Paulo 2012. CETESB-Companhia de Tecnologia de Saneamento Ambiental, São Paulo, Brazil. 2013.

[29] Durbin TD, Zhu XN, Norbeck JM. The effects of diesel particulate filters and a low-aromatic, low-sulfur diesel fuel on emissions for medium-duty diesel trucks. Atmos Environ 2003;37:2105-16.

[30] Ban-Weiss GA, Mclaughlin JP, Harley RA, Kean AJ, Grosjean E, Grosjean D. Carbonyl and nitrogen dioxide emissions from gasoline- and diesel-powered motor vehicles. Environ Sci Technol 2008;42:3944-50.

http://dx.doi.org/10.5772/60944

[31] de Melo TCC, Machado GB, Belchior CRP, Colaco MJ, Barros JEM, de Oliveira EJ, et al. Hydrous ethanol-gasoline blends - combustion and emission investigations on a flex-fuel engine. Fuel 2012;97:796-804.

[32] Martins AA, Rocha RAD, Sodre JR. Cold start and full cycle emissions from a flexible fuel vehicle operating with natural gas, ethanol and gasoline. J Nat Gas Sci Eng 2014;17:94-8.

[33] Randazzo ML, Sodre JR. Exhaust emissions from a diesel powered vehicle fuelled by soybean biodiesel blends (B3-B20) with ethanol as an additive (B20E2-B20E5). Fuel 2011;90:98-103.

[34] Pérez-Martínez PJ, Miranda RM, Nogueira T, Guardani ML, Fornaro A, Ynoue R, et al. Emission factors of air pollutants from vehicles measured inside road tunnels in São Paulo: case study comparison. Int J Environ Sci Technol 2014;11:2155-68.

[35] Martins LD, Andrade MF, D Freitas E, Pretto A, Gatti LV, Albuquerque EL, et al. Emission factors for gas-powered vehicles traveling through road tunnels in Sao Paulo, Brazil. Environ Sci Technol 2006;40:6722-9.

[36] Heck RM, Farrauto RJ. Automobile exhaust catalysts. Appl Catal a-Gen 2001;221:443-57.

[37] Lapuerta M, Armas O, Rodriguez-Fernandez J. Effect of biodiesel fuels on diesel engine emissions. Prog Energ Combust 2008;34:198-223.

[38] Guarieiro LLN, Pereira PAD, Torres EA, da Rocha GO, de Andrade JB. Carbonyl compounds emitted by a diesel engine fuelled with diesel and biodiesel-diesel blends: Sampling optimization and emissions profile. Atmos Environ 2008;42:8211-8.

[39] Rodrigues MC, Guarieiro LLN, Cardoso MP, Carvalho LS, da Rocha GO, de Andrade JB. Acetaldehyde and formaldehyde concentrations from sites impacted by heavy-duty diesel vehicles and their correlation with the fuel composition: Diesel and diesel/biodiesel blends. Fuel 2012;92:258-63.

[40] de Andrade JB, Andrade MV, Pinheiro HLC. Atmospheric levels of formaldehyde and acetaldehyde and their relationship with the vehicular fleet composition in Salvador, Bahia, Brazil. J Brazil Chem Soc 1998;9:219-23.

[41] Pinto JP, Solci MC. Comparison of rural and urban atmospheric aldehydes in Londrina, Brazil. J Brazil Chem Soc 2007;18:928-36.

[42] Nogueira T, Souza KF, Fornaro A, Andrade MF, de Carvalho LRF. On-road emissions of carbonyls from vehicles powered by biofuel blends in traffic tunnels in the metropolitan area of Sao Paulo, Brazil. Atmos Environ 2015;108:88-97.

[43] Gentner DR, Worton DR, Isaacman G, Davis LC, Dallmann TR, Wood EC, et al. Chemical composition of gas-phase organic carbon emissions from motor vehicles and implications for ozone production. Environ Sci Technol 2013;47:11837-48.

[44] Finlayson-Pitts BJ, Pitts Jr. JN. Chemistry of the Upper and Lower Atmosphere: Theory, Experiments, and Applications. Academic Press, San Diego, 2000.

[45] Seinfeld JH, Pandis SN. Atmospheric Chemistry and Physics of Air Pollution, 2nd edn. John Wiley, New York, 1998.

[46] WHO. World Health Organization Air quality guidelines for particulate matter, ozone, nitrogen dioxide and sulfur dioxide - Global update 2005 - Summary of risk assessment, WHO/SDE/PHE/OEH/06.02, available on (http://whqlibdoc.who.int/hq/2006/WHO_SDE_PHE_OEH_06.02_eng.pdf?ua=1). 2005.

[47] Mohanraj R, Azeez PA, Priscilla T. Heavy metals in airborne particulate matter of urban Coimbatore. Arch Environ Con Tox 2004;47:162-7.

[48] Xiu G, Zhang D, Chen J, Huang X, Chen Z, Guo H, et al. Characterization of major water-soluble inorganic ions in size-fractionated particulate matters in Shanghai campus ambient air. Atmos Environ 2004;38:227-36.

[49] Petzold A, Ogren JA, Fiebig M, Laj P, Li SM, Baltensperger U, et al. Recommendations for reporting "black carbon" measurements. Atmos Chem Phys 2013;13:8365-79.

[50] U.S Environmental Protection Agency. Available on http://www.epa.gov/pm/.

[51] Bojes HK, Pope PG. Characterization of EPA's 16 priority pollutant polycyclic aromatic hydrocarbons (PAHs) in tank bottom solids and associated contaminated soils at oil exploration and production sites in Texas. Regul Toxicol Pharm 2007;47:288-95.

[52] Vasconcellos PdC, Artaxo PE, Ciccioli P, Cecinato A, Brancaleoni E, Frattoni M. Determinação dos hidrocarbonetos saturados e policíclicos aromáticos presentes no material particulado da atmosfera amazônica. Quimica Nova 1998;21:385-93.

[53] Pereira Netto AD, Moreira JC, Dias AEXO, Arbilla G, Ferreira LFV, Oliveira AS, et al. Evaluation of human contamination with polycyclic aromatic hydrocarbons (PAHs) and their nitrated derivatives (NHPAs): a review of methodology. Quimica Nova 2000;23:765-73.

[54] Yan J, Wang L, Fu PP, Yu H. Photomutagenicity of 16 polycyclic aromatic hydrocarbons from the US EPA priority pollutant list. Mutat Res - Gen Tox En 2004;557:99-108.

[55] Szulejko JE, Kim K-H, Brown RJC, Bae M-S. Review of progress in solvent-extraction techniques for the determination of polyaromatic hydrocarbons as airborne pollutants. TrAC Trend Anal Chem 2014;61:40-8.

[56] Pandey SK, Kim K-H, Brown RJC. A review of techniques for the determination of polycyclic aromatic hydrocarbons in air. TrAC Trend Anal Chem 2011;30:1716-39.

[57] Giechaskiel B, Maricq M, Ntziachristos L, Dardiotis C, Wang XL, Axmann H, et al. Review of motor vehicle particulate emissions sampling and measurement: from smoke and filter mass to particle number. J Aerosol Sci 2014;67:48-86.

[58] Chow JC. Measurement methods to determine compliance with ambient air-quality standards for suspended particles. J Air Waste Manage 1995;45:320-82.

[59] McMurry PH. A review of atmospheric aerosol measurements. Atmos Environ 2000;34:1959-99.

[60] Sanchez-Ccoyllo OR, Ynoue RY, Martins LD, Astolfo R, Miranda RM, Freitas ED, et al. Vehicular particulate matter emissions in road tunnels in Sao Paulo, Brazil. Environ Monit Assess. 2009;149:241-9.

[61] Guarieiro ALN, Santos JVD, Eiguren-Fernandez A, Torres EA, da Rocha GO, de Andrade JB. Redox activity and PAH content in size-classified nanoparticles emitted by a diesel engine fuelled with biodiesel and diesel blends. Fuel 2014;116:490-7.

[62] de Abrantes R, de Assuncao JV, Pesquero CR, da Nobrega RP, Bruns RE. Comparison of emission of dioxins and furans from gasohol- and ethanol-powered vehicles. J Air Waste Manage 2011;61:1344-52.

[63] de Abrantes R, de Assuncao JV, Pesquero CR, Bruns RE, Nobrega RP. Emission of polycyclic aromatic hydrocarbons from gasohol and ethanol vehicles. Atmos Environ 2009;43:648-54.

[64] de Abrantes R, de Assuncao JV, Pesquero CR. Emission of polycyclic aromatic hydrocarbons from light-duty diesel vehicles exhaust. Atmos Environ 2004;38:1631-40.

[65] da Silva MF, de Assuncao JV, Andrade MD, Pesquero CR. Characterization of metal and trace element contents of particulate matter (PM10) emitted by vehicles running on Brazilian fuels hydrated ethanol and gasoline with 22% of anhydrous ethanol. J Toxicol Env Health Part A 2010;73:901-9.

[66] Correa SM, Arbilla G. Aromatic hydrocarbons emissions in diesel and biodiesel exhaust. Atmos Environ 2006;40:6821-6.

[67] Brito J, Rizzo LV, Herckes P, Vasconcellos PC, Caumo SES, Fornaro A, et al. Physical-chemical characterisation of the particulate matter inside two road tunnels in the Sao Paulo Metropolitan Area. Atmos Chem Phys 2013;13:12199-213.

[68] Mkoma SL, da Rocha GO, Regis ACD, Domingos JSS, Santos JVS, de Andrade SJ, et al. Major ions in PM2.5 and PM10 released from buses: the use of diesel/biodiesel fuels under real conditions. Fuel 2014;115:109-17.

[69] da Rocha GO, Lopes WA, de Paula Pereira PA, Vasconcellos PDC, Oliveir FS, Carvalho LS, et al. Quantification and source identification of atmospheric particulate polycyclic aromatic hydrocarbons and their dry deposition fluxes at three sites in Salvador Basin, Brazil, impacted by mobile and stationary sources. J Brazil Chem Soc 2009;20:680-92.

[70] Martins LD, da Silva CR, Solci MC, Pinto JP, Souza DZ, Vasconcellos P, et al. Particle emission from heavy-duty engine fuelled with blended diesel and biodiesel. Environ Monit Assess. 2012;184:2663-76.

[71] Tavares M, Pinto JP, Souza AL, Scarminio IS, Solci MC. Emission of polycyclic aromatic hydrocarbons from diesel engine in a bus station, Londrina, Brazil. Atmos Environ 2004;38:5039-44.

[72] Ueda AC, Tomaz E. Vehicular emissions inventory from metropolitan region of Campinas (MRC), SP, Brazil. Quimica Nova 2011;34:1496-500.

[73] IEMA. 1º Diagnóstico da Rede de Monitoramento da Qualidade do Ar no Brasil, Instituto de Energia e Meio Ambiente, available on (http://www.forumclima.pr.gov.br/arquivos/File/Rosana/Diagnostico_Qualidade_do_Ar_Versao_Final_Std.pdf). 2014.

[74] Carvalho VSB, Freitas ED, Martins LD, Martins JA, Mazzoli CR, Andrade MdF. Air quality status and trends over the metropolitan area of São Paulo, Brazil, as a result of emission control policies. Environ Sci Policy 2015;47:68-79.

[75] Anderson LG. Ethanol fuel use in Brazil: air quality impacts. Energ Environ Sci 2009;2:1015-37.

[76] Nogueira T, Dominutti PA, de Carvalho LRF, Fornaro A, Andrade MF. Formaldehyde and acetaldehyde measurements in urban atmosphere impacted by the use of ethanol biofuel: metropolitan area of Sao Paulo (MASP), 2012-2013. Fuel 2014;134:505-13.

[77] Martins LD, Andrade MD, Ynoue RY, de Albuquerque EL, Tomaz E, Vasconcellos PD. Ambiental volatile organic compounds in the megacity of Sao Paulo. Quim Nova 2008;31:2009-13.

[78] Correa SM, Martins EM, Arbilla G. Formaldehyde and acetaldehyde in a high traffic street of Rio de Janeiro, Brazil. Atmos Environ 2003;37:23-9.

[79] Correa SM, Arbilla G, Martins EM, Quiterio SL, Guimaraes CD, Gatti LV. Five years of formaldehyde and acetaldehyde monitoring in the Rio de Janeiro downtown area - Brazil. Atmos Environ 2010;44:2302-8.

[80] Kohse-Höinghaus K, Oßwald P, Cool TA, Kasper T, Hansen N, Qi F, et al. Verbrennungschemie der Biokraftstoffe: von Ethanol bis Biodiesel. Angewandte Chemie 2010;122:3652-79.

[81] Leplat N, Dagaut P, Togbe C, Vandooren J. Numerical and experimental study of ethanol combustion and oxidation in laminar premixed flames and in jet-stirred reactor. Combust Flame 2011;158:705-25.

[82] Grosjean D, Grosjean E, Moreira LFR. Speciated ambient carbonyls in Rio de Janeiro, Brazil. Environ Sci Technol 2002;36:1389-95.

[83] Nguyen HTH, Takenaka N, Bandow H, Maeda Y, de Oliva ST, Botelho MMF, et al. Atmospheric alcohols and aldehydes concentrations measured in Osaka, Japan and in Sao Paulo, Brazil. Atmos Environ 2001;35:3075-83.

[84] Colon M, Pleil JD, Hartlage TA, Guardani ML, Martins MH. Survey of volatile organic compounds associated with automative emissions in the urban airshed of Sao Paulo, Brazil. Atmos Environ 2001;35:4017-31.

[85] de Miranda RM, Andrade MD, Fornaro A, Astolfo R, de Andre PA, Saldiva P. Urban air pollution: a representative survey of PM2.5 mass concentrations in six Brazilian cities. Air Qual Atmos Health 2012;5:63-77.

[86] Andrade MF, de Miranda RM, Fornaro A, Kerr A, Oyama B, de Andre PA, et al. Vehicle emissions and PM2.5 mass concentrations in six Brazilian cities. Air Qual Atmos Health 2012;5:79-88.

[87] Mateus VL, Monteiro ILG, Rocha RCC, Saint'Pierre TD, Gioda A. Study of the chemical composition of particulate matter from the Rio de Janeiro metropolitan region, Brazil, by inductively coupled plasma-mass spectrometry and optical emission spectrometry. Spectrochim Acta Part B: Atom Spectr. 2013;86:131-6.

[88] Alvim DS, Gatti LV, dos Santos MH, Yamazaki A. Studies of the volatile organic compounds precursors of ozone in Sao Paulo city. Eng Sanit Ambient 2011;16:189-96.

[89] Martins LD, Andrade MF. Emission scenario assessment of gasohol reformulation proposals and ethanol use in the metropolitan area of Sao Paulo. Open Atmos Sci J 2008;2:166-75.

[90] Martins MHRB, Anazia R, Guardani MLG, Lacava CIV, Romano J, Silva SR. Evolution of air quality in the Sao Paulo Metropolitan Area and its relation with public policies. Int J Environ Pollut 2004;22:430-40.

[91] Orlando JP, Alvim DS, Yamazaki A, Correa SM, Gatti LV. Ozone precursors for the Sao Paulo metropolitan area. Sci Total Environ 2010;408:1612-20.

[92] Sanchez-Ccoyllo OR, Martins LD, Ynoue RY, Andrade MDF. The impact on tropospheric ozone formation on the implementation of a program for mobile emissions control: a case study in Sao Paulo, Brazil. Environ Fluid Mech 2007;7:95-119.

[93] Sanchez-Ccoyllo OR, Ynoue RY, Martins LD, Andrade MD. Impacts of ozone precursor limitation and meteorological variables on ozone concentration in Sao Paulo, Brazil. Atmos Environ 2006;40:S552-S62.

[94] Albuquerque TTD, Andrade MD, Ynoue RY. Characterization of atmospheric aerosols in the city of So Paulo, Brazil: comparisons between polluted and unpolluted periods. Environ Monit Assess 2012;184:969-84.

[95] Harley RA, Russell AG, Mcrae GJ, Cass GR, Seinfeld JH. Photochemical modeling of the Southern California air-quality study. Environ Sci Technol 1993;27:378-88.

[96] Mcrae GJ, Seinfeld JH. Development of a 2nd-generation mathematical-model for urban air-pollution.2. Evaluation of model performance. Atmos Environ 1983;17:501-22.

[97] Mcrae GJ, Seinfeld JH. Development of a 2nd-generation model for urban air-pollution - reply. Atmos Environ 1984;18:1038.

[98] Salvo A, Geiger FM. Reduction in local ozone levels in urban Sao Paulo due to a shift from ethanol to gasoline use. Nat Geosci 2014;7:450-8.

[99] Conselho Nacional do Meio Ambiente (CONAMA). Resolution Nº3 de 06/28/1990. 1990.

[100] Souza PAd, Mello WZd, Mariani RL, Sella SM. Caracterização do material particulado fino e grosso e composição da fração inorgânica solúvel em água em São José dos Campos (SP). Quimica Nova 2010;33:1247-53.

[101] Allen AG, Cardoso AA, da Rocha GO. Influence of sugar cane burning on aerosol soluble ion composition in Southeastern Brazil. Atmos Environ 2004;38:5025-38.

[102] Bourotte C, Curl-Amarante AP, Forti MC, Pereira LAA, Braga AL, Lotufo PA. Association between ionic composition of fine and coarse aerosol soluble fraction and peak expiratory flow of asthmatic patients in Sao Paulo city (Brazil). Atmos Environ 2007;41:2036-48.

[103] Loyola J, Arbilla G, Quiterio SL, Escaleira V, Minho AS. Trace de-38.

[104] Miranda R, Tornaz E. Characterization of urban aerosol in Campinas, Sao Paulo, Brazil. Atmos Res 2008;87:147-57.

[105] Agudelo-Castañeda DM, Teixeira EC. Seasonal changes, identification and source apportionment of PAH in PM1.0. Atmos Environ 2014;96:186-200.

[106] Teixeira EC, Agudelo-Castañeda DM, Fachel JMG, Leal KA, Garcia KDO, Wiegand F. Source identification and seasonal variation of polycyclic aromatic hydrocarbons associated with atmospheric fine and coarse particles in the metropolitan area of Porto Alegre, RS, Brazil. Atmos Res 2012;118:390-403.

[107] Allen AG, da Rocha GO, Cardoso AA, Paterlini WC, Machado CMD, de Andrade JB. Atmospheric particulate polycyclic aromatic hydrocarbons from road transport in southeast Brazil. Transport Res Part D: Transport Environ 2008;13:483-90.

[108] Bourotte C, Forti MC, Taniguchi S, Bicego MC, Lotufo PA. A wintertime study of PAHs in fine and coarse aerosols in Sao Paulo city, Brazil. Atmos Environ 2005;39:3799-811.

[109] Teixeira EC, Mattiuzi CDP, Agudelo-Castañeda DM, De Oliveira Garcia K, Wiegand F. Polycyclic aromatic hydrocarbons study in atmospheric fine and coarse particles using diagnostic ratios and receptor model in urban/industrial region. Environ Monit Assess 2013;185:9587-602.

Chapter 19

An Overview of Biodiesel Production in Mexico

Gisela Montero, Margarita Stoytcheva,
Marcos Coronado, Conrado García, Jesús Cerezo,
Lydia Toscano, Ana M. Vázquez and José A. León

Additional information is available at the end of the chapter

http://dx.doi.org/10.5772/59400

1. Introduction

In Mexico, the primary energy is constituted of 88.5% oil and natural gas, 3.6% coal, 1% nuclear and 6.9% renewable energy as is illustrated in Figure 1. Renewable energy is comprised by 1.7% geothermal, solar, wind; 1.3% hydro and 3.9% biomass [1].

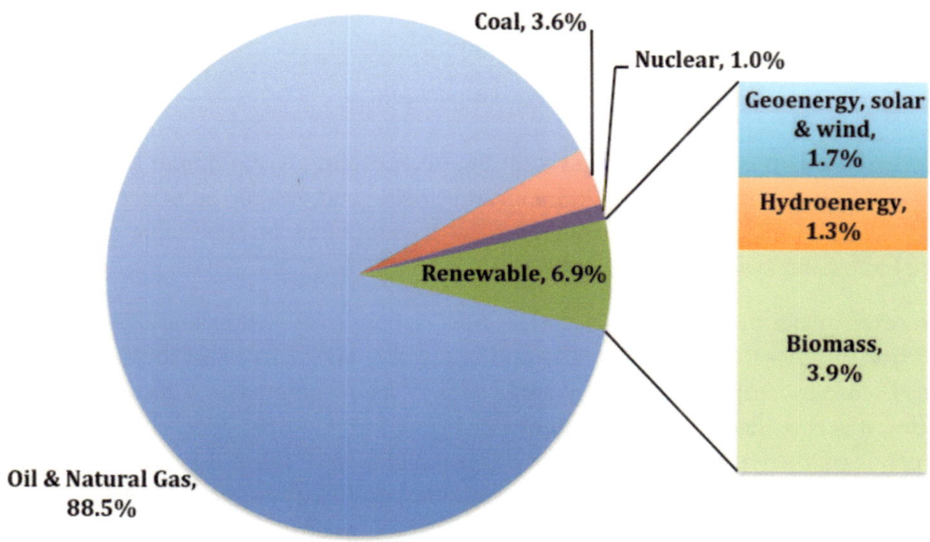

Figure 1. Primary energy of Mexico.

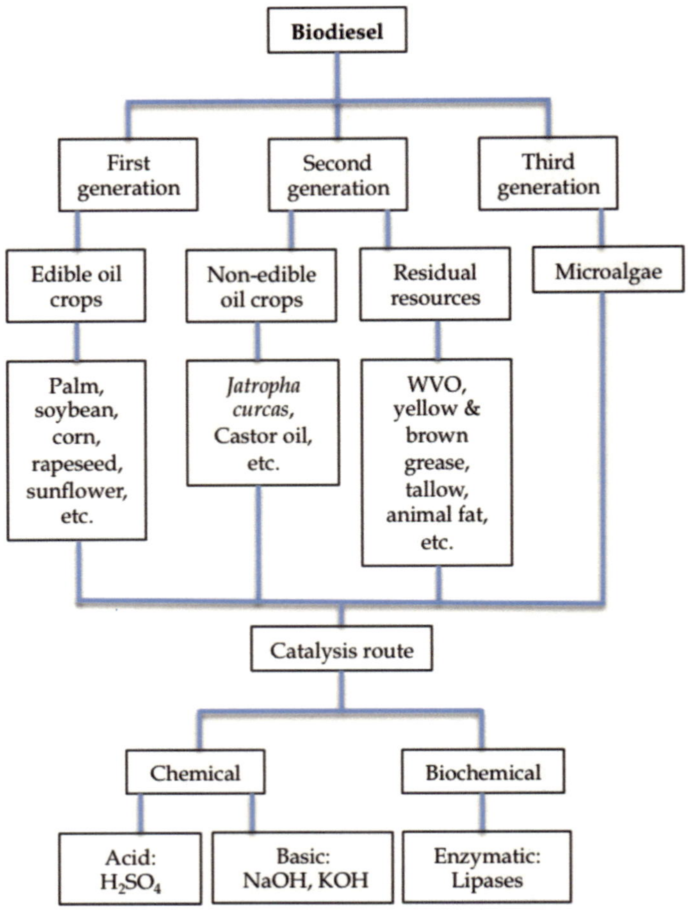

Figure 2. Classification of biodiesel and catalysis routes for its production.

The dependence on fossil fuels joined with the deteriorating environment caused by greenhouse gas (GHG) emissions, climate change, and rising and volatile oil prices are the energy challenges whose answer can be given by the development of renewable energy. Hence, Mexico has established an ambitious target to reduce 30% of GHG emissions by 2020 [2].

In the search of alternative to the fossil fuels, Mexico has developed technologies for biodiesel production. Biodiesel is an eco-friendly and renewable alternative fuel to diesel and can be obtained from oil crops or waste vegetable oils (WVO) generated as a result of productive activities. It is classified as first, second or third generation biofuel depending the feedstocks by which is produced. It is known as first generation biofuel when it is obtained from competing food resources e.g. sunflower, corn, safflower, canola, and soybean. Second generation when it is produced from waste biomass or non-edible energy crops e.g. WVO, yellow and brown grease, tallow, *Ricinus Communis*, *Jatropha Curcas L.*, and finally third generation biofuel from microalgae. The Figure 2 shows the classification of biodiesel according to the different biomass raw materials and the catalysis routes to produce the biofuel.

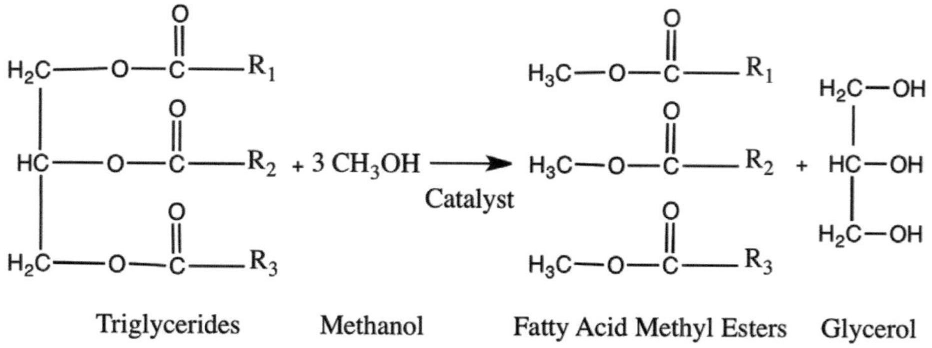

Figure 3. Transesterification reaction to obtain biodiesel.

In 2005, it was published in Mexico the Law of Sustainable Development of Sugarcane for the utilization of agro-energy, particularly ethanol as fuel and to oxygenate gasoline from initial honey and molasses, and exploiting sugarcane bagasse for electric cogeneration and syngas production [3]. Law for the Promotion and Development of Bioenergy (LPDB) was published, in order to contribute to energy diversification and sustainable development in 2008. This law considers promoting bioenergy production inputs, from agricultural activities, forestry, algae, biotechnological and enzymatic processes, without jeopardizing food security and sovereignty of the country [4]. Based on these regulations, Mexico initiated a series of actions to rule the internal market for biofuels and reduce the GHG emissions.

Biodiesel is among the bioenergetics considered. The leading standard setting organization ASTM International, formerly ASTM (American Society of Testing and Materials) defines biodiesel as a fuel comprised of mono-alkyl esters of long chain fatty acids [5]. Biodiesel is produced by transesterification of vegetable oils or animal fat, with a short chain alcohol in the presence of a catalyst, according to the following reaction shown in Figure 3 [6].

The main reason to convert the oil or fat into biodiesel is to reduce its viscosity and to obtain similar properties to diesel. While biodiesel is a lipid-based fuel, diesel is a mix of paraffinic, olefinic and aromatic hydrocarbons derived from the processing of crude oil.

Mexico has a great potential for biodiesel production because it has high biodiversity, intensive agriculture activity and waste biomass resources not exploited productively. For example, it was estimated that the potential of biodiesel production from WVO is between 7.8 PJ and 17.7 PJ nationally [7].

The development of the biodiesel industry in Mexico shows its first steps, relying on first and second generation biofuel technologies. Currently, universities, government agencies and research centers are conducting biodiesel research and development focused on process optimization, new catalysts, new raw materials for biodiesel production and the impacts of biodiesel on materials. The ongoing investigations imply the production of first, second and third generation biodiesel. Also, biorefineries projects at pilot level are conducted. Therefore, this chapter presents an overview of biodiesel production in Mexico and its current efforts for the development of this industry.

2. Biodiesel production plants at industrial scale in Mexico

At present Mexico has six industrial biodiesel production plants located in the states of Chiapas, Michoacán and Nuevo León, as illustrated in Figure 4, which were designed to process palm oil, *Jatropha Curcas L.*, castor oil, WVO and animal tallow and convert it into first or second biodiesel generation. Table 1 summarizes the information of the installed biodiesel production plants in Mexico.

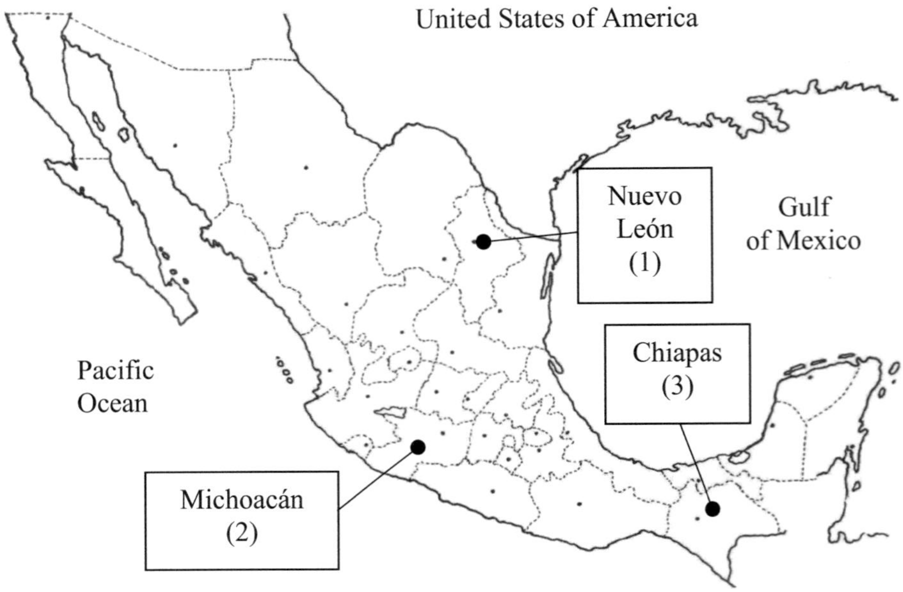

Figure 4. Biodiesel production plants in Mexico.

In 2004, the first biodiesel production plant in Mexico initiated activities in the state of Nuevo León. The installed capacity was 50,000 L/day, to process beef tallow and WVO. It operated in full capacity until 2009 and part of 2010, when it had a selling contract with Mexican Petroleum to supply biodiesel and use it as lubricity enhancing additive for diesel [8]. The biodiesel production plant ceased operations because the company that acquired the biodiesel finished his purchase contract. Due to the lack of market, the biodiesel plant stopped producing the biofuel. Then, the facilities were divided into two modules, one to produce open fuel flame and other for asphalt production [9].

In 2006, the company Moreco, located in Michoacán, launched a project to collect WVO and yellow grease for subsequent conversion into biodiesel. Currently, Moreco collects such waste in restaurants, produces biodiesel and supply it to several companies in Morelia for consumption as fuel in vehicles [10].

The biodiesel production facility located in Lázaro Cárdenas, Michoacán, called Bioenermex opened in 2007. The initial goal was to generate approximately 9,000,000 L of biodiesel per year, in addition to employing a thousand people. Nowadays, the plant is not in operation [11-13].

Location	Capacity (L/day)	Feedstock	Beginning of operation
Cadereyta, Nuevo León	50,000	Beef tallow WVO	2004
Morelia, Michoacán	Not reported	WVO	2006
Lázaro Cárdenas, Michoacán	24,600	*Jatropha Curcas*	2007
Tuxtla Gutiérrez, Chiapas	2,000	*Jatropha Curcas* African palm oil WVO	2009
Puerto Chiapas	8,000	*Jatropha Curcas* African palm oil WVO	2009
Puerto Chiapas	20,000	*Jatropha Curcas* African palm oil WVO	2009

Table 1. Information of biodiesel production plants in Mexico.

In the case of Chiapas, the development of the biodiesel program was called Chiapas Bioenergético and included the establishment of crops, oil extraction and the construction and operation of biodiesel production plants. The raw materials selected to supply these plants were oil from *Jatropha Curcas L.*, palm oil and WVO. The installed production capacity of biodiesel in Chiapas is 30,000 L/day, distributed in a plant located in Tuxtla Gutiérrez of Swedish technology, producing 2,000 L/day; another in Puerto Chiapas consists of two modules of Colombian-Mexican technology, producing 8,000 L/day and one module of English technology, producing 20,000 L/day. The plants began operations in 2009 and closed due to an insufficient supply of raw materials [8, 14].

According to Vega [14], the development of bioenergy in Mexico faces some barriers, including: a) the lack of a program to introduce biofuels for transport, b) standards of quality and sustainability of biofuels under the LPDB were not issued, c) funds for bioenergy program of the Secretariat of Agriculture, Livestock, Rural Development, Fisheries and Food, were used for other purposes.

Currently, biodiesel plants in Chiapas, Nuevo León and Michoacán are out of operation, however, Mexico continues to develop technologies in research centers and universities, for the production of biodiesel in higher yields [15, 16]. Several feedstocks including algae and other catalysts such as enzymes are being investigated.

3. Non-industrial experiences of biodiesel production in Mexico

In the state of Oaxaca, since 2004, José Vasconcelos University has a pilot plant for experimenting with the production of biodiesel from WVO, with a capacity of 45,000 L/year, which is used as B20 in a bus at the same university [12].

In 2005, Monterrey Institute of Technology built a first pilot plant for research purposes, to produce biodiesel from WVO, with the capacity to produce approximately 240,000 L/year [17].

In 2006, it was carried out a project entitled "Potential and Feasibility of Using Bioethanol and Biodiesel for Transport in Mexico", to determine the feasibility of producing such biofuels. In this project, it was assessed the production of biodiesel from rapeseed, soya, jatropha, sunflower and safflower oils, and the use of animal fat and WVO. The results suggest that farm input costs represent between 59% and 91% of biodiesel production costs, and as a result, animal tallow and WVO are an opportunity for biofuels production [18, 19].

In 2007, the Engineering Institute of the Autonomous University of Baja California processed the first batch of biodiesel from WVO. It counts with two reactors to process a total of 416 L of biodiesel per batch. Besides the use of WVO, biodiesel from castor oil and inedible oilseeds from Mexicali Valley has been obtained. Subsequently, there have been developed research projects on topics related to the inventory of residual feedstocks to produce biodiesel, the performance of internal combustion engines operating with diesel-biodiesel blends, as well as the study of the behavior of materials exposed to diesel, biodiesel and mixtures of both [6, 15, 16, 19-21]. Research focused on the production of biodiesel via enzymatic has been conducted as well [22-25].

In 2011, a National Laboratory of Biofuels was inaugurated in Chiapas as a result of the joint efforts by the government, the National Council for Science and Technology (CONACYT) and the Tuxtla Gutiérrez Institute of Technology. Currently, analytical services for the biodiesel production chain are offered [26].

In 2012, the Autonomous University of Puebla began initial production tests to produce 18,000 L/year of biodiesel approximately, despite reporting a production capacity of up to 72,000 L/year. The raw material used was castor oil. The production was intended to supply trucks of the university transport system. The funds were granted by CONACYT and the state government [27].

In 2012, the Pumabus of Ciudad Universitaria (CU) began using biodiesel. This occurred as a result of studies conducted by students at the National Autonomous University of Mexico (UNAM), who determined that more than 10,720 L of WVO were annually discarded by the CU restaurants [28].

In 2013, with an investment of 1.6 million pesos, a plant for biodiesel production in Zacatecas, located in the Autonomous University of Fresnillo began operations. However, the expectations were not met, since it was expected to link production and business sectors, but to date only serves for students to do practice and internship. The goal is to produce 100,000 L of biodiesel weekly [29].

The International Energy Agency (IEA) as a result of a study on "Advanced biofuels in developed and developing countries" mentioned that Mexico has a high level of human resources in the biofuels domain. Up to the date, there are several universities, government agencies and research centers such as CIATEC, CIATEJ, UAG, UABC, CIBNOR, CEPROBI, CIBA, CICY, UNAM, INECOL, CINVESTAV, CICESE, UADEC, SAGARPA, INIFAP and UANL, among others, are working on a wide range of investigation lines and research projects related to the pretreatment of WVO by chemical and enzymatic process for biodiesel production, research and development on microalgae, study of new feedstocks for second and third generation biodiesel production, biotechnological processes and new biocatalysts development for biodiesel production [30]. Further, there are research in progress on the development of biorefineries at pilot scale plant, to optimize the use of biomass resources for the production of biofuels e.g. biodiesel and bioturbosine from *Jatropha Curcas L.* and microalgae.

The CONACYT and Secretariat of Energy (SENER) have supported financially many projects of consortium institutions on biodiesel research and development, and biorefinery systems through national funds programs. One of them is an ongoing project that is conducted with the objective to generate knowledge frontier at laboratory and pilot scale of a biorefinery integral system for biogas production, biodiesel from microalgae and hydrogen from algae biomass residues, by using domestic wastewater. New marine microalgae strains with the highest energy potential and vegetable oil yield are being studied [31].

4. Biodiesel legal framework in Mexico

Considering the importance and the interest of bioenergy worldwide and in Mexico, on February 1st 2008 in the Official Federation Journal was published the decree to enact the LPDB. It aims the promotion and development of bioenergy in order to contribute to energy diversification and sustainable development as conditions possible to ensure support for the Mexican agriculture and provides the basis for [4]:

- **i.** Promoting biofuels production inputs, from agricultural, forestry, algae, biotechnological and enzymatic processes in the Mexican agriculture, without compromising food security and sovereignty of the country.
- **ii.** Develop the production, marketing and efficient use of bioenergy to contribute to the reactivation of the rural sector, generating employment and a better quality of life for the population of high marginalization.
- **iii.** Promote regional and rural development.
- **iv.** Seek to reduce air pollutants and GHG emissions, using international instruments contained in the Treaties signed by Mexico.
- **v.** Coordinate actions among Federal, State, Federal and Municipal District, social and private sectors for the development of bioenergy.

There is also the Regulation of the LPDB, which states that the SENER is responsible for issuing permits for production, storage, pipeline distribution and marketing of bioenergetics [32]. These permissions are granted for activities and necessary transformation processes of biomass fuels from organic matter in agriculture, livestock, forestry, forestry activities, aquaculture, algae-culture, fisheries waste, household waste, commercial waste, industrial waste, microorganisms and enzymes as well as their derivatives.

In accordance with the definitions in the LPDB, the Inter-Secretariat Commission for the development of bioenergy, as the entity of the highest level to promote the development of biofuels in Mexico was created [33]. The Inter-Secretariat Commission is comprised by the following secretariats:

- Secretariat of Energy (SENER).
- Secretariat of Agriculture, Livestock, Rural Development, Fisheries and Food (SAGARPA).
- Secretariat of Environment and Natural Resources (SEMARNAT).
- Secretariat of Economy (SE).
- Secretariat of Finance and Public Credit (SHCP).

The main functions of the Inter-Secretariat Commission are planning, participating in the creation of national and sectoral development plans; setting rules and guidelines to encourage public and private participation in the development of biofuels; monitoring of actions taken; and setting priorities for the allocation of public spending for the promotion and development of biofuels.

Diverse programs for the introduction of bioenergetics and bioenergy sustainable production feedstocks have been created. However, the results at the date have not been scaled industrially with success.

5. Biodiesel associations in Mexico

At present, there are few non-government associations related to biofuels in Mexico: the Mexican Network of Bioenergy (REMBIO) and the National Association of Biofuels Producers (ANPB).

6. Conclusion

The development of the biodiesel industry in Mexico at industrial scale has taken its first steps evidenced by the installation of six production plants. These plants were designed to produce first and second generation biodiesel. It has been made possible by the legal framework on biofuels. However, these industrial biodiesel production developments have been unsuccessful. Some of them were due to the lack of assurance of the availability of raw materials, others by the lack of agreement between existing legislation on biofuels and public policies that promote and encourage the production and consumption of biodiesel. Biodiesel productions from first and second generation in Mexico are assimilated and tested technologies in production plants. Several research-centers have projects in development regarding the third generation biodiesel from microalgae. Mexico has high-level human resources, capacities and capabilities about biodiesel production. However, the government has focused mainly its efforts on energy policy oriented to optimization of oil and natural gas industry, and to a lesser extent to renewable energy, including biodiesel. Mexico should focus its efforts on second and third generation biodiesel without jeopardizing food security.

Acknowledgements

The authors thank to the Institute of Engineering of Universidad Autónoma de Baja California and Consejo Nacional de Ciencia y Tecnología for their support in the development of the present work.

Author details

Gisela Montero[1*], Margarita Stoytcheva[1], Marcos Coronado[1], Conrado García[2], Jesús Cerezo[1], Lydia Toscano[2], Ana M. Vázquez[3] and José A. León[1]

*Address all correspondence to: gmontero@uabc.edu.mx

1 Instituto de Ingeniería, Universidad Autónoma de Baja California, Calle de la Normal S/N, Col. Insurgentes Este, Mexicali, B.C., México

2 Instituto Tecnológico de Mexicali, Academia de Química y Bioquímica, Av. Tecnológico S/N Col. Elías Calles, Mexicali, B.C., México

3 Escuela de Ingeniería y Negocios Guadalupe Victoria, Universidad Autónoma de Baja California, Carretera Estatal No. 3 Col. Gutiérrez, Mexicali, B.C., México

References

[1] SENER. Secretaría de Energía. Balance Nacional de Energía, 2012. http://www.sener.gob.mx/portal/default.aspx?id=1433 (accessed 02 May 2014).

[2] DOF. Diario Oficial de la Federación. Ley General de Cambio Climático, 2012. http://www.diputados.gob.mx/LeyesBiblio/pdf/LGCC.pdf (accessed 14 July 2014).

[3] DOF. Diario Oficial de la Federación. Ley de Desarrollo Sustentable de la Caña de Azúcar, 2005. http://www.diputados.gob.mx/LeyesBiblio/pdf/LDSCA.pdf (accessed 17 August 2014).

[4] DOF. Diario Oficial de la Federación, 2008. Ley de Promoción y Desarrollo de los Bioenergéticos. http://www.diputados.gob.mx/LeyesBiblio/pdf/LPDB.pdf (accessed 30 June 2014).

[5] Stoytcheva M., Montero G., Toscano L., Gochev V., Valdez B. The Immobilized Lipases in Biodiesel Production. In: Stoytcheva M. (ed.) Biodiesel-Feedstocks and Processing Technologies. InTech; 2011. p. 397-410. Available from: http://www.intechopen.com/books/biodiesel-feedstocks-and-processing-technologies/the-immobilized-lipases-in-biodiesel-production (accessed 30 July 2014).

[6] Coronado M., Montero G., Valdez B., Stoytcheva M., Eliezer A., García C., Campbell H., Pérez A. Degradation of nitrile rubber fuel hose by biodiesel use. Energy 2014; 68: 364-9. DOI: 10.1016/j.energy.2014.02.087.

[7] Sheinbaum-Pardo C., Calderón-Irazoque A., Ramírez-Suárez M. Potential of biodiesel from waste cooking oil in Mexico. Biomass and Bioenergy 2013; 56: 230-8.

[8] Prehn M., Cumana I. La Bioenergía en México: Estudios de caso nº 1, 2010. http://www.rembio.org.mx/2011/Documentos/Cuadernos/CT1.pdf (accessed 17 May 2014).

[9] El Economista. Fracasa proyecto de biodiesel, 2011. http://eleconomista.com.mx/estados/2011/07/18/fracasa-proyecto-biodiesel (accessed 23 June 2014).

[10] MORECO. Clientes Moreco, 2010. http://www.moreco.com.mx/clientes/ (accessed 03 May 2014).

[11] Olmedo F. Inauguran primera planta productora de biodiesel en Michoacán, 2007. http://www.biodisol.com/biocombustibles/inauguran-primera-planta-productora-de-biodiesel-en-michoacan/ (accessed 28 May 2014).

[12] Sarmiento R. Primeros desarrollos de producción de biodiesel en México, 2008. http://www.energiaadebate.com/Articulos/Julio2008/RocioSarmientoJulio2008.htm (accessed 06 June 2014).

[13] MORECO. Visita a planta de biodiesel en L.C. Michoacán, 2014. http://www.moreco.com.mx/nosotros/ (accessed 30 June 2014).

[14] Vega J. Bioenergía en México: Posibilidades, Avances, Retos, 2012. http://www.cmic.org/comisiones/sectoriales/medioambiente/Varios/The%20green%20expo%202012/Energ%C3%ADa/presentaciones/26sept/2.bioenergiamexico.pdf (accessed 23 August 2014).

[15] Vázquez A., Montero G., Sosa J., García C., Coronado M. Economic Analysis of Biodiesel Production from Waste Vegetable Oil in Mexicali, Baja California. Energy sci-

ence and technology 2011; 1(1): 87-3. www.cscanada.net/index.php/est/article/viewFile/j.est.1923847920110101.007/1294

[16] Toscano L., Montero G., Stoytcheva M., Campbell H. Preliminary assessment of biodiesel generation from meat industry residues in Baja California, Mexico. Biomass Bioenergy 2011; 35(1): 26-1. DOI: 10.1016/j.biombioe.2010.10.031.

[17] Bernal A. Perfil Comercial: Compilación de la Jatropha Curcas, 2008. http://seder.col.gob.mx/seder2012/comercializacion/perfiles/Jastropha.pdf (accessed 15 May 2014).

[18] SENER. Secretaría de Energía. Potenciales y Viabilidad del Uso de Bioetanol y Biodiesel para el Transporte en México, 2006. http://www.bioenergeticos.gob.mx/descargas/SENER-BID-GTZ-Biocombustibles-en-Mexico-Estudio-completo.pdf (accessed 15 August 2014).

[19] Montero G., Stoytcheva M., García C., Coronado M., Toscano L., Campbell H., Pérez A. Vázquez A. Current status of biodiesel production in Baja California, Mexico. In: Montero G. (ed.) Biodiesel quality, emissions and by-products. InTech; 2011. p. 137-152. Available from: http://www.intechopen.com/books/biodiesel-quality-emissions-and-by-products/current-status-of-biodiesel-production-in-baja-california-mexico (accessed 09 May 2014).

[20] Coronado M., Montero G., Eliezer A., García C., Cerezo J., Pérez L., Ayala J. Materials technological challenges for the biodiesel industry development in Mexico. In: Méndez-Vilas A. (ed.) Materials and processes for energy: communicating current research and technological developments. Formatex Research Center; 2013. p. 279-288. Available from: http://www.formatex.info/energymaterialsbook/book/279-288.pdf (accessed 29 May 2014).

[21] Montero, G., Coronado, M., Campbell, H., Cerezo, J., Lambert, A., Valenzuela, E. Biomass Wastes: An Energy Option for Baja California, México. Journal of Power and Energy Engineering 2014; 2:146-0. DOI: 10.4236/jpee.2014.24021.

[22] Toscano L., Gochev V., Montero G., Stoytcheva M. Enhanced production of extracellular lipase by novel mutant strain of aspergillus niger, Biotechnology & Biotechnological Equipment 2011; 25(1):2243-7. DOI: 10.5504/bbeq.2011.0019.

[23] Toscano L., Montero G., Stoytcheva M., Gochev V., Cervantes L., Campbell H., Zlatev R., Valdez B., Pérez C., Samaniego M. Lipase production through solid-state fermentation using agro-industrial residues as substrates and newly isolated fungal strains, Biotechnology & Biotechnological Equipment 2012. HTTP://DX.DOi.OrG/10.5504/BBeQ.2012.0145.

[24] Toscano L., Montero G., Cervantes L., Stoytcheva M., Gochev V., Beltrán M. Production and partial characterization of extracellular lipase from trichoderma harzianum by solid-state fermentation, Biotechnology & Biotechnological Equipment 2013; 27(2): 3776-1. HTTP://DX.DOi.OrG/10.5504/BBeQ.2012.0140.

[25] Toscano L., Montero G., Stoytcheva M., Cervantes L., Gochev V. Comparison of the performances of four hydrophilic polymers as supports for lipase immobilisation, Biotechnology & Biotechnological Equipment 2014; 28(1):52-0. DOI: 10.1080/13102818.2014.901684.

[26] SENER. Polo tecnológico nacional para el desarrollo de investigación y servicios analíticos en biocombustibles, 2011. http://www.renovables.gob.mx/res/Viernes%2020/3%20-%20Rocio%20Meza.pdf (accessed 06 September 2014).

[27] Foro Consultivo Científico y Tecnológico. Contará la BUAP con planta de producción de biodiesel, 2014. http://www.foroconsultivo.org.mx/innovacion.gaceta/component/content/article/224--innovadores-/303-contara-la-buap-con-planta-de-produccion-de-biodisel (accessed 01 August 2014).

[28] UNAM-DGCS. Donan planta de biodiesel a la Facultad de Ingeniería de la UNAM, 2012. http://www.dgcs.unam.mx/boletin/bdboletin/2012_670.html (accessed 07 July 2014).

[29] Investigación y Desarrollo. Sin "clientes" laboratorio de biodiesel en Zacatecas, 2014. http://www.invdes.com.mx/ciencia-mobil/5074-sin-clientes-laboratorio-de-biodiesel-en-zacatecas (accessed 23 July 2014).

[30] Sandoval G. Biocombustibles avanzados en México, 2011. http://www.rembio.org.mx/2011/Documentos/Cuadernos/CT2.pdf (accessed 02 September 2014).

[31] CIBNOR. Centro de Investigaciones Biológicas del Noroeste, 2014. http://www.cibnor.mx/es/investigacion/acuicultura/lineas-y-proyectos-de-investigacion/proyectos-linea-iv/852-biorrefineria-para-la-produccion-de-biogas-biodiesel-e-hidrogeno-a-partir-de-microalgas-y-aguas-residuales-domesticas (accessed 06 September 2014).

[32] DOF. Diario Oficial de la Federación. Reglamento de la Ley de Promoción y Desarrollo de los Bioenergéticos, 2009. www.diputados.gob.mx/LeyesBiblio/regley/Reg_LPDB.doc (accessed 28 August 2014).

[33] SAGARPA. Secretaría de Agricultura, Ganadería, Desarrollo Rural, Pesca y Alimentación. http://www.bioenergeticos.gob.mx/index.php/programas/marco-institucional.html (accessed 04 September 2014).

Section 6

Biorefinery Processes

Chapter 20

Biorefinery Systems as an Element of Sustainable Development

Krzysztof Biernat and Paulina Luiza Grzelak

Additional information is available at the end of the chapter

http://dx.doi.org/10.5772/60448

1. Introduction

As a result of the review of implementation and research work carried out in the EU countries, aiming to intensify the use of renewable energy processes, it turned out that these works are scattered and do not bring the desired effect, both in terms of environmental and savings of energy conversion. Because of that, it created a vision of the industry based on raw materials of biological origin, which also qualified waste substances from the primary and secondary processes of exploitation and processing of biomass as defined in Directive 28 [1]. The realization of this vision should lead to a transition toward the so-called "post-oil" society through a clear decoupling economic growth from resource depletion and environmental impact.

After consultation carried out in the member states set out the need for separation of a new branch of industry defined as an industry based on raw materials of biological origin ("Bio-based Industries"), which should seek to optimize the use of land and food security through sustainable, efficient (effective), raw and largely limiting the amount of waste, industrial processing of European renewable raw materials in a wide range of products of biological origin, such as:

- Advanced transport fuels
- Chemicals
- Materials
- Food and feed ingredients
- Energy [2]

Thanks to that, "the bio-industry," which is the main component of the EU economy referred to as "bio-economy" ("bioeconomy"), will play an important role in stimulating sustainable growth and increasing Europe's competitiveness by reindustrialization and the revitalization of rural areas, providing tens of thousands of jobs in the field of research, development, and production over the next decade [3].

The Bioeconomy Program for Europe is going to be an evolutionary program. Expected to develop so-called value chains, the implementation of which will ultimately lead to the creation of so-called biorefinery that a comprehensive and zero-waste will be recycled biomass. The most important technological challenges, political and market, therefore will be prior to commercialization of innovative solutions to full scale. These challenges cannot be overcome by an individual company or dispersed industry, so it is necessary to approach the whole system of management system biomass [4].

This is important because of the need to reverse the current trend of significant bioeconomic investments in non-European regions, where conditions seem to be more attractive. The long-term research and innovation jointly financed by public and private entities can help solve this problem. This process will be implemented through the creation and implementation of appropriate and developed value chains, which will lead to reducing the risk of investment in demonstration projects on the implementation of innovative processes.

As part of the preparatory work for the start-up of the scope of the European bio-economy, there was a plan developed for Strategic Innovation and Research Agenda (SIRA). This document proposes a coherent set of actions that should be implemented through established "Biobased Industry Consortium" (BIC), namely:

- Implementation of projects aimed toward the integration and implementation of technology and scientific results and the introduction of technology on a commercial scale by implementing demonstration and flagship projects
- Implementation of development projects aimed at filling the gaps in research and technological innovation
- Supporting projects taking challenges cross-sectors [5]

Schematically, the areas covered by value chains are shown in Fig.1.

As it can be seen from the schematic products, semi-finished and all residues of the process as a result of the implementation of the objectives set in the value chains should be directed to biorefinery systems, in order to complete the transformation into energy carriers and biochemicals for various purposes.

2. The exchange of mass and energy in waste management

Circulation of matter and energy flow are the two main environmental laws describing the basic rules for the functioning of ecosystems on Earth. Life on our planet is possible, thanks to

http://dx.doi.org/10.5772/60448

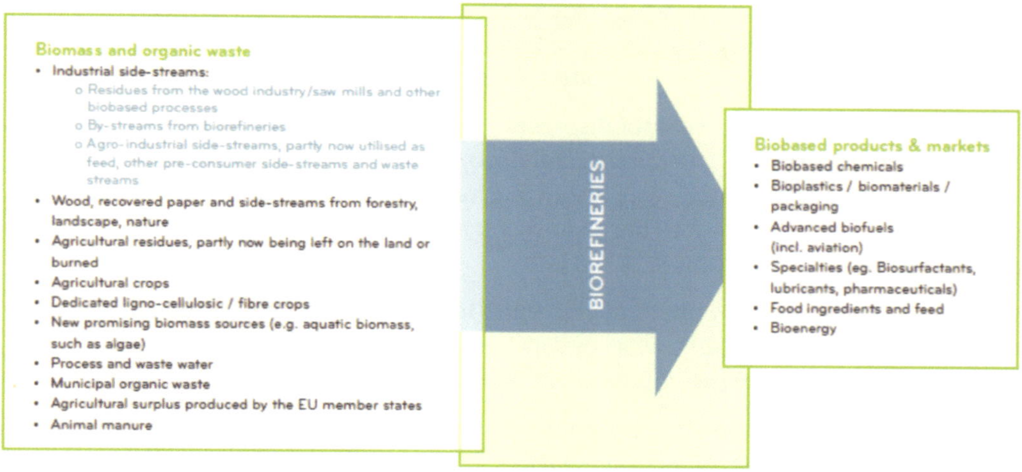

Figure 1. Value chains in bioeconomy [5]

the constant influx of solar energy. This energy flows through the ecosystem and partly returns to space in the form of thermal energy emitted by the Earth. Only a small portion of solar energy is accumulated in living organisms, and they even lose most of it forever. However, despite the exchange of energy, the Earth does not mention the matter with the environment. Earth is thus a closed system. The amount of matter on Earth is constant and circulates in the ecosystem. Under the influence of solar radiation, simple ingredients such as water and carbon dioxide are synthesized biomass, which forms a complex structure such as plants and other biological life forms, which in turn ensure the survival of other more complex forms of biological life and after their death through the action of decomposers and the way the carbon cycle and the nitrogen return to its original state [6].

Behavior of matter and energy is the subject of thermodynamics. The first law of thermodynamics states that energy can neither be created nor destroyed. It can only change its form. This means that everything which is delivered for processing must be accumulated in products in the technosphere[1] or leave them as waste in solid, liquid, or gas form. In addition, the system cannot get more energy than it was put into the system. Every economic activity of man is connected inextricably with the generation of waste, and it should be noted that all industrial products become waste sooner or later. The increasing amount of waste causes obvious environmental problems [7, 8].

All processes in the environment are irreversible processes, in other words, the real processes. In nature, the reversible processes are unique. This is because the reverse process cannot undergo energy dissipation that occurs in the actual process, even, for example, in effect of friction. Similarly, the industrial installations are open systems that interact with the environment together with its products. The energy introduced into the system is transformed from

1 Sphere of human intervention in nature, spreading in the environment

a less useful form, which is more organized, in a more useful form, which is less organized. This interaction can be described by the change in entropy[2], that is state function, which is postulated by the second law of thermodynamics [9].

Entropy is a measure of the degree of disorder of the system, a measure of the "quality" of the energy stored in the system. One version of the second law of thermodynamics states that "the arrangement is reduced (the degree of disorder increases) in any spontaneous process." Energy and matter cannot be destroyed, but their quality is changed. Together they seek a greater mess, so their entropy increases [6, 10, 11].

According to the second law of thermodynamics for a longer period of time, entropy always increases. Each physical system, which is capable of free energy dissipation, does so in such a way that the entropy increases and the amount of used energy decreases [12, 13].

The first law of thermodynamics is a universal law of nature, from which there are no exceptions. The second law of thermodynamics is based on the concept of probability, due to the fact that it is possible when the entropy decreases. There are unlikely but possible situations. The second law of thermodynamics indicates the most likely direction of the incident and does not exclude the other less probable incidents.

Fossil natural resources used by man naturally are ordered matter of a certain structure. The products of human consumption are waste, in which the degree of order of matter is much lower, despite the fact that the elements present in the waste as well as natural resources are the same.

Entropy changes can occur either within the system or may be discharged to the environment. Export of entropy outside of the system is caused precisely by the fact that the processes are irreversible that causes irreversible increase in entropy which is the reduction in the degree of ordering. Reduction of changes in the form of entropy, for example, in the course of the technological process reduces the energy expenditure in obtaining the same effect as a final product, resulting in minimizing the losses in the process and consequently a reduction in the degree of impact on the environment. This can be done, e.g., on the way to limit the number of unit processes in technology, keeping in mind that the total entropy of unit processes will always be higher than the sum process entropy [6, 11].

The concept of entropy resulting in the way of irreversible processes occurring in the natural environment is due to spontaneity of overlapping of these processes. Spontaneous processes are irreversible processes, extending with a predetermined rate in such direction to achieve a state of balance. So it is not possible to restore the state before the occurrence of irreversible process while maintaining all the parameters of both the system and the environment. For example, if the process is reversed and the parameters of the system will return to the initial values, the environment will not return to the previous state because of the entropy emission into the environment as a result of reactions in the system.

Analysis of processes for entropic changes enables optimization of technological processes used in the industry, as well as optimizing the use of raw materials and energy in order to

---

2 A measure of the degree of order – the higher entropy of the system, the lower the degree of order.

obtain specific products. The essence of these considerations is the production of entropy, which determines the degree of energy loss in the process, and so is the energy efficiency of the process. As it was mentioned earlier reduction of entropy production as a result of optimization technology, the substrates used and the energy introduced into the process reduce their impact on the environment. This is possible through the pursuit of the process to reversible process conditions and thus to obtain the equilibrium parameters. However, the pursuit causes inconveniences in the form of diminution of other important parameters of installation, such as enlargement of device size, which in many cases leads to the elimination of some of this type of technologies.

Therefore, the goal of reducing the impact on the environment should be carried out differently. It should be noted that the export of entropy, that is, the energy lost from the process, is carried out through the emission of unused waste heat and/or through the generation of waste. Also, the products of the process, after the step of its use, become waste. Various types of systems comprising external and internal recycles of a flow of mass and energy in the form of heat may be constructed to limit the production of all the waste from the process. This type of system, congruent with the concept of biorefinery, is shown in Fig. 2.

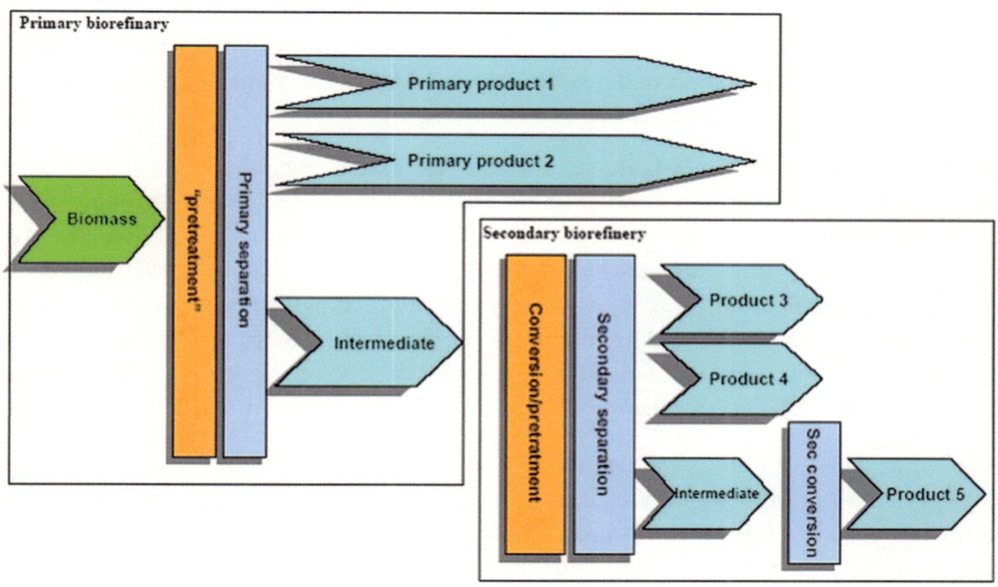

Figure 2. Scheme of biorefinery system containing recycles of mass and/or energy flow [14]

The stream of biomass (matter) and stream of energy flow into the biorefinery system, which are used and processed in the way of biorefinery processes, which results in obtaining useful products and intermediates, which at this stage are treated as process waste. However, by the internal recycle, these intermediates are used as starting materials in subsequent biorefinery processes. The same happens to the energy that is inserted into the process and allows the incident and which may also be the product of the appropriate process. Generally, part of the energy emitted in the form of heat is waste from the process, which with properly optimized technology can be used as the feed energy flow to a subsequent process. From the example of this biorefinery system of the second row can be seen clearly that by optimizing the use of reactants and waste streams in the process, it is possible to reduce the formation of entropy, which prevents the loss of energy by providing greater energy efficiency of the system [15].

It should be noted that the internal and external recycle systems can be used not only at the level of the process but also at the level of the whole plant, as well as at the level of product user or user of energy produced and on the level of region and state. The effect of this will be reduction of the entropy of the system, reducing the consumption of energy and matter, increasing the efficiency of the process, and finally reducing the impact on the environment.

In the analysis of the energy efficiency of processes, of very help is the function of exergy. The precursors of exergy analysis were G. Gouy and A. Stodola, who formulated the law defining the loss of ability to perform work resulting from a thermodynamic irreversibility. According to the definition of the concept proposed by L. Reikerta, "exergy represents the minimum amount of work to be done, to produce from the commonly occurring components of the surrounding nature a desired substance with the required parameters, using the surrounding nature as a heat source, which is worthless in terms of thermodynamic," which in simple can be captured in the words, "exergy is the maximum capacity of the substance, which is an energy carrier, for the work in relation to the environment." Most simply speaking, the exergy is a measure of the quality of various forms of energy [16, 17].

Determination of loss of exergy is the main task of the exergy analysis, because each exergy loss causes a reduction of useful effects of the process or increases the consumption of fuels for its occurrence. Exergy losses can be divided into internal and external. Internal losses are caused by the irreversible processes extending inside the system (control cover of process). External losses result from the discharge of waste products with positive exergy into the environment, and the loss should preferably be expressed directly using the exergy of waste products. Calculation result of internal losses does not depend on the reference level adopted in the calculation of the external losses [18, 19].

However, any loss of exergy should have economic justification. Admission to the loss of exergy is necessary in order to reduce capital expenditures. For example, the flow of heat without loss of exergy would require an infinitely large surface of heat flow. So exergy analysis indicates on the possibilities of improving the thermal process. According to Jan Szargut, this analysis does not decide on the advisability of the improvements, which should be controlled by means of economic analysis. We cannot agree with this statement, because not only economic considerations will determine the attractiveness of technological solutions but also reduction of energy consumption and increase of the energy efficiency of the process will provide a means to achieving the goal of protecting the environment by reducing the impact on them through the process [20].

Analysis of entropy based on the calculation of exergy process allows for optimization of the manufacturing process and proper assessment of its impact on the environment, taking into account all the components and stages of the production process, starting from the raw material constituting substrates in the process and ending on the disposal of waste from the process and waste resulting from the use and consumption of the final product [21, 22, 23].

Therefore, the comparison of the possible use of exergy or entropy emissions in different technological processes allows the systematization and classification of the different technologies in this field. This capability is particularly important in the case of methods of processing and disposal of waste. Among others, the studies showed that the most favorable thermal process in the waste utilization is pyrolysis. This process is ahead of the processes of gasification and combustion for energy. Also, in waste management, this classification is important. Waste actually represents the energy that is correlated with the value of primary energy resources, from which the product was formed and then the waste was produced. Using this energy should be the most efficient and should last as long as possible, and hence, its dissipation should take place as long as possible at the time. This is possible only through the repeated use of the energy contained in the products, intermediate products and wastes [24].

3. Sustainability in the course of processing waste

Sustainable development requires an assessment of the extent to which natural resources are in sufficient quantity and availability, to meet those needs while reducing the amount of waste (both process and also final, generated during the processing and usage of these resources).

The degree of loading of the Earth, which is the environmental capacity of the Earth, is the limited size and is determined by three interrelated factors: the size of the human population, the activity of each human being, and the impact on the environment, which is a result of this activity. The relationship of these factors and their influence on the degree of loading of the Earth is defined by the formula (1)

$$ELL = \left[HI_1 \cdot A_1 \cdot (\pm E_1)\right] + \left[HI_2 \cdot A_2 \cdot (\pm E_2)\right] + \left[HI_3 \cdot A_3 \cdot (\pm E_3)\right] + \ldots + \left[HI_{7\,mld} \cdot A_{7\,mld} \cdot (\pm E_{7\,mld})\right] \quad (1)$$

where:

EL – loading **l**evel of the Earth

HI – **h**uman **i**ndividual

A – the **a**ctivity which is manifested by the human individual

E – the impact on the **e**nvironment of the human individual through the activity

At the present time, there is a continuous and exponential growth of the human population on Earth observed and thus human activity aimed at creating wealth – as a result, omnipresent consumerism is growing globally. In results, there is greater impact on the environment through the use of its resources. To lower the burden on the environment, the value of human activity and the impact of each of the human individual through this activity should be reduced. Human activity cannot be reduced, but this type of activity can be develop on the way of environmental education. In the same way, you can indirectly seek to reduce the environmental impact and the burden of every human activity [25, 26, 27].

There are two ways to reduce the environmental load. The first of these is to limit the use of resources, which is called dematerialization, which is gradually reducing usage of resources in technological processes. The next step should be the determined counteraction character of the life of a consumer society in developing countries and developed countries. Another way to limit the use of natural resources, including energy ones, is to replace the current raw materials for other unknown resources, by-products, or wastes from other processes, which is called transmaterialization [7, 28].

Both of the abovementioned proceedings road are independent, but only their parallel application can provide the most tangible effect of actions to reduce the impact of human activities, including industrial environment. The Earth and its biosphere are thermodynamic systems exchanging with the environment energy only. The components of the biosphere subsystems, however, are open systems, exchanging with their environment both energy and matter. One of the subsystems, so-called man-made technosystem, collects raw materials from the environment and excretes waste. Systematic expansion of the technosystem and its impact not only contributes to an imbalance in the biosphere but also causes a real threat to its future existence [29, 30].

In the context of this situation arises question about the possibility of avoiding further destruction of the biosphere. Following Johansson, there are two possibilities for remedying this problem.

The first one is the complete separation of the technosystem of the biosphere, creating a closed technosystem. This method would involve major technological challenges for the modification of existing technologies for closing circuits of waste, which will entail a significant increase in energy consumption. Note also that technosystem separated from the biosphere would have to thus eliminate human intervention [7].

The second method proposed by Johansson is the adaptation of the technosystem to the environment in such a way that its action plays the strategy and action, which uses nature. In this solution, however, certainly limitation of many existing technological capabilities would occur, including reduction in the size and density of the local industrial activities, which, at this stage of the development of civilization, have already been exceeded in many cases. Implementation of this strategy would involve fundamental reorganization of the existing technological structure which would entail the search for new technological solutions [7].

Due to the fundamental problem of modern society as of a linear flow of matter from its source as raw materials to the waste stage and the accumulation of products of human activity in this form in order to maintain good environmental status as long as possible, our society must in its actions be inspired by nature, reducing the negative environmental impact to a minimum. A society that wishes to survive must maintain the activity causing disorder within order, possible to achieve by the energy captured from the sun. In addition, waste (degraded material

http://dx.doi.org/10.5772/60448

with a high degree of disorder, which comes from industries and households) should be converted back to natural resources [31].

From above it follows that the least favorable from the energy point of view is depositing waste in landfills, where primary energy accumulated in the waste is dissipated during their long storage, without its prior use. In this case, the relatively long time of the energy dissipation is ensured, but the benefit of the use of this energy and hence saving in the use of primary energy is equal to zero. Burning of waste in incinerators is a better solution for the landfill, because it allows you to recover some of the energy from waste, and this energy can be used in other processes. However, the processing of the same material by pyrolysis and gasification is even more preferred environmentally and economically in view of the minimization of oxides and heavy metal migration into the exhaust gases, as well as because of the higher efficiency of cogeneration systems. In addition, in these processes is formed synthesis gas or, in the pyrolysis process, post-pyrolysis oil, which both can be repeatedly used for the production of energy in various forms, including the ability to convert it to transport fuel. Hence, the more energy is consumed for the processing of raw materials, the more waste is created, because of the identity of mass and energy. In nature, there is a balance between the processes of formation of ordered matter, as of organic compounds in living organisms and the production of waste. Nature produces no waste and does not deplete resources, because in nature, nothing is really a waste – something that for one organism means waste is feed for another. In other words, everything is used in the life cycle [8, 32].

4. Biorefinery systems

One of the ways to mitigate the negative effects of local ecosystems is the conversion of biomass and organic waste into a different type of chemical substances or biomaterials and energy to fully exploit the value of the biomass, creating the so-called added value and minimizing the quantity of naturally produced substance or waste. This integrated approach reflects the concept of biorefineries and is gaining more and more attention in many parts of the world.

Similarly to the conventional refinery which produces energy and chemical products from petroleum, biorefineries will produce a variety of industrial products from biomass. These products are both LVHV type (low value and large volume), such as transport fuels and high-volume chemicals and other materials, and HVLV type (high value and low volume), as specialized chemicals and cosmetics, for example. In some types of plant, there also can be produced food and animal feed.

Energy sources are the main driving force behind the development of this type of installation, but sometimes when biorefineries will become more advanced and complex systems, the development of other products such as installations also occurs.

Biorefinery systems are nothing more than a kind of open systems, where part of the input streams is biomass, waste, and energy flow. Within the system there is a series of processes resulting in, for example, the energy exchange with the environment in the form of heat and work. An output streams of biorefinery systems constitute a number of products such as fuels, chemicals (both highly valuable, which are obtained in small quantities, and of low value, obtained in large quantities), feed and food products, polymers, and other materials, as well as the energy produced in cogeneration or trigeneration (heat, electricity, and cooling) and processed waste. Please note that these wastes are waste only for the specific biorefining process. For another manufacturing process, these can be a substrate.

Biorefinery systems, imitating in its actions a living organism as open systems, in contrast to conventional petroleum refinery, may therefore constitute one of the elements of sustainable development [33].

4.1. The concept of biorefinery

Repeatedly attempting to define the concept of biorefineries, and as a consequence, a number of incomplete or differing definitions have been established. After attempts to deduce from them the most important elements and characteristics of this new branch of industry, a comprehensive definition of the biorefinery has been established, according to which, it is an integrated "bio-industry," which uses a variety of technologies in order to obtain products such as chemicals, biofuels, food, feed ingredients, biomaterials (including fibers), and heat and energy, focusing on maximizing the added value, taking into account the three pillars of sustainability: environment, economy, and society. According to the definition of the International Energy Agency (IEA), biorefinery is a way for sustainable biomass processing in a wide range of bio-products of food, feed, chemicals, and biomaterials and bioenergy products such as biofuels, electricity, and heat [34].

By definition, biorefinery is a complex technological system that combines biomass conversion processes and the further processing of products of this conversion to fuels and chemicals – final or intended for further processing. Therefore, biorefinery is equivalent to crude oil processing plants (Figs. 3 and 4), where the substrate is crude oil, natural gas, or other fossil energy resources. These resources are processed in petrorefinery processes on a variety of products, mainly fuel, electricity, heat, chemicals, and various kinds of materials. The substrated in biorefineries are organic materials such as wood, energy crops, grass, and organic waste, which are processed in biorefinery processes – which are similar to the refinery processes – used in conventional petroleum refineries. Products from biorefinery are also the fuels and cogeneration or trigeneration energy, chemicals, and materials as well as food and animal feed. The basic petrorefinery scheme is shown in Fig. 3; Fig. 4 shows a general schematic diagram of a biorefinery.

As can be seen from a comparison of those two patterns, petrorefinery and biorefinery are related systems, processing a substrate or a plurality of substrates into a product or series of products by means of one or more technological processes which, as mentioned earlier, can be used both in a single and the second type of installation.

As mentioned, it should be noted that both the raw materials and also products of the biorefinery should be a much smaller threat to the environment, mainly emission of greenhouse gases. Hence, industrial biorefineries should constitute the most important element of

Figure 3. The basic scheme of the classic oil refinery

Figure 4. Ideological scheme of biorefinery

new industrial sectors, based on renewable energy sources (raw materials), offsetting at least part of the progressive deficiency of existing media such as oil, coal, and natural gas [35].

Biorefinery can be considered as a tool for the implementation of sustainable development in the processes of energy use and waste of natural resources. According to the concept of sustainable development, this type of installation procuring energy is the most optimal solution that simultaneously takes into account the continuous technological development, production of so-called "clean" energy and other products while reducing greenhouse gases and harmful compounds. This technology is almost "waste-free," which uses the existing potential of biomass waste, which currently is not used at all or is used in an irrational way [35].

4.2. Biorefinery processes

Depending on the raw material and the desired product, biorefineries use a variety of conversion technologies of raw biomass to commercial sources. These processes frequently include fermentation, gasification, and transesterification. New and less traditional methods are still in the research area, especially in the development of synthetic biofuels, such as liquid biofuels from biomass (BtL – biomass to liquid). Other substances, except fuels, produced in biorefinery processes, such as chemicals or other materials, are not as popular as energy biorefinery products and are at a much lower level of development in terms of trade with respect to fuels derived from these plants.

The fundamental biorefinery processes used after pretreatment of the biomass material include enzymatic hydrolysis, fermentation, fast pyrolysis, and hydrothermal processing (HTU – *hydrothermal upgrading*), called hydrothermal liquefaction or hydrothermal pyrolysis, with possible further hydrodeoxydation process (HDO – hydrodeoxydation) [36].

Regardless of the complexity of the biorefinery processes, the general scheme of the biorefinery can be presented as it is shown in Fig. 5.

Figure 5. General biorefinery diagram [37]

4.2.1. Biomass pretreatment

Pretreatment is required in order to break the crystalline structure of cellulosic biomass, to make it accessible to the enzymes, which may be combined with the cellulose and hydrolyze the carbohydrate polymers into fermentable sugars. The purpose of the pretreatment is a pre-extraction of hemicellulose, lignin degradation, and release of the cellulose from the plant cell walls. Pretreatment is considered as one of the most costly stages of cellulosic bioethanol production but also has a large potential for its improvement and reduction of costs through continuous research and development.

Many pretreatment technologies have been developed and evaluated for different biomass materials. However, each of these pretreatment methods has its advantages and disadvantages, and one method is not suitable for all types of raw materials.

The most commonly used methods of pretreatment of biomass in order to apply it in biorefinery processes are alkaline pretreatment, pretreatment with hot water, and pretreatment with the dilute acid. Alkaline pretreatment process is carried out using dilute sodium hydroxide (NaOH), ammonia, or lime. This process aims to improve the capacity of the fermentation of cellulose. Pretreatment with hot water is a process called autohydrolysis. This process is intended to hydrate the cellulose before the actual processing. The last type of biomass pretreatment process is a dilute acid pretreatment. This process takes place with the participation of sulfuric acid of 0.5–1.0 % concentration, and it is designed to effectively remove and recover most of the hemicellulose from the processed biomass.

4.2.2. Enzymatic hydrolysis

The enzymatic hydrolysis is based on conversion of carbohydrate polymers to monosaccharides. Although various processes for the conversion of biomass to ethanol were studied, the enzymatic hydrolysis of cellulose provides the ability to improve the technology, so that the ethanol from biomass could be competitive in relation to other fuels in terms of both quality and economically.

Pre-prepared lignocellulosic material is subjected to enzymatic hydrolysis. This process involves the reaction for converting complex sugars to simple compounds – cellulose to glucose and hemicellulose into pentoses (xylose and arabinose) and hexoses (glucose, galactose, and mannose). Conversion processes of cellulose and hemicellulose are catalyzed by cellulase and hemicellulase. Cellulases play a significant role, because they catalyze the decomposition of cellulose into fermentable sugars. The enzymes involved in the hydrolysis of cellulose include endoglucanases, exoglucanases, and β-glucosidase. Endoglucanase randomly catalyze the decomposition of the internal bonds of the cellulose chain, whereas exoglucanases attack chain ends, releasing cellobiose molecules (disaccharide, which is not present in plants alone, transient degradation product of cellulose).

Enzymatic hydrolysis is a process often before fermentation, so it has to decompose cellulose and hemicellulose into fermentable monosaccharides. A solution of simple sugars can be fermented with microbes. However, some plant sugars, such as sugar beet or sugar cane, can be directly used in the fermentation process, without necessity to undergo the process of enzymatic hydrolysis [38].

4.2.3. Fermentation

Lignocellulosic biomass subjected to fermentation process requires separation of hemicellulose and cellulose material from the non-fermentable lignin, which is linked by strong covalent cross bonds. This is done using a pretreatment with an acid, alkali, or steam. Lignin, as waste from the bioethanol production, can be used as fuel and subjected to further processes of combustion or co-firing, in order to obtain energy.

Fermentation of C6 sugars, such as starch or sucrose, requires the use of organisms such as baker's yeast. In contrast, fermentation of C5 sugars – decomposed hemicellulose – requires special organisms, which are capable of fermenting xyloses. Currently, there is a need for more efficient and robust microorganisms that are resistant to higher temperatures and pressures. For example, recent studies carried out to improve the properties of the yeast strain showed that they may contribute to the production of more biofuels from cellulosic plant material by fermentation of all five types of plant sugars: galactose, mannose, glucose, xylose, and arabinose [36].

4.2.4. Fast pyrolysis

Fast pyrolysis is a process of thermal decomposition of biomass to liquid bio-oil, comprising carbohydrates and oxygen content of approx 35–40 %. Through successive hydrogenation and hydrodeoxydation processes or gasification, bio-oil can be converted to a specific hydrocarbon.

The use of fast pyrolysis process, as well as the properties of the thus produced biodiesel, is currently under investigation. However, there is the view that this process can significantly reduce the cost of the gasification process with respect to the direct use of solid biomass in a gas generator.

Fast pyrolysis process requires only one reactor, which involves relatively low costs. The relatively high temperature, - approx. 450–500 °C, the short residence time (approx. one second) of the load at this temperature at atmospheric pressure and resulting high yield of oil, also represents a clear advantage of this process. However, this process is also characterized by a considerable degree of non-selectivity and the formation of many products, including a large amount of soot. The feed used in the fast pyrolysis process must be drained initially, which generates costs and energy consumption and quality of the obtained fuels by the relatively poor [39].

4.2.5. Hydrothermal conversion

Another process of converting biomass in biorefineries is hydrothermal conversion process HTU (*hydrothermal upgrading*), which its diagram is shown in Fig. 6. Originally, this process was used for the conversion of coal to liquid fuels in conventional refining plant. Hydrothermal conversion is the process of biomass depolymerization carried out at high temperature, after which may optionally take place a catalytic hydrodeoxydation process HDO (hydrodeoxygenation). The temperature of this process is lower than in the fast pyrolysis process (300–400 °C), but the residence time of the feedstock in this temperature is approx. 0.5–1 h while ensuring relatively high operating pressure of approx. 5–20 MPa.

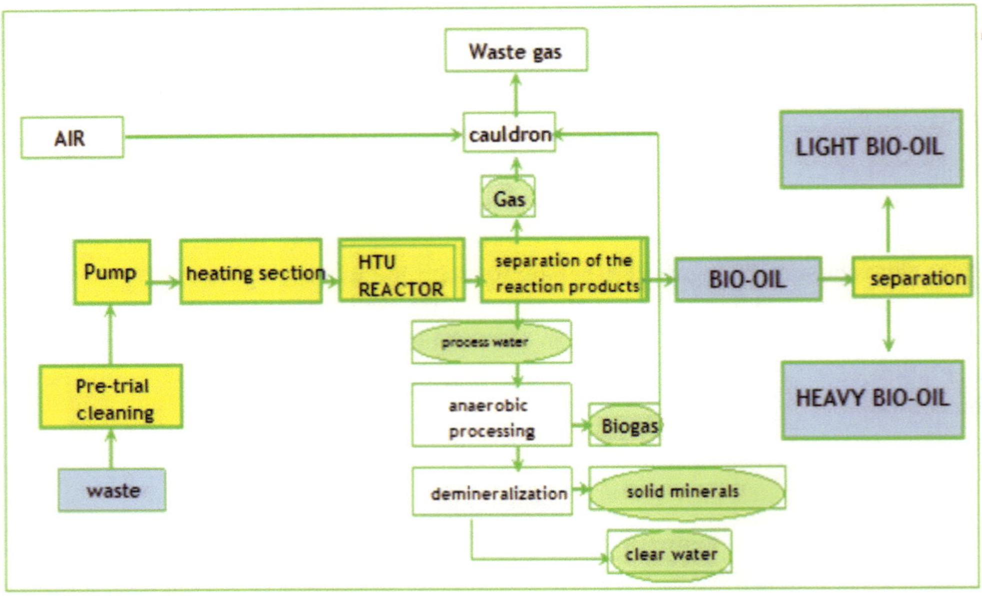

Figure 6. Diagram of HTU process [40]

In contrast to the pyrolysis in the HTU process, it is not necessary to pre-dry the biomass, so the process is particularly suitable for processing natural wet biomass.

HTU process takes place in an aqueous environment, where there are complex reaction sequences. The process involved reducing gas and a catalyst, in order to maximize the extraction of oil and its quality. HTU process is characterized by the high quality of the resultant fuel with low water content.

http://dx.doi.org/10.5772/60448

HTU process has a relatively low yield of oil, approx. 20–60 % dry weight of the stock. Also, the need for high pressure works against the financial side of undertaking due to the increasing demand for energy as a feed and the need for appropriate instrumentation [41, 42].

4.3. Division of biorefinery installations

Biorefineries can be divided into three types, based on technological advancement. The first type is analogous installation to existing installations of conventional oil refinery (petrorefineries). Then in the whole process, one type of raw material which is processed by a single technology is used, thereby providing one primary and main product. Figure 7 shows a schematic diagram of the first type of biorefinery [43].

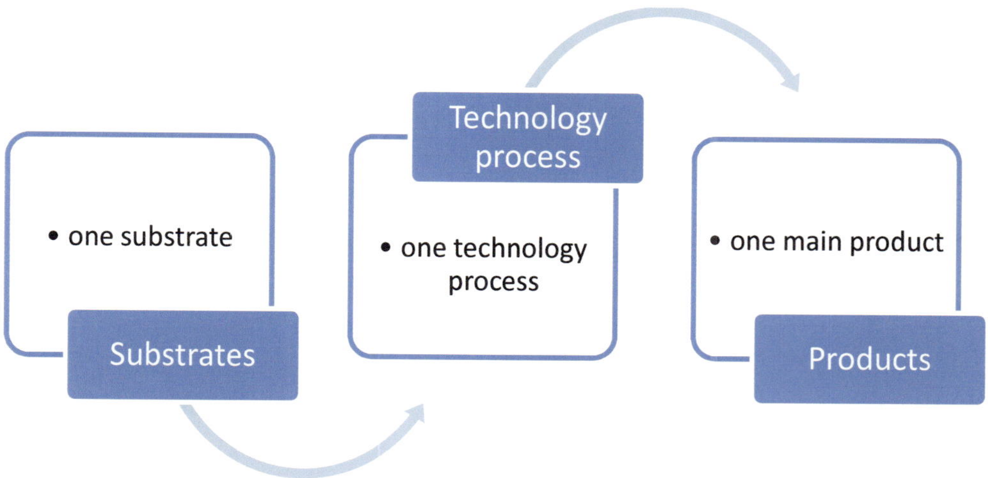

Figure 7. A schematic diagram of the first type of biorefinery (type I)

In another type of biorefinery one substrate is processed through a number of processes to many "major" products, that are equally important from the point of view of investment. A schematic diagram of this type of installation is shown in Fig. 8.

The last type of biorefinery installation is the most advanced system, which uses multiple substrates as input, processed through multiple technologies to a number of main products that are equivalent from the point of view of the investment. A schematic diagram of this type of installation is shown in Fig. 9 [43].

A very important role in the concept of biorefinery installation plays is the possibility of multiple processing of organic waste and more. Biorefinery systems – the first, second, and third type – may use various types of technologies and thus can process various types of biomass. Hence, input substances, regardless of their type, can be processed in the selection of the appropriate installation using appropriate processing technologies. Biomass processed in the so-called "primary" biorefinery systems is often transformed in an incomplete way,

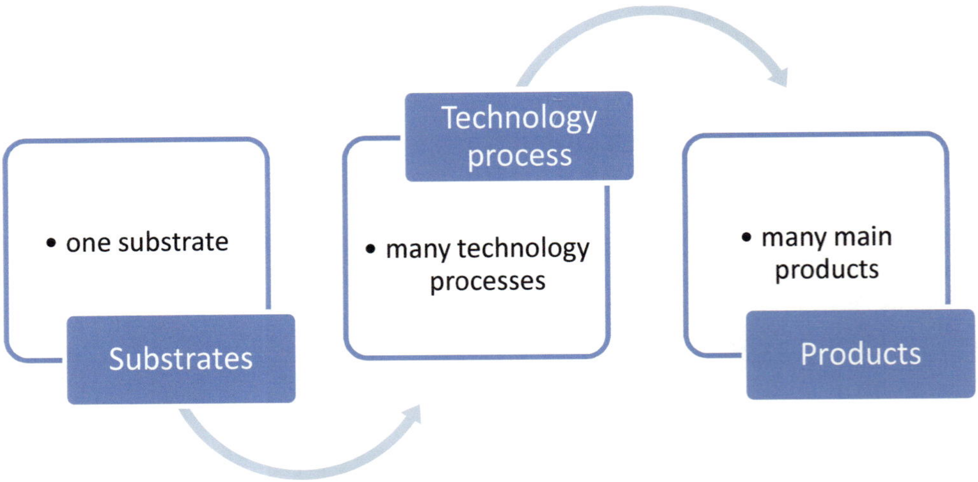

Figure 8. A schematic diagram of the second type of biorefinery (type II)

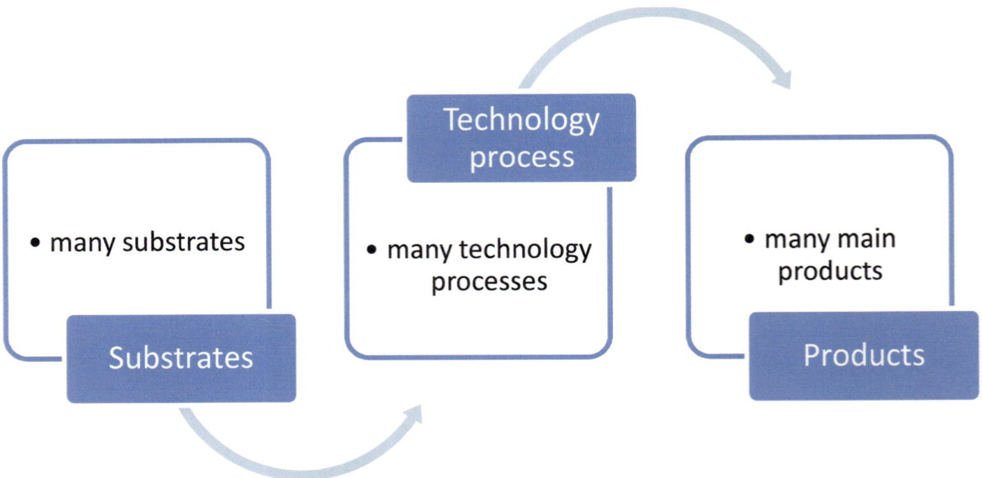

Figure 9. A schematic diagram of the third type of biorefinery (type III)

which should be understood that there are obtained some parts of the products and also are formed intermediates suitable for further processing. In this case, used are so-called secondary biorefinery installations, in which intermediates are processed by the same or other biorefinery processes. The flow of raw material through a series of biorefinery installations is applied until the total conversion of biomass and waste formation, which is not suitable for further processing [44].

Figure 10 shows the environmental biorefinery cycle, taking into account the carbon cycle in nature.

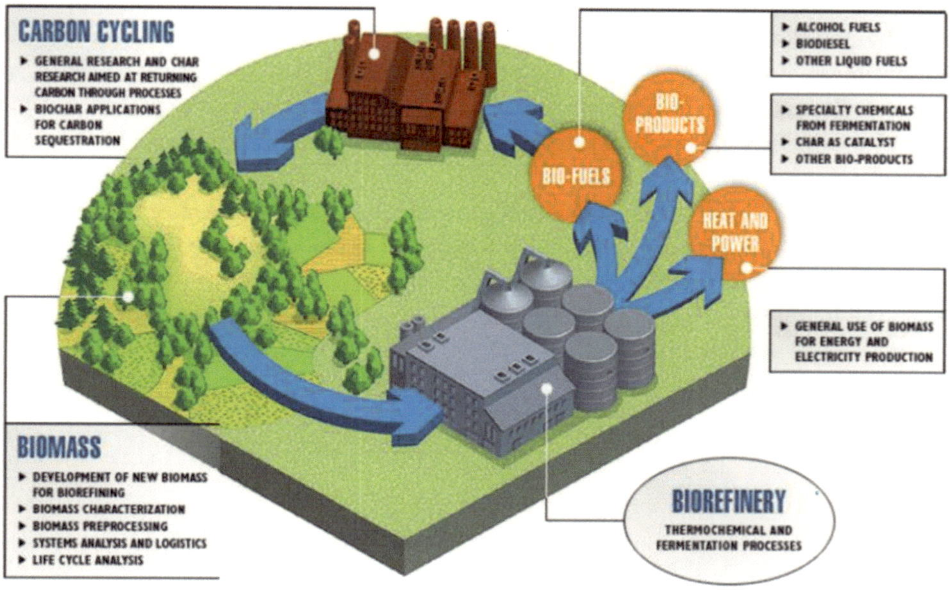

Figure 10. Biorefinery cycle – natural carbon cycle [45]

Biomass in the form of waste, energy crops, etc., shall be prepared and processed (decomposed) to simple compounds (sugars, proteins, carbohydrates, and fats) for the subsequent processes in the biorefinery. To the biorefinery plant gets adequately prepared feedstock from biomass, which is subjected to thermochemical and fermentation biorefinery processes. As a result of biorefinery processes, there are many products created: heat and electricity in co-generation; bio-products such as specialty chemicals, bio-carbon, and others; and also biofuels such as alcohol fuels, bio-oil (biodiesel), and other liquid fuels.

Considering the emissions from vehicles in the WtW cycle, meaning from the source to the wheels (well to wheel), there is the question of the carbon cycle in nature. According to this, there are undertaken research aimed at maximum reduction of carbon dioxide emissions into the atmosphere and return of carbon to nature through various processes, also the use of the resultant bio-carbon in sequestration processes.

Part of the emitted carbon dioxide throughout the lifecycle is absorbed by plants through photosynthesis process, which enables the development and creation of new plants and, consequently, new biomass for biorefinery processes.

The raw materials used in biorefineries are very diverse renewable raw materials, starting from agricultural products such as corn, wheat, and barley grains, agricultural crops such as oilseeds, waste from the agro-food and forestry industry as agro waste, wood chips and deforestation forest products, as well as special energy crops such as switchgrass or willow. Recently biorefinery raw materials also include organic waste, mainly municipal waste and any waste biomass.

As can be seen in Fig. 10, biorefinery raw materials can be a plant specifically designed for that purpose, by-products of processing of other substances, and products from different industries. Biorefinery processes also require power, assuming that the processes need to be exothermic – the energy emitting to the environment. So as biorefinery products is obtained the desired products of all kinds, as well as the energy. A side effect may also be the residual substances (waste). These substances are residues, which in the present state of knowledge cannot be further processed reasonably. It is assumed that the optimal biorefinery should be completely waste-free, because these substances should be used even as internal energy carriers. Therefore, research efforts have led to a comprehensive concept of biorefineries; the diagram is shown in Fig. 11.

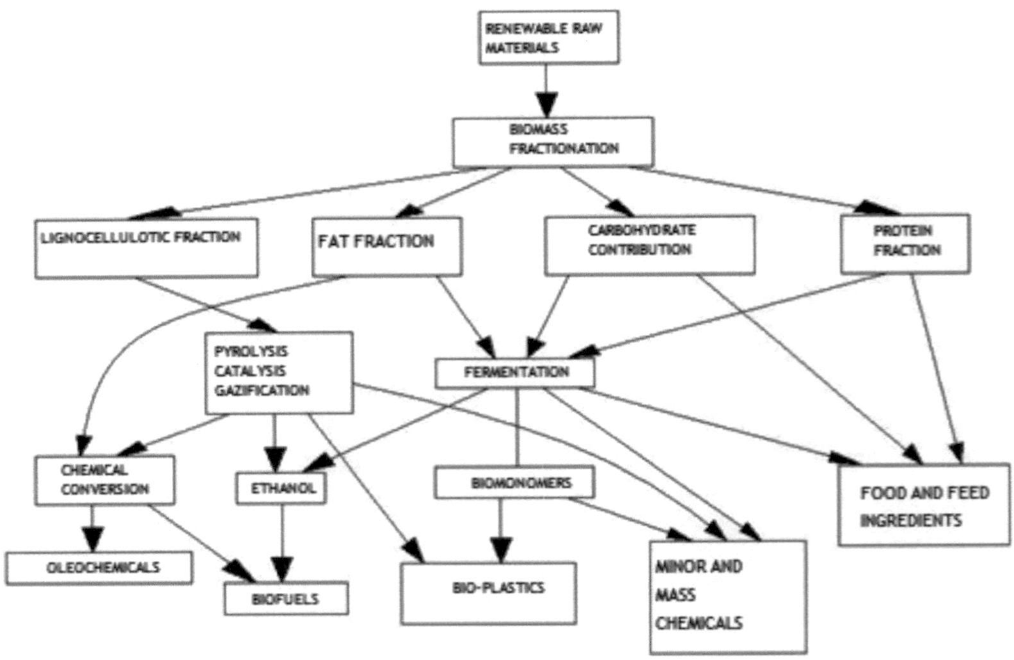

Figure 11. Comprehensive concept of biorefinery (developed on the basis of [46])

As is apparent from the diagram, the first step of biorefinery processes is biomass fractionation. The result of this process is a fraction of the cellulose, fats, protein, and carbohydrates. For each of the fractions, there are different ways of processing and hence various types of possible products, such as oleochemicals, biofuels, "bio"-plastics, chemicals, and food and feed ingredients.

The biomass can be converted into many useful forms of energy by several processes. There are two basic platforms of biorefining processes: "sugar" and "thermochemical." Both systems can produce chemicals and fuels, including methanol, ethanol, and polymers.

"Sugar" platform is based on the breakdown of biomass to an aqueous sugar solution with the use of chemical and biological agents. Fermenting sugars can be further processed in ethanol (produced by fermentation), aromatic hydrocarbons (through a process of dehydration), or liquid paraffins (by processing the aqueous phase). The residues – mainly lignin – can be used to produce electricity (through co-firing) or can be dedicated to the production of other products (e.g., etherified gasoline).

In the "thermochemical" platform, biomass is converted to synthesis gas via the gasification process to bio-oils by the pyrolysis and hydrothermal conversion (HTC). Bio-oils can be further refined to receive a liquid fuel such as methanol, gasoline and diesel fuel, and other chemical compounds.

The comprehensive biorefinery concept, shown in Fig. 11, assumes the universality and thus the energy cost-effectiveness (economic) of multiple processes. Unfortunately, in today's state of knowledge, those processes are mostly in the research phase (laboratory). Hence, due to the technological possibilities and availability of raw materials, there are four basic biorefinery systems in pre-industrial research phase:

- Biorefinery with the feed of the whole plant
- Biorefinery with the feed of non-edible parts of plants
- Biorefinery with lignocellulosic feedstock
- A so-called "two-platform" biorefinery

The biomass can be converted into many useful forms of energy by several processes. There are two basic biorefining platforms: "sugar" and "thermochemical". Both systems can obtain chemicals and fuels, including methanol, ethanol, and polymers.

"Sugar" platform is based on the breakdown of biomass to an aqueous sugar solution by using of chemical and biological agents [33].

4.3.1. Biorefineries based on feed of the whole plant

Biorefineries, where crops are feed material are entirely termed "agrorefineries." Common crops so far, mainly their grain, were intended as a basic raw materials for agro-food industry, which is the basis of food security needs of humanity. While in industrialized countries, there is a surplus of food, in third-world countries there is a permanent lack of food. Thus, the earmarking of raw materials for energy purposes (nonfood) is seen as inhumane. For these reasons, the EU began to develop technologies for biofuels and bioliquids from nonfood raw materials, specifying these fuels as second-generation biofuels. Regardless, it was found that first-generation biofuels, which are made from food raw material (ethanol distillers, FAME from rapeseed, sunflower, etc.), do not change the balance of greenhouse gases and in some cases even worsen it.

In turn, further processing of nonfood use plants, and therefore useless in the agri-food industry is very beneficial for social reasons, because of the possibilities of degraded area development for food crops.

The technology assumes the breakdown of the entire crop plants into edible parts and straw. From the edible parts by biotechnology and chemical and also physicochemical methods can be obtained starch derivatives and flour, which can be considered as final products or continue to process them into fuels, chemicals, polymers, and other materials, which are the final products in this process. This process generates heat and power in cogeneration and waste. Straw, which constitutes a second portion of the raw material, is treated by biotechnology and chemical methods into the lignocellulosic material, which is further processed to the final products of the biorefinery process with production of energy in cogeneration and waste.

Figure 12 shows a general illustrative flowchart of biorefinery, where the raw material feed is whole plants.

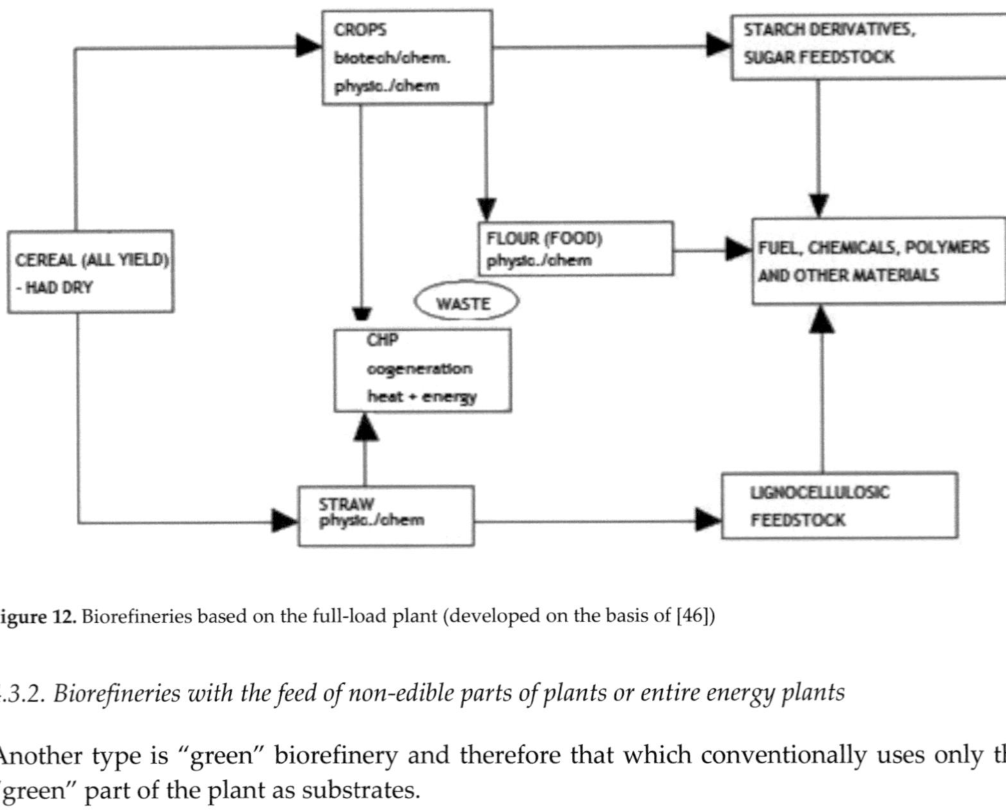

Figure 12. Biorefineries based on the full-load plant (developed on the basis of [46])

4.3.2. Biorefineries with the feed of non-edible parts of plants or entire energy plants

Another type is "green" biorefinery and therefore that which conventionally uses only the "green" part of the plant as substrates.

The feed is naturally wet biomass, green grass, alfalfa, clover, immature corn. etc. Within the scope of the raw materials also includes redundant or useless for breeding and food industry plants or their parts (silage).

Green biomass (the green parts of the plant) by means of extrusion is converted into liquid and solid substances. Fluids are subjected to biotechnological processes, physical and biochemical, which leads to the formation of proteins and soluble sugars. By contrast, the solids

http://dx.doi.org/10.5772/60448

by the hydrothermal, enzymatic, and thermal methods are converted into cellulose and lignocellulose. The products of these two paths are processed for animal feed, fuels, polymers, chemicals, and other materials. These processes generate energy in cogeneration and waste.

Figure 13 shows a general diagram of a biorefinery platform, where the feed is the green parts of plants.

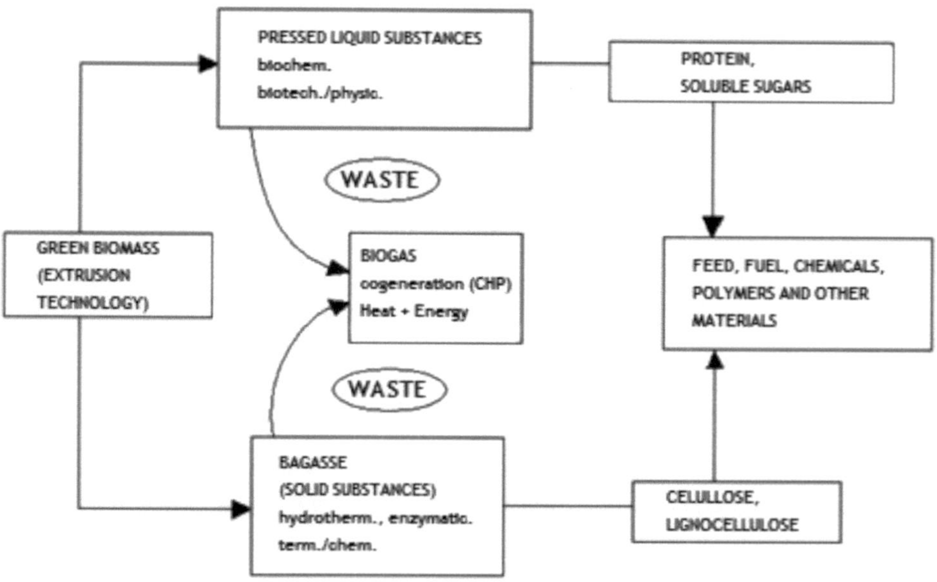

Figure 13. Biorefinery with a load of green plant parts (developed on the basis of [46])

4.3.3. Biorefinery with lignocellulosic feedstock (lignorefineries)

Lignorefinery is a plant, which substrates are rich in organic substances of lignocellulose. This type of materials are, for example, naturally dry biomass, wood, straw, redundant or useless fodder maize, and cellulose-containing biomass.

Excellent raw material including this type biorefinery is waste from many industries, such as the forest industry, wood, paper, furniture, etc.

Firstly, lignocellulosic feed is separated into lignin, hemicellulose, and cellulose. In the biorefinery processes in the case of lignorefinery apply various types of auxiliary substances such as enzymes or yeast. Lignin in a chemical way is processed to obtain lignin raw material, and cellulose is converted into sugar raw material by biotechnology and chemical processes. From the products of these two processes and hemicellulose can be obtained fuel, chemicals, polymers, and other materials with energy in cogeneration and waste [47].

A diagram of lignorefinery is shown in Fig. 14.

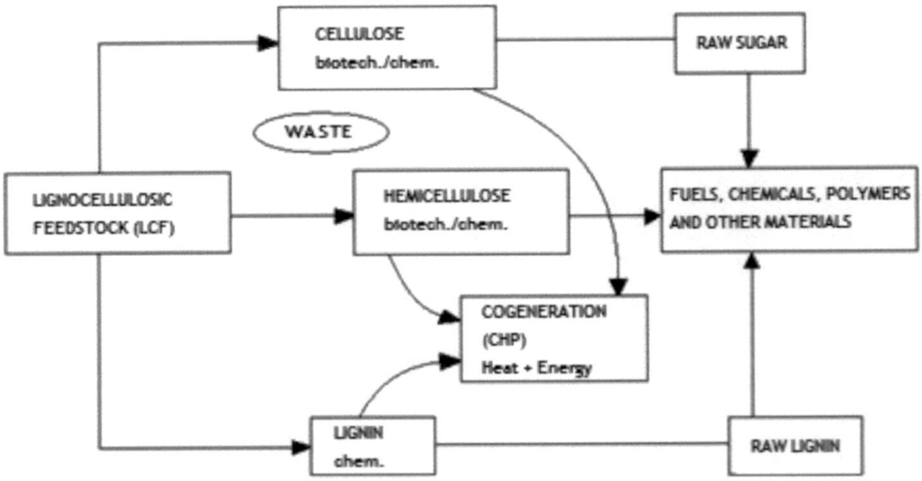

Figure 14. Lignocellulosic biorefinery – lignorefinery (developed on the basis of [46])

4.3.4. "Two-platform" biorefineries

The idea of "two-platform" biorefinery is to obtain a synthesis of gas and sugars and so the parallel production of fuels based on renewable raw materials and other products, in terms of production in two technology platforms.

Raw materials used in this type of biorefinery is widely understood biomass consisting of any substances, mostly waste from various industries, such as agriculture, forestry, marine, food industry, public utilities, etc.

Biomass is divided into two contractual "platforms" – sugar and syngas platform. Sugar raw materials are converted through chemical reactions with the separation of waste, while the synthesis gas is conditioned by thermal and chemical methods. The end products of these processes are fuels, chemicals, polymers, and other materials. In these processes, the energy produced in cogeneration is produced.

Figure 15 shows a flowchart of a "two-platform" biorefinery.

4.3.5. Oleorefinery concept

There is also a variety of agrorefinery called oleorefinery. As the name suggests, it uses oilseeds as raw material such as canola, sunflower, olive trees, and soybeans. Typically, these seeds are used for food production. Therefore, at present, when the first-generation biofuels go into oblivion, oleorefinery may have only use these crops from degraded areas, which cannot be used in food industry. Besides oilseeds as a substrate can also be used animal fats and waste. Oleorefinery products may be biodiesel and oleochemical products such as phytosterols or

http://dx.doi.org/10.5772/60448

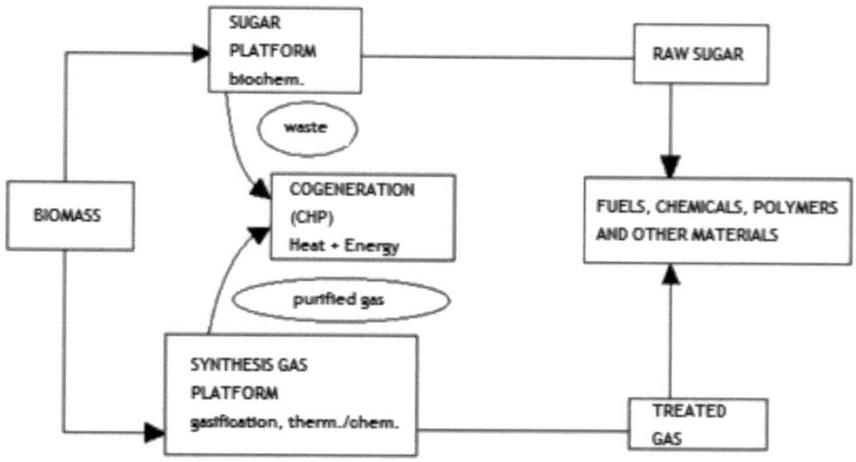

Figure 15. "Two-platform" biorefinery (developed on the basis of [46])

sphingomyelin. This process also generates energy, which in part can be used as the energy delivered to a subsequent process.

Figure 16 shows a general flowchart of oleorefinery.

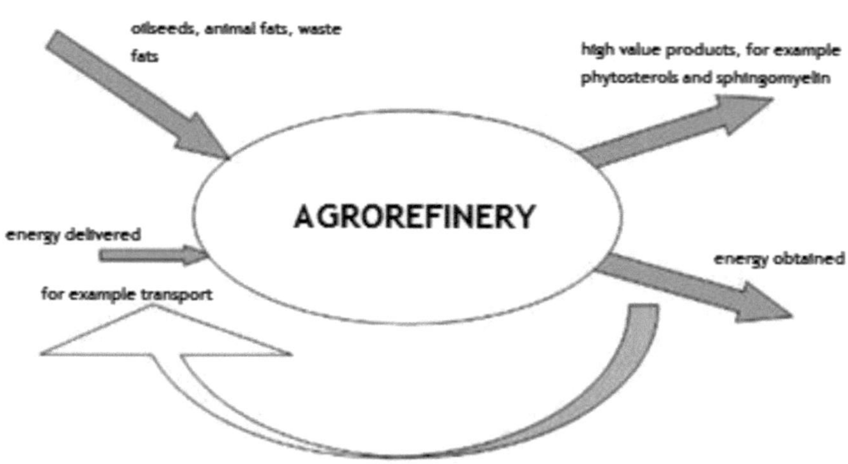

Figure 16. The general scheme of oleorefinery [37]

4.3.6. Prospective biorefinery concept

There was developed a future-oriented concept of biorefinery based on the raw waste, which the diagram is shown in Fig. 17. That biorefinery practically implements processes in the field of WtL ("wastes to liquid"). Feed for this type of installation can provide all kinds of waste oils: waste used frying oils, animal fats, grease, animal waste, etc. Oil fractions by lipid extraction and refining are converted into crude oil and waste.

The crude oil is converted by transesterification to the methyl ester or glycerol, by hydrogenation into the liquid hydrocarbon (biodiesel), or by chemical and enzymatic modification in all kinds of oleochemical products, such as fatty acids, alcohols, fatty esters, fatty ketones, dimer acids, glycerine, etc. Wastes are converted by gasification into synthesis gas, resulting in a yield of energy and heat in cogeneration. Syngas may be further converted into transport fuels or chemicals by catalytic processes [48].

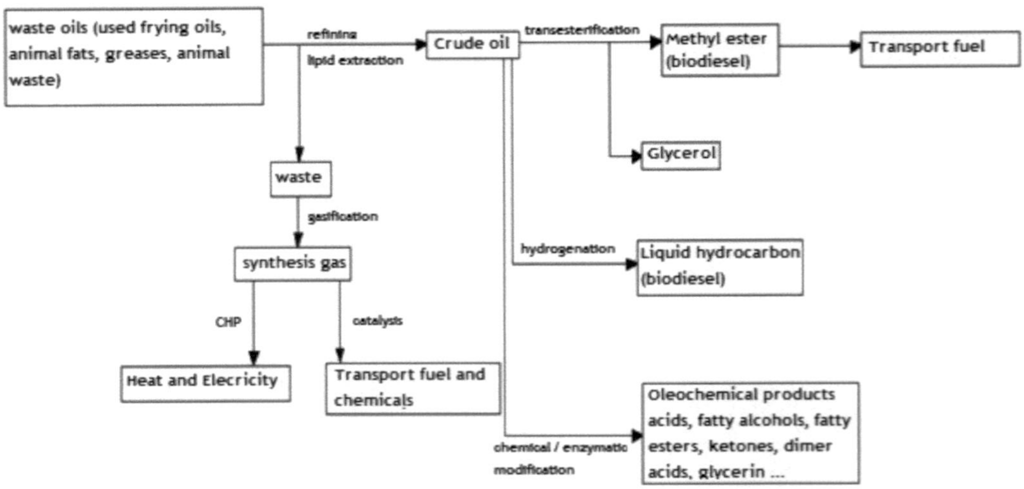

Figure 17. Biorefinery perspective based on the raw waste (developed on the basis of [46])

5. The existing biorefinery systems in the world

Progress in research on new, more efficient complex biomass processing methods in technological and environmental way led to the development of new technologies to enable the construction of biorefinery systems. Detailed information in this field are presented in the International Energy Agency (IEA) report, "IEA Bioenergy Task 42, Biorefining: Sustainable and synergetic processing of biomass into marketable food & feed ingredient, chemical, materials and energy (fuels, power, heat)." IEA report defines biorefinery systems, which include systems that meet the basic functions, that is, the number and the name of the technology platforms, materials, processes, and expected products. Examples of applicable platforms in biorefinery systems are placed in Fig. 18.

http://dx.doi.org/10.5772/60448

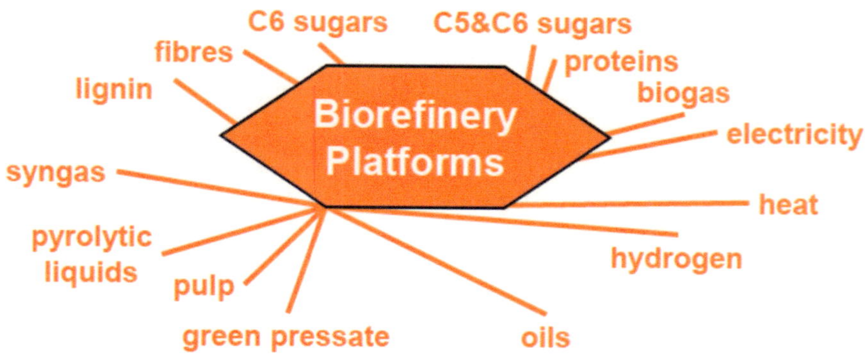

Figure 18. Examples of possible platforms in a biorefinery systems (according to IEA Bioenergy Task 42) [49]

The research in the field of the possibility of obtaining different kinds of products or chemical intermediates from biomass, so-called biochemicals, estimates the broad spectrum of these compounds, as shown in Fig. 19.

Figure 19. Portfolio of chemicals potentially to be produced from biomass (according to IEA Bioenergy Task 42) [49]

The result from the IEA report is that currently in the world, mainly in the pilot or demonstration version, there are 80 plants which can be classified as more or less complex components of biorefinery systems. In terms of some paths proposed in these technology systems, fuller research will be necessary, especially in terms of quality of products planned to be obtained and their reproducibility. In particular, reproducibility and efficiency of individual development paths can be difficult to obtain, and the difficulty is due to the heterogeneity and diversity of biorefinery raw materials. Hence, most of the operating systems in the world are the one-platform systems and two- or three-platform systems. Production processes planned or implemented in these systems are mainly for the obtaining of bioethanol, biomethanol, and less biomethane and a few types of biochemicals, including biodegradable monomers. One of the most interesting biorefinery plants is a pilot plant LanzaTech, New Zealand, in which the fermentation process of compressed syngas obtained by the gasification of waste wood and solid urban municipal waste (MSW) is conducted. Synthesis gas fermentation process is carried out to obtain bioethanol, and regardless of this process, in a separate technology path, from the recovery of CO, CO_2, and H_2 from syngas fermentation process, it is planned to receive a hydrocarbons from C2 to C5, as a potential biofuel components and/or biochemicals, including again the ethanol and acetic acid (ethanoic), C2; isopropanol and acetone (dimethylketone), C3; and 2,3-butanediol (2,3-BDO), butane, isobutane, and succinic acid (1,4-butanedioic), C4, and isoprene, C5. A block diagram of this plant is shown in Fig. 20.

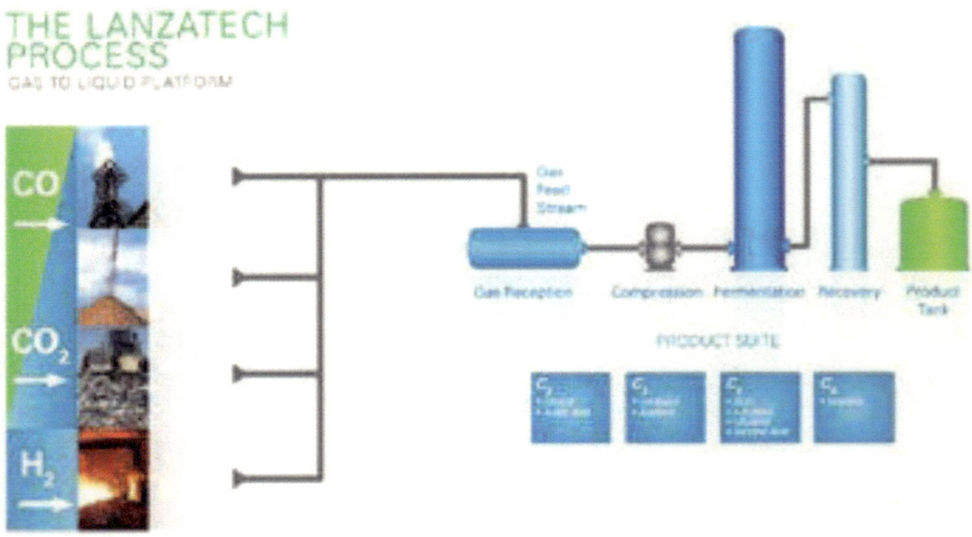

Figure 20. Block diagram of the installation LanzaTech in New Zealand (according to IEA Bioenergy Task 42) [49]

The diagram also shows the power unit which is used to supply heat and power to the plant in the biorefinery system. This block is supplied by different types of raw materials than technological resources, but also coming from renewable sources or waste substances.

Interesting system is proposed by the company Avantium Chemicals B.V. from the Netherlands. In this three-platform system (C5 and C6 sugars and lignin) is proposed to receive a perspective and very interesting furan fuels, polymers and monomers (furan dicarboxylic acid, furan diamine), fine and specialty chemicals (organic acids, solvents, flavors and fragrances), and solid fuels (humans and lignin residues). The raw material for the installation proposed by Avantium is generally defined lignocellulosic material (cellulose, hemicellulose, starch, and sucrose), contained in the various sources. The essence of the proposed process is to develop

http://dx.doi.org/10.5772/60448

the technology for obtaining derivatives of furan with catalytic processes of dehydration/etherification of carbohydrates. These derivatives can be compiled as a replacement for the relevant petroleum production of biofuels, plastics, and biochemicals. At the same time, it is possible to use emerging process residues, with the possible addition of biomass (and also coal) to produce heat and electricity, while technological processes used the resulting excess of steam and heat. As already mentioned, there are a number of streams of raw materials, possible to use in the proposed process. It may be waste from the food industry; corn stover (possibly other straw); grass, including the grass from the city greenery; as well as bagasse and municipal waste (from the households). An important and perspective solution is possible conversion of C6 sugars (glucose, mannose, galactose, and fructose) and C5 sugars (xylose and arabinose) to the derivatives of very promising chemical intermediate which is hydroxymethylfurfural (HMF), which is inter alia an intermediate for the preparation of DMF (2,5-dimethylfuran), the perspective biofuel, and 2,5-furandicarboxylic acid (FDCA), which can be used as a replacement to terephthalic acid (TA) [49].

6. Conclusion

The essence of the biorefinery processes is renewable raw materials, by-product, and waste processing for biofuels and high-value chemical products. Those processes should be led with possibly high effectiveness and in closed circulation of CO_2. Process of biorefinery system creation requires high capital expenditures of technology and installation and needs assurance of appropriate raw material, personnel, and logistic background.

As can be seen from the foregoing considerations, there are a number of concepts of biorefinery development. These concepts are substantially different from each other by type of raw material used in the planned manufacturing processes to yield energy carriers like fuel and a certain group of derivatives which may be used in further chemical processes.

Generally, the biorefinery process should consist of transforming raw materials of organic origin and various origins, so to obtain a mixture of liquid components (mainly hydrocarbons), so-called bio-oil in transformation process.

The next step in the process should constitute processes similar in principle to the oil preparation processes, leading to a distillation treatment (degassing, demulsibility, dewatering, cleaning, etc.).

In the final part of the biorefining system undistilling, refining, and hydrogenation processes of bio-oil in principle are no different from the nonconservative processes realized for crude oil converting. Therefore, biorefinery complexes can be built at the beginning based on the modification of existing oil refineries by expansion of complex installation of feed transformation into mixtures of liquid fractions.

For example, in Poland, there is a few low-efficient oil refineries, which untapped technological, installation, and human potential could be used with biorefinery complex building. The potential can be used with further processing and refine processed biomass to the liquid form.

Planned undertaking widely reduces capital costs and enables obtaining chemical substances and biofuel production, which fulfill modern engines and heat devices qualitative requirements. That is why it was assumed to build an experimental biorefinery using technology potential of little oil refinery in Poland. Figure 21 shows the technological schema of planned biorefinery system.

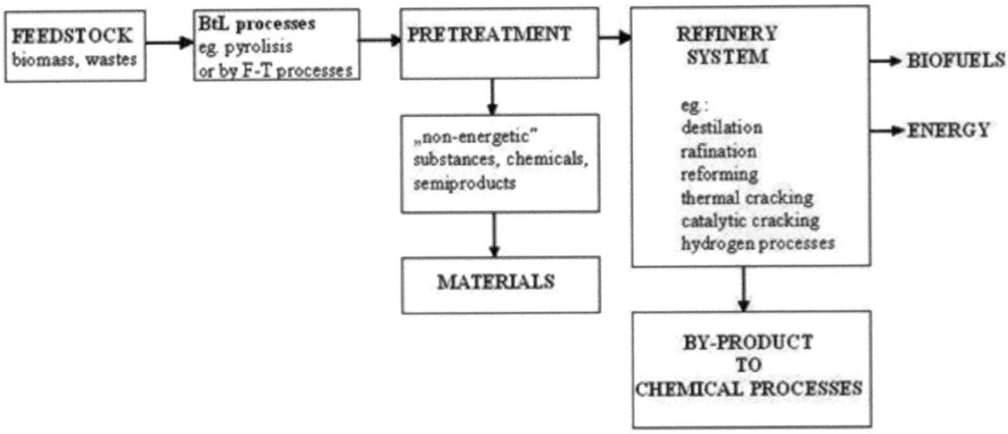

Figure 21. Diagram of biorefinery based on conventional oil refinery [50]

The essence of planned undertaking is elaboration or selection of suitable technologies of biomass liquefaction by pyrolytic process to "bio-oil" phase. It is also possible to use of gasification processes and then selective liquefaction. In this process, there is expected a technology elaboration, where from the liquid fraction will be selected "non-energetic" substances. Those substances can't be used in further processing like biocomponents, but they could become biorefinery products or industry semiproducts for further synthesis. Refined bio-oil will become resource for obtaining and refining process using existing refinery installations like atmospheric and vacuum distillation, rectification, refinement, thermal and catalytic cracking, alternately reforming, and hydrogen process. After processing remains could become raw material for further synthesis.

Development of technology for biomass conversion in the biorefinery system based on the technology processes and systems of conventional oil refinery will require the realization of research processes in the following main areas:

- Optimization of biomass and residues pyrolysis processes or new biomass conversion processes
- Technology of production value-added co-products
- Adaptation of existing production technologies for converting bio-oil into biofuels, also second-generation biofuels

- Technology of converting co-products into marketable products

Author details

Krzysztof Biernat[1,3*] and Paulina Luiza Grzelak[2,3]

*Address all correspondence to: k.biernat@pimot.org.pl

1 Automotive Industry Institute, Warsaw, Poland

2 Motor Transport Institute, Warsaw, Poland

3 Institute for Ecology and Bioethics of Cardinal Stefan Wyszynski University, Warsaw, Poland

References

[1] Directive 2009/28/EC of the European Parliament and of the Council of 23 April 2009 on the promotion of the use of energy from renewable sources and amending and subsequently repealing Directives 2001/77/EC and 2003/30/EC (Text with EEA relevance).

[2] K. Biernat, A. Malinowski, M. Gnat, "The Possibility of Future Biofuels Production Using Waste Carbon Dioxide and Solar Energy", Chapter 5 in the Monograph "Biofuels, Economy, Environment and Sustainability" Edited by Zhen Fang, IntechOpen ISBN 978-953-51-0950-1, Croatia, Feb. 2013.

[3] K. Biernat, "The Possibility of Creating Biorefinery in Polish Bioeconomy", Stakeholder Event on the Bio-based Industries Public-Private Partnership, Warsaw, Poland, 22 November 2013.

[4] The European Bioeconomy in 2030, "Delivering Sustainable Growth by addressing the Grand Societal Challenges", Seventh Framework Program.

[5] Biobased and Renewable Industries for Development and Growth in Europe. Strategic Innovation and Research Agenda (SIRA), BRIDGE 2020, March 2013.

[6] L. Jones, P. Atkins, "Chemistry: Molecules, Matter and Change", W. H. Freeman, 2000.

[7] A. Johansson, "Clean Technology", LEWIS PUBL, 1992.

[8] R. Zarzycki, M. Imbierowicz, M. Stelmachowski, "Wprowadzenie do inżynierii i ochrony środowiska" (*Introduction to environmental engineering and environmental protection*), Wydawnictwo WNT, Warszawa, 2007.

[9] F. Reif, "Fizyka statystyczna" (*Statistical physics*), PWN, Warszawa, 1973.

[10] A. Ślęzak, "Termodynamiczna ocena źródła entropii w układzie zawierającym dwuskładnikowy opatrunek membranowy" (*Thermodynamic evaluation of entrophy source including two-component membrane dressing*), Katedra Zdrowia Publicznego, Politechnika Częstochowska, Inż. Biomed. (2008) praca złożona w redakcji.

[11] P. G. Hewitt, L. A. Hewitt and J. A. Suchocki, "Conceptual physics", Pearson Education, 2010.

[12] L. D. Landau, E. M. Lifszyc, "Fizyka statystyczna" (*Statistical physics*), PWN, Warszawa, 1973.

[13] R. S. Ingarden, A. Jamiołkowski, R. Mrugała, "Fizyka statystyczna i termodynamika" (*Statistical physisc and thermodynamics*), PWN, Warszawa, 1990.

[14] Biorefinery.nl: a joint initiative of Wageningen University and Research Centre (WUR) and the Energy Research Centre of the Netherlands (ECN), supported by SenterNovem - http://www.biorefinery.nl/background-biorefinery/biorefinery-concepts/.

[15] B. Kamm, M. Kamm, "Principles of Biorefineries", Appl. Microbiol. Biotechnol. 64 (2004), 137-145.

[16] J. Szargut, A. Ziębik, "Podstawy energetyki cieplnej" (*The basics of heat energetics*), PWN, Warszawa, 1998.

[17] P. Perrot, "A to Z of Thermodynamics", Oxford University Press, 1998.

[18] J. Szargut, D. R. Morris, F. R. Steward, "Exergy analysis of thermal, chemical and metallurgical processes", Hemisphere Publ. Corp., New York, Washington, London, 1988.

[19] J. Szargut, "Exergy Analysis: Technical and Ecological Applications", WIT-press, Southampton, 2005.

[20] J. Szarawara, "Termodynamika chemiczna stosowana" (*Chemical thermodynamics applied*), Wydawnictwa Naukowo-Techniczne, Warszawa, 2007.

[21] J. Szargut, R. Petela: "Egzergia" (*Exergy*), PWN, Warszawa, 1965.

[22] J. Szargut, "Application of Exergy for the Calculation of Ecological Cost", Bull. Pol. Acad. Sei. Techn., 1986, nr 7-8.

[23] J. Szargut, "Standard Chemical Exergy of Some Elements and Their Compounds, Based upon the Concentration in Earth's Crust", Bull. Pol. Acad. Techn., 1987, nr 1-2.

http://dx.doi.org/10.5772/60448

[24] R. L. Cornelissen, "Thermodynamics and Sustainable Development. The Use of Exergy Analysis and the Reduction of Reversibilit", PhD thesis, University of Twente, Holandia, 1997.

[25] A. Papuziński (red.), "Zrównoważony rozwój. Od utopii do praw człowieka" (*Sustainable development. From utopia to human rights*), Oficyna Wydawnicza Branta, Bydgoszcz, 2005.

[26] C. A. S. Hall, C. J. Cleveland, R. Kaufmann, "Energy and Resource Quality", New York, John Wiley & Sons, 1986.

[27] "Our Common Future", Report of the World Commission for Environment and Development, Warsaw, PEW, 1991.

[28] A. Bogda, C. Kabała, A. Karczewska, K. Szopka, "Zasoby naturalne i zrównoważony rozwój" (*Natural resources and sustainable development*), Wydawnictwo Uniwersytetu Przyrodniczego we Wrocławiu, Wrocław, 2010.

[29] R. Kulik, "Konsumpcjonizm - uwarunkowania i wpływ na świat" (*Konsumpcjonizm - uwarunkowania i wpływ na świat*), Workshop: The phenomenon of consumerism and materialism as risk factors of climate change, Bystra, Poland, 14-15 December 2013.

[30] D. O. Hall, "Food versus Fuel, a World Problem?", in Pric. 10 CD/UNESCO Conf. Photochemical Conversion, Ecole Polytechnique Federal Lausanne, June 1983.

[31] M. Grathwohl, "World Energy Supply: Resources, Technologies, Perspectives", Berlin, Walter de Gruyter, 1982.

[32] J. W. Wandrasz, J. Biegańska, "Odpady niebezpieczne - podstawy teoretyczne", (*Hazardous Waste - the Theoretical Basis*) Wydawnictwo Politechniki Śląskiej, Gliwice, 2003.

[33] S. Fernando, S. Adhikari, C. Chandrapal, N. Murali, "Biorefineries: Current Status. Challenges and Future Direction", Energy Fuels 4 (2006).

[34] A. J. Ragauskas et al., "The Path Forward for Biofuels and Biomaterials", Science 311, 484 (2006).

[35] T. L. Richard, "Challenges in Scaling Up Biofuels Infrastructure", Science 329, 793 (2010).

[36] S. Xiu, B. Zhang, A. Shahbazi, "Biorefinery Processes for Biomass Conversion to Liquid Fuel", Biofuel's Engineeing Process Technology, chapter in book "Biofuel's Engineering Process Technology" edited by Dr. Marco Aurelio dos Santos Bernardes, InTech, Augusr 2011.

[37] "Narodowa Strategiczna Agenda Badawcza w Zakresie Biopaliw" (*National Strategic Research Agenda in the area of Biofuels*), elaboration of Polish Technology Platform for Biofuels, Editor: K. Biernat, Warsaw, 2008.

[38] K. Świątek, M. Lewandowska, M. Świątek, W. Bednarski, "Doskonalenie warunków hydrolizy enzymatycznej polisacharydów zawartych w słomie rzepakowej" (*Improv-

ing conditions for enzymatic hydrolysis of polysaccharides contained in the rape straw), Zeszyty Problemowe Postępów Nauk Rolniczych. 570, 2012 z, 107-116.

[39] D. Meier et al., "State-of-the-art of Fast Pyrolysis in IEA Bioenergy Member Countries", Renewable and Sustainable Energy Reviews 20 (2013), 619-641.

[40] J. E. Namber, F. Goudriaan, "HTU® - Diesel from biomass", Waszyngton DC, 31 August 2005.

[41] J. Schindler, W. Weindorf, "Fuels for Transportation Derived from Renewable Energy Sources", Hyforum 2000, Munich, 11-12 October 2000.

[42] B. E. Kampman, L. C. den Boer, H. J. Croezen, "Biofuels under Development: An Analysis of Currently Available and Future Biofuels, and a Comparison with Biomass Application in Other Sectors", Delft, May 2005.

[43] R. W. R. Zwart, "Biorefinery – The Worldwide Status at the Beginning of 2006", BioRef 0603. www.biorefinery.nl, July 2006.

[44] R. Van Ree, E. Annevelink, "Status Report Biorefinery 2007", www.biorefinery.nl. November 2007.

[45] Biorefining and Carbon Cycling Program at the University of Georgia, http://www.biorefinery.uga.edu/.

[46] W. Soetaert, "Defining Biorefineries and Different Concepts", BioreFuture 2009, Biorefinery Euroview BIOPOL, Brussels, 30 March 2009.

[47] L. Fagernäs, Y. Solantausta, "Lignocellulosic Biorefinery", BioreFuture 2009, Biorefinery Euroview BIOPOL, Brussels 30 March 2009.

[48] P. Söderholm, R. Lundmark, "The Development of Forest-based Biorefineries: Implication for Market Behavior and Policy", Forest Product Journal 1 February 2009.

[49] IEA Bioenergy Task 42 Biorefinery.

[50] Biorefinery Joint Call INFO DAY. Information Booklet. Brussels 16 September 2008.

Chapter 21

Conversion of Biomass to Bio-Oil in Sub- and Supercritical Water

Armando T. Quitain, Chan Yi Herng, Susana Yusup,
Mitsuru Sasaki and Yoshimitsu Uemura

Additional information is available at the end of the chapter

http://dx.doi.org/10.5772/59947

1. Introduction

Biomass-derived energy is widely considered to be the most potential and feasible among many forms of available energy on earth. Unlike any other renewable energy sources, such as geothermal, solar and wind which only generate heat and power, biomass can be converted to solid, liquid and gaseous fuels [1]. However, the choice of conversion methods suitable for a particular type of biomass and the target fuels becomes a challenge because of the complex three-dimensional (3D) structure connecting the main biomass components (i.e. lignin, cellulose and hemicellulose). This structure gives strength to the materials, and offers resistance that protects them against chemicals and microbial attack making it very difficult to hydrolyze. An effective dissociation of these components and their subsequent separation are needed for the production of high value products from lignocellulosic biomass [2]. One promising approach introduced recently is "hydrothermal liquefaction method" or commonly referred to as "hydrothermal upgrading" in many parts of Europe, wherein biomass is liquefied using water as a solvent at elevated temperatures and pressures.

2. Hydrothermal liquefaction (or upgrading) of biomass

Researches related to hydrothermal liquefaction of biomass have gained significant momentum during the last decade. Search for published literatures under the keywords "hydrothermal liquefaction" using SCOPUS database resulted into heavy turn outs of about 257 published articles in the recent 5 years as compared to only 40 articles during the previous 5-year term. Several review articles of the method have also been published recently including the most comprehensive review of the technology summarized by Peterson et al. [3]. Reviews on its application to various biomass components such as protein, carbohydrates, lignin and fats have also been reported [4]. The most recent applications to the liquefaction of microalgae and marine biomass have also been summarized [5–8]. This hot-compressed water has also been compared with other hydrolysis methods [9] and shows it to be very attractive economically and environmentally because of higher sugar recovery even in the absence of acid and chemical catalysts.

The increasing popularity of this method is due to the deemed tremendous opportunities that biomass of any sources can offer as energy carrier, combined with the promising properties of water at high temperature and high pressure, especially under sub- and supercritical conditions.

3. Water under sub- and supercritical conditions

Water under sub- and supercritical conditions behaves totally different to that at ambient conditions, and its properties have been extensively reviewed by G. Brunner in his recent publication [10], which is basically based on the work by M. Modell in the 70s. To summarize, water with a dielectric constant of about 80 at ambient condition changes its properties in near critical (or subcritical) conditions, which can be used as a solvent for ionic or polar species. But, as the temperature approaches the critical temperature (Tc=373°C, Pc=22.1 MPa) and beyond (supercritical), the dielectric constant dramatically decreases to less than 10 (similar to that of methylene chloride) [11], behaving like a non-polar solvent that can dissolve and degrade a variety of non-polar organic compounds (please refer to Fig. 1 for the typical phase diagram of a substance).

Other than the dielectric constant, which indicates polarity, the ion product also exhibits changes in the properties of water. Water dissociates to produce hydrated hydrogen (hydronium ion) and hydroxyl ions. Under ambient conditions, the product of their concentrations is about 10–14 $(mol/L)^2$. The ion product also changes dramatically with increasing temperature, and the maximum product is obtained at a temperature of about 250°C, under saturated vapor pressure. Having low dielectric constant and high ion product, water under near critical (subcritical) conditions become a suitable solvent for hydrolysis of organic compounds, even without any additives. With these properties, water serves as a potential solvent and a catalyst for organic reactions.

The review of G. Brunner [10] also showed several physicochemical properties of water under the above-mentioned conditions including density, viscosity and solubility of some gases and hydrocarbons. Sophisticated techniques for the measurement of those properties under high temperature and high pressure conditions were also presented. A micro-cell for potentiometric pH measurements of aqueous solutions in supercritical state has been reported [12]. An apparatus utilizing an optical cell under flow conditions was also developed to measure solubility, salt deposition and Raman spectroscopic studies in aqueous solutions near the critical point [13]. Mixing of solutions under supercritical water conditions had been a

http://dx.doi.org/10.5772/59947

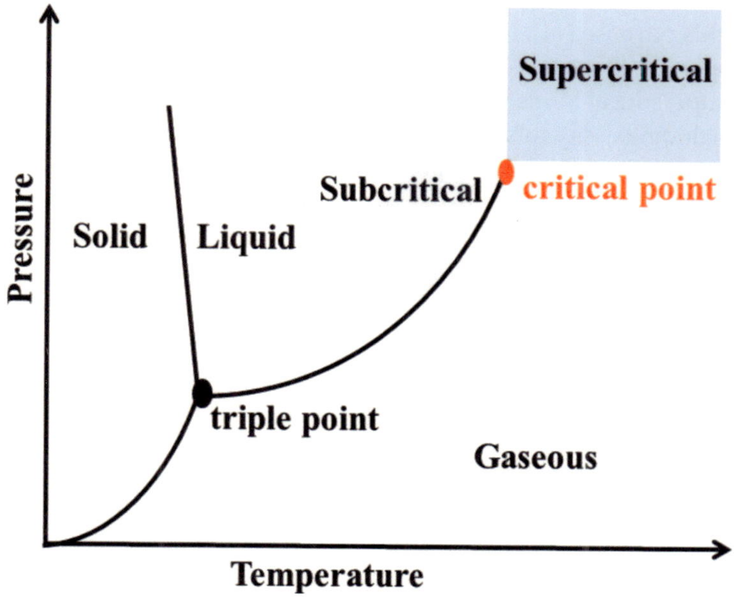

Figure 1. Typical phase diagram of any substance (for H_2O: Tc=373°C, Pc=22.1 MPa).

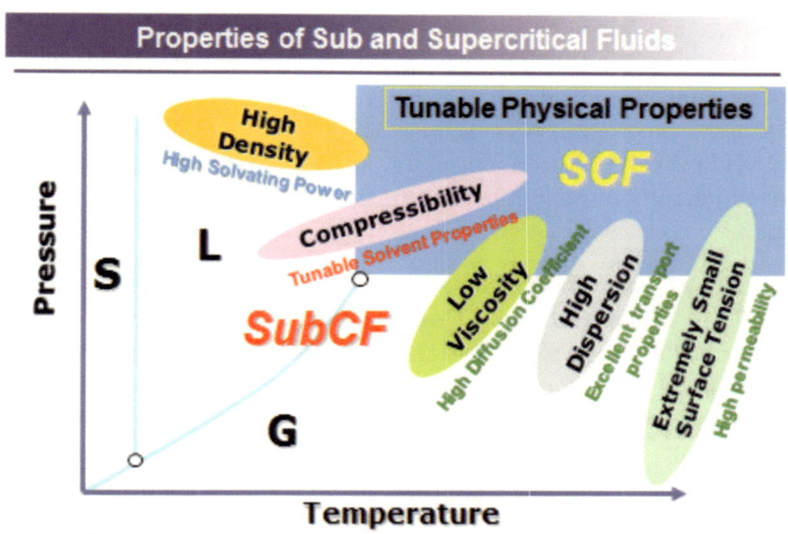

	H_2O	CO_2	EtOH	Benzyl Alcohol (BZA)
Tc (°C)	374	31.1	241	403
Pc (MPa)	22.1	7.4	6.1	4.6

Figure 2. Summary of tunable physical properties of sub- and supercritical water and comparison of critical temperatures and pressures of some commonly used supercritical solvents.

challenging and hot topic, a device for direct observation of channel-tee mixing has been developed for this purpose [14]. Further, a reactor for the studies of in-situ X-ray scattering studies of nanoparticle formation in supercritical water system was also developed [15]. The advantages of supercritical fluids with regard to its physicochemical properties are summarized in Fig. 2, including easily tunable physical properties which make the water suitable for highly selective extraction and reaction processes involving biomass samples.

4. Potential biomass feedstocks

Biomass being considered for bio-oil production includes woody biomass, grass and most recently different types of micro- and macroalgae. The characteristics of these feedstocks will be introduced here giving more emphasis on oil palm biomass which is selected as one of the main materials investigated in this work.

4.1. Oil palm biomass

Malaysia is one of many agricultural countries in the world, producing main agricultural crops such as oil palm, rubber, cocoa, rice and coconut [16]. Hence, the amount of production of agricultural wastes and residues is staggering with 47762 ktons of biomass generated as of year 2007. These biomass feedstocks have been long identified as a source of renewable energy which could reduce the dependency of fossil fuels as the main source of energy supply and multiple fine chemicals. The abundance of various biomass feedstocks has provided a good potential for their utilization. Thus, this has been extensively investigated more especially in looking for the best technology applicable for this conversion into either fuels or fine chemicals more efficiently, effectively and economically. Among the many available biomass feedstocks, those that are abundantly present in the Southeast Asian region are mostly derived from oil palm plantation.

Oil palm plantation in Malaysia is the most important agricultural crop, producing biomass close to about 37.0 million tons in 2008. This biomass consists of 22% empty fruit bunch (EFB), 13.5% fruit press fiber (FPF) and 5.5% shell. This high amount of oil palm biomass wastes could be utilized in the production of bio-oil, a promising future renewable energy source. In this research, the oil palm wastes such as empty fruit bunch (EFB), mesocarp fiber (MCF) and palm kernel shell (PKS) will be utilized as feedstocks for the hydrothermal liquefaction process to produce bio-oil using sub- and supercritical water technologies as depicted in Fig. 1. Fig. 3 shows images of the samples and the lignocellulosic composition with respect to the lignin, hemicellulose and cellulose contents. The EFB has lower lignin content at about 18%. The PMF and PKS have almost similar composition consisting of about 30–33% lignin, 22% hemicellulose and 24% cellulose. These changes in composition could have significant effect on its liquefaction. Table 1 summarizes the ultimate analysis and high heating value (HHV) of the samples. The EFB has the lowest carbon content at 43.6%, but has the highest oxygen content at 50.2%. The HHV of EFB and PMF are almost the same at 16.3 MJ/kg, while PKS has about 17.5 MJ/kg of heat value.

Empty fruit brunch (EFB) Palm mesocarp fiber (PMF) Palm kernel shell (PKS)

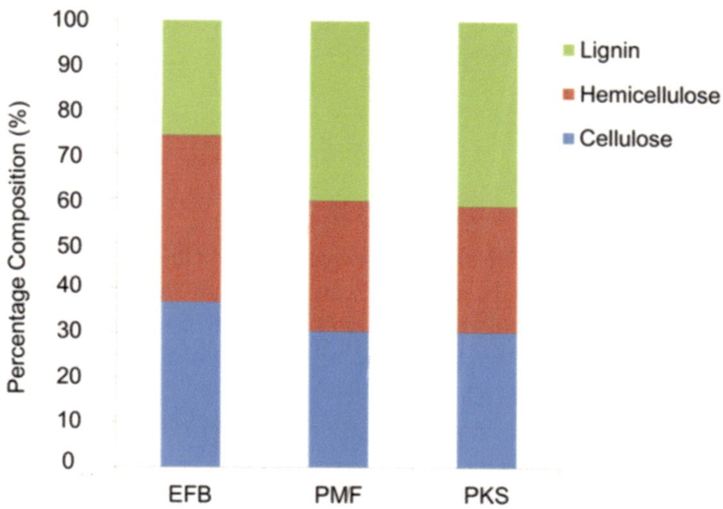

Figure 3. Lignocellulosic composition of Malaysian oil palm biomass samples.

Element		EFB	PMF	PKS
Carbon	(C) (%)	43.6	46.3	47.8
Hydrogen	(H) (%)	4.0	4.7	4.1
Nitrogen	(N) (%)	2.0	1.4	0.5
Sulfur	(S) (%)	0.2	0.2	0.2
Oxygen (by difference)	(O) (%)	50.2	47.4	47.6
High Heating Value (HHV) (MJ/kg)	16.3	16.5	17.5	

Table 1. Ultimate analysis and HHV of various Malaysian oil palm biomass samples

4.2. Algal biomass (microalgae or macroalgae)

Algae are considered to be primitive plants lacking in roots, stems and leaves [17]. This inherent structure of these type of plants makes it very suitable for energy production [18]. Thus, its use as feedstocks for bio-oil production is gaining popularity in recent years due also to the presence of some functional components including oils, proteins and carbohydrates [8]. Among the various types of algae, microalgae are the most potential energy source because cultivation and production of these biofeedstocks do not match those for food production [5]. The concept of bio-oil conversion using these materials has been introduced in 1970s, but mainly focused on direct thermochemical method. Recently, liquefaction under hydrothermal condition has also received increasing attention because of the advantages the method can offer for rapid reaction, no pre-drying of samples and no restrictions due to lipid contents [8]. Hydrothermal liquefaction (HTL) converts biomass feedstocks into four phases including the biocrude oil (or simply referred to as bio-oil), aqueous products, solid residue and gaseous products.

5. Objectives of the study

This work focused on the conversion of biomass (mainly oil palm biomass) to bio-oil by liquefaction under hydrothermal conditions using sub- and supercritical water. The effects of operating parameters such as temperature and pressure on the liquefaction were studied. The obtained bio-oil was then characterized using various analytical methods such as TG-DTA and GC-MS analyses. Other approaches like the use of catalysts to enhance liquefaction rates and selectivity of the products, and coupling of microwave irradiation will also be introduced.

6. Experimental methodology

6.1. Chemicals and reagents

Most of the chemical reagents used in this study, such as toluene for separation of bio-oil from the solvent, were purchased from Wako Chemicals (Japan).

6.2. Experimental methods

The apparatus for hydrothermal experiments (AKICO Co. Japan) shown in Fig. 4 consists of a batch-type Inconel reactor (about 8.8 cm^3 inner volume) and a heating electric furnace. Mechanical stirring of reactor is provided with a cyclic horizontal swing span of 2 cm fixed at a frequency of 60 cycles per minute. The amount of solvent used in each experiment depends on the desired reaction pressure. Only water was used as solvents for liquefaction experiments. Critical temperatures (Tc) and pressures (Pc) for water are Tc = 375°C, Pc = 22.1 MPa, respectively. It took about 15 min to reach the set reaction temperature of 210 to 300°C. The reaction time was varied from 0.5 to 8 h. After the reaction time elapsed, batch reactor was quenched

http://dx.doi.org/10.5772/59947

in a water bath for rapid cooling. The degradation products were then separated for analyses. The degradation rate (DR) was then calculated from the amount of resin removed from the sample on the basis of the amount of resin originally present, and reported in %.

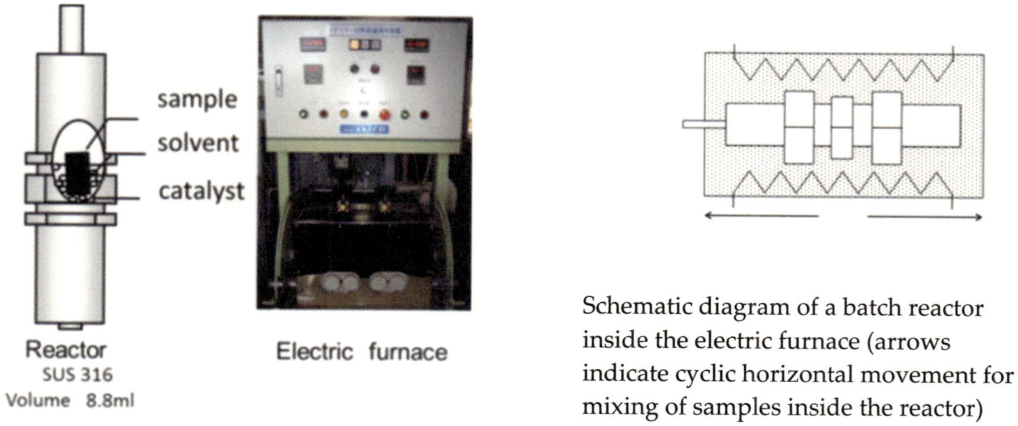

Figure 4. Schematic diagram of the reactor and actual apparatus for solvothermal experiments (Tmax=500°C, Pmax=50MPa).

Biomass samples were loaded into the reactor (8.8 mL) with distilled water at a sample to water ratio of 1:10. The reactor was inserted into the heating chamber preheated at the desired temperature, and then was allowed to react for 1 h. After 1 h had elapsed, the heater was turned off and the reactor was quenched in water at room temperature. After sufficiently cooling the reactor, its content was washed and rinsed with water and toluene. The toluene solution was subjected under vacuum in a rotary evaporator set at T=60°C to remove toluene and obtained the bio-oil for further analyses.

6.3. Gas Chromatography-Mass Spectrometry (GC-MS) analysis

The obtained bio-oil was analyzed by gas chromatography-mass spectrometry (GC-MS, Hewlett Packard-6890 series, Palo Alto, CA), coupled with a mass selective detector (Hewlett Packard-5973). The GC conditions were as follows:

Oven Temperature: 40°C (3 min) → 230°C → 300°C (3 min), rate = 5°C /min

Injector/Detector Temperature: 250°C, Injection Volume: 0.5 μL

Carrier Gas: Helium (split ratio of 12: 1, flow rate of 24 mL min^{-1})

Ionizing energy: 70 eV

The compounds corresponding to the peaks appearing in the chromatograms resulting from the GC-MS analyses were identified by comparing the mass spectra of each peak with those stored in the database (the NIST library).

6.4. Thermogravimetry-Differential Thermal Analysis (TG-DTA) analysis

To determine the temperature-dependent mass behavior of biomass, thermogravimetry-differential thermal analysis (TG-DTA) was performed using EXSTAR 6000 TG/DTA6300 apparatus (Seiko Instruments. Inc., Japan). The temperature range investigated was from 35 to 500°C at a heating rate of 20°C /min under N_2 gas flow.

7. Results and discussion

7.1. Effect of temperature and pressure

Table 2 summarizes the effect of temperature and pressure on sub- and supercritical hydrothermal liquefaction of oil palm biomass at reaction time of 1 h and sample-to-water ratio = 1 : 10. From all the runs, the bio-oil yields of PMF were higher than that of EFB. This can be explained by the differences in composition of cellulose, hemicellulose and lignin of the biomass samples. PMF has relatively higher lignocellulosic content (75.9%) and lignin content (30.6%) compared to EFB (72.1% and 18.6, respectively).

At 25 and 30 MPa, the yield of bio-oil decreased with increasing temperature, most likely due to possible secondary decomposition of biomass components at higher temperatures, causing gasification and charring. At 35 MPa, increasing the temperature from 360°C to 390°C resulted into an increase in bio-oil yield. However, increasing the temperature further to 450°C decreased the yield. High pressure increases water density, hence enhancing solubility of the materials and its decomposition. However, higher temperature causes secondary decomposition of biomass and recombination of some free radicals, leading to gas and char formation.

Reaction Conditions	T (°C)	P (MPa)	Bio-oil Yield (wt%)		
			EFB	PMF	PKS
Subcritical water	330	25	15.72 ± 1.05	16.40 ± 0.21	22.83 ± 0.56
	330	30	16.96 ± 1.18	15.72 ± 0.81	25.62 ± 1.87
	330	35	15.68 ± 0.68	14.79 ± 0.21	23.67 ± 0.27
	360	25	25.31 ± 0.44	22.71 ± 0.83	26.55 ± 1.29
	360	30	26.02 ± 0.44	23.22 ± 0.54	27.54 ± 0.70
	360	35	22.76 ± 0.28	21.75 ± 0.01	23.44 ± 1.24
Supercritical water	390	25	**37.39 ± 0.67**	**34.32 ± 1.87**	**38.53 ± 1.46**
	390	30	28.17 ± 0.35	27.57 ± 1.03	31.16 ± 0.81
	390	35	30.16 ± 0.98	24.07 ± 1.04	29.35 ± 0.71

Table 2. Effect of temperature and pressure on sub- and supercritical hydrothermal liquefaction of oil palm biomass (reaction time = 1 h, sample-to-water ratio = 1 : 10)

http://dx.doi.org/10.5772/59947

The increase in pressure has changing impacts on the bio-oil yield. Increasing the pressure increases the water density and also the solubility of the target compounds. These properties combined with inherent highly diffusive characteristics of supercritical fluids can enhance the ability of water to penetrate more easily into the sample matrix, hence improving extraction and decomposition of target compounds in biomass.

However, in some cases the increase in solvent density due to increase in pressure could result into cage effect for the C-C bonds present in the molecules of the target biomass components. This in effect inhibits the cleavage of C-C bonds, thus lowering hydrolysis rate resulting into low yields of liquefied products. Hence, as observed from the obtained data in this study, variation of pressure may or may not have positive results in sub- and supercritical hydrothermal liquefaction of oil palm biomass. This observation merits more detailed investigation in the future.

7.2. Relation of obtained results with the lignocellulosic composition of samples

As shown in the results of TG-DTA analysis of the samples in Fig. 5 and Fig. 6, respectively, hemicellulose degrades at a lower temperature range of 210–330°C due to the less uniform structure and low-degree crystallinity. Cellulose is a naturally occurring polymer consisting of linearly linked monomer units of glucose; hence, it has a higher degree of crystallinity and degrades at 300–375°C. Lignin is a highly crossed-linked polyphenolic aromatic polymer having no ordered repeating units. As such, lignin has the highest thermal stability and decomposed at 150–1000°C.

Due to higher lignin content, higher temperature and pressure (450°C and 30 MPa) are needed to degrade PMF and PKS compared to that of EFB (390°C and 25 MPa). PKS produces highest bio-oil yield at optimum condition, followed by PMF and EFB due to decreasing lignin content in the order of PKS > PMF > EFB.

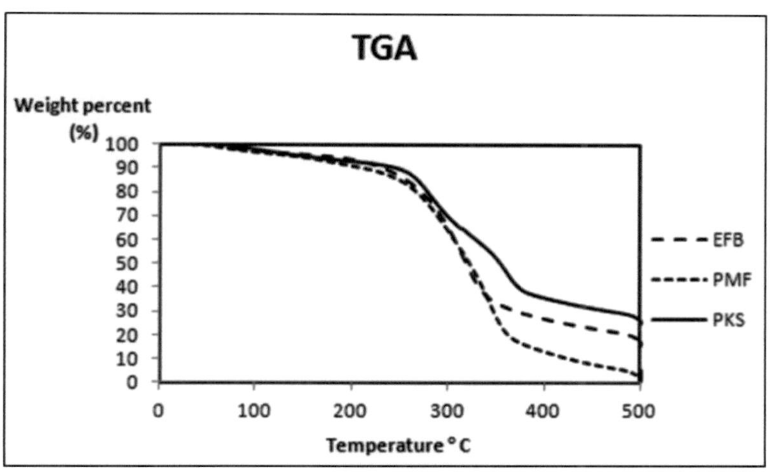

Figure 5. Results of thermogravimetric analysis of various oil palm biomass samples.

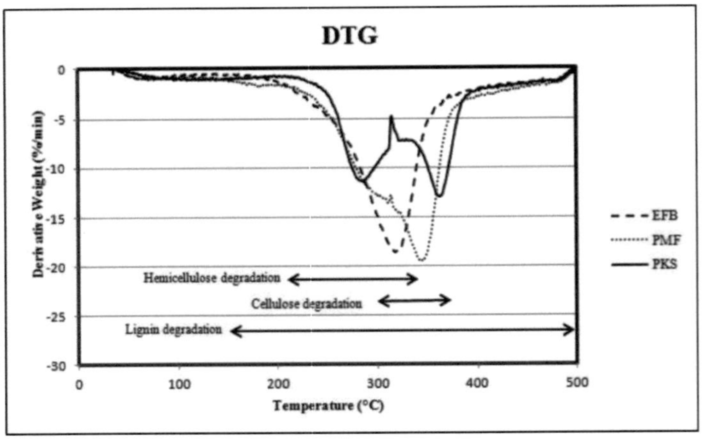

Figure 6. Results of thermogravimetric analysis of various oil palm biomass samples.

7.3. Results of GC-MS analyses of the obtained bio-oils

The typical GC-MS chromatograms of the bio-oil obtained from EFB, PMF and PKS at the optimum condition in this study (T=390°C, P=25 MPa) are shown in Fig. 7. The bio-oil obtained was typically dark brown, highly viscous liquid with distinct smoky odor. The chromatograms were almost identical, consisting of mostly phenolic and benzene compounds as depicted in Table 3. These compounds were likely obtained mainly from decomposition of lignin and cellulosic components of the biomass. Slight differences in composition for PKS as compared to other samples were observed as can be seen in the appearances of peaks 12 to 16. These peaks were identified as phenolic (peak 12), 1,2,3-trimethoxybenzene (peak 13), ester compound (peak 14), ketone (peak 15) and 1-(2,5-dimethoxyphenyl)-propanol (peak 16). These results agree with chemical compositions reported in literature consisting of several hundreds of organic compounds, such as acids, alcohols, aldehydes, esters, ketones, phenols and lignin-derived oligomers [19]. It is also possible that some of these products are undesirable; thus, post-treatment of the obtained biocrude oil by removal of these unwanted compounds or by upgrading has been suggested. Production of these unwanted molecules can also be avoided by using catalysts suitable for selective degradation of the lignocellulosic components into the target compounds in addition to the enhancement of liquefaction rates.

Peak No.	Retention Time (min)	Compound	Percentage Peak Area (%)		
			EFB	PMF	PKS
1	18.44	Ethylbenzene	5.12	5.30	3.93
2	18.77	Benzene, 1,3-dimethyl-	5.55	-	-
3	20.03	2-cyclopenten-1-one, 2-methyl	3.42	-	-
4	22.38	Phenol	38.95	41.74	32.30

Peak No.	Retention Time (min)	Compound	Percentage Peak Area (%)		
			EFB	PMF	PKS
5	24.93	2,3-dimethylcyclopent-2-en-1-one	13.04	5.28	3.08
6	25.20	Phenol, 2-methyl-	-	-	2.26
7	26.72	Mequinol / Phenol, 2-methoxy	11.14	16.28	16.13
8	30.08	Phenol, 2-methoxy-4-methyl	-	4.11	4.40
9	32.69	1,2-benzenediol, 4-methyl-	11.97	-	-
10	32.70	Phenol, 4-ethyl-2-methoxy	-	9.64	8.78
11	34.73	Phenol, 2,6-dimethoxy	10.80	17.66	14.66
12	35.26	Phenol, 2-methoxy-4-[[[2-(4-hydroxyphenyl)ethyl]ami	-	-	2.17
13	37.28	1,2,3-trimethoxybenzene	-	-	2.50
14	39.08	Dodecanoic acid, methyl ester	-	-	3.30
15	39.30	Ethanone,1-(2,6-dihydroxy-4-methoxyphenyl)-	-	-	3.69
16	41.40	1-(2,5-dimethoxyphenyl)-propanol	-	-	2.80

Table 3. Major chemical components of the obtained bio-oil identified using GC-MS and NIST library

8. Use of catalysts to enhance liquefaction rates

The use of catalysts is indispensable and important to enhancement of liquefaction rates of biomass even under hydrothermal conditions. This will shorten reaction time even under mild conditions, thus reducing the required energy and processing cost. Some of the catalysts used for this purpose includes both homogeneous (alkali, alkali salts, etc.) and heterogeneous catalysts (metal oxides, etc.).

Akhtar et al. [20] has investigated liquefaction of empty palm fruit bunch (EPFB) under subcritical water conditions using various alkalis (such as NaOH, KOH and K_2CO_3). Catalytic performance and biomass to water ratio suitable for high EPFB conversion, liquid hydrocarbons yield and lignin degradations were screened in a batch reactor operating at 270°C and 2 MPa in a reaction time of 20 min. Results showed that addition of alkali had positive effect on conversion, liquefaction and lignin degradation, increasing the rates by almost two-folds compared to that without any catalysts. The reactivity of the alkalis was in the order of K_2CO_3 > KOH > NaOH. The composition of bio-oil also depended on the type of catalysts, obtaining the highest yield of phenols using K_2CO_3 (1.0 M), while much of ester compounds were obtained using 1.0 M NaOH.

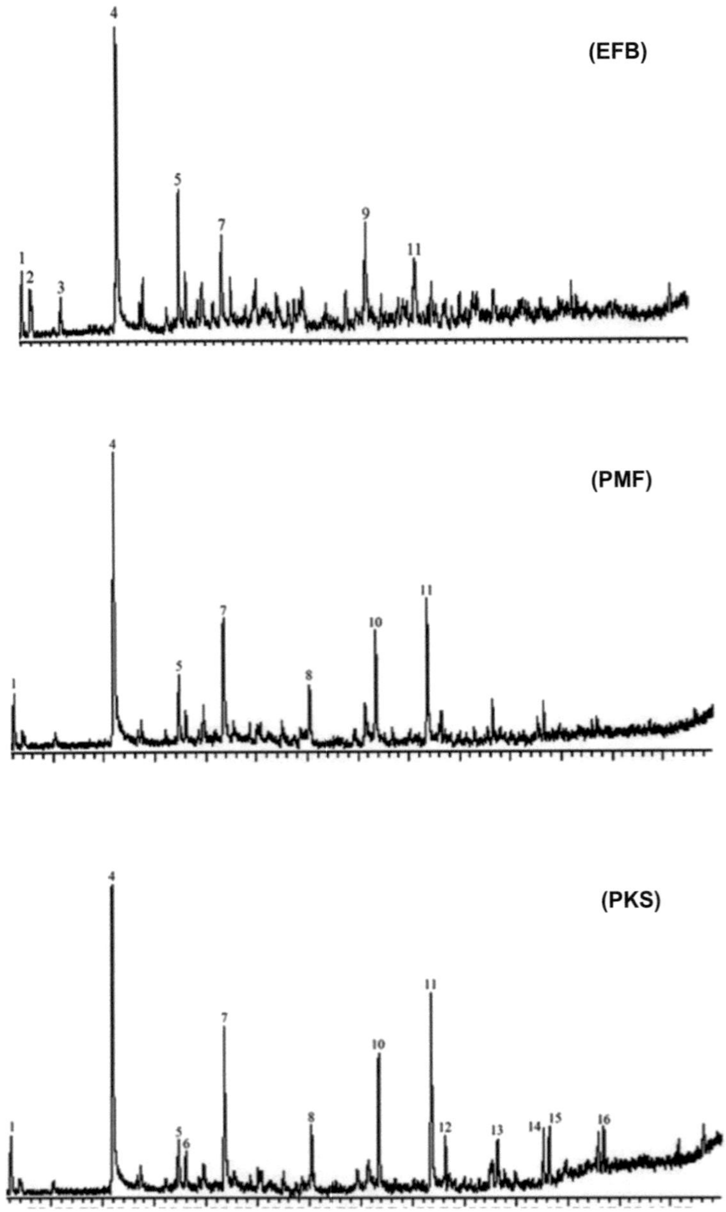

Figure 7. Typical GC-MS chromatograms of bio-oil obtained from EFB, PMF and PKS at the optimum condition (T = 390°C, P=25 MPa).

Various catalytic processes for production of biofuels from palm oil and oil palm biomass were also summarized by Chew and Bhatia [21]. This includes reviews on catalytic processes of palm oil to produce biodiesel, and cracking to produce high-grade biofuels. This review also discusses biomass gasification to produce hydrogen and syngas, and its conversion to liquid fuels by Fischer-Tropsch synthesis (FTS), including upgrading of liquid/gas fuels obtained from liquefaction/pyrolysis of biomass. Several catalysts including zeolites such as ZSM-5 and aluminosilicates were reported to enhance biomass conversion and upgrading; however, application to hydrothermal liquefaction was not reported.

The use of rare earth modified zeolite, Ce/HZSM-5 has been proposed by Xu et al. [22] for hydrothermal liquefaction of microalgae. They found that this novel type of catalysts exhibited good potential and benefits for the preparation of bio-oil from microalgae (e.g. *Chlorella pyrenoidosa*) with high efficiency.

Several other catalysts have been screened for hydrothermal catalytic processing of pretreated algal oil that was produced from the hydrothermal liquefaction of Chlorella pyrenoidosa [23]. The activities of 5% Pt/C, 5% Pd/C, 5% Ru/C, 5% Pt/C (sulfided), Mo_2C, MoS_2, alumina, CoMo/ γ-Al_2O_3 (sulfided), Ni/SiO_2-Al_2O_3, HZSM-5, activated carbon, and Raney-Ni for hydrothermal hydrodeoxygenation and hydrodenitrogenation of the pretreated algal oil at 400°C were investigated. It was reported that Ru/C showed the best performance for deoxygenation, and Raney-Ni was the most suitable catalyst for denitrogenation. The upgraded oil from the Ru/C catalyzed reaction contained the highest hydrocarbon content, the highest fraction of material boiling below 400°C, and the highest higher heating value (45.1 MJ/kg). All of the metal catalysts produced freely flowing upgraded oil. The combination of Ru/C and Raney Ni, which showed very good deoxygenation and denitrogenation of the oil, produced upgraded oil that retained 86% of the heating value in the original pretreated oil. This upgraded oil was also produced in a higher yield (77.2 wt.%) and with lower gas (9.4 wt.%), coke (15.6 wt.%) and water-soluble products yields (2.1 wt.%) than from either catalyst alone. The authors also suggested a two-step biocrude treatment strategy (non-catalytic treatment followed by catalytic upgrading) and a two-component catalyst bed in the second step for the hydrothermal catalytic upgrading of algal biocrude.

9. Coupling with microwave irradiation

Microwave technology utilizes electromagnetic waves for heating as a result of friction produced by oscillation of molecules as they realign with microwave upon absorption. Microwaves operate in between infrared radiation and radiofrequencies of 30 GHz to 300 MHz, corresponding to wavelengths of 1 cm to 1 m. Strict government regulations only allow domestic and microwave apparatus to operate at either 122.2 cm (2.45 GHz) or 33.3 cm (900 MHz) to avoid interference with RADAR transmissions and telecommunications [24].

Using microwave, rapid heating occurs as a result of heat being generated within the material, unlike the conventional method where heating of material is performed by conduction from heat sources outside of the vessel. Many microwave-assisted reactions are accelerated due to this rapid internal heating, resulting to tremendous increase in reaction rates compared to the conventional methods. Thus, even at shorter reaction times, higher yields and selectivity of target compounds are expected. Some other reactions not possible with the conventional heating methods had also been reported to proceed by microwave irradiation. Kingston and

Haswell had summarized available information related to the fundamentals of microwave chemistry in sample preparations and other applications [25].

Microwave non-thermal effects on reaction have also been reported to occur, thus even at milder conditions, dramatic increase in the yield can be obtained. Hoz et al. reported in a review of related topics some evidences and postulates on the existence of this phenomenon, but doubts still remain for many researchers [26]. Jacob et al. has reported some interesting results on specific microwave effects on thermal and non-thermal interaction of microwaves with materials [27]. Some of the topics discussed include "hot spots" or localized heating, molecular agitation, improved transport properties of materials with microwaves, and some evidences on reaction rate enhancements as result of these improved properties of materials. A lot of mechanisms of activation of materials are thought to be possible with microwave interaction; thus, it is difficult to simply treat microwave heating as similar to that of the conventional.

The above-mentioned thermal and non-thermal effects of microwave irradiation offer enormous benefits to the treatment of biomass for synthesis of biofuels including energy efficiency, development of a compact process, rapid heating and instant on-off process (instant heating-cooling process), among many other possible advantages.

Microwave-based pretreatment approach utilizes both thermal and non-thermal effects generated by an extensive intermolecular collision as a result of realignment of polar molecules with microwave oscillations. Compared to conventional heating, electromagnetic field generated by microwave has the ability to directly interact with the material to produce heat, thereby accelerating chemical, physical and biological processes. The advantages of employing microwave rather than the conventional heating include reduction of process energy requirements, selective processing and capability for instantaneous starting and ceasing of the process. This also offers enormous benefits such as energy efficiency due to rapid and selective heating, and the possibility for developing a compact process.

When microwave is applied to the treatment of lignocellulosic biomass, the unique feature of selectively heating the more polar part will result in an improved disruption of the recalcitrant structures of lignocellulose. With the non-thermal effects, electromagnetic field enhances the destruction of crystalline structures and changes the super molecular structure of lignocellulosic material thereby improving its reactivity.

Microwave pretreatment is also an energy-efficient and environmentally benign technology that aids in the transport of chemicals into the substrates. The project team from the US Department of Energy in partnerships with research institutes including the Oak Ridge National Laboratory [28] has showed that by opening the cellular microstructures of wood, for example, microwave pretreatment could permit pumping chemicals for easy access of even large sections (10 cm long x 10 cm diameter) of hardwoods. The project team has demonstrated that, for both hardwood and softwood chips, microwave pretreatment can decrease both H-factor and chemicals required to pulp hardwoods and softwoods by greater than 40% with acceptable quality. The steam pressure generated inside the wood breaks the pit membranes and vessel cell walls, thereby enhancing the woods permeability to chemicals and process liquors.

http://dx.doi.org/10.5772/59947

Other than the lignocellulosic biomass, the use of microwave for pretreatment of samples for a more efficient oil extraction and pretreatment of free fatty acids for biodiesel conversion has also been proposed.

10. Summary and future directions

This research focuses on the feasibility of converting oil palm biomass wastes to bio-oil under sub- and supercritical water. The effects of reaction temperature and pressure on the efficiency of the liquefaction of EFB, MCF and PKS were investigated. Based on the obtained results, the optimum liquefaction condition for EFB, PMF and PKS could be obtained at supercritical conditions of water. EFB gave the highest bio-oil yield at 390°C and 25 MPa, whereas for PMF and PKS, the highest bio-oil yields were at 450°C and 30 MPa. Lignocellulosic contents of the biomass affect their optimum liquefaction condition and the yield of bio-oil. Temperature and pressure have also different impacts on the bio-oil yield. Compared to pyrolysis, the proposed hydrothermal liquefaction method using sub- and supercritical fluid utilizes the water originally present in the feedstock, thus avoiding the energy-intensive drying procedures.

Future directions will cover evaluation of global kinetics of the liquefaction rate based on the reaction equilibrium data collected and use of various catalysts to enhance liquefaction rates. The major contribution expected from this research will be a new technology as applied to the liquefaction of biomass using sub- and supercritical fluid.

Acknowledgements

This work was partly supported by JSPS Bilateral Exchange Program – Open Partnership Joint Research Projects with UTP (Malaysia). YUTP Research Grant for CYH stay in Kumamoto University to carry out the experimental works is also being acknowledged.

Author details

Armando T. Quitain[1*], Chan Yi Herng[2], Susana Yusup[2], Mitsuru Sasaki[3] and Yoshimitsu Uemura[4]

*Address all correspondence to: quitain@kumamoto-u.ac.jp

1 Graduate School of Science and Technology, Kumamoto University, Japan

2 Chemical Engineering Department, Universiti Teknologi PETRONAS, Malaysia

3 Institute of Pulsed Power Science, Kumamoto University, Japan

4 Centre for Biofuel and Biochemical Research, Universiti Teknologi PETRONAS, Malaysia

References

[1] Nilsen MH, Antonakou E, Bouzga A, Lappas A, Mathisen K, Stocker M. Investigation of the effect of metal sites in the Me-Al-MCM-41 (Me = Fe, Cu or Zn) on the catalytic behavior during the pyrolysis of wooden based biomass. Microporous and Mesoporous Materials 2007; 105, 189–203.

[2] Hamzeh Y, Ashori A, Khorasani Z, Abdulkani A, Abyaz A. Pre-extraction of hemicelluloses from bagasse fibers: Effects of dry-strength additives on paper. Industrial Crops and Products 2013; 43(5), 365–371.

[3] Peterson AA, Vogel F, Lachance RP, Fröling M, Antal MJ Jr., Tester JW. Thermochemical biofuel production in hydrothermal media: A review of sub- and supercritical water technologies. Energy & Environmental Science 2008; 1, 32–65.

[4] Toor SS, Rosendahl L, Rudolf A. Hydrothermal liquefaction of biomass: A review of subcritical water technologies. Energy 2011; 36, 2328–2342.

[5] Barreiro DL, Prins W, Ronsse F, Brilman W. Hydrothermal liquefaction (HTL) of microalgae for biofuel production: State of the art review and future prospects. Biomass and Bioenergy 2013; 53, 113–127.

[6] Brennan L, Owende P. Biofuels from microalgae – A review of technologies for production, processing, and extractions of biofuels and co-products. Renewable and Sustainable Energy Reviews 2010; 14, 557–577.

[7] Ruiz HA, Rodriguez-Jasso RM, Fernandes BD, Vicente AA, Teixeira JA. Hydrothermal processing, as an alternative for upgrading agriculture residues and marine biomass according to the biorefinery concept: A review. Renewable and Sustainable Energy Reviews 2013; 21, 35–51.

[8] Tian C, Li B, Liu Z, Zhang Y, Lu H. Hydrothermal liquefaction for algal biorefinery: A critical review. Renewable and Sustainable Energy Reviews 2014; 38, 933–950.

[9] Yu Y, Lou X, Wu H. Some recent advances in hydrolysis of biomass in hot-compressed water and its comparisons with other hydrolysis methods. Energy and Fuels 2008; 22, 46–60.

[10] Brunner G. Near critical and supercritical water. Part I. Hydrolytic and hydrothermal processes. Journal of Supercritical Fluids 2009; 47, 373–381.

[11] Haar LJ, Gallagher S, Kell GS. National Bureau of Standards/National Research Council steam tables. Hemisphere Publishing Corp, Bristol, PA (1984)

[12] Sue K, Ouchi F, Arai K. Microcell for potentiometric pH measurements of supercritical aqueous solutions. Journal of Supercritical Fluids 2006; 39, 271–276.

[13] Hurst WS, Hodes MS, Bowers WJ Jr., Bean VE, Maslar JE, Griffith P, Smith KA. Optical flow cell and apparatus for solubility, salt deposition and Raman spectroscopic

studies in aqueous solutions near the water critical point. Journal of Supercritical Fluids 2001; 22, 157–166.

[14] Aizawa T, Masuda Y, Minami K, Kanakubo M, Nanjo H, Smith RL. Direct observation of channel-tee mixing of high-temperature and high-pressure water. Journal of Supercritical Fluids 2007; 43, 222–227.

[15] Bremholm M, Jensen H, Brummerstedt Iversen S, Brummerstedt Iversen B. Reactor design for in situ X-ray scattering studies of nanoparticle formation in super-critical water syntheses. Journal of Supercritical Fluids 2008; 44, 385–390.

[16] Sumathi S, Chai SP, Mohamed AR. Utilization of oil palm as a source of renewable energy in Malaysia. Renewable and Sustainable Energy Reviews 2008; 12, 2404–2421.

[17] Lee RE. Phycology. Cambridge University Press, New York (1980).

[18] Khan SA, Rashmi, Hussain MZ, Prasad S, Banerjee UC. Prospects of biodiesel production from microalgae in India. Renewable and Sustainable Energy Reviews 2009; 13 (9), 2361–2372.

[19] Xiu S, Shahbazi A. Bio-oil production and upgrading research: A review. Renewable and Sustainable Energy Reviews 2012; 16, 4406–4414.

[20] Akhtar J, Kuang SK, Amin NS. Liquefaction of empty palm fruit bunch (EPFB) in alkaline hot compressed water. Renewable Energy 2010; 35 (6), 1220–1227.

[21] Chew TL, Bhatia S. Catalytic processes towards the production of biofuels in a palm oil and oil palm biomass-based biorefinery. Bioresource Technology 2008; 99 (17), 7911–7922.

[22] Xu Y, Zheng X, Yu H, Hu X. Hydrothermal liquefaction of *Chlorella pyrenoidosa* for bio-oil production over Ce/HZSM-5. Bioresource Technology 2014; 156, 1–5.

[23] Bai X, Duan P, Xu Y, Zhang A, Savage PE. Hydrothermal catalytic processing of pretreated algal oil: A catalyst screening study. Fuel 2014; 120, 141–149.

[24] Mingos DMP, Baghurst DR. Applications of microwave dielectric heating effects to synthetic problems in chemistry, In: Microwave-Enhanced Chemistry, H. M. (Skip) Kingston & S. J. Haswell, (Ed.), pp. 3–50. American Chemical Society, Washington, DC, ISBN 0-8412-3375-6 (1997).

[25] Kingston HM (Skip), Haswell SJ. Microwave-enhanced chemistry: fundamentals, sample preparation and applications. American Chemical Society, Washington, DC, ISBN 0-8412-3375-6 (1996).

[26] De la Hoz A, Diaz-Ortiz A, Moreno A. Critical review: Microwaves in organic synthesis. thermal and non-thermal microwave effects. Chemical Society Reviews 2005; 34, 164–178.

[27] Jacob J, Chia LHL, Boey FYC. Review: Thermal and non-thermal interaction of microwave irradiation with materials. Journal of Material Science 1995; 30, 5321–5327.

[28] Compere A, Parent D, et al. Increasing yield and quality of low-temperature, low-alkali kraft cooks with microwave pre-treatment. US Department of Energy, Energy Efficiency and Renewable Energy Reports (2007).

Chapter 22

Glycerol as a Raw Material for Hydrogen Production

Sandra Imaculada Maintinguer,
Rafael Rodrigues Hatanaka and
José Eduardo de Oliveira

Additional information is available at the end of the chapter

http://dx.doi.org/10.5772/60013

1. Introduction

Fossil fuels continue to be the primary global energy sources, suppling approximately 80% of global demand. However, such energy sources can cause the greenhouse effect with the generation of harmful gases such as COx, NOx, SOx, CxHx and other organic compounds. These pollutants are released into the atmosphere as a result of fossil fuel combustion [1]. Renewable technologies, such as biofuels, present possible alternative sources of energy that are carbon neutral.

Approximately 68% of global biodiesel supply is produced by five countries: Brazil, Germany, USA, France and Argentina. Brazil is among the largest producers and consumers of biodiesel in the world, with a production capacity of 8539 million liters [2]. Several factors favour the cultivation of different plants for biodiesel production in Brazil, especially the climatic condition, and the availability of arable land. The increase in the production of biodiesel makes Brazil a strategically important country for the whole world, due to the depletion of already known fossil energy reserves.

Biodiesel is a biofuel obtained by the transesterification of raw materials such as animal fats and vegetable oils. In Brazil, sources like soybeans and palm are economically attractive for biodiesel production due to their abundance. Approximately 80% of the biodiesel produced in Brazil is derived from soybean [3]. In Brazil, the addition of 2% biodiesel in diesel fuel has been mandatory since 2008, this amount was increased to 6% in July 2014 and then increased to 7% in November 2014 [4].

The large surpluses of glycerol that are generated in this process require new commercial uses to be identified. Brazil will become a major producer and consumer of biodiesel due to strong strategy for biofuel production. This is possible due to exceptional conditions for the cultivation of oilseeds for oil extraction. The selection of feedstock depends strongly on potentialities of each region [5].

When biodiesel is produced from vegetable oils and animal fats through transesterification process, high amounts of waste are created. This waste chiefly consists of crude glycerol, which has limited commercial value, unless expensive purification processes are performed. Even when purified, there is such a large global overproduction that traditional markets find difficult to absorb it. Each ton of biodiesel produced generates approximately 100 kg of crude glycerol. However, the amount of crude glycerol generated in biodiesel production can vary from 1% to 85% (v/v), depending on operating conditions of the transesterification plants. Thus, to overcome these issues it is necessary to discover new uses for this significant residue.

Ethanol has been primarily produced from sugarcane in Brazil since 1975, encouraged by the implementation of the National Program for Alcohol (1975–1985). Brazil produces (in 2011/2012) nearly 571 million tons of sugarcane, which is processed by sugar mills to produce 36.9 million tons of sugar and 22.9 billion liters of ethanol [6]. The ethanol could potentially be used in the esterification process of biodiesel [7] to develop cheaper and more environmentally friendly processes. Several studies have been developed to obtain hydrogen gas, as a renewable energy source that can be generated from waste glycerin, a byproduct of biodiesel production.

In this sense, this chapter presents a comparative study on biological hydrogen production from crude glycerol, the microorganisms involved in the biological processes for hydrogen production and some of the strategies applied in the literature for the improvement of these systems. It will contribute to innovation in research into the reuse of crude glycerol, in a sustainable manner, thus leading to potential cost reduction and clean energy generation.

2. Glycerol: A byproduct of biodiesel production

Glycerin is the principal byproduct of biodiesel production, with waste streams containing at least 95% glycerol; however its purity can vary depending on the efficiency of the production process [8].

Glycerol is generated by the process for obtaining biodiesel from vegetable oils or animal fats. This process often uses the addition of catalysts, such as sodium hydroxide, and alcohols, such as methanol or ethanol, with reactors maintained under heat and agitation (Figure 1).

However, during the transesterification process a high volume of glycerol is produced as a byproduct: for 90 m^3 of produced biodiesel, approximately 10 m^3 of glycerin is generated [8]. Pure glycerol can be used in many different applications, mainly in textile, chemical, pharmaceutical and food industries. However, to use it in these applications is necessary a degree of purity higher than 95% [9]. To achieve this, the crude glycerin must be submitted to a purification process, often resulting in high financial costs.

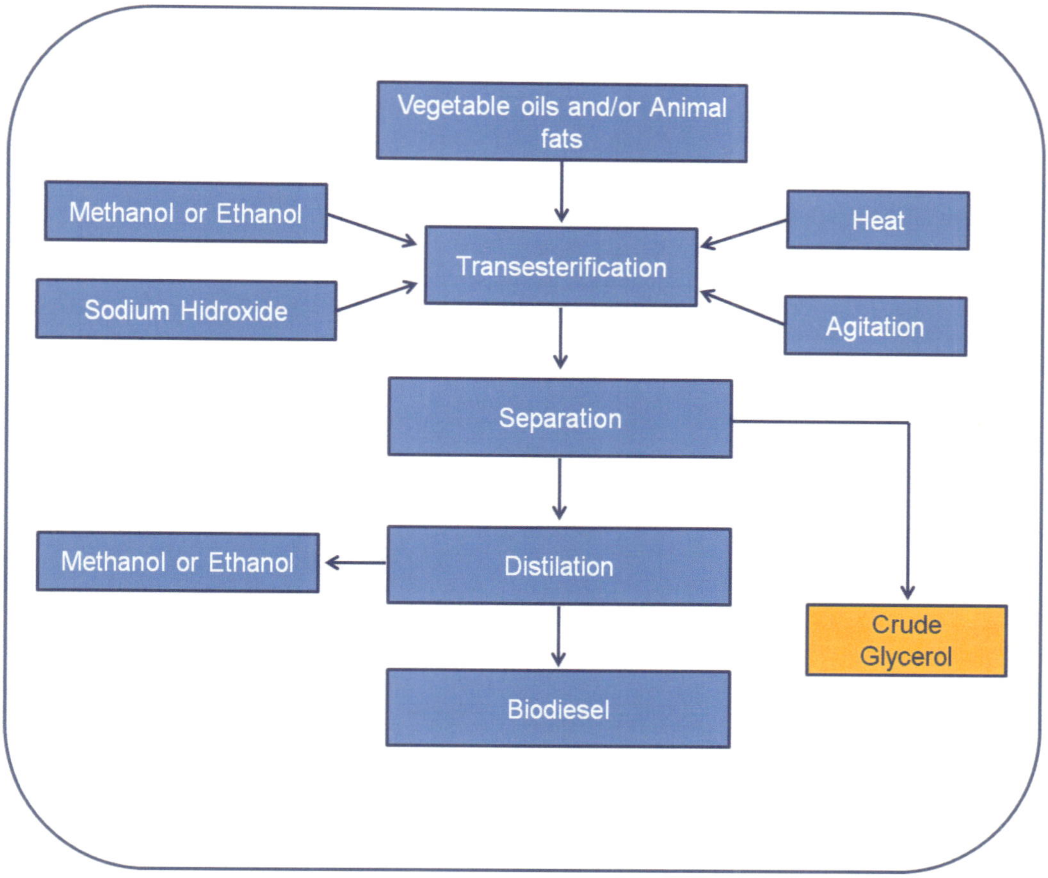

Figure 1. Representation of Biodiesel production by Transesterification Process (adapted from [10]).

Due to the significant amount of unwanted glycerol generated in the biodiesel manufacturing process [11], traditional markets have not found the capacity to absorb it, as described above. Depending on the purity, the main uses for glycerol are direct burning for energy production, inputs for various industrial segments and as raw material in animal feed [8].

3. The impurities present in crude glycerol

The main impurities present in crude glycerol are soap, free fatty acids, methanol, unreacted triglicerides, diglycerides and monoglycerides [10]. The exact composition of the waste glycerol often differs depending on the manufacturing process employed (Table 1). Silva et al. [12] observed methanol, water and sodium chloride, in samples of crude glycerol during the biodiesel production in a Brazilian plant with soybean oil as raw material.

Crude glycerol should be pre-treated before being used as feedstock. Methanol or ethanol can be removed using heating or distillation processes. Soap is another impurity present in the crude glycerol and can be removed by precipitation from the liquid medium through pH adjustment. Sodium ions can be removed from crude glycerol by neutralization with addition of phosphoric acid and lime in excess, in order to crystallize/precipitate hidroxyapatite [17]. However, these treatments are costly and not economically justifiable. Alternatively, there are many studies covering the use of crude glycerol for bio-hydrogen production without pre-treatments [11]. These bioconversions of the crude glycerol may be suitable and economically attractive alternatives to the industrial processes.

Source of biodiesel	**Glycerol content w/w**	**Product Yield**	**Impurities**	**Ref.**
Nittany Biodiesel, State College, PA	69.5%	0.41 m^3H_2. m^3 d^{-1}	MeOH, NaOH and sodium methylate	[13]
Integrity Biofuels, Indiana, USA	67.5±3.2%	Phytase (1125 U ml^{-1} supernatant)	$NaSO_4$, MeOH, Water,	[14]
Virginia Biodiesel Refinery (West Point, VA, USA)	84%	Docosahexaenoic acid (4.91 g l^{-1})	Soap, FFA, MeOH, Mono, Di or Triglycerides	[15]
Biodiesel factory, Portugal	86%	710.0 cm^3H_2 dm^3 medium	OMNG, ash and methanol	[16]

OMNG - Organic Matter Not Glycerol; FFA – Free Fat Acids; MeOH – Methanol

Table 1. Impurities present in the crude glycerol during the biodiesel production

4. Applications of crude glycerol in biological processes

Glycerol can be consumed by microorganisms in biological processes to generate byproducts with added value, such as ethanol, 1,3-propanediol (PD), H_2 and organic acids, among others. PD may be used in industrial applications such as polymers, cosmetics, foods, adhesives, lubricants, laminates, solvents, antifreeze and pharmaceuticals [9]. Ethanol has been used in the pharmaceutical industry, solvent, cleaning products and personal hygiene. In Brazil, its use is remarkable mainly due to the sugarcane producing. In 2011 over 27 billion liters of ethanol was produced in Brazil from sugarcane. Most of it was destined for use as a fuel[18], mainly hydrid vehicles that can be driven by mixtures of gasoline and ethanol [19]. H_2 has been utilized as a reactant in the chemical and petroleum industries during the production of ammonia, petroleum processing and methanol [2].

The energy content of the pure glycerol is 19.0 MJ/kg, however for crude glycerol it is 25.30 MJ/kg, possibly due to presence of methanol and biodiesel [10]. Such high energy content of crude glycerol indicates its potential to be an effective carbon source for hydrogen, PD and ethanol bioproduction.

The microorganisms involved in hydrogen production may be classified in four groups: strictly anaerobic, facultative anaerobic, aerobic and phototrophic [22], for example, green algae, cyanobacteria, phototrophic bacteria and fermentative bacteria. However, higher yields of H_2 generation are obtained by fermentation processes. The main fermentative bacteria known to produce hydrogen include *Enterobacter* sp., *Bacillus* sp., *Clostridium* sp., *Klebsiella* sp. and *Citrobacter* sp.[9,21]. The process of dark fermentation from crude glycerol may be followed by fotofermentation [9] because phototrophic bacteria grow with organic acids (the possible metabolites from fermentation) and they may produce more hydrogen [22].

Glycerin may have different yields of hydrogen per mole of organic substrate, depending on the route of the fermentation used and the exact composition of the end products. According reference [9], the end products of the fermentation process may be acetic acid (equation 1); butyric acid (equation 2), butanol (equation 3) and ethanol (equation 4). Furthermore, as reported by several authors, it may also generate PD (equation 5),[23,24]. Generation of acetic acid (equation 1) and butyric acid (equation 2) are accompanied by higher yields of H_2, as observed in the fermentation of sugars [25].

$$C_3H_8O_3 + H_2O \rightarrow CH_3COOH + CO_2 + 3H_2 \quad (1)$$

$$2C_3H_8O_3 \rightarrow C_4H_8O_2 + 2CO_2 + 4H_2 \quad (2)$$

$$2C_3H_8O_3 \rightarrow C_4H_{10}O + 2CO_2 + H_2O + 2H_2 \quad (3)$$

$$C_3H_8O_3 \rightarrow C_2H_6O + CO_2 + H_2 \quad (4)$$

$$2C_3H_8O_3 \rightarrow CH_3COOH + C_3H_8O_2 + CO_2 + 2H_2 \quad (5)$$

Glycerol can also produce lactate (equation 6) and succinate (equation 7) [26].

$$C_3H_8O_3 \rightarrow C_3H_6O_3 + H_2 \quad (6)$$

$$2C_3H_8O_3 \rightarrow C_4H_6O_4 + 2CO_2 + 5H_2 \quad (7)$$

Fermentation processes of hydrogen production using anaerobic acidogenic bacteria have been extensively described by several authors [27-31]. Additionally there are several key intermediate products created during the fermentation of glycerol: mainly organic acids, such as acetic acid and butyric acid and alcohols, such as ethanol and PD [9].

The following metabolites are obtained from the fermentation of glycerol: dihydroxyacetone, succinic acid, citric acid, docosahexaenoic acid, propionic acid, hydrogen, ethanol, and PD.

Figure 2 demonstrate that during the oxidative metabolism of glycerol, pyruvate is formed as an intermediate.

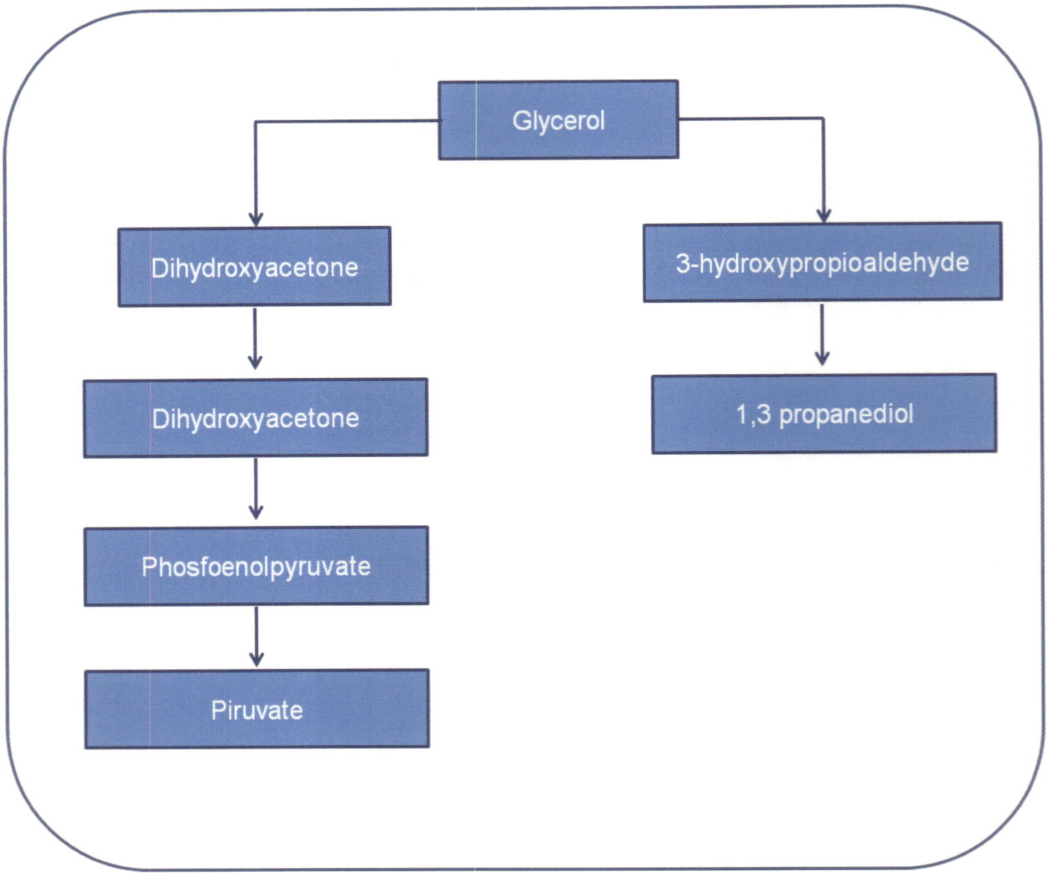

Figure 2. Glycerol metabolism during anaerobic fermentation (adapted from [10])

The production of PD is achieved through a reductive pathway in the anaerobic fermentation of glycerol. However, production of H_2 and other metabolites (ethanol, butanol, acetone, acetate, butyrate and lactate) compete with the production of PD by oxidative pathways [24].

However pyruvate formed during the conversion of glycerol (using the oxidative route) may be employed in various ways by microorganisms. The pyruvate is responsible for the formation of numerous metabolites such as lactate, ethanol, acetone, butanol and butyrate, as demonstrated during glucose metabolism (Figure 3). A similar metabolism using glycerol first produces pyruvate, before conversion to different metabolites and H_2 [10].

In many bacteria there are two biochemical pathways for glycerol metabolism: an oxidative pathway, where H_2 is generated and a reductive pathway leading to PD generation. When both pathways exist in the same microorganism PD production is preferential against H_2 [10].

http://dx.doi.org/10.5772/60013

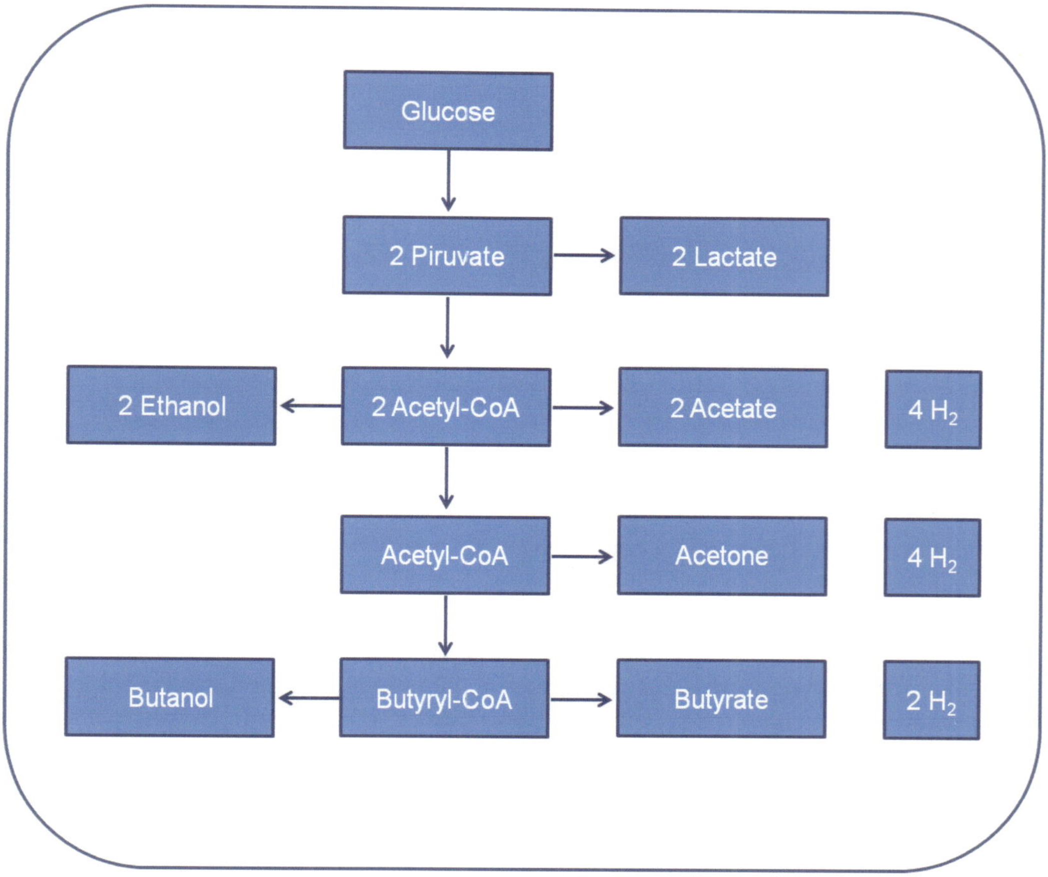

Figure 3. The glucose metabolism to pyruvate and hydrogen (Adapted from [10])

The biotechnological production of H_2, ethanol and PD from glycerol has been demonstrated by several bacteria species. The species of anaerobic bacteria as *Klebsiella* sp., *Enterobacter* sp., *Citrobacter* sp., *Lactobacillus* sp., *Bacillus* sp. [32] and some *Clostridium* sp., have demonstrated the ability to ferment glycerol or mixtures of glycerol and sugars [9]. In addition to the anaerobic microorganisms, nutrients are required in the reaction medium. These allow growth and fermentation of organic substrates leading H_2 production. Complex compounds used as nutrients include: peptone, tryptone, polypeptone, yeast extract, vitamin solutions, among others. Such bacterial species may generate H_2 from various carbon sources, particularly sugars [33] and glycerol.

Table 2 shows some studies about the bioconversion of glycerol using pure cultures or mixed microbial consortia to generate H_2 and other byproducts.

PD and H_2 are the two major products which can be obtained by bioconversion of crude glycerol. A co-culture of anaerobic bacteria, which can simultaneously use PD and produce H_2 via glycerol fermentation, may be a suitable option for crude glycerol bioconversion [10].

However, a combined production process of hydrogen and ethanol provides higher energy yield when compared with hydrogen or ethanol alone [40].

Microorganisms	Hydrogen Yield	Other by products	Ref.
Enterobacter aerogenes HU-101	63 mmol H_2 l^{-1} h^{-1}	0.85 mol ethanol mol^{-1} glycerol	[34]
Klebsiella pneumoniae ATCC 25955	-	PD	[35]
Mixed micro-flora of organic waste or soil	11.5-38.1 ml H_2 g^{-1} COD	PD	[36]
Mixed culture	0.31 mol H_2 mol^{-1} glycerol	0.59 mol PD mol^{-1} glycerol	[37]
Anaerobic digested sludge	0.41 mol H_2 mol^{-1} glycerol	0.784± 0.063 L CO_2L^{-1} media	[38]
Halanaerobium saccharolyticum	0.62 mol H_2 mol^{-1} glycerol	PD, butyrate, ethanol	[39]
Halanaerobium saccharolyticum	1.61 mol H_2 mol^{-1} glycerol	1.11 mol CO_2 mol^{-1} glycerol, acetate	[39]

COD- Chemical Oxygen Demand; PD - Propanediol

Table 2. Bioconversion of glycerol to H_2 and other products

A range of reactor types has been used in hydrogen production that utilizes organic waste materials such as crude glycerol or pure glycerol. They may be simple serum bottles [33], laboratory scale fermenters, pack-bed or up flow reactors [41] (Figure 4). However, it should be noted that most of the studies were performed in batch anaerobic reactors on laboratory scale (Table 3). To the best of our knowledge, there are no studies with pilot scale reactors, this suggests that the research into the bio-production of hydrogen using glycerol is currently in a preliminary phase.

Figure 4. Anaerobic batch reactors applied on laboratory scale

http://dx.doi.org/10.5772/60013

Substrate	Inocula	Reactor	T (°C)	pH	H_2 Yield [a]	Ref.
Crude glycerol	Kefir	Batch	25	5.2	0.22	[42]
Crude glycerol	*Enterobacter aerogenes*	Batch	37	6.8	1.12	[34]
Crude glycerol	Mixture(wheat oil)	Batch	30	6.2	0.31	[13]
Pure glycerol	*Enterobacter aerogenes*	Continuous	37	6.8	0.94	[34]
Pure glycerol	Wastewaters	Continuous	30	8.0	0.05	[41]
Crude glycerol	Soil from blueberry farm	Batch	30	5.5	0.18	[43]
Crude glycerol	Organic Soil	Batch	30	6.0	0.75	[24]
Crude glycerol	*Enterobacter aerogenes*	Batch	37	6.5	0.12	[44]

[a] mol-H_2.mol^{-1} glycerol

Table 3. Different bioreactors applied for bioconversion of glycerol

However there are many promising results of hydrogen generation using different configurations of anaerobic reactors fed with industrial wastewater, sugars, starch and others. The configurations of anaerobic reactors applied for biological H_2 production are AFBR (Anaerobic Fluidized-Bed Reactor), CSTR (continuously stirred tank-reactor), EGBS (Expanded granular sludge bed reactor) and UASB (Up-Flow Anaerobic Sludge Blanket reactor) (Table 4).

Substrate	Inocula	Reactor	H_2 Yield [a]	Ref.
Sucrose	UASB Sludge	Batch	1.2	[25]
Starch	Anaerobic Sludge	CSTR	2.3	[45]
Glucose	UASB Sludge	AFBR	2.45	[46]
Waste water of coffee	Anaerobic Sludge	UASB	1.29	[47]
Glucose and L-arabinose	CSBR and UASB Sludge	EGBS	2.71 $l^{-1}d^{-1}$	[48]

[a] mol-H2 mol-1 substrate

Table 4. Different configurations of anaerobic reactors applied for H_2 production

The UASB reactor is a single tank process where the wastewater enters from the bottom and flow upward (Figure 5). A suspended sludge blanket filters treats the wastewater flows through it and bacteria living in the sludge break down organic matter by anaerobic digestion, transforming it into biogas. Some advantages of this configuration are: the conversion of the organic matter in all reactor areas (bed and sludge blanket); the microorganisms can grow close to the bottom of the reactor in the form of flocks or granules (1 to 5 mm); the mixing of the system is promoted by the upward flow of wastewater and gas bubbles [49,50].

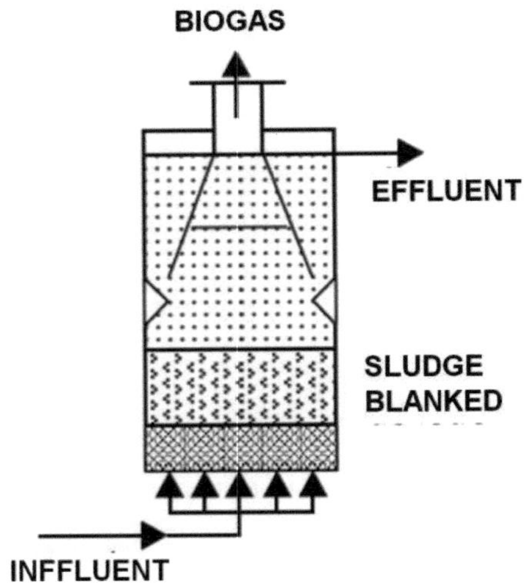

Figure 5. Schematic representation of UASB reactor (Adapted from [50])

The EGBS reactor has a cylindrical structure, packed with inert particles (0.3 to 3.0 mm of diameter) as support for microorganisms to form the biofilm (Figure 6). Several types of materials may be used as support mediums such as sand, coal, PVC, resins, ground tire and PET [51], etc. The biofilm may develop on the particles surface [49].

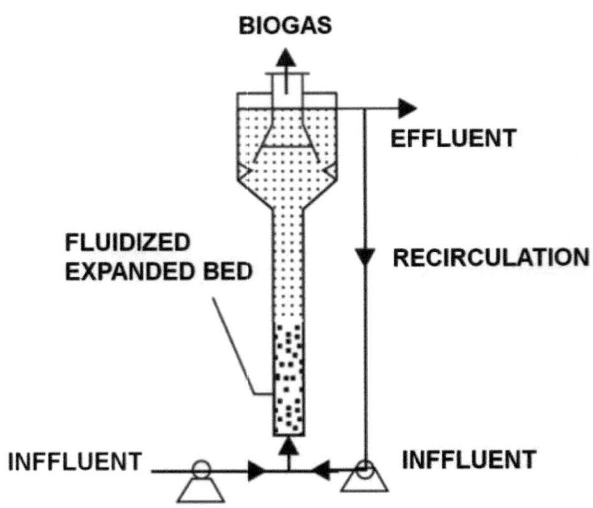

Figure 6. Schematic representation of EGBS reactor (Adapted from [50])

AFBR has the same operating principles of the EGBS reactors, except the particles size (0.5 to 0.7 mm) of the support medium and the expansion rates (Figure 7). The upward velocity of the liquid must be enough high to fluidize the bed until it reaches the point at which the gravitational force is equaled by the upward drag force. A high recirculation rate is necessary and the particles do not stay a fixed position inside the bed [49].

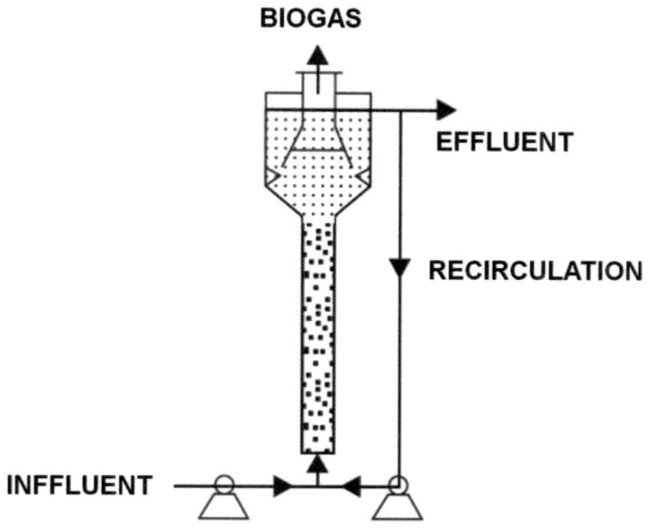

Figure 7. Schematic representation of AFBR reactor (Adapted from [50])

CSTR is known as a mix batch reactor and is an ideal type reactor in chemical engineering, for studies on laboratory scale [50]. CSTR can provide continuous or intermittent flow and comprises the follow steps: (1) filling (input of organic matter and microorganisms); (2) reaction (organic matter come into contact with microorganisms and they will degraded it); (3) sedimentation (settling of anaerobic sludge) and (4) emptying (removal of treated effluent) (Figure 8).

Studies on anaerobic fermentation of glycerol present major advances in pure cultures, such as with *Enterobacter aerogenes*. However, pure cultures do not represent real situations, such as those found in industrial waste [43]. To address this, research has been conducted onto hydrogen generation with mixed cultures obtained from anaerobic microorganisms present in biological treatment system sludge [39]. However, H_2-generating bacteria may be present in addition to bacteria that consume this gas, such as methanogenic archaea.

The fermentative production of hydrogen can be facilitated with methanogen inhibition, since methanogenic archaea use hydrogen in anaerobic biological processes [25]. To inhibit this methane formation process, reagents can be introduced, such as 2-brometanosulfonic acid (BES) and acetylene [2]. Additionally, pH control and heat treatment may provide other effective ways to prevent methanogens [25]. These methods may promote hydrogen production and encourage the growth of endospore-forming bacteria which are tolerant to high temperatures and adverse environmental conditions [21].

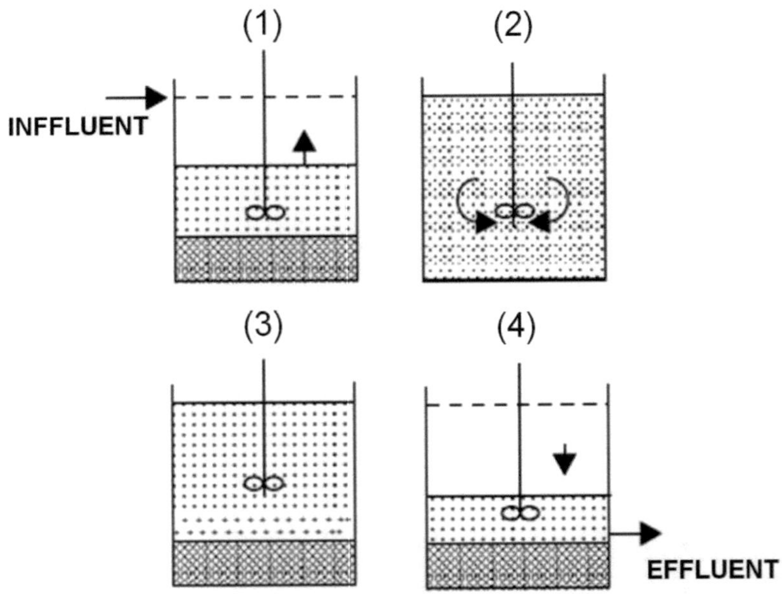

Figure 8. Schematic representation of CSTR reactor: (1) filling; (2) reaction; (3) sedimentation and (4) emptying (Adapted from [50])

When the pH is controlled, organic acids that favor microbial selection and the consequent production of hydrogen gas are formed. Other methods for elimination hydrogen consumers utilize ultra-sonication, acidification, sterilization and freezing/thawing [10]. Sá et al. [52] studied the biological hydrogen production using anaerobic sludge of the sewage treatment system of Rio de Janeiro city, Brazil. The authors (op cit.) applied heat treatment (120 °C for 1 h) upon the sewage to inhibit methanogenesis. Tests in anaerobic batch reactors using glycerin for H_2 production were obtained of 0.80 mol-H_2.mol glycerine^{-1}. Therefore, all such methods need to be verified for crude glycerol fermentation and hydrogen production efficiency.

5. Energetic applicarions and storage of H_2

Hydrogen can be used as energetic source in different systems and technologies, such as vehicular, stationary or portable devices.

In vehicular applications, hydrogen can be used as a supplemental fuel in conventional spark ignition engines without extensive engine modifications, reducing CO and hydrocarbons emissions, improving engine performance characteristics, such as thermal efficiency and specific fuel consumption. The researches show that when H_2 is used as a sole fuel in spark ignition engine, it is more efficient (and cleaner, since its combustion produces only water) than fossil fuel [53].

http://dx.doi.org/10.5772/60013

Another option for vehicular transportation is use hydrogen in fuel cells (Figure 9). Fuel cells are high-efficiency power generation systems that convert hydrogen and oxygen directly into electricity using a low-temperature electrochemical process assisted by catalysts, emitting only water and virtually no pollutant. This system consists mainly of two electrodes (where electrochemical reactions takes place) separated by electrolyte or a membrane. Hydrogen and oxygen (i.e. from air) are fed into the fuel cell. The flow of ions between the electrodes occurs through electrolyte, while the excess electrons flow through an external circuit, providing electrical power [54].

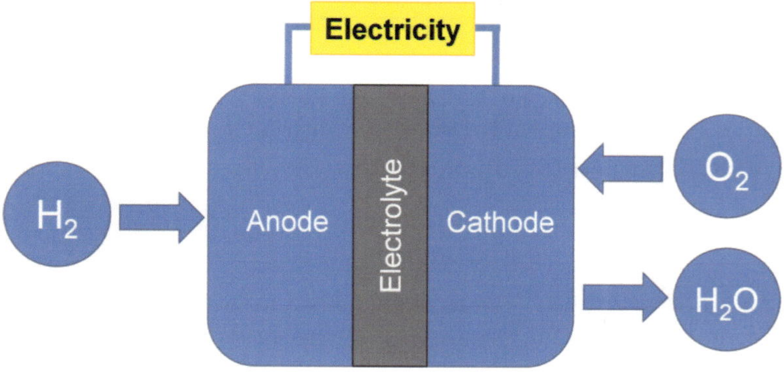

Figure 9. Simplified schematic representation of a fuel cell

Fuel cells convert hydrogen into electricity at high efficiency since they are not subject to the Carnot cycle limitations [55]. In present-day vehicles, a petrol-driven car engine operates at 25% efficiency, in the other hand a hydrogen fuel cell engine can operate at more than 65% [54].

Several types of fuel cells have been developed at different scales and characteristics (Table 5). The Brazilian government (Ministry of Mines and Energy - MME) plans introduce H_2 in the national energy matrix until 2025 not only as fuel for vehicles, but also as a clean source for stationary power generation for energy supply. According the MME plans, after 2020 the hydrogen produced in Brazil should be mostly provided from renewable sources [56].

Many global electronic companies such as Samsung, Sony, Toshiba, Motorola, Panasonic, Fujitsu, NEC, Hitachi and others have been developed patents and prototypes using fuel cells portable systems. Even though the market is not yet mature, the volume of investments of these companies indicates that technology should be consolidated in a few years [57]. However, the widespread utilization of H_2 as an energy source requires solutions to several problems: the gas must be able to be produced from a cheap and renewable source; safe storage and handling of H_2 must be addressed and refueling infrastructure developed.

A significant challenge regarding the large scale use of hydrogen gas is its storage. A storage device is an important part of the hydrogen energy system and it is a serious problem due to high inflammability, adequate safety measures should be taken during the production, storage, and use of H_2 fuel.

Fuel cell type	Operating temperature (ºC)	Eletrical power range (kW)	Eletrical efficiency (%)
Proton exchange membrane	60-110	0.01-250	40-55
Alkaline	70-130	0.1-50	50-70
Direct methanol	60-120	0.001-100	40
Phosphoric acid	175-210	50-1000	40-45
Molten carbonate	550-650	200-100,000	50-60
Solid oxide	500-1000	0.5-2000	40-72

Table 5. Characteristics of various fuel cell types (adapted from [54])

Hydrogen is quite difficult to store or transport with current technology. There are many ways for storing hydrogen fuel; as a gas (hydrogen compressed), a liquid (liquid hydrogen) and chemicals (metal hydride) [53].

Hydrogen compressed in tanks (with similar technologies applied in natural gas compression) is the easiest and cheapest way to store it. These tanks can store hydrogen at a high pressure (about 25 MPa - 35 MPa), but even under these conditions the energy density by volume for hydrogen is lower than for gasoline or diesel as can be seen from Table 6. In liquid form (-253 ºC), the energy density has higher value than hydrogen in compressed form [58]. However, it is necessary to spend more energy to liquefy hydrogen than to compress it (up to 20% of the energy content of hydrogen is required to compress the gas and up to 40% to liquefy it), so the cryogenic process efficiency demands elevated costs [54].

Fuel	Way of storage	Energy density by weight (kWh/kg)	Energy density by volume (kWh/L)
Hydrogen	Gas (20 MPa)	33.3	0.53
	Gas (30 MPa)	33.3	0.75
	Liquid (-253 ºC)	33.3	2.36
	Metal hydrides	0.58	3.18
Natural gas	Gas (20 MPa)	13.9	2.58
	Gas (30 MPa)	13.9	3.38
	Liquid (-162 ºC)	13.9	5.8
Methanol	Liquid	5.6	4.42
Gasoline	Liquid	12.7	8.76
Diesel	Liquid	11.6	9.7

Table 6. Energy density for some fuels (adapted from [58])

http://dx.doi.org/10.5772/60013

Metals and metal alloys can also be used as a storage medium for hydrogen (hydrides form), i.e. chemical hydrides, Ca_2H, LiH, $NaBH_4$, MgH_2, $LiAlH_4$ and H_3NBH_3 have been widely studied as storage materials [53]. The positive aspects of this storage technology are low risk of unwanted losses, low pressures and energy densities greater than liquid and compressed hydrogen. The greatest disadvantage is the weight of these storage systems, about three times heavier than compressed hydrogen tank [58]. Probably, for this reason Toyota chose to use high-pressure tanks to equip the "Mirai" - its first commercial car powered by hydrogen produced on industrial scale [59].

The massive use of energy from hydrogen is expected to be gradual, with production and utilization initially on-site, with the development of new strategies environmentally friendly without the necessity for storage and transportation. For ex, initialy the hydrogen produced in a treatment plant can be used as energy source for its production plant; the use of H_2 for energetic necessity to maintaining a pump system would be appropriated.

6. Conclusion

The use of crude glycerol by biologic processes that generate PD, hydrogen and ethanol should be ensured for large scale production. Therefore, detailed economic studies and the optimization of such processes are interesting subjects for future investigations. New strategies may involve developing a proper market for the bioconversion of crude glycerol, as this would determine the economic viability of obtains clean energy from the glycerol feedstock.

Crude glycerol from the biodiesel manufacturing processes is a potential feedstock for bacterial hydrogen, PD and ethanol production. It can be used as substrate for the production of these bio products instead of other more expensive, carbon sources such as sugars.

A high hydrogen yield is possible when acetic acid is produced as the end product of crude glycerol fermentation. Other similar strategies should be developed for a metabolic route of acetic acid generation during the fermentation of crude glycerol.

Most investigations on crude glycerol bioconversion have been performed in serum bottle scale batch reactors. Only a few studies have been performed in a continuous mode. The continual improvement of investigations into bacterial hydrogen production using the continuous mode is recommended.

Consortia of anaerobic bacteria from environmental sources or pure cultures may be used for bioconversion of crude glycerol to hydrogen PD and ethanol. However using co-cultures may reduce the accumulation of metabolites and improve hydrogen yield. Application of the biological processes to directly convert abundant crude glycerol into higher value products may represent a promising route to achieve economic viability in the biofuels industry.

The widespread utilization of H_2 as an energy source requires solutions to several problems: the H2 must be able to be produced from a cheap and renewable source; refueling infrastructure must be developed for fuel cells. The appropriate utilization of energy from hydrogen in large scale must be initially expected on-site. Further investigations for safe and economic storage of hydrogen are recommended.

Acknowledgements

Authors would like to acknowledge ABES – Associação Brasileira de Engenharia Sanitária e Ambiental for their kind permission to reuse figures from previously published material Mr. Frank Garrik (University of St Andrews, Scotland) for their language review and FUNDUNESP for finantial support.

Author details

Sandra Imaculada Maintinguer, Rafael Rodrigues Hatanaka and José Eduardo de Oliveira

CEMPEQC –Center for Monitoring and Research of the Quality of Fuels, Biofuels, Crude Oil, and Derivatives. Institute of Chemistry. UNESP – São Paulo State University. Araraquara, Brazil

References

[1] Das D, Veziroglu TN. Hydrogen production by biological processes: a survey of literature. International Journal of Hydrogen Energy 2011; 26 13-28.

[2] BIODIESEL BR. Todas as usinas de biodiesel do Brasil. http://www.biodieselbr.com/usinas/ (accessed 23 March 2014)

[3] Sousa LS, Rodarte CV, Oliveira JE, Moura EM. Use of natural antioxidants in soybean biodiesel. Fuel 2014; 134 420–428.

[4] Brasil. Medida Provisória nº 647, de 28 de maio de 2014. Diário Oficial da República Federativa do Brasil, Poder Executivo, Brasília, DF, 29 mai. 2014. Seção 1, p. 1. http://www.planalto.gov.br/ccivil_03/_Ato2011-2014/2014/Mpv/mpv647.htm (accessed 26 September 2014)

[5] Souza ACC, Silveira, JL. Hydrogen production utilizing glycerol from renewable feedstocks—The case of Brazil. Renewable and Sustainable Energy Reviews 2011; 15 1835–1850.

[6] Andrade MF, Colodette JL. Dissolving pulp production from sugar cane bagasse. Industrial Crops and Products 2014; 52 58–64.

[7] Bergmann AJC, Tupinamba DD'A, Costa OYAA, Almeida JRMB, BarretoCCA, Quirino BF. Biodiesel production in Brazil and alternative biomass feedstocks. Renewable and Sustainable Energy Reviews 2012; 21 411–420.

[8] Sequinel R. Caracterização físico-química da glicerina proveniente de usinas de biodiesel e determinação de metanol residual por CG com amostragem por Headspace estático. PhD Thesis. UNESP - Instituto de Química de Araraquara; 2013.

[9] Rossi DM, Costa JB, Souza EA, Peralba MCR, Ayub MAZ. Bioconversion of residual 8 glycerol from biodiesel synthesis into 1,3-propanediol and ethanol by isolated bacteria from environmental consortia. Renewable Energy 2012; 39 223-227.

[10] Sarma SJ, Brar SK, Sydney EB, Bihan YL, Buelna G, Soccol CR. Microbial hydrogen production by bioconversion of crude glycerol: a review. International Journal of Hydrogen Energy 2012; 37 6473-90.

[11] Rossi DM, Costa JB, Souza EA. Peralba MCR, Ayub SD, Zachia MA. Comparison of different pretreatment methods for hydrogen production using environmental microbial consortia on residual glycerol from biodiesel. International Journal of Hydrogen Energy 2011; 36 4814-4819.

[12] Silva CXA, Mota CJA. The influence of impurities on the acid-catalyzed reaction of glycerol with acetone. Biomass and Bioenergy 2011; 35 3547-3551.

[13] Selembo PA, Perez JM, Lloyd WA, Logan BE. High hydrogen production from glycerol or glucose by electrohydrogenesis using microbial electrolysis cells. International Journal of Hydrogen Energy 2009;345373-5381.

[14] Shuiquan T, Lindsay B, Howard L, Zisheng Z. *Pichiapastoris*fermentation for phytase production using crude glycerol from biodiesel production as the sole carbon source. Biochemical Engineering Journal2009;43157-162.

[15] Zhanyou C, Denver P, Zhiyou W, Craig F, Shulin C. A laboratory study of producing docosahexaenoic acid from biodiesel-waste glycerol by microalgal fermentation. Process Biochemistry 2007;421537-1545.

[16] Marques PA, Bartolomeu ML, Thomas MM, Neves LM. Biohydrogen production from glycerol by a strain of *Enterobacter aerogenes*. In: Hypothesis VIII 2009:proceeding of Hypothesis VIII, 1-3Apr 2009; Lisbon, Portugal. URI: http://hdl.handle.net/10400.9/604 (accessed 22 September 2014)

[17] Chuang-Wei C, Mohanprasad AD, Willam RS, Galen JS. Removal of residual catalyst from simulated biodiesel's crude glycerol for glycerol hydrogenolysis to propylene glycol. Industrial & Engineering Chemistry Research 2006; 45791-5.

[18] Wang L, Quiceno R, Price C, Mapas R, Woods J. Economic and GHG emissions analyses for sugarcane ethanol in Brazil: Looking forward. Renewable and Sustainable Energy Reviews 2014; 40 571-582.

[19] Ferreira Filho JB, Horridge M. Ethanol expansion and indirect land use change in Brazil. Land Use Policy 2014. 36 595-604.

[20] Fang HHP, Li C, Zhang T. Acidophilic biohydrogen production from rice slurry. International Journal of Hydrogen Energy 2006; 31 683-692.

[21] Kawagoshi Y, Hino N, Fujimoto A, Nakao M, Fujita Y, Sugimura S, Furukawa K. Effect of seed sludge conditioning on hydrogen fermentation and pH effect on bacterial community relevant to hydrogen production. Journal of Bioscience and Bioengineering 2005; 100(5)524-530.

[22] Loss RA, Fontes ML, Reginatto V, Antônio RV. Biohydrogen production by a mixed photoheterotrophic culture obtained from a Winogradsky column prepared from the sediment of a southern Brazilian lagoon. Renewable Energy 2013; 50 648-654.

[23] Homann T, Tag C, Biebl H, Deckwer WD, Schink B. Fermentation of glycerol to 1,3-propanediol by Klebsiella and Citrobacter strains. Applied and Microbiology Biotechnology 1990; 33(2) 121-126.

[24] Liu B, Christiansen K, Parnas R, Xu Z, Li B. Optimizing the production of hydrogen and 1,3-propanediol in anaerobic fermentation of biodiesel glycerol. International Journal of Hydrogen Energy 2013; 3 3196-3205.

[25] Maintinguer SI, Fernandes BS, Duarte IC, Saavedra NK, Adorno MT, Varesche MB. Fermentative Hydrogen Production by Microbial Consortium. International Journal of Hydrogen Energy 2008; 33 4309-4317.

[26] Zeng AP, Menzel K, Deeckwer WD. Kinetic, dynamic, and pathway studies of glycerol metabolism by *Klebsiella pneumoniae* in anaerobic continuous culture: II analysis of metabolic rates and pathways under oscillation and steady state conditions. Biotechnology and Bioengineering 1996; 56 561-571.

[27] Fan Y, Li C, Lay JJ, Hou H, Zhang G. Optimization of initial substrate and pH levels for germination of sporing hydrogen-producing anaerobes in cow dung compost. Bioresource Technology 2004; 91 189-193.

[28] Khanal SK, Chen W-H, Li L, Sung S. Biological hydrogen production: effects and intermediate products. International Journal of Hydrogen Energy 2004; 29 1123-1131.

[29] Wang C-H, Lin P-J, Chang J-S. Fermentative conversion of sucrose and pineapple waste into hydrogen gas in phosphate-buffered culture seeded seeded with municipal sewage sludge. Process Biochemistry 2006; 41 1353-1358.

[30] Hawkes FR, Hussy I, Kyazze G, Dinsdale R, Hawkes DL. Continuous dark fermentative hydrogen production by mesophilic microflora: Principles and progress. International Journal of Hydrogen Energy 2007; 32 172-184.

[31] Kotay SM, Das D. Microbial hydrogen production with Bacillus coagulans IIT-BT S1 isolated from anaerobic sewage sludge. Bioresource Technology 2007; 98 1183-1190.

[32] Agnieszka D, KatarzynaL, Katarzyna C. Biotechnological production of 1,3-propanediol from crude glycerol. BioTechnologia 2011; 92(1) 92-100.

[33] Maintinguer SI, Fernandes BS, Duarte IC, Saavedra NK, Adorno MT, Varesche MB. Fermentative Hydrogen production with xylose by Clostridium and Klebsiella species in anaerobic batch reactors. International Journal of Hydrogen Energy 2011; 36: 13508-13517.

[34] Ito T, Nakashimada Y, Senba K, Matsui T, NishioN. Hydrogen and ethanol production from glycerol-containing wastes discharged after biodiesel manufacturing process. Journal of Bioscience and Bioengineering 2005; 100 260-265.

[35] Barbirato F, Himmi EH, Conte T, Bories A. 1,3-propanediol production by fermentation: an interesting way to valorizeglycerin from the ester and ethanol industries. Indus Crops Prod 1998; 7 281-9.

[36] Akutsu Y, Lee D-Y, Li Y-Y T Noike. Hydrogen production potentials and fermentative characteristics of various substrates with different heat-pretreated natural microflora. International Journal of Hydrogen Energy 2009;34(13) 5365-5372.

[37] Priscilla AS, Joe MP, Wallis AL, Bruce EL. Enhanced hydrogen and 1,3-propanediol production from glycerol by fermentation using mixed cultures. Biotechnol Bioeng 2009; 104 1098-1106.

[38] Anniina K, Ville S, Matti K. Hydrogen production from glycerol using halophilic fermentative bacteria. Bioresource Technology 2010; 1018671-7.

[39] Seifert K, Waligorska M, Wojtowski M, Laniecki M. Hydrogen generation from glycerol in batch fermentation process. International Journal of Hydrogen Energy 2009; 34(9) 3671-3678.

[40] Zhao C, Karakashev D, Lu W, Wang H, Angelidaki I. Xylose fermentation to biofuels (hydrogen and ethanol) by extreme thermophilic (70°C) mixed culture culture. International Journal of Hydrogen Energy 2010; 35 3415-3422.

[41] Temudo MF, Poldermans R, Kleerebezem R, Van Loosdrechtet MCM. Glycerol fermentation by (open) mixed cultures: A chemostat study. Biotechnology and Bioengineering 2008; 100(6) 1088-1098.

[42] Agnelli JAB, Peixoto G,Zaiat M. Avaliação da produção de hidrogênio a partir de glicerol em reatores em batelada. In: Anais do I Seminário do Projeto Temático, 2011, Produção de bioenergia no tratamento de águas residuárias e adequação ambiental dos efluentes e resíduos gerados, São Carlos, Brasil.São Carlos: Universidade de São Paulo; 2011.

[43] Sharma Y, Parnas R, Li B. Bioenergy production from glycerol in hydrogen producing bioreactors (HPBs) and microbial fuel cells (MFCs). International Journal of Hydrogen Energy 2011; 36 3853-3861.

[44] Reungsang A,Sittijunda S,Angelidaki I. Simultaneous production of hydrogen and ethanol from waste glycerol by Enterobacteraerogenes KKU-S1. International Journal of Hydrogen Energy 2013; 38 1813-1825.

[45] Li S-L, Whang L-M, Chao Y-C, Wang Y-H, Wang Y-F, Hsiao C-J, Tseng I-C, Bai M-D, Cheng S-S. Effects of hydraulic retention time on anaerobic hydrogenation performance and microbial ecology of bioreactors fed with glucose-peptone and starch-peptone. International Journal of Hydrogen Energy 2010; 35 61-70.

[46] Shida GM, Sader LT, Amorim ELC. Sakamoto IK, Maintinguer SI, Saavedra NK, Varesche MBA, Silva EL. Performance and composition of bacterial communities in anaerobic fluidized bed reactors for hydrogen production: Effects of organic loading rate and alkalinity). International Journal of Hydrogen Energy 2012; 37 16925-16934.

[47] Jung K-W, Kim D-H, Shin H-S. Continuous fermentative hydrogen production from coffee drink manufacturing wastewater by applying UASB reactor. International Journal of Hydrogen Energy 2010; 35 13370-13378.

[48] Abreu AA, Alves JI, Pereira MA, Karakashev D, Alves MM, Angelidaki I. Engineered heat treated methanogenic granules: a promising biotechnological approach for extreme thermophilic biohydrogen production. Bioresource Technology 2010; 101 9577-9586.

[49] Chernicharo CAL. Biological Wastewater Treatment Series. Volume Four. Anaerobic Reactors. London - New York:IWA Publishing; 2007.

[50] Campos JR. Tratamento de esgotos sanitários por processo anaeróbio e disposição controlada no solo. Rio de Janeiro: ABES; 1999.

[51] Barros AR, Adorno MAT, Sakamoto IK, Maintinguer SI, Varesche MBA, Silva EL. Performance evaluation and phylogenetic characterization of anaerobic fluidized bed reactors using ground tire and pet as support materials for biohydrogen production. Bioresource Technology 2011; 102 3840-3847.

[52] Vasconcelos de Sá LR, Cammarota MC, Ferreira-Leitão VS. Hydrogen production by anaerobic fermentation - general aspects and possibility of using brazilian agro-industrial wastes. Química Nova 2014; 37(5) 857-867.

[53] Dutta S. A review on production, storage of hydrogen and its utilization as an energy resource. Journal of Industrial and Engineering Chemistry 2014; 20 1148–1156

[54] Eduards PP, Kuznetsov VL, David WIF, Brandon NP. Hydrogen and fuel cells: Towards a sustainable energy future. Energy Policy 2008; 36 4356–4362

[55] Dutton AG. Hydrogen energy technology.TWP 17. Didcot: Tyndall Centre for Climate Change; 2002. http://www.tyndall.ac.uk/sites/default/files/wp17.pdf (accessed 22 September 2014).

[56] Ministério de Minas e Energia. MME: Roteiro Hidrogênio http://www.mme.gov.br/mme/galerias/arquivos/publicacoes/Roteiro_Hidroigenio/01_Texto_Completo_Versxo_Beta_2005.pdf (accessed 22 Sep 2014).

[57] CGEE 2010: Hidrogênio energético no Brasil: subsídios para políticas de competitividade, 2010-2025; Tecnologias críticas e sensíveis em setores prioritários – Brasília: Centro de Gestão e Estudos Estratégicos, 2010 http://www.cgee.org.br/atividades/redirect.php?idProduto=6771 (accessed 22 September 2014).

[58] Cipriani G, Di Dio V, Genduso F, La Cascia D, Liga R, Miceli R, Galluzzo GR. Perspective on hydrogen energy carrier and its automotive applications. International Journal of Hydrogen Energy 2014; 39 8482-8494.

[59] Chang K. Hydrogen Cars Join Electric Models in Showrooms. http://www.nytimes.com/2014/11/18/science/earth/hydrogen-cars-join-electric-models-in-showrooms.html?_r=0 (accessed 26 November 2014).

Chapter 23

Synthesis of Oxygenated Fuel Additives from Glycerol

Jorge Sepúlveda, Mariana Busto, Carlos Vera,
Maraisa Gonçalves, Wagner Carvalho and
Dalmo Mandelli

Additional information is available at the end of the chapter

http://dx.doi.org/10.5772/59868

1. Introduction

The last years, a dramatic increase in the installed capacity for biodiesel fuel has taken place. This is a technically mature biofuel replacement of petrodiesel, with improved properties of cetane number, lubricity, biodegradability and flash point. This increasing biodiesel production has resulted in an excess production of the glycerol by-product (stoichiometrically 10 wt% of the product of a biodiesel plant). The transformation of bio-glycerol into glycerol-ethers and glycerol-esters via etherification and esterification reactions is considered to be a convenient alternative for glycerol utilization. These value-added chemicals have potential uses in many industrial applications. Particularly acetylation of bio-glycerol with acetic acid into glycerol-esters can produce di- and triacetin that have potential for vast quantity utilization as valuable biodiesel and petro fuel additives. In the case of the addition to biodiesel, the in-factory utilization of the product is quite advantageous. In addition, di- and triacetin are used as fuel additives for viscosity reduction. Triacetin meets the specifications of flash point (>374 K) and oxidation stability (6 h at 383 K) required by the standards EN 14214 and ASTM D6751 [1-3].

The esterification of glycerol with acetic acid produces mono-, di- and tri-acetates of glycerol. The mono- and di-acetates are known as monoacetin (2-monoacetyl-1, 3-propanediol or 3-monoacetyl-1, 2-propanediol, MAG) and diacetin (1, 2-diacetyl-3-propanol or 1, 3-diacetyl-propanol, DAG). The scheme of reaction is depicted below:

$$\text{glycerol} + \text{acetic acid} \leftrightarrow \text{monoacetin} + \text{water} \tag{1}$$

$$\text{monoacetin} + \text{acetic acid} \leftrightarrow \text{diacetin} + \text{water} \tag{2}$$

$$\text{diacetin} + \text{acetic acid} \leftrightarrow \text{triacetin} + \text{water} \tag{3}$$

Current glycerol esterification processes are carried out using mineral acids. However, these technologies are not environmentally friendly, and much attention has been put on the development of new techniques that use acidic heterogeneous catalysts. Recently, the synthesis of new supported materials containing immobilized sulfonic acid groups, which behave as active and selective catalysts for esterification, has been reported [4-6]. Other reports can also be found that deal with zeolites, poly vinyl sulfonic resins and niobic acid.

Sulfated zirconia obtained by the sol-gel method, was evaluated in the esterification of glycerol with acetic acid at 328 K; however, leaching of sulfur occurred during the reaction [7]. Propylsulfonic and fluorosulfonic acid functionalized mesostructured silica (SBA-15) was synthesized and have demonstrated excellent catalytic behavior in the acetylation of glycerol with acetic acid [8]. Sulfonation of carbon-based materials also produced a highly active, and stable solid acid catalyst for this reaction [9].

A great attention has been devoted to the conversion of glycerol into oxygenated additives for liquid fuels. In this context, an industrially relevant route for the conversion of glycerol into oxygenated chemicals is the etherification to tert-butyl ethers. Tert-butyl ethers of glycerol with a high content of di-ethers are considered promising as oxygenated additives for diesel fuels (smoke suppressors and pour point depressants for diesel, biodiesel and their mixtures).

It is found however that mono-tert-butyl ethers of glycerol (MBGEs) have a low solubility in diesel fuel and they are soluble in water. However, if the etherification of glycerol produces mainly di- and tri-ethers, the product is readily blended in the fuel, and other restrictions related to the fuel properties controlled by quality standards can also be met. Thus, when di- and tri-tertiary butyl ethers of glycerol are incorporated to standard 30–40% aromatic-containing diesel fuel, emissions of particulate matter, hydrocarbons, carbon monoxide and unregulated aldehydes decrease significantly [10, 11].

The alkylation of glycerol can be performed with many etherifying agents: isobutylene, tert-butyl alcohol and C_4 olefinic petrochemical fractions. Tert-butyl alcohol avoids the need to use solvents to dissolve glycerol; however, water is formed as a by-product that may deactivate the heterogeneous catalysts used. When isobutylene is used, two phases might be present depending on the reaction conditions. The existence of multiple phases may lead to some problems of mass transfer in the reactor.

Many heterogeneous catalysts have been used in the alkylation of glycerol and reported in the scientific literature: acidic ion-exchange resins (mainly Amberlyst15 and 35), acid form wide pore zeolites (e.g., H-Y and H-Beta), sulfonic mesostructured silicas, sulfonated niobia and pillared clays.

Reported homogeneous catalysts for etherification of glycerol are the p-toluenesulfonic acid and sulfuric acid. Glycerol etherification with tert-butyl alcohol (TBA) is an acid catalyzed

http://dx.doi.org/10.5772/59868

reaction, resulting in a mixture of mono-tert-butyl-glycerol (MTBG), di-tert-butyl-glycerol (DTBG) and tri-tert-butyl-glycerol (TTBG). Some unwanted by-products can also be formed that are mainly a result of polymerization reactions.

2. Experimental

2.1. Materials

An Amberlite 15W resin was supplied by Rohm & Haas and used as reference acid catalyst in most tests.

CNR 115 activated carbon (AC sample) was supplied by Norit. This activated carbon was sulfonated by three different procedures: (a) Immersion in hot (373 K) concentrated sulfuric acid (H_2SO_4 98%) for 10 h (ACa sample). (b) Immersion in aqua regia for 20 h at room temperature followed by rinsing with water until neutral pH and drying. Final sulfonation in hot concentrated sulfuric acid like in (a) (ACb sample). (c) Sulfonation with sulfuric acid and naphthalene. 0.3 g of naphthalene were dissolved in 20 ml of tetrahydrofuran at room temperature and 2 g of AC were immersed in this solution for 1 hour with gently stirring. Then the carbon was filtered and dried at 373 K. Finally the naphthalene doped AC sample was sulfonated with hot sulfuric acid like in (a) (ACc sample).

MWNT (multi-walled nanotubes) were supplied by Arkema (Lot number 6068). This MWNT were also sulfonated by three different procedures: (a) Immersion in hot (373 K) concentrated sulfuric acid (H_2SO_4 98%) for 10 h (MWNTa sample). (b) Immersion in aqua regia for 20 h at room temperature followed by rinsing with water until neutral pH and drying. Final sulfonation in hot concentrated sulfuric acid like in (a) (MWNTb sample). (c) Sulfonation in very hot concentrated sulfuric acid (503 K), H_2SO_4 98%) for 10 h (MWNTc sample).

Molybdophosphoric acid ($H_3PMo_{12}O_{40}.7H_2O$) (HPA) was supplied by Merck. HPA/AC was prepared by first treating AC with aqua regia (3 parts of HCl + 1 part HNO_3 + 1 part water H_2O), then rinsing with water and drying. 0.4 g of HPA were dissolved in 15 ml of water and 0.5 ml of HNO_3. AC was added to a solution of 10 ml of water and 0.3 ml of HNO_3. Then, the HPA solution was added to the AC solution while stirring gently. The solution was then kept at room temperature under constant stirring for 12 h. Then the carbon was washed repeatedly with hot water (373 K) and finally filtered and dried [12-14].

2.2. Characterization of the carbon-based catalysts

The functionalized carbon materials were characterized by Raman spectroscopy, thermogravimetry, FT-IR spectroscopy and chemical titration.

Raman spectroscopy analysis of the solid samples were performed at room temperature with a Jobin–Yvon Horiba Labram II micro-Raman system with an excitation laser wavelength of 632 and 514 nm. The incident power was kept well below 3 mW to avoid sample damage or laser-induced heating. For each sample, spectra were acquired at three different spots and averaged, except when large variations were observed.

Thermogravimetric analysis (TGA) was carried out in a N_2 atmosphere with a heating rate of 10 K min^{-1}, from 25 ºC to 800 ºC (TA Instruments, Q500 TGA).

The samples were analyzed by infrared spectroscopy using a Varian 3100 FT-IR Spectrometer. Spectra were acquired by accumulating 100 scans at 4 cm^{-1} resolution in the range of 400–1200 cm^{-1}.

The titration of the acidic sites was performed using 100 mg of catalyst and a back titration method. The sample was first immersed in 10 cm^3 of a 0.1 M NaOH aqueous solution and stirred gently for 1 h. Then the resulting solution was titrated with a 0.1 molar HCl solution.

2.3. Catalytic tests

Esterification of glycerol: the reagents were glycerol (99.5% purity, Sigma-Aldrich) and acetic acid (99.7% purity, Sigma-Aldrich). The reaction was carried out in liquid phase 20 at 80 ºC in a stainless steel PTFE lined autoclave. Typically, the mass composition of the reaction mixture was 2.5 g of glycerol, 10 g acetic acid, i.e. 6:1 acetic acid/glycerol molar ratio, and a constant catalyst mass of 0.1 g. Samples of the reacting mixture were analyzed by gas chromatography in a Varian 3900 chromatograph using a CP-SIL 8 CB column (30 m long, 0.25 mm ID, film thickness 0.25 μ) and a flame ionization detector.

Etherification of glycerol: the reagents were glycerol (99.5% purity, Sigma-Aldrich) and tert-butyl alcohol (99.7% purity, Sigma-Aldrich). The reaction was carried out in liquid phase in a stainless steel PTFE lined autoclave. The stirring rate was maintained at 1200 min^{-1} in order to limit the effects of external mass transfer phenomena. Experiments were performed under different reaction conditions, 70-90 ºC reaction temperature, 2-6 tert-butyl alcohol/glycerol ratio and 1-7 h reaction time. The catalyst concentration was constant, 5% with respect to the glycerol mass. The catalysts were dried before each catalytic test. In a typical run, 5 g of glycerol and 0.2 g of the dry catalyst were used. The reaction products were sampled periodically and analyzed off-line in a Shimadzu 2014 gas chromatograph equipped with a flame ionization detector and a capillary column (J&W INNOWax 19091N-213, 30 m length) using acetonitrile as internal standard.

3. Results and discussion

3.1. Catalyst characterization

Raman spectroscopy is a potentially useful technique to obtain information of carbon materials. Raman results of AC and MWNT exhibited some typical features of carbonaceous materials: The tangential G band around 1580 cm^{-1} and the defect D band around 1320-45 cm^{-1}.

Raman spectra of AC and MWNT are shown in Figure 1. The main difference between AC and MWNT is the intensity band D / band G; AC has a lower ratio than MWNT, indicating that there is a higher concentration of defects (sp^3) on MWNT.

http://dx.doi.org/10.5772/59868

The thermogravimetric analysis (TGA) shows an increased weight loss with respect to the raw materials. The evolution of the weight loss was between 400 and 700 K, and it can be concluded that AC and MWNT have been functionalized with sulfonic group. All the samples showed weight loss between 4 and 9 %.

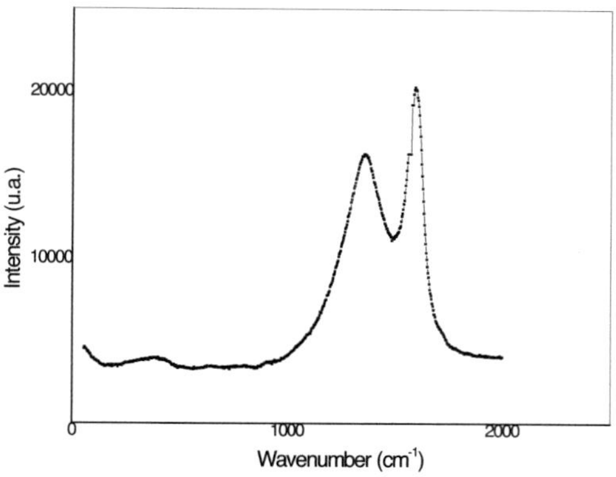

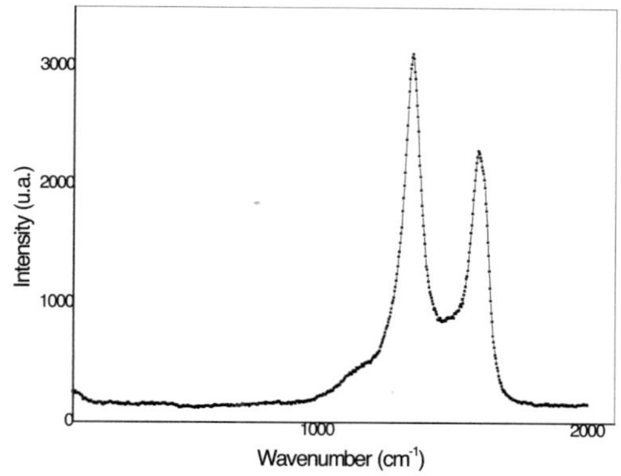

Figure 1. Raman spectra with excitation at 532 nm of AC (left) and MWNT (right).

Table 1 shows the specific surface area of the used catalysts. It can be seen that the acid treatment does not affect the physical properties of the activated carbon (AC). MWNT is also stable and not affected by the acid treatment. Only a shortening of the tubes or a slight variation of the length/diameter ratio might occur. The length of the nanotubes can range from several hundred nanometers to several micrometers, and the diameters from 2 to 100 nm [15, 16]. In the case of the catalyst of the last row of the table (HPA/AC), the HPA content (6.7%, mass basis) was determined by atomic absorption.

Catalyst	*BET area ($m^2\ g^{-1}$)*	*Mean pore radius (Å)*	*Acid content ($mEq\ g^{-1}$)*
Amberlyst 15	45	280	4.8
AC	1970	19	
Aca	1850	20	0.9
Acb			1.2
Acc	1700	22	2.7
MWNT			
MWNTa			0.8
MWNTb			1.2
MWNTc			1.9
HPA	5		
HPA/AC	780	30	

Table 1. Textural and acid properties of the catalysts.

Some authors [17, 18] have posed that glycerol etherification reactions can show intraparticle diffusion problems, especially in the case of the formation of tri-tert-butyl-glycerol. In these reports, zeolites were used with a pore size of less than 10 Å. However, other authors have reported that using a beta zeolite as catalyst, glycerol can be etherified by 70-80% with tert-butyl alcohol with a 40% selectivity to di- and tri-ethers of glycerol [19]. The treatment of an HY zeolite with citric acid and/or nitric acid modifies the micropores and as a result selectivities to di- and tri-ethers of 85% and 58%, respectively, can be obtained when running the reaction at 343 K [11]. Therefore, the acidity strength significantly influences the formation of di- and tri-ethers of glycerol, although the accessibility of the glycerol to the acid sites must be guaranteed.

Table 1 also contains values of the total surface acidity of the catalysts. These concentration of acid sites can be considered as that of the strong sulfonic groups, because of the titrating method used. If other sites of weaker acid strength exist, they were not titrated by this method.

Regarding the sulfonation of carbonaceous materials, many reports have been published with values of surface sulfonic acid sites of 0.3-7 milimols $H^+ g^{-1}$. This big dispersion of values is due to the diverse nature of the starting carbons and the different ways of sulfonation used. Both sulfuric acid and óleum have been used and the temperature, time of sulfonation and method of titration have been varied.

In one approach, the carbon sample was left in contact overnight with fuming sulfuric acid (7% SO_3). In the second approach, the carbon sample was mixed with H_2SO_4 (>98%), and heated during 10 h [9]. SO_3H densities of about 1 mmol g^{-1} were got. In Ref. [20], the final sulfur content found was 3.29%, that corresponds to 1.03 mmol g^{-1}, after sulfonating at 453 K. 4-aminobenzene sulfonic acid at 278 K has also been used and functionalized carbons have been obtained with 1.16-1.86 mmol g^{-1} [21]. Some authors use more sophisticated sulfonation methods, like impregnation with 4-benzene-diazoniumsulfonate in ethanol/water solution in the presence of H_3PO_2. With this method, 1.70 mmol g^{-1} were obtained [22]. At 423 K with fuming sulfuric acid, the carbon with higher concentration had 1.1 mmol g^{-1} [23]. In a similar report at 423 K, a carbon with 7.07 %S was obtained that corresponds to 2.21 mmol g^{-1} [24]. Gomes et al. [25] found that impregnation at 353 K and 423 K with sulfuric acid yielded materials with 1 mmol g^{-1}. Other authors report that sulfonation at 423 K for 15 h yields a material with S content of 0.39 mmol g^{-1} [26] and 2.9-3.9 mmol g^{-1} calculated by titration with NaOH [27].

For the preparation of sulfonated carbon materials, only the sulfonation conditions have been varied. Other authors have focused their research efforts on the variation of the carbon precursor. For example, glucose and cellulose have been reported as carbon precursors, yielding materials with 7.2 and 7.3 mmol g^{-1} were obtained [28, 29].

All catalysts prepared from carbon supports (AC or MWNT) had an amount of acid sites lower than that of the resin. AC impregnated with naphthalene had a higher amount of sulfonic groups due to the incorporation of unsaturated surface carbons.

HPA/AC was also characterized by IR spectroscopy. Figure 2 shows that the four modes of vibration of HPA [12-14] can be found in the spectrum when the catalyst is the bulk one or when it is supported over activated carbon. These bands are attributed to the Keggin anion

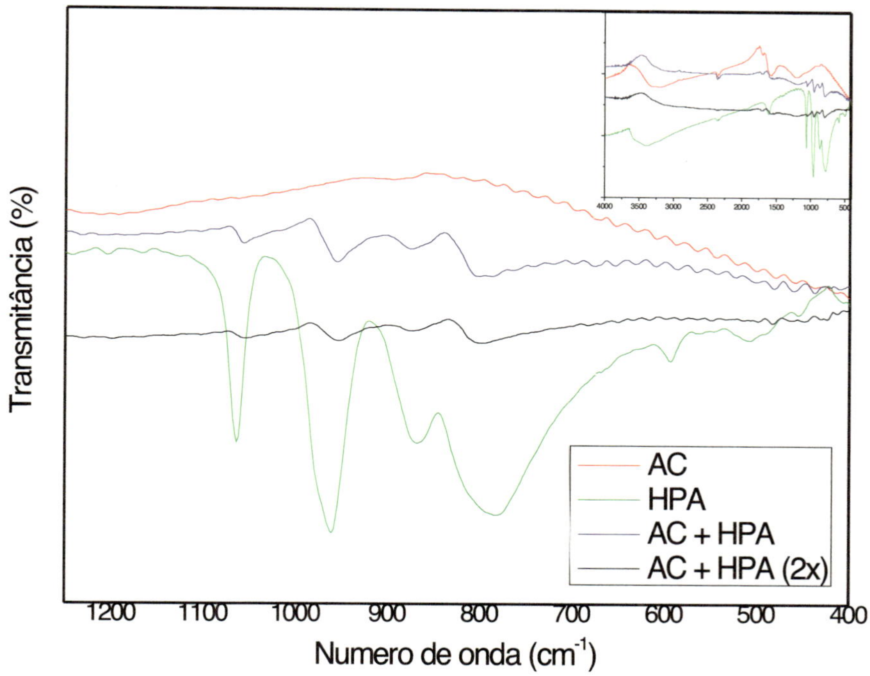

Figure 2. IR spectrum of HPA and HPA/AC.

$[PMo_{12}O_{40}]^{3-}$. The surface area of HPA/AC decreases due to the blocking of the micropores of AC by the heteropolyacid. In a previous work, we have demonstrated [12] that the acid treatments do not alter the total pore volume, though they decrease the fraction of micropores.

3.2. Esterification of glycerol

Some reported results on the esterification of glycerol with acetic acid are detailed in Table 2. One main variable is the molar ratio of acetic acid (AA) to glycerol (AA/Gly). This parameter varies widely from one report to the other. An inspection of the most active catalysts shows that these have an acid strength similar to that of pure sulfuric acid. Niobic acid that has an $H_o < 5.6$ (Hammett scale) is the least active.

It can also be seen that the temperatures used are those needed for reflux or 353-393 K. Variations of up to 100 K do not introduce meaningful changes in conversion, probably because of the low activation energy for this reaction.

The first assays of catalytic activity were performed at room temperature. The conversion of glycerol as a function of time can be seen in Figure 3.

The catalytic activity of the catalysts based on carbon catalysts, either AC or MWNT, was compared with that of the Amberlyst 15 resin. At 20 h of reaction time, the equilibrium was not achieved and the conversion was low. For the resin, conversion was 28% and for the other catalysts was about 20% or less.

Catalyst	AA / Gly	AA / Cat	T, K	X, %	MAG, %	DAG, %	TAG, %	Time, h	Ref.
Amberlyst 15	6/1	79	353	90	75	22	3	8	[30]
STARBON-400-SO_3H	1/1	6	373	98	78	20	2	0.16	[31]
Niobic acid	3/1	0.6	R	30	85	---	---	0.5	[32]
C-SO_3H	1/9	5 %	453	99.6	---	---	50	4	[33]
MP(10)/NbSBA-15-64	9/1	146	423	92	11	50	39	4	[34]
PMo1_NaUSY	16/1	105	R	60	33	62	4	3	[35]
Ag_1PW	10/1	100	393	96.8	48.4	46.4	5.2	0.25	[36]
MgF_2-87	3/1	72	373	94.2	---	60	30	22	[37]

Table 2. Reported values of catalytic activity and selectivity for the esterification of glycerol over several catalysts. MAG: selectivity to monoacetylglycerol; DAG: selectivity to diacetylglycerol; TAG: selectivity to triacetylglycerol; AA/Gly: molar ratio of acetic acid (AA) to glycerol (Gly); AA/Cat: mass ratio of acetic acid to catalyst; X: total conversion of acetic acid; R: reflux temperature. The resin was more active due to the higher concentration of acid sites. This was due to the fact that the resin has twice the concentration of acid sites of the carbon-based catalysts. The selectivity of the resin in these assays was 95% to monoacetin. For HPA/AC, ACc and MWNTc, the selectivity values to monoacetin were 92%, 93% and 94%, respectively. The selectivity to the other di- and tri- ethers was negligible.

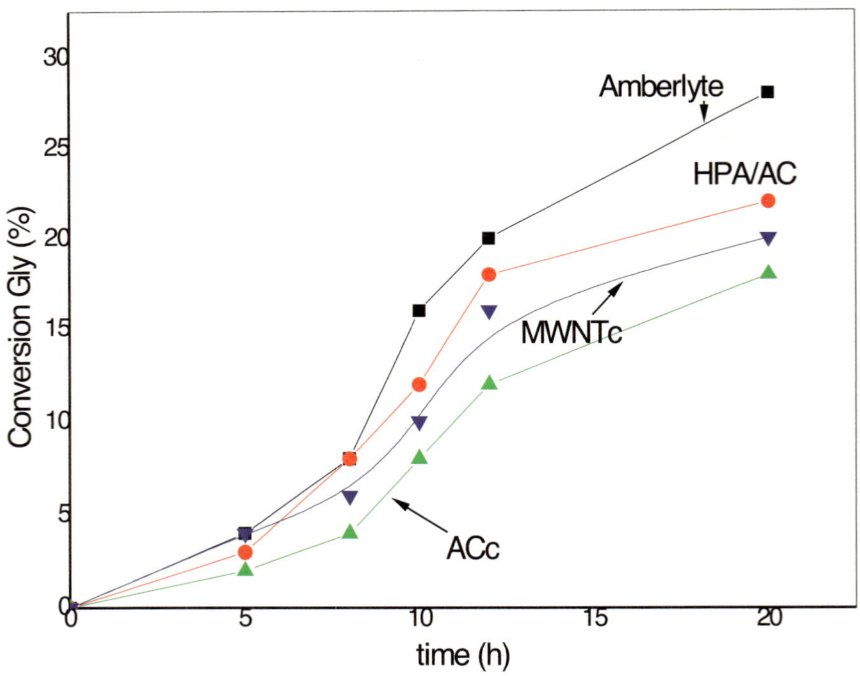

Figure 3. Conversion of glycerol during esterification with acetic acid at 293 K. Gly/AA=6, 0.1 g of catalyst.

Although the conversion achieved in these tests was relatively high, the reaction rate in these tests was fairly low and the thermodynamic equilibrium was not achieved. In these conditions,

http://dx.doi.org/10.5772/59868

monoacetin was the main product and could not be esterified with one or two additional glycerol molecules, as it could be expected from the consecutive reaction scheme depicted earlier.

A blank test was also performed to see the contribution of a non-catalytic path. Conversion was 4%, with a 90% selectivity to monoacetin. This conversion is not negligible in the reaction conditions used. The reactivity could be due to the protonic acid sites of acetic acid, that could autocatalyze the reaction.

Though the reaction can be performed at these conditions, the temperature was varied in order to maximize the glycerol conversion and the yield to di- and tri-acetylated products. The influence of the temperature was assessed by changing the reaction temperature to 373 K, keeping the other variables (AA/Gly molar ratio 6/1, 0.1 g catalyst mass) constant. Results for the catalysts with higher acidity (HPA/AC, MWNTc and ACc) are presented in Figure 4.

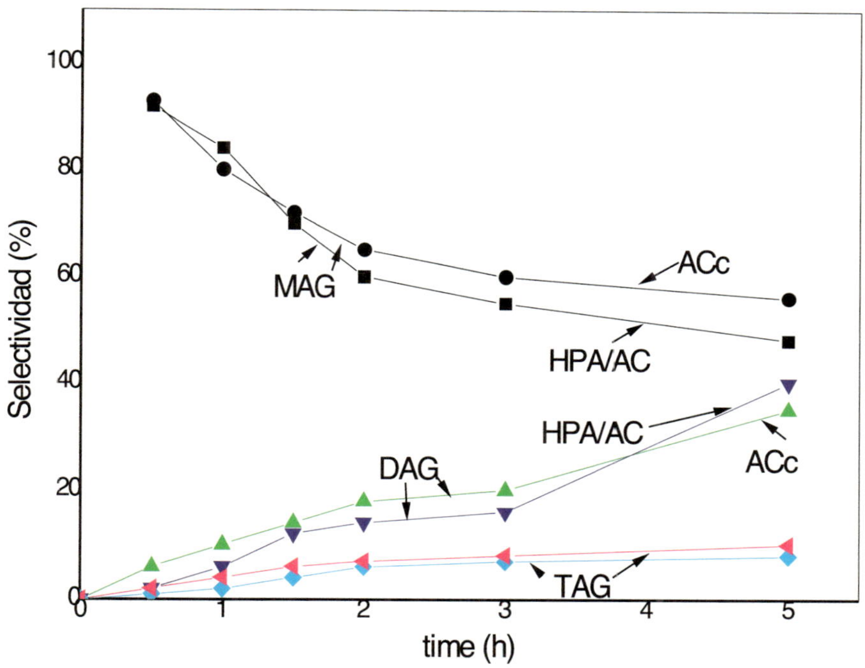

Figure 4. Selectivity in the reaction of esterification of glycerol with acetic acid at 373 K. Gly/AA=6, 0.1 g catalyst mass.

Conversion of glycerol reached 90-96% after 3-4 h of reaction time for all the catalysts tested. These results point to a higher reaction rate at higher temperatures and with favorable thermodynamics [38-41]. The equilibrium conditions are reached in a shorter time and the different selectivities can be inspected. Values of the selectivity to the diacetins (1, 3-DAG and 1, 2-DAG) and triacetin (TAG) can be seen in Figure 4. MAG concentration is decreased with reaction time while DAG and TAG are increased, showing the typical pattern of consecutive reactions.

Catalyst	*AA/Gly*	*X, %*	*MAG, %*	*DAG, %*	*TAG, %*
HPA/CB	1	90	56	37	7
	3	92	52	39	9
	6	92	49	41	10
CBc	1	88	62	32	6
	3	90	60	34	6
	6	94	56	36	8
MWNTc	1	86	64	30	6
	3	88	60	33	7
	6	90	58	32	10
Blank	1	27	66	32	2
	3	35	64	33	3
	6	44	62	36	2

Table 3. Conversion of glycerol and selectivity to acetates of glycerol. Esterification with acetic acid at 100 ºC, 0.1 g catalyst mass, 4 h reaction time. Notation as in Table 2.

HPA/AC showed the best yield to polyethers, with a 92% conversion at 3 h and a DAG+TAG selectivity next to 50%. Catalyst ACc that has the highest concentration of sulfonic sites also has a selectivity near to 45% for the products of interest.

Some additional tests were performed to check the reusability of the catalysts. These tests indicated that the catalysts were reusable and that surface sulfonic acid was not removed by leaching.

The influence of the molar ratio of acetic acid to glycerol (AA/Gly) was also assessed. The temperature was kept at 373 K and the values of AA/Gly used were 1, 3 and 6. The results are included in Table 3. It is especially noticeable that for the blank test at AA/Gly=6, the conversion was about 50%. This points to the autocatalytic role of the protons of acetic acid.

3.3. Etherification of glycerol with tert-butyl alcohol

3.3.1. Effect of reaction temperature

Glycerol was etherified with tert-butyl alcohol with the same series of catalysts used in the previous section. The convenient temperature was first screened using the reaction with HPA and comparing with the results obtained with the resin. HPA catalysts with Keggin structure are widely used as acid catalysts because of their very strong Brönsted acidity. They however have low specific surface area. Variations have been prepared to increase their area by supporting HPA on carbon, silica, zeolites, etc.

Figure 5 shows that the conversion of glycerol is increased when increasing the temperature from 70 to 90 ºC. Although this could be foreseen, there is no agreement with the published reports. Some authors have found that conversion of glycerol at 90 ºC is lower than conversion at 120 ºC [42]. It is also found that the maximum yield occurs at 90 ºC [43]. For the Amberlyst 15 resin, the maximum conversion is achieved between 70 and 90 ºC [44]. These differences can be due to the system not being in equilibrium in some cases and to the occurrence of other reactions. Degradative reaction would produce gases and increase the system pressure, thus modifying the reaction rate and the equilibrium [44].

In our experiments, 80 ºC was selected as optimal temperature for testing the activity and selectivity of the catalysts. First, HPA can be compared to the Amberlyst resin. According to the results of Figure 4, the initial reaction rate on HPA is higher than the reaction on the resin. After some time, the conversion however is lower with HPA, being 60% with the resin and 50% with HPA. At this temperature, a weight loss could also be detected. This could be due to the presence of degradation reactions catalyzed by the strong acid sites. *tert*-Butanol could be dehydrated to isobutylene and glycerol decomposed to acetaldehyde, acrolein, acetic acid, isobutylene, diisobutylene, etc. At 80 ºC, the reaction system with the resin had a weight loss of 3.8% and the system with HPA, 2.1%. In the first system, the amount of water formed was 3.2%; however, 2.3% water was formed when using the HPA catalyst. A blank test was performed in order to check the presence of autocatalytic reactions, but under similar conditions, the conversion of glycerol was negligible.

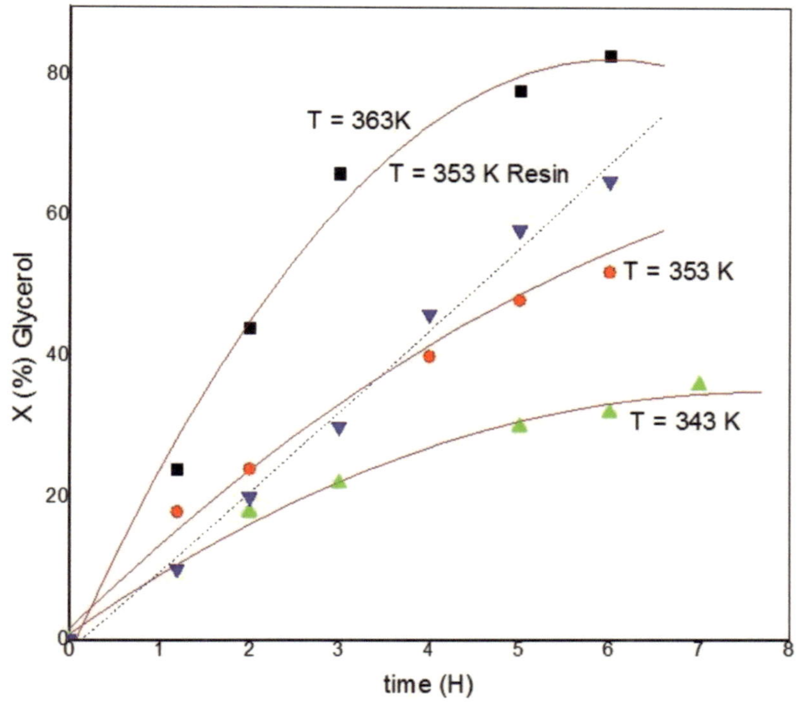

Figure 5. Conversion of glycerol during etherification of glycerol with *tert*-butanol at 80 ºC. *tert*-butanol/Gly=4, 0.2 g catalyst mass.

It can be seen that HPA is an optimal homogeneous catalyst with a great affinity for polar molecules. However, the resin, with its high concentration of active sites, is also very active. In both cases, no deactivation was detected due to the formation of water.

3.3.2. *Influence of catalysts on the selectivity*

Figure 6 shows values of conversion of glycerol and selectivities to the different glycerol ethers, when using catalysts ACc and MWNTc. Both catalysts have Brönsted acidity from sulfonic groups, the only difference being their concentration and their accessibility. In the case of the ACc catalyst, the average size of the pores is 22 Å. In the case of MWNTc, MWNTs have a greater amount of sp^3 carbons, and hence this catalyst should have a greater amount of defects of easier sulfonability. For a system of 5 nanotubes with 10 nm diameter (lower limit) and 10 tubes with 15 nm diameter (upper limit), 1 carbon atom out of 4.3-8.5 total atoms is on the surface. In this sense, 1 out of 6.4 carbon atoms is accessible for reaction with a liquid reagent [45].

The conversion of glycerol and the selectivity to different ethers was studied at 80 ºC and a *tert*-butyl alcohol/Gly molar ratio of 4. It is expected that high temperatures promote undesirable reactions. These could be the dehydration of *tert*-butyl alcohol to isobutene or the decomposition/reaction of glycerol. Usually, these side reactions produce an increase of the pressure in the system. There is in consequence an optimal temperature that some authors report as 75-85 ºC. Lower temperatures would not provide a sufficiently high reaction rate. Higher temperatures would lower the selectivity to the desired products [46]. Figure 5 shows values of the selectivity to the different glycerol ethers as a function of time, for a reaction temperature of 80 ºC.

The conversion of glycerol after 10 h reaction time was 50-60% on the ACc and MWNTc catalysts. Both catalysts were less active than the Amberlyst resin. This was attributed to their lower concentration of acid sites. With respect to the selectivity and according to the mechanism of consecutive reactions for this system, 5 glycerol ethers can be expected: 3-*tert*-butoxy-1, 2-propanediol, 2-*tert*-butoxy-1, 3-propanediol, 1, 3-di-*tert*-butoxy-2-propanol, 1, 2-di-*tert*-butoxy-3-propanol and tri-*tert*-butoxy-propane. Despite the high number of possible products, the selectivity was mostly to the monoether and to a lower extent to the di-ether. Only traces of the tri-ether could be detected. Several authors report that the acid strength of the catalyst should be increased [18] in order to improve the selectivity to di- and tri-ethers, or the temperature should be increased [41], or the water product eliminated from the reaction medium to help shift the thermodynamic equilibrium [43]. However, none of these modifications permits increasing the concentration of tri-ethers.

With respect to the MWNTc catalyst, we can safely say that the accessibility of the reactants to the acid sites is good, given the open structure of this catalyst. Another aspect to analyze is that of the surface density of acid sites. Once the glycerol monoether is formed, the availability of another active site nearby could permit the onset of a second etherification. Fluorinated beta zolite [18] permits the etherification of glycerol at 75 ºC with 75% conversion and 37% to di+tri ethers, but with a small amount of tri-ethers. Better results of higher catalytic activity and better selectivities to di- and tri-ethers have been published and concern the use of isobutylene as

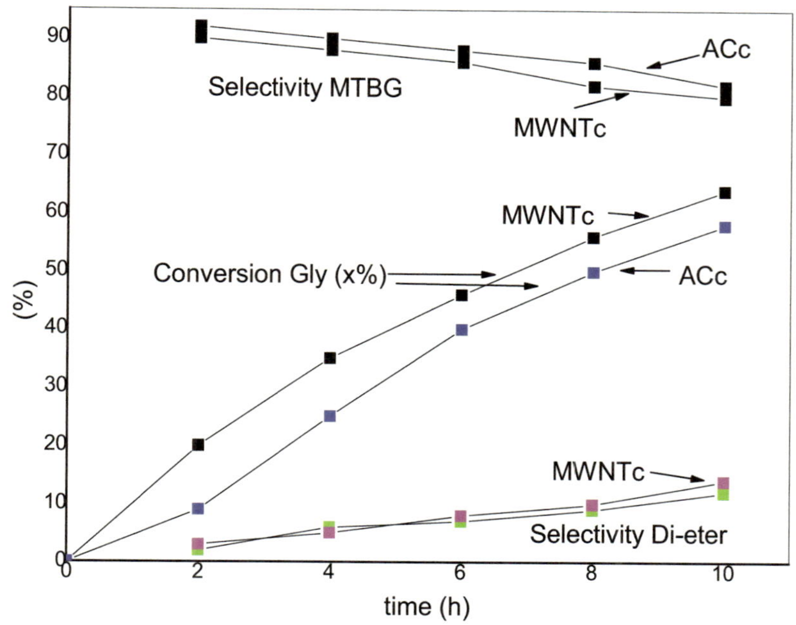

Figure 6. Conversion and selectivity as a function of time during etherification of glycerol with *tert*-butanol at 80 ºC over different catalysts. *tert*-Butanol/Gly=4, 0.2 g catalyst mass.

etherifying agent. Under similar conditions of temperature and Gly/etherifying agent ratio, better results are obtained when using isobutylene than *tert*-butyl alcohol [18].

3.3.3. Effect of the molar ratio of tert-butyl alcohol to glycerol

The sulfonated carbons were also tested in the reaction of tert-butylation of glycerol at varying molar ratios of *tert*-butyl alcohol to glycerol. Only the most acidic catalysts were used. The temperature was kept at 80 ºC and the catalyst mass used was 4% with respect to glycerol. The results are included in Table 4.

When a molar ratio in the range 2-6 and the resin Amberlyst 15 are used, the glycerol conversion increases as a function of the molar ratio from 68% to 82%. The reaction reaches equilibrium faster, although the selectivity is not significantly altered. Only a little increase of the selectivity to DTBG was observed in the case of the carbon-based catalysts. It can be seen that increasing the TBA/Gly molar ratio does not produce large modification in the activity or selectivity. This is probably a result of the similar acid strength of the catalysts due to the similar nature of the acid site (SO_3H). A change in activity or selectivity probably needs an increase of the acid strength, which might be obtained with another kind of acid sites. Some authors [19] observed that an increase in the Brönsted acid strength improved the yield of di- and tri-ethers of glycerol.

The effect of water on the activity and selectivity of these reactions must also be mentioned. In order to favor the consecutive reactions of glycerol over zeolite catalysts, Frusteri [44] removed the water from the reaction system after 6 hours. He obtained an increase of the DTBG yield from 28.5% to 41.5% and demonstrated that the difficulty in obtaining tri-ethers is related to the presence of water.

Catalyst	*TBA/Gly molar ratio*	*Gly conversion (%)*	*MTBG*	*DTBG*	*TTBG*
	2	48	84	16	---
HPA/AC	4	56	82	18	2
	6	62	80	17	3
	2	52	85	15	---
ACc	4	60	84	16	2
	6	68	80	18	2
	2	50	84	16	---
MWNTc	4	58	82	15	3
	6	58	79	19	2
	2	68	80	20	---
Amberlyst 15	4	75	80	19	1
	6	82	78	18	4

Table 4. Conversion of glycerol and selectivity to tert-butyl ethers of glycerol at 80 ºC, 8 h reaction time and different TBA/Gly molar ratios. TBA: *tert*-butyl alcohol. MTBG: mono *tert*-butyl ether of glycerol. DTBG: di *tert*-butyl ether of glycerol. TTBG: tri *tert*-butyl ether of glycerol.

4. Conclusions

Carbon supports such as activated carbon and MWNTs can be irreversibly sulfonated to yield materials with sulfonic groups surface concentrations of about 3 milimols per gram.

Sulfonation of activated carbon does not yield materials with meaningful concentration values of surface acid sites unless a graphitic precursor such as naphthalene is coimpregnated during sulfonation.

MWNTs are more efficiently sulfonated at 230 ºC. In this case, only 1/6 of the surface carbon atoms are exposed.

Esterification of glycerol with acetic acid can be carried out at room temperature, but in these conditions, the selectivity is biased toward monoacetin because the esterification rate is low and monoacetin cannot be acetylated. The selectivity to TAG and DAG is increased at higher temperatures (e.g., 60% at 100 ºC).

Etherification of glycerol with tert-butyl alcohol is a complex reaction with many undesired by-products being produced. All catalysts tried were effective for converting glycerol, but obtaining selectivities higher than 70-80% to di- and tri-ethers is still a challenge.

http://dx.doi.org/10.5772/59868

Acknowledgements

The authors are grateful to Universidade Federal do ABC and Universidad Nacional del Litoral for supporting this research work.

Author details

Jorge Sepúlveda[1*], Mariana Busto[1], Carlos Vera[1], Maraisa Gonçalves[2], Wagner Carvalho[2] and Dalmo Mandelli[2]

*Address all correspondence to: jsepulve@fiq.unl.edu.ar

1 INCAPE, Institute for Research on Catalysis and Petrochemistry (FIQ, Universidad Nacional del Litoral-CONICET), Argentina

2 Center for Natural and Humane Science (Universidade Federal do ABC), Santo André-SP, Brazil

References

[1] Dosuna-Rodríguez I, Gaigneaux EM, Catalysis Today 2012; 195: 14-21.

[2] García E, Laca M, Perez E, Garrido A, Peinado J, Energy & Fuels 2008; 22: 4274–4280.

[3] Melero JA, Gemma V, Morales G, Paniagua M, Bustamante J, Fuel 2010; 89: 2011–2018.

[4] Kim I, Kim J, Lee D, Applied Catalysis A: General 2014; 482: 31–37.

[5] Melero JA, Morales G, Iglesias J, Paniagua M, Hernández B, Penedo S, Applied Catalysis A: General 2013; 466: 116– 122.

[6] Gonzalez MD, Cesteros Y, Llorca J, Salagre P, Journal of Catalysis 2012; 290: 202–209.

[7] Dosuna-Rodriguez I, Adriany C, Gaigneaux EM, Catalysis Today 2011; 167: 56–63.

[8] Melero JA, van Grieken R, Morales G, Paniagua M, Energy & Fuels 2007; 21: 1782-1791.

[9] Galhardo T, Simone N, Goncalves M, Figueiredo F, Mandelli D, Carvalho W, ACS Sustainable Chemistry and Engineering 2013; 1: 1381-1389.

[10] Gonzalez MD, Cesteros Y, Salagre P, Applied Catalysis A: General 2013; 450: 178–188.

[11] [11] Xiao L, Mao J, Zhou J, Guo X, Zhang Sh, Applied Catalysis A: General 2011; 393: 88–95.

[12] Sepulveda JH, Yori JC, Vera CR, Applied Catalysis A: General 2005; 288: 18–24.

[13] Ferreira P, Fonseca IM, Ramos AM, Vital J, Castanheiro JE, Catalysis Communications 2011; 12: 573-576.

[14] Schwegler MA, Vinke P, van der Eijk M, van Bekkum H, Applied Catalysis A: General 1992; 80: 41-57.

[15] Bikiaris D, Vassiliou A, Chrissafis K, Paraskevopoulos KM, Jannakoudakis A, Docoslis A, Polymer Degradation and Stability 2008; 93: 952-967.

[16] Valcarcel M, Simonet B, Cardenas S, Suarez B, Anal. Bional. Chem. 2005; 382: 1783-1790.

[17] Viswanadham N, Saxena SK, Fuel 2013; 103: 980–986.

[18] González MD, Salagre P, Mokaya R, Cesteros Y, Catalysis Today 2014; 227: 171–178.

[19] González MD, Salagre P, Linares M, García R, Serrano D, Cesteros Y, Applied Catalysis A: General 2014; 473: 75–82.

[20] Song X, Fu X, Zhang Ch, Huang W, Zhu Y, Yang J, Zhang Y, Catalysis Letters 2012; 142: 869-874.

[21] Geng L, Yu G, Wang Y, Zhu Y, Applied Catalysis A: General 2012; 427: 137-144.

[22] Liu R, Wang X, Zhao X, Feng P, Carbon 2008; 46: 1664-1669.

[23] Janaun J, Ellis N, Applied Catalysis A: General 2011; 394: 25–31.

[24] Shu Q, Zhang Q, Xu G, Nawaz Z, Wang D, Wang J, Fuel Processing Technology 2009; 90: 1002–1008.

[25] Gomes HT, Miranda SM, Sampaio MJ, Figueiredo JL, Silva AMT, Faria JL, Applied Catalysis B: Environmental 2011; 106: 390– 397.

[26] Peng L, Philippaerts A, Ke X, Van Noyen J, De Clippel F, van Tendeloo G, Jacobs PA, Sels BF, Catalysis Today 2010; 150: 140–146.

[27] Valle-Vigón P, Sevilla M, Fuertes AB, Applied Surface Science 2012; 261: 574– 583.

[28] Macia-Agullo JA, Sevilla M, Diez MA, Fuertes AB, Chem. Sus. Chem. 2010; 3: 1352-1354.

[29] Suganuma S, Nakajima K, Kitano M, Yamaguchi D, Kato H, Hayashi S, Hara M, Solid State Science 2010; 12: 1029-1034.

[30] Kim I, Kim J, Lee D, Applied Catalysis B: Environmental 2014; 148–149: 295–303.

[31] Luque R, Budarin V, Clark JH, Macquarrie DJ, Applied Catalysis B: Environmental 2008; 82: 157–162.

[32] Goncalves VLC, Pinto BP, Silva JC, Mota CJA, Catalysis Today 2008; 133–135: 673–677.

[33] Sánchez JA, Hernández DL, Moreno JA, Mondragón F, Fernández JJ, Applied Catalysis A: General 2011; 405: 55-60.

[34] Trejda M, Stawicka K, Dubinska A, Ziolek M, Catalysis Today 2012; 187: 129-134.

[35] Ferreira P, Fonseca IM, Ramos AM, Vital J, Castanheiro JE, Catalysis Communications 2009; 10: 481–484.

[36] Zhu Sh, Gao X, Dong F, Zhu Y, Zheng H, Li Y, Journal of Catalysis 2013; 306: 155–163.

[37] Troncea SB, Wuttke S, Kemnitz E, Coman SM, Parvulescu VI, Applied Catalysis B: Environmental 2011; 107: 260–267.

[38] Jagadeeswaraiah K, Balaraju M, Prasad PSS, Lingaiah N, Applied Catalysis A: General 2010; 386: 166–170.

[39] Reddy PS, Sudarsanam P, Raju G, Reddy BM, Catalysis Communications 2010; 11: 1224–1228.

[40] Khayoon MS, Hameed BH, Applied Catalysis A: General 2012; 433–434: 152–161.

[41] Ferreira P, Fonseca IM, Ramos AM, Vital J, Castanheiro JE, Applied Catalysis B: Environmental 2009; 91: 416–422.

[42] Celdeira PA, Goncalves M, Figueiredo FCA, Dal Bosco SM, Mandelli D, Carvalho WA, Applied Catalysis A: General 2014; 478: 98–106.

[43] Goncalves M, Souza VC, Galhardo TS, Mantovani M, Figueiredo FCA, Mandelli D, Carvalho Wa, Ind. Eng. Chem. Res. 2013; 52: 2832–39.

[44] Frusteri F, Arena F, Bonura G, Cannilla C, Spadaro L, Di Blasi O, Applied Catal A: General 2009; 367: 77-83.

[45] Voiry D, Valles C, Roubeau O, Penicaud A, Carbon 2011; 49: 170-5.

[46] Melero JA, Vicente G, Morales G, Paniagua M, Moreno JM, Roldan R, Ezquerro A, Perez C, Applied Catalysis A: General 2008; 346: 44–51.

Chapter 24

Phosphorus Removal and Recovery from Digestate after Biogas Production

Hongjian Lin, Jing Gan, Aravindan Rajendran,
Cristiano Eduardo Rodrigues Reis and Bo Hu

Additional information is available at the end of the chapter

http://dx.doi.org/10.5772/60474

1. Introduction

1.1. Anaerobic digestion

Anaerobic digestion (AD) is a commercially available industrial process that generates biogas (roughly consisting of 60% of CH_4 and 40% of CO_2) and breaks down organic materials by anaerobic microbes. It is a process that can greatly reduce the amount of organic matter which might otherwise be destined to be land filled or burnt in an incinerator, both scenarios generating strong public concerns. The use of AD is suitable for most types of organic wastes such as livestock manure, waste paper, grass clippings, municipal waste, food and fruit/ vegetable processing waste etc. AD provides benefits including substantial odor reduction, production of a renewable energy source (biogas), reduction of greenhouse gas (GHG) emissions, potential pathogen reduction, minimization of solid waste for disposal, and enhanced nutrient management (Borowitzka 1999). Different groups of microorganisms are working together as a food chain to degrade the organic materials to produce methane as the final product. Briefly, insoluble organic material is hydrolyzed to produce simple soluble materials such simple sugar, amino acid and long chain fatty acid. Acidogenic bacteria degrade them to produce volatile fatty acid (VFA) and hydrogen, which is called acidogenesis. Then, acetogenic bacteria produce acetate from VFA and solvents in acetogenesis. There is a group of acetogenic bacteria which can synthesize acetate from hydrogen and carbon dioxide, referred as homoacetogenesis. And finally methanogens use acetate or hydrogen to produce methane as the final product. There are also other bacteria groups involved in the AD, for example sulfate reduction bacteria. Of all the current bioenergy options, AD is a well-established technology in Europe with large scale systems developed primarily in countries such as German and Denmark. In the US, large dairy farms are the leading users of AD in agriculture. EPA's AgSTAR program estimates that 137 dairy farms and 23 swine operations are using anaerobic digestion in the US. The current rate of anaerobic digestion deployment contrasts with AgStar's estimates of 2,645 dairy and 5,596 swine operations that could use anaerobic digestion with current designs.

While AD is the widely recommended technology for bioenergy production and nutrient management for animal wastes, the literature on AD is vastly focused on the influence of reactor configuration and manure characteristics based on four goals: solids destruction, biogas production, odor reduction, and pathogen reduction [1]; and few has been done on regards to P dynamics and mechanisms of transformation during AD. It is commonly believed that organic phosphate is degraded during the anaerobic degradation and therefore converted to inorganic phosphate. Gerritse and Vriesema [2] worked on the fractionation of P into organic and inorganic portions, as well as phase distribution, i.e., dissolved vs. particulate P, in liquid cow manure samples, either raw and digested. Their results confirm the hypothesis that most P found in manure is inorganic (around 90%), and overall, the inorganic:organic ratio does not change significantly after digestion – having found inorganic P at 86% of total P in their digestate samples. Similar qualitative conclusions are found in other literatures [3-5]. In regards to AD on phosphorus availability, discussions are very discrepant among studies. While some state that this degradation process increases the nutrient availability for plants [6] or that it does not have direct effects [7], some state that AD has potentially the opposite influence, i.e., it decreases P availability for plants [8-10]. It is known that pH influences the solubility of P and micronutrients; e.g., raising the pH moves the chemical equilibrium toward the formation of dissociated phosphate ion, which facilitates the precipitation of such ion as insoluble Ca and Mg phosphates. Some other concerns, such as the binding form of other elements, as Fe, may be regulated by AD [11]. Also, during AD, the fraction of dissolved P becomes mineralized and it becomes associated with suspended solids [12]. Also, the water-extractable P-fraction decreases substantially during AD. Struvite formation, which will be further discussed in this chapter, is very likely to be formed and crystallized, due to a combination of mineralization of P, N, and Mg during AD, being regulated by many ionic species found in the digestate media, e.g., Ca^{2+}, K^+, CO_3^{2-} [13]. P loss, i.e., loss of phosphorus due to leaching, retention, etc, during AD are also in a wide range of results in the literature, being reported as smaller than 10% [14] or as high as 36% [15]. This is due to many factors, since AD systems are different among them with different operational conditions, which also include partial retention in the digesters during P precipitation [16]. Möller and Müller [7] suggested some losses could also occur via leaching and runoff after field application, even though no data has been published by then.

Since AD does not show any significant effects on phosphorus (P) removal [17], AD effluent still contain a high level of phosphorus (either organic or inorganic phosphate) that, when directly discharged, has a potential to cause various environmental issues, for instance, eutrophication, which severely damages aquatic ecological systems. In terms of manure management, large portions of the phosphorus can be removed by the solid–liquid separation of manure, which was proposed mostly in alignment with the AD of dairy manure. The

http://dx.doi.org/10.5772/60474

phosphorus is relatively enriched in the solid phase while the liquid phase primarily only contains the soluble phosphate. The solid-liquid separation can be installed before the AD process and separating the solids from the liquid manure makes the liquid easier to pump and handle. However, it may significantly decrease the amount of organic materials available for the biogas production. The solid-liquid separation can also be installed after the AD process, where the solids can be used as the animal bedding materials. Numerous technologies have been developed to effectively and efficiently separate the solid and the liquid portions of the manure [18-23]; however, it is still considered as a pretreatment or as a first stage decontamination in highly polluted effluents with high content of particulate phosphorus. Even with the solid-liquid separation, the liquid waste still contains high concentration of phosphorus, while the phosphorus content of the solids is very low and the transportation of manure solids is still limiting the applications of phosphorus in a wider region.

As the animal industry is developing to larger production and operation to meet the increasing food and meat consumption as a result of the population growth and higher living standard [24], animal wastes, mainly animal manure in liquid, slurry, or solid forms, need more appropriate handling and utilization. Research on AD is boosted to a new level because of the recent research highlights on bioenergy and biofuel, so studies on methods for phosphorus removal and recovery from AD effluent needs investigation of the same level of attentions. In fact, recovery and recycle of P from AD effluent would offer a sustainable way of producing P fertilizer compared to the current approach that P is unsustainably mined from phosphate rock which according to some estimation that the reserve would be depleted within a century [25]. This book chapter will elucidate the mechanisms, processes and performances of some of the currently available P recovery technologies for manure and AD effluent, including chemical, electrochemical and biological methods.

1.2. Phosphorus in manure

Phosphorus (P) is one of the most abundant elements in the Earth's crust and it occurs in a large variety of forms, either in organic or inorganic forms, and also as monomeric (phosphates) or as constituent part of macromolecules (polyphosphates). Its discovery dates back to 1669, by Hennig Brand of Hamburg, through the distillation of urine. Its history proceeds to further characterization of some phosphorus compounds and production of phosphoric acid in the 18th century by Boyle. It was during the first half of the nineteenth century that some scientists, especially Liebig and Lawes, made very significant advances in the science of plant nutrition, and the first studies on the utilization of phosphates as fertilizers were recorded [26]. Some remarkable work in the 20th century in the field of biochemistry has been developed upon the understanding of phosphorus in biological systems, such as the discovery of adenosine triphosphate (ATP) in 1929; the concept of high energy phosphate bonds in 1941; and the elucidation of the molecular structure of nucleic acids (DNA and RNA) by Crick and Watson in 1953. These findings, and other outstanding results, led to the understanding that phosphorus plays a vital role in living processes. Phosphorus is usually not found free in nature and mostly occurs in the fully oxidized state as phosphate; and phosphates can be classified according to their molecular structure. The first attempt to classify them was introduced by

Graham, in 1833, in which he proposed the division into orthophosphates, pyrophosphates, and metaphosphates [26].

Phosphorus is present in agricultural residues and wastes, for instance, animal manure and litter. These residues are used as inexpensive fertilizer to improve soil quality. Soil tests have been conducted to estimate how much nutrient may be available for plant uptake during growth (Corbridge, 1985). For phosphorus characterization in soil, the values are aimed at identifying the labile P fraction, i.e., the fraction that is readily available for plants. From these observations, significant advances have been achieved over the last 65 years. The utilization of manure as fertilizer brought a different scenario for the analysis, due to the rich phase of phosphorus present in several samples reported in the literature [27]. Analytical techniques for phosphorus have transformed from simple gravimetric and volumetric titration methods to advanced new applications of 2D-NMR, chromatographic, spectroscopic, and microscopic methods. However, for several studies, understanding the chemical behavior of phosphorus is more relevant than predicting its molecular properties, like on sediments, soils, and residual materials. On these studies, the different phosphorus forms are usually categorized within their capability of being recovered by some physiochemical methods. The inorganic fraction are usually categorized under the following groups: (i) adsorbed by exchange sites; (ii) associated with iron, aluminum and manganese oxides; (iii) associated with carbonate; (iv) associated with calcium as apatite; and (iv) bound in a crystalline mineral form. The organic fraction is divided into: (i) labile organic substances; (ii) organic phosphorus associated with humic and fulvic acid; (iii) acid-soluble components; and (iv) residuals consisting of phosphate esters and phosphonates [28, 29]. The method of phosphorus fractionation relies on sequentially extracting compounds from a sample with selective solvents, that are able to isolate P pools of different solubility and of different chemical behavior. The major drawback of such analysis is that, it is unable to isolate discrete chemicals, though sophisticated methods of fractionation is used. Hence, extractants are usually designated to solubilize groups of minerals defined as P associated with Fe, Al, Ca, or even residual forms [30].

On regards to the variety and solubility of manure phosphorus, it must be understood that manure is a complex system and there are numerous interactions between the organic and inorganic phases within its matrix. It has been stated that manure relies on a sensitive and balanced dynamic equilibrium where minor changes, such as through chemical, physical or biological processes, affect the matrix as a whole [31]. Researchers defined the characteristics of four types of manures based on the different phosphorus contents and their characterization: i) Swine; ii) Beef and Dairy; iii) Chicken and Turkey; and iv) other species [27]. SEM images of swine manure samples revealed the presence of $MgNH_4PO_4 \cdot 6H_2O$ (struvite) and trace amounts of $MgKPO_4 \cdot 6H_2O$ (K-struvite); and found that these forms of struvite were in chemical equilibrium with beta-tricalcium phosphate (beta-TCP), and $CaHPO_4 \cdot 2H_2O$ (brushite) [32]. It was also found that swine manure has significant portions of brushite and $AlPO_4 \cdot 2H_2O$ (variscite) [33]. The majority of analysis shows that struvite and brushite are commonly present in swine manure [27]. Similar to swine manure, cattle manure analysis provides a range of phosphate minerals; struvite and $CaHPO_4$ (dicalcium phosphate, DCP) were found to be the main mineral forms of manure inorganic phosphorus as detected by SEM and X-ray diffraction [1, 31]. It was reported that [33] dairy manure predominantly has struvite

http://dx.doi.org/10.5772/60474

and brushite, like swine manure. Chicken and turkey manure, shows significant amounts of brushite, beta-TCP, DCP, and struvite, but they vary with different studies [27]. Studies on other manures, such as horse, deer, sheep, and goat, have been reported as being constituted by a mix of struvite and at least one form of Ca-phosphate mineral, usually brushite [26].

Land application of manure has been a more sustainable practice to provide an alternative source for nutrients in order to improve agricultural crop production. Over the past few decades, livestock production has undergone an industrial revolution, resulting in the large-scale generation of livestock manure [34]. Recent developments in corn ethanol production are also transforming the feed industry for wide applications of corn ethanol coproducts such as dry distiller's grains with solubles (DDGS) and corn gluten feeds (CGF) in animal diets. These new feeding materials are causing an increase in P excretion in animal manure [35]. There are multiple barriers to the land application of manure for P utilization and uptake by plants. First, the application is limited to the site close to the livestock, due to the low nutrient content (less than 1% of P in dry volatile solids, and solid content of swine manure around 6%) and subsequent high transportation cost. With the increasing size of livestock farms, especially in the areas where animal farming is highly concentrated, tremendous amounts of surplus manure must be discharged while the land in the surrounding area is oversaturated with P. Second, land application of animal manure is limited by its composition. For example, nutrients such as nitrogen (N) and P are present in swine manure in N:P ratios ranging from 1:1 to 2:1, while the N:P ratios needed by crops are between 3:1 to 15:1 [36, 37]. Therefore, when manure, especially swine manure, is applied to supply the crops demand for N, it results in the overdose of P. P-based application of manure is proposed as a new practice; however, this practice will not only result in under-application of N in most cases, but also will require more land to apply the same volume of manure [38, 39]. Over-application of P leads to its accumulation in soils; in turn, soils with high levels of P have been linked to environmental problems such as eutrophication of water bodies [40]. The land application of manure, especially swine manure, is considered an important contributor of P entering surface waters. Finally, once commercial chemical P fertilizer and manure are applied to soil, a large portion of soluble inorganic P is rapidly converted into insoluble forms by adsorption to the surface of soil particles, reacting with soil cations (such as calcium, iron, and aluminum), or immobilized into organic P by microorganisms in soil [41, 42]. Although the total P content in soil (average 0.05% w/w) is sufficient for plant growth, only 0.1% of the total P in soil is available to plants [43]. Overall, phosphorus separation from animal manure and digestate is a critical step in the sustainable utilization of the nutrient and in the healthy development of livestock industry. Some common phosphorus separation methods for liquid manure and digestate are discussed in the following sections.

2. Coagulation

2.1. Chemical coagulation/flocculation

Many commercialized processes for phosphorus removal from wastewater use chemical coagulation/flocculation methods by dosing divalent or trivalent metal cations (e.g.,

$[Al(H_2O)_6]^{3+}$ and $[Fe(H_2O)_6]^{3+}$, abbreviated as Al^{3+} and Fe^{3+} for convenience) via chemicals such as ferric chloride, ferric sulfate, aluminum sulfate, alum $[KAl(SO_4)_2 \cdot 12H_2O]$ and poly-aluminum chloride [44]. Coagulation process removes not only phosphorus in the form of phosphate, but also dissolved organic matter, colloids and particulates which are prone to be coagulated [45]. Coagulation is different from crystal precipitation process (crystallization) in which by adding lime or magnesium compounds to wastewater precipitates of calcium phosphate (e.g., hydroxyapatite) or struvite are formed and gravitationally separated from liquid fraction [8]. But when multivalent metal cations of coagulants are added, only a portion of phosphate is precipitated with these ions to serve to the phosphate removal from liquid. Further information about precipitation is described in the section of "Struvite precipitation" in this chapter. It is suggested by some authors that the real mechanisms in the coagulation/flocculation are complex, and it may include effects induced by other processes such as hydrolysis, complexation, crystallization, precipitation, adsorption (Figure 1), reduction-oxidation, other types of interactions (e.g., ligand competition) among different ions, etc. It is not surprising that a study with emphasis on theoretical aspect of a process may have little indication of the real significance in experimental observations [46].

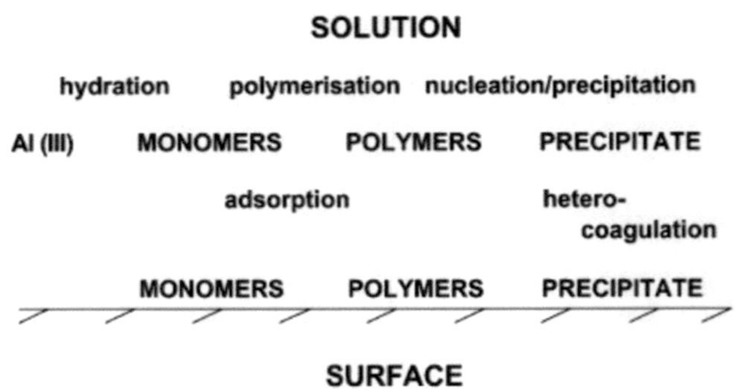

Figure 1. Schematic illustration of the charge neutralization by aluminum species [45].

The stability of colloidal system in wastewater is maintained principally by two ways: the coverage of the negative charge on particulate surface so that a repelling electrostatic force counteracts with van der Waals force and separates particulates from each other; and the hydration of the surface layers of colloids [46]. Destabilization of colloidal systems is the first step for coagulation. The efficacy of aluminum and iron coagulants principally originates from their ability to form multi-charged poly-nuclear complexes that enhance their adsorption capability [46]. Hydrated metal ions (Al^{3+} and Fe^{3+}) with one or more hydroxyl ions are observed to substantially improve absorptivity and coagulation, but it is not clear through what mechanisms the hydrolysis and the poly-nuclear complexation improve adsorption (sweep flocculation) [45, 46]. The above processes for aluminum species are illustrated in Figure 1, while ferric species follow a similar way. The formation of insoluble amorphous metal hydroxide precipitate also provides an important way as adsorbents for phosphate to attach [44, 47]. Generally, the hydrolyzed cations generated from dosed chemicals (applied in solid or solution forms depending on the chemicals used) will provide positive charges in bulk liquid and neutralize the surface charge of colloids, causing an increased attraction among colloids and destabilizing the colloidal system. The destabilized particles are followed by flocculation process through metal hydroxo complexes to form larger agglomerates, eventually forming larger particles of flocs. This mechanism can be especially useful because it is well accepted that the majority of manure phosphorus already exists in the colloidal form rather than dissolved phosphate anions, so crystallization of phosphate salts may be unnecessary prior to coagulation. Precipitation of insoluble phosphate salts, e.g., $FePO_4$ and $Fe_5(PO_4)_2(OH)_9$ when ferric is added, is another significant factor contributing to phosphate removal [48]. When chemical dosing is high, insoluble metal hydroxides are precipitated from liquid, which will also enmesh particulate materials by a sweep action. After flocculation the wastewater can be directly subjected to some physical separation processes, such as floating, gravitational sedimentation, screw press, or filtration.

Both Fe^{3+} and Al^{3+} salts (sulfate and chloride salts in solid or solution form) are widely used for coagulation for phosphorus removal, and the dose in liquid manure should reach 2 to 3 for the molar ratio of metal to phosphorus in order to achieve over 95% removal. Aluminum salts usually outperform iron salts at anaerobic condition because part of ferric is reduced to ferrous which is less effective in coagulation. A study evaluated the phosphorus removal by adding iron salts to simulated cattle manure (dissolving back dry solids in water): ferric salts were found much more effective than ferrous salts, and ferric chloride was more effective than ferric sulfate. Adding calcium oxide (CaO) removed additional amount of phosphorus [49]. Ferric chloride, ferric sulfate, aluminum chloride, and aluminum sulfate were assessed in jar test for phosphorus removal from liquid dairy manure [50]. Compared to the removal efficiency of 50% by natural sedimentation, 100 mg-Fe/L by ferric chloride slightly reduced the removal by 2%. Further increase of Fe (ferric chloride) to 200 mg/L, 300 mg/L, 400 mg/L, 500 mg/L, and 600 mg/L achieved removal efficiencies of 71%, 82%, 89%, 89%, and 92%, respectively. Ferric sulfate addition achieved very close removal efficiency for phosphorus, and removal efficiency was slightly improved from 82% to 88% when ferric was increased from 300 mg/L to 600 mg/L. Aluminum chloride obtained much better phosphorus removal: 80%, 85%, and 99% efficiency at 100, 200, and 300 mg-Al/L. Aluminum sulfate was comparable to the chloride salt: 60%, 75%, 89%, 95% and 100% removals were achieved at 59, 119, 179, 239, and 300 mg-Al/L [50]. 200 mg/L of flocculant (polyacrylamide) dosing with ferric (300 mg/L) and aluminum (180 mg/L) achieved over 98% of phosphorus removal. The total cost of P removal (84%) and transportation of 5 mile was calculated to be $4.09/m^3 ($0.02/gal). More data based on liquid cattle/dairy manure expressed in molar units are presented in Table 1.

Another study compared the coagulation/flocculation performance in terms of phosphorus removal from liquid swine manure by coagulants/flocculants of $FeCl_3$, $FeCl_2$, $FeClSO_4$, poly aluminum chloride (PAC), and sodium aluminate solutions [54]. The manure was liquid/solid separated and treated by activated sludge in a reactor for 30 days. The resulting total phosphorus was 322 mg/L. The ranges for the chemical dosing (mmol-metal/L) and removal efficiencies were listed as follows: $FeCl_3$, 3.3 to 16.3, 39% to 95%; $FeCl_2$, 2.1 to 10.6, 17% to 78%; $FeClSO_4$, 3.3 to 16.7, 26% to 91%; poly aluminum chloride (PAC), 4.5 to 23, 23% to 91%; and sodium aluminate, 5.7 to 28.7, 14% to 34%. $FeCl_3$ had the best performance of 95% removal, and the corresponding molar ratio of Fe to total-P was 1.57. Based on extrapolated data, in order to achieve total phosphorus down to the 2 mg/L from 322 mg/L, 27 mmol/L of $FeCl_3$ should be dosed to manure, corresponding to a molar ratio of 2.6 [54]. Performance variance in different studies (Table 1) indicates that further investigation is needed for optimal phosphorus removal and a reduced chemical cost.

Coagulant	**Fe/Al dose**	**Total solids**	**Initial total P**	**Final total P**	**Ratio of Fe/Al to total P**	**Reference**
	mmol/L	%	mg/L	mg/L	Molar ratio	
$FeCl_3$	5.0	1	104	12.6	1.5	[51]
$Fe_2(SO4)_3$	4.4	1	104	16.4	1.3	[51]
$FeCl_3$	5.4	0.9	110	19.8	1.5	[50]
$FeCl_3$	10.7	0.9	110	8.8	3.0	[50]
$Fe_2(SO4)_3$	5.4	0.9	110	19.8	1.5	[50]
$Fe_2(SO4)_3$	10.7	0.9	110	13.2	3.0	[50]
$AlCl_3$	3.7	0.9	110	22.0	1.0	[50]
$AlCl_3$	11.1	0.9	110	1.1	3.1	[50]
$Al_2(SO4)_3$	6.6	0.9	110	12.1	1.9	[50]
$Al_2(SO4)_3$	11.1	0.9	110	0.0	3.1	[50]
$Al_2(SO4)_3$	8.0	1.6	106	8.4	2.3	[52]
$FeCl_3$	16.0	1.6	106	29.6	4.7	[52]
$Al_2(SO4)_3$	5.9	1.1	143	25.0	1.3	[53]
$Al_2(SO4)_3$	11.7	1.1	143	4.4	2.5	[53]
$FeCl_3$	3.4	1.1	143	25.9	0.7	[53]
$FeCl_3$	6.7	1.1	143	16.0	1.5	[53]

Table 1. Performance of chemical coagulation on total phosphorus removal from liquid dairy manure

2.2. Electrochemical Coagulation (electrocoagulation; EC)

Electrochemical coagulation (EC) is an alternative to chemical coagulation. The main mechanism responsible for coagulation is similar in electrocoagulation and chemical dosing, except the self-generation of metal cations by anode oxidation. The electrocoagulation offers some advantages over chemical dosing: it has simple equipment requirement and can be readily automated; reduces the chemical cost by using cheaper materials; gas bubbling provides gentle mixing that promotes coagulation and helps form bigger flocs; and gas bubbling carries some particles up to the top of liquid in a way of flotation, which may be easily separated. So electrocoagulation is not only an alternative to the conventional way, but also a promising method due to its effectiveness and low cost [55, 56], which has been used for phosphate removal from drinking water [57], turbidity reduction [58], and wastewater remediation [59, 60].

The application and performance of EC for animal manure treatment has been reported in several recent publications [61-63]. EC was explored to remove solids from digested and screw pressed manure [62]. Before the pre-processed manure entered the EC system, the screw pressed digestate contained 4.2% TS and 3.0% VS. The EC effluent had a decreased TS and VS to 0.89% and 0.28%, corresponding to a removal efficiency of 79% and 89%, respectively. A parallel EC yielded a comparable results in effluent, with only 0.62% TS and 0.13% VS. Electrode materials, configurations, operating conditions, and phosphorus removal were not reported in the literature. Another field EC test reported total and dissolved phosphorus removal from lagoon effluent which was chemically pretreated and centrifuged before entering EC unit [61]. The dissolved and total phosphorus was reduced from 0.15 and 7 mg/L to 0.01 and 2.1 mg/L respectively, with limited description of the EC reactor configuration and operating conditions in the literature [61]. The preprocessing of lagoon effluent may generate effluent that did not reveal the total capability of EC for P removal. 304 stainless steel was used as EC electrodes for dairy manure treatment [63] with 1 A (at a voltage of about 6 V) for 500 mL manure. Within the first 30 min, 83% TP was removed but the ensuing current application seemed not to substantially improve removal efficiency. TSS removal reached 88% after 20 min EC operation. In a continuous mode, 53% to 78% TP was removed at current density of 3 A to 5.5 A after 30 min operation. Based on these results, EC can be a method to treat liquid discharge from dairy farm, but more studies must be conducted before its field application in order to articulate the interfering factors, appropriate electrode materials and configuration, and operating conditions.

3. Struvite precipitation

The crystallization technology was originally developed primarily because of the more stringent phosphorus removal requirements to further decrease the phosphorus (P) level of effluent in wastewater treatment plants and produce more valuable and sustainable end-products, such as struvite (magnesium ammonium phosphate, $MgNH_4PO_4 \cdot 6H_2O$) and calcium phosphate (e.g., hydroxyapatite, $Ca_5(PO_4)_3OH$) [17]. Struvite is a mineral crystalline substance consisting of equal molar magnesium, ammonium, and phosphate, and it can potentially be an excellent slow-release fertilizer that has numerous potential uses in agriculture and horticulture. Its precipitation reaction can be expressed as Eq. (1):

$$Mg^{2+} + NH_4^{+} + HPO_4^{2-} + 6H_2O \rightarrow MgNH_4PO_4 \cdot 6H_2O \downarrow + H^{+} \quad (1)$$

Precipitation of struvite often occurs in wastewater when ammonium, phosphate, and magnesium ions exceed the struvite solubility limit. The solubility product constant of struvite is $10^{-13.26}$ [64]. Struvite precipitation from digested manure has attracted tremendous attention recently as a method of P removal and recovery [65-68], although the reusability of this precipitated phosphate salt is debatable [66, 69], especially when it is applied to the soil with higher pH due to the low solubility of the struvite in the alkaline conditions [70]. Manure sources to apply this technology include calf manure [71], swine wastewater [72-80], poultry manure wastewater [81], and dairy manure [82-85]. Struvite precipitation can be separated into two stages: nucleation and growth. Nucleation occurs when constituent ions combine to form crystal embryos. Then, they quickly form a larger nucleus of crystals and crystal growth continues until equilibrium is reached [86]. Struvite precipitation is influenced by type of reactors, pH, temperature, Mg:P ratio, chemicals added, reaction time and the presence of other ions in solution such as calcium.

3.1. Crystallizer

Several types of crystallizer have been developed to remove phosphorus from manure wastewater by struvite precipitation. Bowers and Westerman [87] developed a cone-shaped fluidized-bed for struvite precipitation from swine wastewater. This design provided a high ratio of crystal surface area to reactor volume and it struvite seed to promote the growth of struvite crystals within the reactor. Field tests with pH adjustment and magnesium amendment demonstrated dissolved reactive phosphorus (DRP) removal efficiencies ranging from 70% to 82%, and total phosphorus (TP) reductions ranging from 63% to 80% [88]. However, high content of solids in the influent slurry could lead to washout of fine bed particles, resulting in lower phosphorus removal. The solids also could interfere with the bed particles and inhibit the struvite crystal growth. Shepherd et al. [89] developed a bench-scale continuous flow air sparged tank reactor (ASTR) for struvite precipitation in swine manure slurries. This system used air sparging for both pH adjustment and mixing, and used a peristaltic pump to continuously inject $MgCl_2$ for struvite precipitation. The bench-scale ASTR system provided DRP reductions of 78% and 95% in swine manure slurry collected from a shallow under floor pit collection system and from a concrete storage tank with a permeable cover, respectively. However, separation of precipitated struvite for TP reduction was not achieved with an up-flow clarifier operated in continuous flow mode. Then, they tested the performance of a pilot-scale ASTR-hydrocyclone system [90]. The pilot-scale ASTR-hydrocyclone system provided a 92% reduction of DRP in manure slurry from a swine finishing facility concrete storage tank and a 91% reduction of DRP in manure slurry collected from a swine finishing facility deep-pit under floor collection system. The pilot-scale ASTR-hydrocyclone system removed 18% of TP in swine manure from a concrete storage tank and 9% to 14% of TP in swine manure slurry from a deep-pit under floor collection system. The low TP recovery was attributed to the hydrocyclones inability to provide effective struvite separation as operated. Although ASTR was simple in design, fabrication, and operation, economics analysis indicated that this ASTR-hydrocyclone system in swine finisher manure slurries was not currently economically viable. The third design presented the reactor for removing and recovering phosphorus from swine wastewater with dual functions of crystallization through aeration and separation of formed struvite by settling [78]. The swine wastewater was fed to the aeration column of the reactor continuously. Air was aerated to diffuser units via a stainless steel air tube. The crystals formed in the aeration column were settled and withdrawn from the bottom of the reactor. The maximum yield of struvite was 171 g/m^3 swine wastewater, and the purity of the recovered struvite was approximately 95% without washing [79].

3.2. Effect of pH

pH is an important role for struvite production because struvite plays slightly soluble under neutral and alkalinic conditions but readily soluble in acid [91]. At 25°C, the solubility of struvite is 0.018 g/100 mL in water, and increases to 0.033 g/100 mL in 0.001N HCl and 0.178 g/100 mL in 0.01N HCl [92]. Therefore, optimized struvite precipitation in manure slurries generally requires increasing the slurry pH. The minimal pH required for effective precipitation of struvite from anaerobically digested cattle manure effluents is about 8 and the operation pH should be controlled between 8.5 and 9.5 [83]. Miles and Ellis [76] reported that the optimum pH was about 9.5 for struvite precipitation in anaerobically digested piggery wastes. The experiments using anaerobic swine lagoon liquid for struvite precipitation showed that that the minimum concentrations of DRP occurred between pH 8.9 and 9.25 at all Mg:P ratios [77]. Burns et al. [74] also found that 96%-98% of the DRP was removed from swine manure slurries at a pH of 8.6 with a 10-minute reaction time and $MgCl_2$ addition. However, a higher operation pH concurrently increases the caustic consumption and the cost of neutralizing the final treated effluents, even though it promotes better phosphate removal efficiency. Air sparging to elevate pH may be an advantageous approach for P recovery at larger scales because it would avoid liabilities from alkali additions [84]. Suzuki et al. [93] confirmed that aeration was effective for elevating the pH of swine wastewater to 8.5, and 65% removal of orthophosphate was achieved.

3.3. Effect of magnesium salts

The $Mg^{2+}/PO_4^{3-}/NH_4^+$ molar ratio is 1:1:1 in struvite based on the molecular formula ($MgNH_4PO_4 \cdot 6H_2O$). The concentration of soluble P in animal manure is normally much higher than that of magnesium ion but much lower than ammonium nitrogen, which results in that magnesium ion becomes the limiting constituent for struvite precipitation from manure. Without magnesium addition, only 20–30% of soluble phosphorus was removed from centrifuged digested cattle manure effluent at pH 7.8–9.0 as both struvite and calcium phosphates due to the low levels of the initial Mg^{2+} and Ca^{2+} in the effluent [83]. In order to increase phosphate removal efficiency through struvite precipitation, more Mg^{2+} would be needed. Phosphate removal efficiency greater than 80% was achieved when the molar ratio of Mg^{2+}/PO_4^{3-} in swine wastewater was between 1.1 and 1.6 [74, 77]. However, the requirement of higher Mg^{2+}/PO_4^{3-} molar ratio was also observed by other studies [83]. The phosphate removal efficiency just increased to 50% at a Mg^{2+}/PO_4^{3-} molar ratio larger than 5 and reached 73% at a Mg^{2+}/PO_4^{3-} molar ratio up to 22:1. Struvite formation could be hindered by high Ca^{2+} concentration, ionic strength, alkalinity, suspended solids content, or a combination of these [85, 93]. Huchzermeier and Tao [85] reported that Ca^{2+} inhibited the struvite formation by blocking active struvite growth sites and competing for orthophosphate to form calcium phosphates. They also found that calcium can be effectively removed from anaerobically digested dairy manure through precipitation of calcium carbonate at pH 9 to 10 while retaining magnesium and orthophosphate.

Technically, many magnesium sources that produce Mg^{2+} ions can be used for the struvite precipitation process. $MgCl_2$, $MgSO_4$, MgO, $Mg(OH)_2$ and other low-cost sources of magnesium, such as bittern and seawater, were tried as magnesium amendment for struvite precipitation [83, 94, 95]. Magnesium chloride ($MgCl_2 \cdot 6H_2O$) was considered a good source of Mg^{2+} for struvite formation and reduced the reaction time that was required to dissolve Mg^{2+} into solution for struvite formation [73]. Burns et al. [74] indicated that $MgCl_2$ can force the precipitation of P from swine manure. Zeng and Li [83] also found that $MgCl_2$ and $MgSO_4$ provided the higher P removal efficiency, likely due to their highly water-soluble property. The advantage of using $MgCl_2$ is its faster dissolution in water and hence a shorter reaction time. However, it decreases pH because it is slightly acidic. A pH of 8.5 or greater would be required for effective struvite precipitation [96]. Use of MgO or $Mg(OH)_2$ can supply Mg^{2+} and raise solution pH, and they are normally less expensive than $MgCl_2$ and $MgSO_4$. Although MgO and $Mg(OH)_2$ are both poorly soluble in water, MgO is still a good magnesium source for struvite precipitation and it was better than $Mg(OH)_2$ for struvite precipitation from digested swine wastes after they incurred insolubility problems with $Mg(OH)_2$ [76]. For struvite precipitation from digested cattle manure, MgO was slightly worse than $MgCl_2$ and $MgSO_4$ but much better than $Mg(OH)_2$ and $MgCO_3$. Due to its insolubility, $MgCO_3$ was the least effective for struvite precipitation from manure among the salts used [83]. A fine particle size and vigorous agitation of the reaction solution were needed when using MgO or $Mg(OH)_2$ for struvite precipitation.

3.4. Hydroxyapatite precipitation

Except struvite, small amount of other phosphate precipitates are also formed, which include K-struvite ($KMgPO_4 \cdot 6H_2O$), calcium phosphates, and magnesium phosphates. Zeng and Li [83] reported that less than 1.5% K-struvite was formed in precipitated solids although the original digested cattle manure effluent contains a high level of potassium. Burns et al. [74] detected brushite ($CaPO_3(OH) \cdot 2H_2O$) in precipitates from struvite precipitation of swine manure slurries due to the Ca^{2+} in the slurries. Magnesium phosphates possibly as $MgHPO_4 \cdot 3H_2O$ (newberyite), $Mg_3(PO4)_2 \cdot 8H_2O$ (bobierrite), and $Mg_3(PO_4)_2 \cdot 22H_2O$ were formed [83].

As a matter of fact, phosphorus removal in high Ca^{2+} content wastewater also can be achieved by direct precipitation of calcium phosphate (hydroxyapatite, $Ca_5(PO_4)_3OH$). For sludge side streams of wastewater treatment plant, phosphorus removal efficiencies by hydroxyapatite crystallization ranged from 75% to 85% with additional artificial crystal seed material, consisting of calcium silicate hydrate (tobermorite crystals, manufactured by mixing siliceous and calcareous raw materials, pelletizing, and autoclaving) [97]. However, few studies have been done to remove phosphorus via hydroxyapatite crystallization from manure slurries. When Harris and coworkers [84] recovered phosphorus from flushed dairy manure wastewater by precipitation, they found that phosphorus can be recovered from flushed dairy manure wastewater not only in the form of struvite, but also calcium phosphate. Presence or formation of carbonates inhibits hydroxyapatite formation and solution pH value, again, is a

http://dx.doi.org/10.5772/60474

key factor influencing the precipitation process. Addition of $MgSO_4$ can suppress carbonate precipitation and enhance Ca phosphate precipitation at elevated pH (>9) [84].

3.5. Commercialization and application of struvite precipitation

Struvite precipitation is currently the most commercially adopted technology for phosphorus recovery from wastewater as fertilizer. Modification on waste streams' constituents or operating conditions, such as chemical dosing, temperature elevation, air stripping and reactor innovation, is made to improve struvite crystallization and precipitation, and therefore, to increase P removal and recovery efficiency. Based on internet search, we found that there are numeral commercially available processes which have been developed and marketed. These processes include AirPrex™ developed by Berlin Water (Wassmannsdorf, Germany), NuReSys® by Akwadok Company (Waregem, Belgium), Pearl® with WASSTRIP (patent pending) by Ostara Nutrient Recovery Technologies Inc. (Vancouver, Canada), Crystalactor® by Royal HaskoningDHV Company (Amersfoort, the Netherlands), PHOSNIX by Unitika Ltd (Osaka, Japan), and Quich Wash™ by Renewable Nutrients LLC (Raleigh, NC). Struvite precipitation technology dominates the market compared to other types of technologies such as incineration, enzyme hydrolysis, critical point oxidation, and adsorption.

Phosphorus recovery from raw manure or anaerobically digested manure via struvite precipitation has been widely studied, and high phosphorus removal efficiencies were both obtained from the two kinds of manure. However, just few researches investigated the effect of anaerobic digestion on struvite precipitation. Moody et al. (2009) found that anaerobic digestion of swine slurry increased orthophosphate (PO_4^{3-}) in solution by 25% (from 1.26 g/L to 1.59 g/L), but the amount of Mg^{2+} increased by 254% (from 88.3 mg/L to 313.3 mg/L), which indicated that anaerobic digestion have no significant effect on reactive phosphate releasing, but it significantly increased the amount of available Mg^{2+}. Increasing the solution Mg^{2+} concentration means less amendment of magnesium salt and low cost. In addition, PO_4^{3-} removal efficiency of struvite precipitation was increased by 36% with anaerobic digestion pretreatment.

4. Enhanced Biological Phosphorus Removal (EBPR)

Biological nutrient process (BNP) refers to modified activated sludge processes where contaminants, biochemical oxygen demand (BOD), nitrogen and phosphorus can be simultaneously removed from the bulk wastewater and accumulated to the waste sludge [98]. In this method, phosphorus removal is usually realized by enhanced biological phosphorus removal process (EBPR), a process that recirculates activated sludge in alternating anaerobic and aerobic conditions and enriches or selects the microbial strains synthesizing polyphosphate (polyphosphate accumulating organisms, PAO). It has been applied to municipal wastewater treatment, but is also acknowledged that in large-scale applications the process experiences reactor upset and failure from time to time due to environmental or biological reasons such as nitrate overloading, high rainfall, and microbial competition [99]. Selecting and maintaining suitable operational conditions to avoid reactor upset can be important to the process. One limitation for EBPR is the insufficient carbon source in wastewater, or relatively low soluble COD-to-phosphorus ratio so PAOs does not sequester enough PHAs or energy for phosphorus accumulation. Manure before or even after anaerobic digestion usually contains a high level of organic compounds and a large portion of it is readily biodegradable such as acetate, propionate and butyrate. So manure can at least be a good substrate for providing carbon source for PAOs; in other words, EBPR can be an alternative method for phosphorus removal from liquid manure [100]. Some of the previous studies using animal manure in fact achieved good results in phosphorus removal via EBPR. Note that EBPR process usually is not the final step closing the phosphorus removal but a pretreatment method, although phosphorus is concentrated in sludge. The reason is that poly-phosphate in the sludge tends to be hydrolyzed, released and leaked to liquid phase. There are several other processes that can be used to further treat the EBPR sludge and phosphorus-accumulated media, e.g., coagulation and crystallization.

4.1. Microbiology and mechanisms

EBPR has been used in wastewater treatment plants for a long period of time, and most fundamental studies were based on activated/recycled sludge from these treatment plants. Briefly, the anaerobic/aerobic cycling selects and enriches PAOs, and PAOs take up phosphate from wastewater at the aerobic stage (process diagram illustrated in Figure 2). In the PAOs well developed sludge, acetate and other short-chain fatty acids are taken up at the anaerobic stage by PAO cells and transferred to acetyl-CoA with ATP consumption from poly-P hydrolysis. This process is accompanied with release of cations such as K^+ and Mg^{2+} as well as $H_2PO_4^-$ to wastewater. After several steps of transformation and polymerization, acetyl-CoA is transformed to poly-β-hydroxyalkanoates (PHAs), mainly poly-β-hydroxybutyrate (PHB) and poly-β-hydroxyvalerate (PHV). The reducing power for acetyl-CoA transformation at anaerobic condition comes from NADH in the degradation of internal carbohydrate, via full [102] or split TCA cycles [103], but also found via the EMP pathway (a glycolysis pathway) from consumption of internal carbohydrates mainly glycogen [104]. As said before, hydrolysis of poly-P may also contribute energy source as ATP at the anaerobic stage, for acetate transport across cell membranes. Multiple of these pathways may co-exist depending on sludge sources and environmental conditions, which needs clarification by future studies [99]. At aerobic condition, the degradation of PHAs leads to the generation of acetyl-CoA and propionyl-CoA, both of which enter TCA cycle as carbon and energy source for biomass growth, phosphate uptake and poly-P generation, and glycogen generation.

Another important microbial composition of activated sludge is glycogen-accumulating organisms (GAOs). GAOs consume external carbon source at anaerobic stage, but at aerobic stage they do not uptake phosphate from environment and there is no poly-P accumulation occurring; instead, carbon and energy from PHAs hydrolysis is mostly used for biomass growth and glycogen synthesis. Therefore, there is a substrate competition between PAOs and GAOs for acetate and other short-chain fatty acids utilization. This relationship dominates the performance of phosphate removal in EBPR process [99], and is affected by environmental

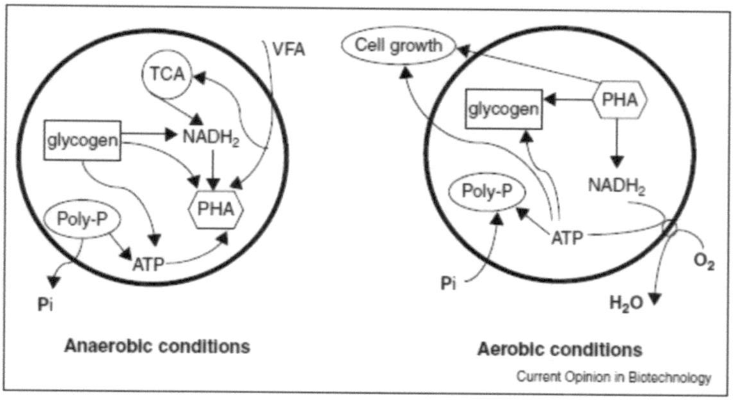

Figure 2. Schematic diagrams of the anaerobic and aerobic PAO metabolism in EBPR process [101].

conditions for selective advantages between two groups, such as types and concentrations of carbon sources, organic loading rate, nutrients level and their ratio to carbon sources, pH, temperature, etc. Generally speaking, at the ambient pH range (6 to 8), a relatively higher pH is found advantages for phosphate removal and for PAOs selection over GAOs. At the ambient temperature range (5 to 35 °C), a lower temperature is favored by PAOs over GAOs for phosphate removal. Relative abundance of PAOs over GAOs may be increased by a relative lower dissolved oxygen levels (1.5 to 3 mg/L) at the aerobic stage [99].

Phosphate uptake is also possible by denitrifying PAOs which use nitrate as electron acceptor in the anoxic condition, corresponding to oxygen in the aerobic condition. This pathway is promising because it removes both nitrogen and phosphate and it reduces sludge production due to less energy release compared to aerobic condition [99]. There is no exclusive relationship between the concepts of general PAOs and denitrifying PAOs. Multiple commercial processes have been established for the combined biological nitrogen and phosphate removal in wastewater treatment plants, such as A_2O, modified-UCT, five-stage Bardenpho, and DEPHANOX with either pre- or post-denitrification [99].

4.2. EBPR with animal manure as substrate

At the anaerobic stage, VFAs are substrates for PHAs regeneration inside PAOs cells, and the accumulated PHAs can be degraded for biomass growth and phosphate uptake at the aerobic stage. However, municipal wastewater contains only a small amount of VFAs, which reduces the suitability of EBPR process. Acetate may be added to wastewater but it increases overall chemical cost. So supplementing agro-food wastewater can be a good option, because it is considered waste and cost-effective. EBPR potential test, which evaluates the phosphorus release from activated sludge in certain medium composition in batch mode, was developed for assessing the feasibility of implementing EBPR for a substrate [105, 106]. Based on this test, it was found that tomato processing and milk bottling industry wastewater had short-term enhancement for EBPR, while wastewater from cheese industry, slaughter houses, beet sugar processing, and winery processing didn't show improvement [107].

The spatial requirement on condition variation (anaerobic/aerobic) in conventional flow systems for EBPR can be transformed to temporal variation by using sequencing batch reactors (SBRs) [108, 109]. This approach is widely adopted for treating liquid manure (liquid swine and dairy manure) [110, 111] because it saves space and operating cost over the conventional alternate flow systems, thus suitable for use in animal farms. SBRs consist several phases in its operation, typically including feeding, anaerobic, aerobic, settling, and withdrawal. Operating conditions such as medium composition, ORP, HRT, SRT, length of a cycle, length of each phase, pH, and temperature will have an impact on SBR performance. Surplus phosphorus uptake from swine manure and poly-P accumulation in biomass was observed in liquid swine manure treatment in SBRs, with 93% of COD removal, 88%-93% of total nitrogen removal and 95% of phosphate removal [112]. With non-diluted swine wastewater that initially contained 1500 mg-N/L of ammonium and 144 mg-P/L of phosphate, a removal efficiency of 99.7% for nitrogen and 97.3% for phosphate was achieved in SBR operated with 3 cycles/day at temperature of 30 °C, SRT of 1 day and HRT of 11 days [110]. Subjecting the digested manure to EBPR may encounter lack of VFAs at the anaerobic stage of EBPR, so non-digested liquid swine manure can be supplemented to AD effluent as additional carbon source. This combined substrate achieved good nutrient removal by EBPR in SBR, 99.8% of nitrogen and 97.8% of phosphate removal from an initial content of 900 mg/L of ammonia and 90 mg/L of phosphate [111]. A 16 L SBR was constructed for treating liquid dairy manure by EBPR [113], and with 6-fold dilution, 98% of phosphate removal (from 33 mg/L to 0.5 mg/L of phosphate) was obtained, while the removal was dropped to 70% with 5-fold dilution. It was suggested by the authors that the higher strength dairy manure contained inhibitory components to EBPR, and high content of acetate was thought as the major reason but was not conclusive. Another study fed EBPR SBR with dairy industry wastewater, and phosphate concentration was decreased from 29.8 - 43.6 mg/L to less than 1 mg/L, demonstrating a highly effective phosphate removal [114]. The EBPR sludge showed a high proportion (31.4% to 38.3%) of *Rhodociclus*-related bacteria, an indicator of PAOs population, in the total microbial cell counts, and phosphorus accounted for 5.4% to 10.1% of the MLVSS [114]. Another SBR study reported 59% of phosphate removal with an initial phosphate level of 37.4 mg/L in dairy manure, generating sludge with 2.6% of phosphorus in the MLVSS [115].

5. Phytoremediation

Phytoremediation could be a low-cost clean up technology for wastewater treatment and also for the P removal, as the green algae and plants offer the finest eco-friendly option for environment remediation. The major hitch with the phytoremediation is the slow growth rate of the plant species and its survival capacity in the non-ideal environment. But, algae and aquatic plants offer a realistic time frame for the nutrient recovery from eutrophic water and other contaminated waters rather than the P removal from the soils, and also require less than one-tenth of the area to recover phosphorus compared to terrestrial crops [116]. The characteristic advantages of the system has made the algal ponds and macrophyte wetlands more popular for environmental applications including herbicides [117], heavy metals [118] and antibiotics [119] removal, and is already been explored widely for varieties of wastewater types including animal manure [120]. Anaerobic treatment of manure results in digestate which requires further treatment before discharge, especially for the removal of P and N. Growing microalgae or macrophytes on the anaerobic effluent could be a commendable option considering the valuable byproducts and process efficiency.

5.1. Algae for phosphorous recovery

Anaerobically treated swine manure has proved to be good medium for algae growth [120-122]. Microalgae with its rapid growth potential, better adaptation to various ecological habitats and as an important feedstock for third generation biofuel, can be a best suited strain to grow in the waste waters for its valuable biomass and the nutrient recovery. It is reported that the major bottleneck to commercialization of algal fuels is the supply of N and P nutrients [123, 124]. Many recent research focuses on a suitable nutrient management strategy to use wastewater (industrial, municipal, dairy, food wastewater and digested dairy manure) as a nutrient supplement for cultivation of oil-rich green microalgae growth and recycling of nutrients [121, 122, 125-127]. The anaerobic digested effluent after decomposing organic waste to produce biogas has been used to grow algae for nutrient recovery. The effluent has relatively lower carbon levels for algae because of microbial utilization during anaerobic digestion [121]. Pretreatment steps like dilution to avoid inhibition and sterilization to prevent the contamination may be required for certain algae systems [121]. Raw and anaerobically digested swine manure has been treated widely by (1) suspended algae in (i) high rate pond systems [128-131], (ii) mixed algae systems [132] and (iii) mixed algae-bacterial systems [133] or (2) by immobilized algae [120], e.g., algal turf scrubber units [134-136]. The harvested algae can be a good high-grade protein supplement for animal feed and also can be used as a slow-release fertilizer [136] which can be directly sprayed as suspension in farm land or stored for future use [137].

Different species of freshwater micaralgae have been tested for nutrient removal from municipal wastewater and manure. The nutrient removal capacity of a *Chlorella sp.* from a highly concentrated municipal wastewater stream generated from activated sludge thickening process (raw and autoclaved medium) was tested by Li and coworkers [125]. After 14-day batch culture, algae could remove ammonia, total nitrogen, total phosphorus, and chemical oxygen demand (COD) by 93.9%, 89.1%, 80.9%, and 90.8%, from raw medium respectively. It was concluded that the system could be successfully scaled up, and continuously operated at 50% daily harvesting rate, providing a net biomass productivity of 0.92 g-algae/(L day) [6]. The immobilized and free cell cultures of two nanoplanktonic algal species, *Scenedesmus intermedius Chod.* and *Nannochloris sp.* isolated from different sources of pig manure was used to study the growth rate, phosphorus and nitrogen uptake from anaerobically treated manure [120]. P and N uptake rates for *S. intermedius* were 0.014 and 0.012 mg P h^{-1} and 0.022 and 0.009 mg N h^{-1} for free and immobilized cells respectively; and rates for *Nannochloris sp.* were 0.006

and 0.009 mg P h^{-1} and 0.011 and 0.006 mg N h^{-1} for free and immobilized cells. It was observed that the isolated species were more efficient in nutrient recovery than the commercially available strains [120], as the isolated strains were better acclimatized to the prevailing conditions. The anaerobically digested swine manure from a farm digester was used to culture *Chlorophyceae, Chlorella sp., Scenedesmus obliquus,* and a cyanobacterium, *Phormidium bohneri,* to evaluate the inorganic nitrogen and orthophosphate removal efficiency. *Chlorella sp.* performed well in batch cultures wheres *P. bohneri* in semi-continuous conditions [138]. Benthic freshwater algae was also used to recover nutrients from dairy manure [134] in algae growth chambers operated in semi-batch mode by continuously recycling wastewater and adding manure inputs daily. It was found that, when compared to a conventional corn/rye rotation, such benthic algae production rates would require 26% of the land area requirements for equivalent N uptake rates and 23% of the land area requirements on a P uptake basis [134]. Besides microalgae species, some filamentous fungal species also showed some potential to combine with AD to remove and recover the phosphorus [139].

Harvesting microalgae from treated wastewater is cost intensive, therefore becoming the key to remove and recover the phosphorus. The attached algae cultures for the nutrient removal from the manure waste water was evaluated [140]. It was found that, depending on different culture conditions, the attached algal culture removed 61–79% total N and 62–93% total P from dairy manure. Overall, the attached algal culture removed 62–90% of total phosphorus, 62–87% of soluble phosphorus, and 43–80% of orthophosphate from dairy manure. The economic assessment of algal turf scrubber technology for treatment of dairy manure effluent showed that economic balance would become more favorable if values from algae as a byproduct and nutrient trading credits can be realized [141].

5.2. Macrophytes for phosphorous recovery

Aquatic macrophytes are the conspicuous plants that dominate wetlands, shallow lakes, and streams, playing a vital role in healthy ecosystems. Total nitrogen and total phosphorous removal in treatment wetlands can range from 3–98% to 31–99% respectively [142, 143] with an average removal of about 50% [144]. Studies have showed that vascular aquatic plants have acceptable animal feed qualities, ability to remove nutrients from water, and high production rates [145]. Macrophytes constitute a diverse assemblage of taxonomic groups and are often separated into four categories based on their habit of growth: floating unattached, floating attached, submersed, and emergent [146]. Macroscopic flora includes the aquatic angiosperms (flowering plants), pteridophytes (ferns), and bryophytes (mosses, hornworts, and liverworts). Macrophytes based nutrient removal technology has the merits of (1) high productivity of several large-leaf floating plants; (2) high nutritive value of floating plants relative to many emergent species; and (3) ease of stocking and harvesting [147]. Also, the harvested floating macrophytes biomass can potentially be used for composting, soil amendments, anaerobic digestion with methane production, being processed for animal feed, and could be mixed with separated manure solids to increase the amount of nutrients available for exporting off the farm [148]. The biomass can be a good resource of starch, and utilized for the production of value-added products such as fuel ethanol [149]. For example, *Spirodela polyrrhiza* grown on anaerobically treated swine wastewater had a starch content of 45.8% (dry weight) and enzymatic hydrolysis of the duckweed biomass yielded a hydrolysate with a reducing sugar content corresponding to 50.9% of the original dry duckweed biomass, yielding 25.8% ethanol (dry weight) after fermentation by yeast [150]. Duckweed (*Lemnaceae*) has been widely used to recover the nutrients from pig effluents, because of its tolerance to high nutrient levels and preferred absorption of ammonium [149, 151], and can also grow all seasons in areas with warm climates and doubles its biomass within two days under the optimal conditions [152]. Screened duckweed strains, *Lemnaceae* that grew well on the anaerobically treated swine wastewater in laboratory and greenhouse experiments were tested for nutrient recovery under field conditions. Under nutrient abundant conditions in waste water, duckweed takes the nutrients for its growth and store the nutrients in its tissue for future nutrient limited conditions for a significant period of time, and the nutrient reserve in duckweed biomass has been found the key to the kinetics of duckweed growth [153]. Plants take up nutrients while growing and if not harvested, decompose in wetlands returning nutrients to the ecosystems.

6. Conclusions and future remarks

With the increasing size of livestock farms, especially in the area where livestock raising is highly concentrated, the surplus digested manure applied on soil increases P concentration in agricultural runoff, causing environmental problems like eutrophication. Phosphorus removal and recovery from digested manure reveals its importance in livestock raising area. Coagulation and electrocoagulation methods have been used for P removal from either digested or undigested animal manure. Compared to the municipal wastewater treatment, the dosing of multivalent cations is more intensive for manure treatment, e.g., the molar ratio of metal to P is mostly more than 3. This can be a result of the presence of high solids content and high level of carbonate/bicarbonate, which may consume additional portion of the added metal coagulants. Aluminum salts work better than the ferric counterparts in the anaerobic condition of manure media. Electrocoagulation avoids the direct chemical dosing by releasing metal ions through sacrificing metal anode, so is less chemically intensive but consumes additional electric energy. Struvite precipitation is the most commercially available method to recover the phosphorus from manure as fertilizer. The anaerobic digestion of manure seems to be beneficial to the struvite precipitation while more detailed studies are needed. The key issue related to this process is the bioavailability of phosphate in these precipitates, including struvite and hydroxyapatite, to the plant growth. Animal manure contains high level of carbon source in the form of VFAs which can be the ultimate reducing power for PAOs. Therefore, manure can be treated with EBPR process without additional carbon dosing. The high level of ammonium may be simultaneously removed by enriching denitrifying PAOs community with suitable process design. An integrated process combining anaerobic digester and algae cultivation / macrophyte growth can also be an eco-friendly and sustainable process to reduce nutrient loss to environment and to produce valuable biomass.

Overall, these currently available methods for phosphorus removal and recovery are primarily designed for industrial and municipal wastewater treatment, where there is incentive related to the operations. However, the phosphorus removal from the manure is economically challenging because the end products of phosphorus recovery do not justify the cost of the removal process. It is beneficial to combine the AD process with the phosphorus removal and recovery so that the overall techno-economic feasibility of the process can be significantly improved.

Acknowledgements

The authors gratefully acknowledge the funding support from MNDRIVE Bioremediation, UMN BTI Synthetic Ecology Program, Minnesota Agricultural Experiment Station (MAES) and CAPES Foundation of the Brazilian Ministry of Education (grant BEX13252135).

Author details

Hongjian Lin, Jing Gan, Aravindan Rajendran, Cristiano Eduardo Rodrigues Reis and Bo Hu*

*Address all correspondence to: bhu@umn.edu

Department of Bioproducts and Biosystems Engineering, University of Minnesota, USA

References

[1] Güngör, K. and K. Karthikeyan, *Phosphorus forms and extractability in dairy manure: A case study for Wisconsin on-farm anaerobic digesters.* Bioresource Technology, 2008. 99(2): p. 425-436.

[2] Gerritse, R. and R. Vriesema, *Phosphate distribution in animal waste slurries.* The Journal of Agricultural Science, 1984. 102(01): p. 159-161.

[3] Barnett, G., *Phosphorus forms in animal manure.* Bioresource Technology, 1994. 49(2): p. 139-147.

[4] Dou, Z., et al., *Laboratory procedures for characterizing manure phosphorus.* Journal of Environmental Quality, 2000. 29(2): p. 508-514.

[5] Sharpley, A. and B. Moyer, *Phosphorus forms in manure and compost and their release during simulated rainfall.* Journal of environmental quality, 2000. 29(5): p. 1462-1469.

[6] Massé, D., F. Croteau, and L. Masse, *The fate of crop nutrients during digestion of swine manure in psychrophilic anaerobic sequencing batch reactors.* Bioresource technology, 2007. 98(15): p. 2819-2823.

[7] Möller, K. and T. Müller, *Effects of anaerobic digestion on digestate nutrient availability and crop growth: a review.* Engineering in Life Sciences, 2012. 12(3): p. 242-257.

[8] Hjorth, M., et al., *Solid–liquid separation of animal slurry in theory and practice. A review.* Agronomy for sustainable development, 2010. 30(1): p. 153-180.

[9] Christensen, M.L., M. Hjorth, and K. Keiding, *Characterization of pig slurry with reference to flocculation and separation.* Water research, 2009. 43(3): p. 773-783.

[10] Burton, A.L., *Phenotypic evaluation and genetic basis of anatomical and architectural root traits in the genus Zea.* 2010: The Pennsylvania State University.

[11] Callander, I. and J. Barford, *Precipitation, chelation, and the availability of metals as nutrients in anaerobic digestion. I. Methodology.* Biotechnology and bioengineering, 1983. 25(8): p. 1947-1957.

[12] Field, J., et al., *Fertilizer recovery from anaerobic digesters.* Transactions of the ASAE, 1984. 27(6): p. 1871-&.

[13] Sommer, S. and S. Husted, *The chemical buffer system in raw and digested animal slurry.* The Journal of Agricultural Science, 1995. 124(01): p. 45-53.

[14] Schievano, A., et al., *On-field study of anaerobic digestion full-scale plants (Part I): An on-field methodology to determine mass, carbon and nutrients balance.* Bioresource technology, 2011. 102(17): p. 7737-7744.

[15] Salminen, E. and J. Rintala, *Anaerobic digestion of organic solid poultry slaughterhouse waste–a review.* Bioresource Technology, 2002. 83(1): p. 13-26.

[16] Marcato, C.-E., et al., *Bioavailability of Cu and Zn in raw and anaerobically digested pig slurry.* Ecotoxicology and environmental safety, 2009. 72(5): p. 1538-1544.

[17] Morse, G., et al., *Review: phosphorus removal and recovery technologies.* Science of the total environment, 1998. 212(1): p. 69-81.

[18] Xia, M.J., et al., *Passive solid-liquid separation of anaerobically digested dairy manure using two-stage filtration.* Biosystems Engineering, 2012. 111(4): p. 392-397.

[19] Mukhtar, S., M.S. Borhan, and J. Beseda, *Evaluation Of A Weeping Wall Solid-Liquid Separation System For Flushed Dairy Manure.* Applied Engineering in Agriculture, 2011. 27(1): p. 135-142.

[20] Walker, P.M. and C.A. Wade, *Comparison Of The Effectiveness And Economic Costs Of Two Production Scale Polyacrylamide Assisted Solid/Liquid Separation Systems For The Treatment Of Liquid Swine Manure.* Applied Engineering in Agriculture, 2010. 26(2): p. 299-305.

[21] Walker, P.M., C.A. Wade, and T.R. Kelley, *Evaluation of a polyacrylamide assisted solid/liquid separation system for the treatment of liquid pig manure.* Biosystems Engineering, 2010. 105(2): p. 241-246.

[22] Garcia, M.C., et al., *Enhanced solid-liquid separation of dairy manure with natural flocculants.* Bioresour. Technol., 2009. 100(22): p. 5417-5423.

[23] Meunier, N., et al., *Solid/Liquid Separation of Pig Manure by Biological Flotation: Pilot-Scale Study.* Journal of Environmental Engineering-Asce, 2009. 135(9): p. 869-875.

[24] He, Z. and H. Zhang, *Applied Manure and Nutrient Chemistry for Sustainable Agriculture and Environment*. 2014: Springer.

[25] Cordell, D., J.-O. Drangert, and S. White, *The story of phosphorus: Global food security and food for thought.* Global environmental change, 2009. 19(2): p. 292-305.

[26] Corbridge, D.E.C., *Phosphorus-an outline of its chemistry, biochemistry and technology.* 1985: Elsevier Science Publishers BV.

[27] Pagliari, P.H., *Variety and Solubility of Phosphorus Forms in Animal Manure and Their Effects on Soil Test Phosphorus,* in *Applied Manure and Nutrient Chemistry for Sustainable Agriculture and Environment*. 2014, Springer. p. 141-161.

[28] Barbanti, A., et al., *Critical aspects of sedimentary phosphorus chemical fractionation.* Journal of Environmental Quality, 1994. 23(5): p. 1093-1102.

[29] McKelvie, I.D., *Separation, preconcentration and speciation of organic phosphorus in environmental samples.* Organic phosphorus in the environment, 2005: p. 1-20.

[30] Leikam, D.F. and F.P. Achorn, *Phosphate fertilizers: production, characteristics, and technologies, in Phosphorus: Agriculture and the Environment.,* J.T. Sims and A.N. Sharpley, Editors. 2005, American Society of Agronomy, Inc.: Madison, WI. p. 23-50.

[31] Fordham, A. and U. Schwertmann, *Composition and reactions of liquid manure (Gülle), with particular reference to phosphate: II. Solid phase components.* Journal of Environmental Quality, 1977. 6(2): p. 136-140.

[32] Bril, J. and W. Salomons, *Chemical composition of animal manure: a modelling approach.* Netherlands Journal of Agricultural Science, 1990. 38(3A): p. 333-351.

[33] Ajiboye, B., et al., *Phosphorus speciation of sequential extracts of organic amendments using nuclear magnetic resonance and X-ray absorption near-edge structure spectroscopies.* Journal of environmental quality, 2007. 36(6): p. 1563-1576.

[34] Kumar, R.R., B.J. Park, and J.Y. Cho, *Application and Environmental Risks of Livestock Manure.* Journal of the Korean Society for Applied Biological Chemistry, 2013. 56(5): p. 497-503.

[35] Stein, H.H. and G.C. Shurson, *BOARD-INVITED REVIEW: The use and application of distillers dried grains with solubles in swine diets.* Journal of Animal Science, 2009. 87(4): p. 1292-1303.

[36] Edwards, D.R. and T.C. Daniel, *Environmental impacts of on-farm poultry waste disposal: a review.* Bioresource Technology, 1992. 41(1): p. 9-33.

[37] Naylor, S.J., R.D. Moccia, and G.M. Durant, *The Chemical Composition of Settleable Solid Fish Waste (Manure) from Commercial Rainbow Trout Farms in Ontario, Canada*. North American Journal of Aquaculture, 1999. 61(1): p. 21-26.

[38] Ndegwa, P.M., et al., *Enhanced phosphorus removal from swine-nursery manure in aerated batch reactors*. Transactions of the Asae, 2003. 46(3): p. 797-803.

[39] Lory, J.A., et al., *Feasibility and costs of phosphorus application limits on 39 U.S. swine operations*. J Environ Qual, 2004. 33(3): p. 1114-23.

[40] Pagliari, P.H. and C.A.M. Laboski, *Investigation of the Inorganic and Organic Phosphorus Forms in Animal Manure*. Journal of Environmental Quality, 2012. 41(3): p. 901-910.

[41] Bolan, N.S., *A Critical-Review On The Role Of Mycorrhizal Fungi In The Uptake Of Phosphorus By Plants*. Plant and Soil, 1991. 134(2): p. 189-207.

[42] Vadas, P.A., R.D. Harmel, and P.J.A. Kleinman, *Transformations of soil and manure phosphorus after surface application of manure to field plots*. Nutrient Cycling in Agroecosystems, 2007. 77(1): p. 83-99.

[43] Chai, B., et al., *Isolation and phosphate-solubilizing ability of a fungus, Penicillium sp from soil of an alum mine*. Journal of Basic Microbiology, 2011. 51(1): p. 5-14.

[44] De-Bashan, L.E. and Y. Bashan, *Recent advances in removing phosphorus from wastewater and its future use as fertilizer (1997–2003)*. Water research, 2004. 38(19): p. 4222-4246.

[45] Duan, J. and J. Gregory, *Coagulation by hydrolysing metal salts*. Advances in Colloid and Interface Science, 2003. 100: p. 475-502.

[46] Bratby, J.R., *Coagulation and flocculation: with an emphasis on water and wastewater treatment*, in *Coagulation and flocculation: with an emphasis on water and wastewater treatment*. 1980, Uplands Press.

[47] Duffy, S.J. and G.W. vanLoon, *Characterization of amorphous aluminum hydroxide by the ferron method*. Environmental science & technology, 1994. 28(11): p. 1950-1956.

[48] Rodriguez, M.E. and A.G. del Puerto, *Separation of phosphorus from pig slurry using chemical additives*. Applied engineering in agriculture, 2005. 21(4): p. 739-742.

[49] Barrow, J., et al., *Effects of Fe and Ca additions to dairy wastewaters on solids and nutrient removal by sedimentation*. Applied engineering in agriculture (USA), 1997.

[50] DeBusk, J.A., et al., *Chemical Phosphorus Removal for Separated Flushed Dairy Manure*. 2008. 24(4).

[51] J.T., B., et al., *Effects of Fe and Ca additions to dairy wastewaters on solids and nutrient removal by sedimentation*. Appl. Eng. Agric., 1997. 13: p. 259-267.

[52] Karthikeyan, K.G., et al., *Chemical Treatment of Dairy Manure Using Alum, Ferric Chloride and Lime, in 2002 ASAE Annual Meeting*. 2002: Chicago, Illinois.

[53] Sherman, J.J., H.H. Van Horn, and R.A. Nordstedt, *Use of flocculants in dairy wastewaters to remove phosphorus.* Appl. Eng. Agric., 2000. 16: p. 445-452.

[54] Meers, E., et al., *Physico-chemical P removal from the liquid fraction of pig manure as an intermediary step in manure processing.* Water, Air, and Soil Pollution, 2006. 169(1-4): p. 317-330.

[55] Cañizares, P., et al., *Coagulation and electrocoagulation of wastes polluted with dyes.* Environmental science & technology, 2006. 40(20): p. 6418-6424.

[56] Lacasa, E., et al., *Electrochemical phosphates removal using iron and aluminium electrodes.* Chemical Engineering Journal, 2011. 172(1): p. 137-143.

[57] Vasudevan, S., et al., *Studies on the removal of phosphate from drinking water by electrocoagulation process.* Industrial & Engineering Chemistry Research, 2008. 47(6): p. 2018-2023.

[58] Merzouk, B., et al., *Removal turbidity and separation of heavy metals using electrocoagulation–electroflotation technique: A case study.* Journal of hazardous materials, 2009. 164(1): p. 215-222.

[59] Ahmad, A., S. Ismail, and S. Bhatia, *Optimization of coagulation-flocculation process for palm oil mill effluent using response surface methodology.* Environmental science & technology, 2005. 39(8): p. 2828-2834.

[60] Tchamango, S., et al., *Treatment of dairy effluents by electrocoagulation using aluminium electrodes.* Science of the total environment, 2010. 408(4): p. 947-952.

[61] Lazenby, L.A., *Evaluation of selected new technologies for animal waste pollution control.* 2006, Texas A&M University.

[62] Hall, H.R. and C.L. Hansen, *Changes in Settling Characteristics and Fate of Solids in Dairy Manure Using Anaerobic Digestion, Mechanical Liquid-Solid Separation, and Electrocoagulation, in 2006 ASABE Annual International Meeting.* 2006: Portland, Oregon.

[63] Jiang, A., et al., *Phosphorous Recovery Technology in Conjunction with Dairy Anaerobic Digestion, in Climate Friendly Farming: Research Report 2010 - 001.* 2010, CSANR.

[64] Ohlinger, K., T. Young, and E. Schroeder, *Predicting struvite formation in digestion.* Water Research, 1998. 32(12): p. 3607-3614.

[65] Fernandes, G.W., et al., *Chemical phosphorus removal: a clean strategy for piggery wastewater management in Brazil.* Environmental Technology, 2012. 33(14): p. 1677-1683.

[66] Huchzermeier, M.P. and W.D. Tao, *Overcoming Challenges to Struvite Recovery from Anaerobically Digested Dairy Manure.* Water Environment Research, 2012. 84(1): p. 34-41.

[67] Zhang, W.X. and A. Lau, *Reducing ammonia emission from poultry manure composting via struvite formation.* Journal of Chemical Technology and Biotechnology, 2007. 82(6): p. 598-602.

[68] Zeng, L. and X.M. Li, *Nutrient removal from anaerobically digested cattle manure by struvite precipitation.* Journal of Environmental Engineering and Science, 2006. 5(4): p. 285-294.

[69] Gasser, M.O., et al., *Plant-Available and Water-Soluble Phosphorus in Soils Amended with Separated Manure Solids.* Journal of Environmental Quality, 2012. 41(4): p. 1290-1300.

[70] Ackerman, J.N., et al., *Evaluation of manure-derived struvite as a phosphorus source for canola.* Canadian Journal of Plant Science, 2013. 93(3): p. 419-424.

[71] Schuiling, R. and A. Andrade, *Recovery of struvite from calf manure.* Environmental Technology, 1999. 20(7): p. 765-768.

[72] Moody, L., R. Burns, and K. Stalder, *Effect of anaerobic digestion on manure characteristics for phosphorus precipitation from swine waste.* Applied Engineering in Agriculture, 2009. 25(1): p. 95.

[73] Burns, R., et al., *Laboratory and in-situ reductions of soluble phosphorus in swine waste slurries.* Environmental Technology, 2001. 22(11): p. 1273-1278.

[74] Burns, R., et al., *Optimization of phosphorus precipitation from swine manure slurries to enhance recovery.* Water Science & Technology, 2003. 48(1): p. 139-146.

[75] Maqueda, C., J.L. Pérez Rodríguez, and J. Lebrato, *Study of struvite precipitation in anaerobic digesters.* Water Research, 1994. 28(2): p. 411-416.

[76] Miles, A. and T.G. Ellis, *Struvite precipitation potential for nutrient recovery from anaerobically treated wastes.* Water Science & Technology, 2001. 43(11): p. 259-266.

[77] Nelson, N.O., R.L. Mikkelsen, and D.L. Hesterberg, *Struvite precipitation in anaerobic swine lagoon liquid: effect of pH and Mg: P ratio and determination of rate constant.* Bioresource Technology, 2003. 89(3): p. 229-236.

[78] Suzuki, K., et al., *Recovery of phosphorous from swine wastewater through crystallization.* Bioresource technology, 2005. 96(14): p. 1544-1550.

[79] Suzuki, K., et al., *Removal and recovery of phosphorous from swine wastewater by demonstration crystallization reactor and struvite accumulation device.* Bioresource Technology, 2007. 98(8): p. 1573-1578.

[80] Wrigley, T., K. Webb, and H. Venkitachalm, *A laboratory study of struvite precipitation after anaerobic digestion of piggery wastes.* Bioresource technology, 1992. 41(2): p. 117-121.

[81] Yetilmezsoy, K. and Z. Sapci-Zengin, *Recovery of ammonium nitrogen from the effluent of UASB treating poultry manure wastewater by MAP precipitation as a slow release fertilizer.* Journal of Hazardous Materials, 2009. 166(1): p. 260-269.

[82] Massey, M., et al. *Struvite production from dairy wastewater and its potential as a fertilizer for organic production in calcareous soils. in International Symposium on Air Quality and Waste Management for Agriculture. CD-Rom Proceedings of the 16–19 September 2007, Conference (Broomfield, Colorado), USA. ASABE Publication.* 2007.

[83] Zeng, L. and X. Li, *Nutrient removal from anaerobically digested cattle manure by struvite precipitation.* Journal of Environmental Engineering and Science, 2006. 5(4): p. 285-294.

[84] Harris, W.G., et al., *Bench-scale recovery of phosphorus from flushed dairy manure wastewater.* Bioresource technology, 2008. 99(8): p. 3036-3043.

[85] Huchzermeier, M.P. and W. Tao, *Overcoming challenges to struvite recovery from anaerobically digested dairy manure.* Water Environment Research, 2012. 84(1): p. 34-41.

[86] Ohlinger, K.N., et al., *Kinetics effects on preferential struvite accumulation in wastewater.* Journal of Environmental Engineering, 1999. 125(8): p. 730-737.

[87] Bowers, K. and P. Westerman, *Design of cone-shaped fluidized bed struvite crystallizers for phosphorus removal from wastewater.* Transactions of the ASAE, 2005. 48(3): p. 1217-1226.

[88] Bowers, K. and P. Westerman, *Performance of cone-shaped fluidized bed struvite crystallizers in removing phosphorus from wastewater.* Transactions of the ASAE, 2005. 48(3): p. 1227.

[89] Shepherd, T.A., et al., *Development of a bench-scale air sparged continuous flow reactor for struvite precipitation from two different liquid swine manure storage systems.* Applied Engineering in Agriculture, 2009. 425(430): p. 425.

[90] Shepherd, T.A., et al., *Performance of a pilot-scale air sparged continuous flow reactor and hydrocyclone for struvite precipitation and removal from liquid swine manure.* Applied Engineering in Agriculture, 2009. 25(2): p. 257.

[91] Rahman, M.M., et al., *Production of slow release crystal fertilizer from wastewaters through struvite crystallization–A review.* Arabian Journal of Chemistry, 2014. 7(1): p. 139-155.

[92] Le Corre, K., et al., *Phosphorus recovery from wastewater by struvite crystallization: A review.* Critical Reviews in Environmental Science and Technology, 2009. 39(6): p. 433-477.

[93] Suzuki, K., et al., *Removal of phosphate, magnesium and calcium from swine wastewater through crystallization enhanced by aeration.* Water Research, 2002. 36(12): p. 2991-2998.

[94] Lee, S., et al., *Removal of nitrogen and phosphate from wastewater by addition of bittern.* Chemosphere, 2003. 51(4): p. 265-271.

[95] Shin, H. and S. Lee, *Removal of nutrients in wastewater by using magnesium salts.* Environmental technology, 1998. 19(3): p. 283-290.

[96] Stratful, I., M. Scrimshaw, and J. Lester, *Conditions influencing the precipitation of magnesium ammonium phosphate.* Water Research, 2001. 35(17): p. 4191-4199.

[97] Moriyama, K., et al., *Development of artificial seed crystal for crystallization of calcium phosphate.* Environmental technology, 2001. 22(11): p. 1245-1252.

[98] Chunjie Xia, J.Z.a.B.H. *Microbial oil accumulation via pelletized cultivation of Mucor circinelloides. in 33rd Symposium on Biotechnology for Fuels and Chemicals.* 2011. Seattle, WA.

[99] Oehmen, A., et al., *Advances in enhanced biological phosphorus removal: from micro to macro scale.* Water research, 2007. 41(11): p. 2271-2300.

[100] Bernet, N. and F. Béline, *Challenges and innovations on biological treatment of livestock effluents.* Bioresource technology, 2009. 100(22): p. 5431-5436.

[101] Yuan, Z., S. Pratt, and D.J. Batstone, *Phosphorus recovery from wastewater through microbial processes.* Current opinion in biotechnology, 2012. 23(6): p. 878-883.

[102] Martín, H.G., et al., *Metagenomic analysis of two enhanced biological phosphorus removal (EBPR) sludge communities.* Nature biotechnology, 2006. 24(10): p. 1263-1269.

[103] Pramanik, J., et al., *Development and validation of a flux-based stoichiometric model for enhanced biological phosphorus removal metabolism.* Water Research, 1999. 33(2): p. 462-476.

[104] Pereira, H., et al., *Model for carbon metabolism in biological phosphorus removal processes based on in vivo.* 1996, C-13-NMR labelling experiments. Water Reseach, 1996 (30): 2128-2138.

[105] Kang, S.J., et al., *A new procedure for screening feasibility of biological phosphorus removal for a wastewater.* Water Science & Technology, 1991. 23(4-6): p. 595-602.

[106] Park, J., et al., *A biological phosphorus removal potential test for wastewaters.* Water environment research, 2001: p. 374-382.

[107] Fernández, F., et al., *Agro-food wastewaters as external carbon source to enhance biological phosphorus removal.* Chemical Engineering Journal, 2011. 166(2): p. 559-567.

[108] Pijuan, M., et al., *Enhanced biological phosphorus removal in a sequencing batch reactor using propionate as the sole carbon source.* Biotechnology and bioengineering, 2004. 85(1): p. 56-67.

[109] Singh, M. and R. Srivastava, *Sequencing batch reactor technology for biological wastewater treatment: a review.* Asia-Pacific Journal of Chemical Engineering, 2011. 6(1): p. 3-13.

[110] Obaja, D., et al., *Nitrification, denitrification and biological phosphorus removal in piggery wastewater using a sequencing batch reactor.* Bioresource technology, 2003. 87(1): p. 103-111.

[111] Obaja, D., S. Mace, and J. Mata-Alvarez, *Biological nutrient removal by a sequencing batch reactor (SBR) using an internal organic carbon source in digested piggery wastewater.* Bioresource technology, 2005. 96(1): p. 7-14.

[112] Bortone, G., et al., *Nitrification, denitrification and biological phosphate removal in sequencing batch reactors treating piggery wastewater.* Water Science & Technology, 1992. 26(5-6): p. 977-985.

[113] Ogejo, J.A., et al., *Enhanced Biological Phosphorus Removal for Liquid Dairy Manure, in ASABE Annual International Meeting.* 2008: Providence, Rhode Island.

[114] Ahn, C.H., J.K. Park, and J. Wang, Changes in polyhydroxy-alkanoates (PHAs) during enhanced biological phosphorus removal with dairy industrial wastewater. Journal of Environmental Engineering, 2009. 135(11): p. 1213-1220.

[115] Qureshi, A., et al., *Dairy manure treatment, digestion and nutrient recovery as a phosphate fertilizer.* Journal of Environmental Science and Health Part B: Pesticides, Food Contaminants, and Agricultural Wastes, 2006. 41(7): p. 1221-1235.

[116] Shilton, A.N., N. Powell, and B. Guieysse, *Plant based phosphorus recovery from wastewater via algae and macrophytes.* Current opinion in biotechnology, 2012. 23(6): p. 884-889.

[117] Knuteson, S., T. Whitwell, and S. Klaine, *Influence of plant age and size on simazine toxicity and uptake.* Journal of environmental quality, 2002. 31(6): p. 2096-2103.

[118] Padmavathiamma, P.K. and L.Y. Li, *Phytoremediation technology: hyper-accumulation metals in plants.* Water, Air, and Soil Pollution, 2007. 184(1-4): p. 105-126.

[119] Gujarathi, N.P., B.J. Haney, and J.C. Linden, *Phytoremediation potential of Myriophyllum aquaticum and Pistia stratiotes to modify antibiotic growth promoters, tetracycline, and oxytetracycline, in aqueous wastewater systems.* International journal of phytoremediation, 2005. 7(2): p. 99-112.

[120] Jimenez-Perez, M., et al., *Growth and nutrient removal in free and immobilized planktonic green algae isolated from pig manure.* Enzyme and Microbial Technology, 2004. 34(5): p. 392-398.

[121] Wang, L., et al., *Anaerobic digested dairy manure as a nutrient supplement for cultivation of oil-rich green microalgae< i> Chlorella</i> sp.* Bioresource technology, 2010. 101(8): p. 2623-2628.

[122] Levine, R.B., M.S. Costanza-Robinson, and G.A. Spatafora, *< i> Neochloris oleoabundans</i> grown on anaerobically digested dairy manure for concomitant nutrient removal and biodiesel feedstock production.* Biomass and Bioenergy, 2011. 35(1): p. 40-49.

[123] Chisti, Y., *Constraints to commercialization of algal fuels.* Journal of biotechnology, 2013. 167(3): p. 201-214.

[124] Yang, J., et al., *Life-cycle analysis on biodiesel production from microalgae: water footprint and nutrients balance.* Bioresource technology, 2011. 102(1): p. 159-165.

[125] Li, Y., et al., *Characterization of a microalga< i> Chlorella</i> sp. well adapted to highly concentrated municipal wastewater for nutrient removal and biodiesel production.* Bioresource technology, 2011. 102(8): p. 5138-5144.

[126] Wang, L., et al., *Semi-continuous cultivation of Chlorella vulgaris for treating undigested and digested dairy manures.* Applied biochemistry and biotechnology, 2010. 162(8): p. 2324-2332.

[127] Farooq, W., et al., *Two-stage cultivation of two< i> Chlorella</i> sp. strains by simultaneous treatment of brewery wastewater and maximizing lipid productivity.* Bioresource technology, 2013. 132: p. 230-238.

[128] Costa, R., W. Medri, and C. Perdomo, *High-rate pond for treatment of piggery wastes.* Water Science & Technology, 2000. 42(10): p. 357-362.

[129] Goh, A., *Production of microalgae using pig waste as a substrate.* Algal Biomass Technologies. Cramer, Berlin, 1986: p. 235-244.

[130] Olguín, E.J., et al., *The effect of low light flux and nitrogen deficiency on the chemical composition of< i> Spirulina</i> sp.(< i> Arthrospira</i>) grown on digested pig waste.* Bioresource technology, 2001. 77(1): p. 19-24.

[131] Olguin, E., et al., *Production of Spirulina sp. in sea water supplemented with anaerobic effluents in outdoor raceways under temperate climatic conditions.* Applied microbiology and biotechnology, 1997. 48(2): p. 242-247.

[132] Asmare, A., B. Demessie, and G. Murthy, *Investigation of Microalgae Co-Cultures for Nutrient Recovery and Algal Biomass Production from Dairy Manure.* 2014.

[133] González, C., et al., *Efficient nutrient removal from swine manure in a tubular biofilm photo-bioreactor using algae-bacteria consortia.* 2008.

[134] Wilkie, A.C. and W.W. Mulbry, *Recovery of dairy manure nutrients by benthic freshwater algae.* Bioresource Technology, 2002. 84(1): p. 81-91.

[135] Kebede-Westhead, E., C. Pizarro, and W.W. Mulbry, *Treatment of dairy manure effluent using freshwater algae: elemental composition of algal biomass at different manure loading rates.* Journal of agricultural and food chemistry, 2004. 52(24): p. 7293-7296.

[136] Kebede-Westhead, E., C. Pizarro, and W.W. Mulbry, *Treatment of swine manure effluent using freshwater algae: production, nutrient recovery, and elemental composition of algal biomass at four effluent loading rates.* Journal of applied phycology, 2006. 18(1): p. 41-46.

[137] Becker, E., Micro-algae as a source of protein. Biotechnology advances, 2007. 25(2): p. 207-210.

[138] De la Noüe, J. and A. Basseres, *Biotreatment of anaerobically digested swine manure with microalgae*. Biological wastes, 1989. 29(1): p. 17-31.

[139] Ye, Y., *Biogas Production and Nutrient Recovery from Waste Streams, in Bioproducts and Biosystems Engineering*. 2013, University of Minnesota: St Paul.

[140] Johnson, M.B. and Z. Wen, *Development of an attached microalgal growth system for biofuel production*. Applied microbiology and biotechnology, 2010. 85(3): p. 525-534.

[141] Pizarro, C., et al., *An economic assessment of algal turf scrubber technology for treatment of dairy manure effluent*. ecological engineering, 2006. 26(4): p. 321-327.

[142] Spieles, D.J. and W.J. Mitsch, *The effects of season and hydrologic and chemical loading on nitrate retention in constructed wetlands: a comparison of low-and high-nutrient riverine systems*. Ecological Engineering, 1999. 14(1): p. 77-91.

[143] Steer, D., et al., *Efficiency of small constructed wetlands for subsurface treatment of single-family domestic effluent*. Ecological Engineering, 2002. 18(4): p. 429-440.

[144] Verhoeven, J.T. and A.F. Meuleman, *Wetlands for wastewater treatment: opportunities and limitations*. Ecological engineering, 1999. 12(1): p. 5-12.

[145] Boyd, C.E., Vascular aquatic plants for mineral nutrient removal from polluted waters. Economic Botany, 1970. 24(1): p. 95-103.

[146] Dodds, W.K., *Freshwater ecology: concepts and environmental applications*. 2002: Academic press.

[147] Boyd, C., *Utilization of aquatic plants, in Aquatic vegetation and its use and control*. 1982, UNIPUB. p. 107-15.

[148] Sooknah, R.D. and A.C. Wilkie, *Nutrient removal by floating aquatic macrophytes cultured in anaerobically digested flushed dairy manure wastewater*. Ecological Engineering, 2004. 22(1): p. 27-42.

[149] Xu, J. and G. Shen, *Growing duckweed in swine wastewater for nutrient recovery and biomass production*. Bioresource technology, 2011. 102(2): p. 848-853.

[150] Xu, J., et al., *Production of high-starch duckweed and its conversion to bioethan*ol. Biosystems engineering, 2011. 110(2): p. 67-72.

[151] Shen, G., et al., *Nitrogen removal pathways in shallow-water duckweed-based wastewater treatment systems*. J. Ecol. Rural Environ, 2006. 22(1): p. 42-47.

[152] Landesman, L., et al., *Modeling duckweed growth in wastewater treatment systems*. Livestock Research for Rural Development, 2005. 17(6): p. 61.

[153] Cheng, J.J. and A.M. Stomp, *Growing duckweed to recover nutrients from wastewaters and for production of fuel ethanol and animal feed*. Clean–Soil, Air, Water, 2009. 37(1): p. 17-26.

Chapter 25

Enhancing Biomass Utilization for Bioenergy — Crop Rotation Systems and Alternative Conversion Processes

Ronald Hatfield

Additional information is available at the end of the chapter

http://dx.doi.org/10.5772/59883

1. Introduction

With ever increasing global populations there is a rising demand for energy to support even modest changes in lifestyle. It has been recognized for some time now that with decreasing oil reserves on a global scale there is a need for alternative energy sources. Many of our needs for energy utilizing electricity can be met by alternatives to petroleum and coal-based power generation. Of particularly high potential is the efficient utilization of solar energy. According to Lewis and Nocera [1], the earth receives approximately 7000 times more energy from the sun than is utilized by all of mankind. There are several technologies that are being utilized, ranging from photovoltaic to focusing mirrors to super heat fluids for steam generation in the production of electricity. The continued development of these technologies, along with other types such as wind-driven turbines, geothermal, hydroelectric, and ocean wave motion for electricity production, will greatly lessen the demand on petroleum-based energy. However, a critical need is liquid fuels for transportation. The movement of people and goods over great distances is a vital part of the world economy.

Part of the answer may still lie in the utilization of solar energy; not in a direct manner to power vehicles (cars, trucks, trains, and airplanes), but what it has been doing for billions of years in providing energy to growing plants. Conversion of plant biomass to energy or the production of bio-based liquid fuels (biofuels) has received greater attention in the last couple of decades. Although there is a tremendous amount of potential energy stored in the total plant biomass as it goes through its normal life cycle, much of the current technology has focused on the utilization of grains (corn, cereals, and soybeans) or sugars from storage organs of specialty plants (sugar cane, sugar beets). This has allowed a rapid ramping up of liquid fuel production in the form of ethanol. The technology needed for this production was not something that required a lot of development, but was basically a matter of scale. After all the brewing industry has been utilizing this process for centuries. For corn grain and cereals, it is a matter of converting starch to glucose, a simple enzymatic process followed by the fermentation of glucose by yeast to ethanol. In the case of sugar cane or sugar beets, the same technology was already being utilized to efficiently remove the sugar (sucrose) from plant biomass and easily convert to sugars fermentable with yeast [2]. Even for the production of plant-derived biodiesel, the grains from oil-producing crops are pressed to release oils in which the fatty acids can be methyl- or ethyl–esterified, producing a suitable diesel alternative. Biodiesel lags well behind other types of biofuel production systems and seems to be focused primarily on the utilization of waste products from the food industry[3].

With current scenarios, the ethanol industry will have to compete with increasing demands on grains for feed and food [2]. A concern has been the diversion of land from food production to energy production and rightly so with increasing world populations. With this in mind, much attention has been directed to the conversion of cellulosic biomass to liquid fuels. This subject has been highly reviewed in the past few years, addressing a wide range of concerns and potential advantages. It is clear that crop residues will play a key role in meeting the projected total biomass needed to provide the amount of liquid fuel to meet the goal of replacing 30% of U.S petroleum consumption by 2030 [2]. Dedicated biofuel crops such as switchgrass and fast-growing poplar also figure prominently into meeting this goal. It is envisioned that the dedicated energy crops could be grown on marginal lands poorly suited for the high capacity needs of feed and food [4]. Recently Schmer et.al.,2008 [5]demonstrated that switchgrass grown in areas considered to be margin cropland could be an effective source of biomass for biofuels. It has been proposed that establishment of low input man made prairies could be an economical way of producing biomass for biofuels [6]. Although this could be a way to supply some of the required biomass it may fall well short of the amount needed per acre to make it a practical enterprise for harvest and transportation. Well-managed switchgrass plots on marginal croplands supplied higher estimated ethanol yields per acre (93% greater than poor management) [5]. Genetic improvement is a critical component to establish switch-grass as a major biomass source that can meet the demands for more biofuels [7]. It should be kept in mind that biofuel programs must fit into an agricultural system that maximizes the production potential of each acre of farmland while protecting the environment. In this respect switchgrass on marginal croplands could also provide a nutrient sink for nitrogen waste from animal production. Switchgrass needs little nitrogen input but as with any crop production increases with the application of nitrogen [5]. Well-managed switchgrass plots could extend the useful life of croplands no longer fit for typical row crop production. Perennial grasses such as switchgrass can provide runoff protection as buffer strips along streams and rivers to keep nutrients out of waterways and lakes, thus providing dual benefits.

Although there have been a wide range of crop residues proposed to contribute to the total biomass needed for biofuel production, corn stover would be the largest contributor. It has been estimated that corn stover would contribute as much as 20% of the total biomass requirement [2]. One of the concerns of removing crop residues is the long-term impact upon soils. Removing large portions of the residues leaves the soil surface vulnerable to wind and water erosion. Guidelines have been proposed for leaving sufficient biomass on the fields to keep this from becoming too much of a problem [8]. In addition, removing large portions of the biomass leads to a depletion of the soil organic carbon levels [9]. If sufficient amounts were left in place to meet these demands, this in turn would limit the amount of biomass for biofuel production [10]. With anticipated small profit margins, especially in the early going, there will be a temptation to remove more of the biomass, leaving the soils vulnerable to erosion and risking soil organic carbon depletion. Once these soils have reached high depletion levels, productivity will be severely restricted and returning them to better productivity will be a monumental task. Switching these lands to crops such as switchgrass that can do well in marginal soils would help the biofuels industry, but some of the most productive farmland for food and feed would be lost. This would most certainly sharpen the debate over land use for biofuels vs. food. No matter the approach it is clear multiple scenarios will need to be investigated to meet biomass for biofuel needs in a sustainable manner. The driving force behind future directions should be one of maintaining our existing high production lands while capturing increased value from lands that are should not be in continuous crop production. The challenge moving forward is to develop farming systems that are both economic and environmentally sustainable while meeting the increasing demands of food, feed, fiber, and now bioenergy. There is no doubt that crop residues, especially corn stocks, play a major part in making this vision a reality but as already pointed out it is walking a fine line between productivity and maintaining soil health.

2. The role of crop rotations

At one time crop rotations utilizing nitrogen fixing legumes were much more prevalent on the landscape due to the cost and availability of commercial fertilizers. With the availability of commercial fertilizers there was no longer a need for utilizing legume forages that are particularly good at fixing nitrogen to be used for subsequent crop production. In the most productive regions in the United States particularly the Midwest Breadbasket there is economic pressure to produce monocultures of crops such as corn. This is made possible due to the relatively cheap source of commercial nitrogen-based fertilizer [11] and to the development of pesticides and herbicides. The Haber-Bosch process to produce ammonia requires large amounts of energy and appropriate catalysts to complete the transformation of hydrogen and nitrogen into ammonia. The commercialization of this process has been referred to as the detonator for the world population explosion because lands could now produce much higher levels of food to support increased populations [12]. Although this has allowed increased grain production the cost of nitrogen fertilizers has increased nearly 8 to 14 fold from a low in early 1970s to 2013 (USDA-REE statistics, http://www.ers.usda.gov/dataproducts/ferti izer-use-and-price.aspx#.VDwPcOe9i-Q). Much of the increased cost of nitrogen based commercial fertilizers has been driven by rising energy costs not only for production of anhydrous ammonia but also for transportation. As fossil based fuels continue to become in greater demand and at some point become limiting the price of fertilizers will continue to go up (See fertilizer price trends USDA-REE statistics) putting greater pressure on the value of crops produced on each acre of land. An alternative is to find other methods of increasing soil fertility. In farming regions where animal production is an integral part of the farming system, animal waste provides a valuable nutrient source (e.g., dairy production). Although a good source of nitrogen based nutrients for crops, good management is critical to maintaining nutrient availability for crop production and preventing excessive soil erosion.

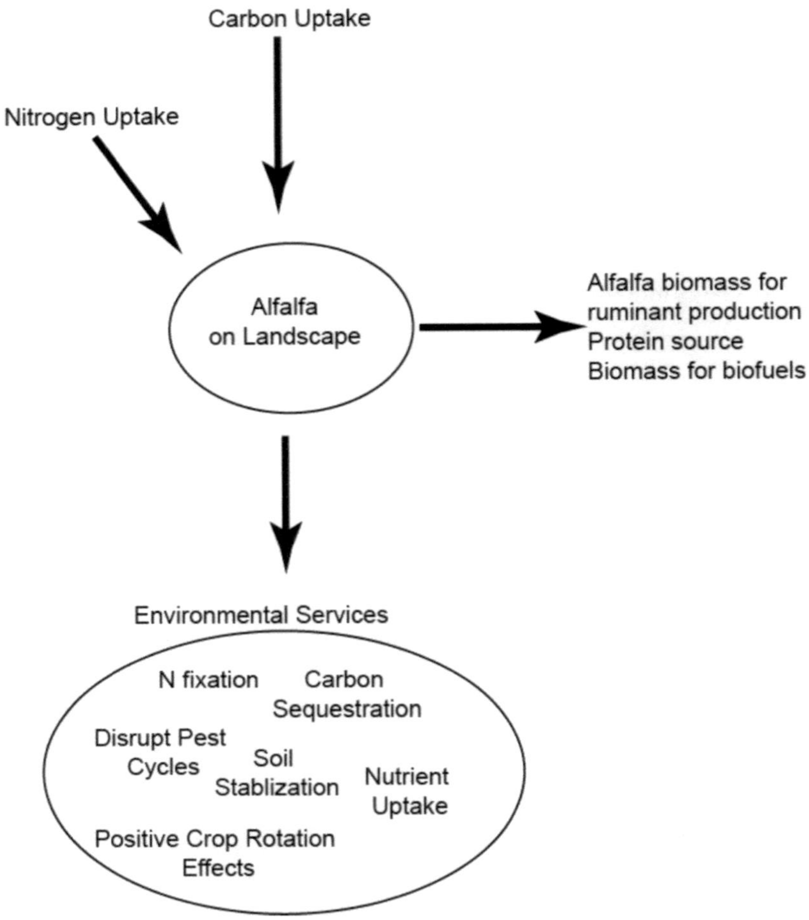

Figure 1. Diagram of alfalfa production with environmental and economic impacts. Alfalfa as a rich source of protein in its leaves can have multiple uses in terms uses as animal feedstuff. The high fiber stem fraction could be used for bioenergy production. There are also many benefits to the environment by including alfalfa into crop rotations to allow sustainable production systems.

Production of forage legumes in rotation with row crops provides opportunities for increasing nitrogen for crop production while stabilizing and improving the environment (Figure 1). In 2010, a workshop (organized by National Alfalfa & Forage Alliance, Pioneer, USDA-Agricultural Research Service, and the National Corn Growers Association) was held to discuss the feasibility and benefits of establishing alfalfa-corn rotations to meet food and feed demands, as well as providing biomass for biofuel production (proceedings available online: www.alfalfa-forage.org). Workshop attendees evaluated the feasibility of using crop rotations to maintain soil fertility while providing sufficient biomass for biofuel production. Jung reported [13] alfalfa (*Medicago sativa* L.) is a deep-rooted perennial legume forage typically used as a feed source for ruminant animal production. Because of its high capacity to fix nitrogen, there is no need for the addition of nitrogen fertilizer for its own growth. Nitrogen stored in the roots after two years of growth would be sufficient to supply approximately 75% of the next two years of corn production [13]. This result would have several positive environmental impacts: 1) decreased greenhouse gas emissions from reduced dependence upon commercial fertilizers; 2) reduced soil erosion; 3) reduced nutrient run-off; and 4) improved carbon sequestration [13]. A potential advantage of such a rotation system would be the accumulation of soil organic carbon if proper soil/plant management was put into place [14] (Figure 1). However, Baker [15] cautions that assessing changes in soil organic carbon is not easy in a rotation system due to the relatively short duration of the alfalfa in its rotation sequence especially in the early years of adaption of such a farming system. Having the organic matter incorporated into the soil already in the form of extensive root systems eliminates the need for soil tillage to assist in moving organic matter in crop residue to the soil biome.

Accumulation of fixed nitrogen in alfalfa is substantial (152 kg N ha^{-1} over a range of environments and soil types) [16]. This decreases the need for application of commercial fertilizer that is dependent upon fossil fuels in the form of methane for production. As a perennial legume, alfalfa's early spring growth as well as late fall growth provides cover for soils when row crops would be planted and after harvest when soils are most vulnerable to erosion. This does not remove the need for good management practices during the corn production part of the cycle; the severity is greatly reduced over a continual corn or corn-soybean rotation. According to Vadas et.al., [17] alfalfa-corn rotations for bioenergy production can have significant advantages mostly in terms of efficiency of energy production and decreased soil erosion and less nitrogen leaching compared to continuous corn. The bottom line was continuous corn had the greatest production costs but also had the greatest profit potential. This is not assigning a cost to the soil erosion. Scientists at the U.S. Dairy Forage Research Center in conjunction with University of Wisconsin-Madison researchers Grabber, Renz, and Lauer have shown that interseeding alfalfa with corn can double the first-year yields from the alfalfa [18]. Such a practice would insure cover-crop availability once the corn is harvested and would provide a jumpstart on the production of alfalfa the following spring [19]. The use of alfalfa as a cover crop would appear to have some drag on total corn production during the establishment year but alfalfa production would to significantly increased during the first full year of production. Most importantly the soil would be better protected during the last year of corn production and during the alfalfa establishment decreasing soil erosion potential during alfalfa establishment. Additionally since alfalfa is a deep-rooted perennial it can recover nitrogen that has leached beyond the limited root zone of corn, helping prevent further leaching and contamination of ground water.

In the early 90s (1993 to 2000) a pilot program was initiated to test the feasibility of alfalfa-corn rotation for energy production [13]. The alliance involved the University of Minnesota, USDA-Agricultural Research Service, Minnesota Valley Alfalfa Producers, and the DOE. The proposed system utilized dry baled alfalfa from which stems were mechanically separated from the leaves creating two feedstock components; one being the high fiber stems for energy production and the other leaf meal as a high protein fraction. Feeding trials with the alfalfa leaf meal found that it could successfully replace other protein sources such as soybean meal in diets of calves, dairy cows, and feedlot steers [13]. Although the early work indicated feasibility and advantages of alfalfa-corn rotations in a bioenergy production system the project fell apart before it could move to the next stages of testing and the project abandoned. However, these initial results indicated an existing infrastructure for handling alfalfa that could be easily adapted to a biofuel production program.

There is no doubt that rotation of corn and alfalfa would have significant environmental benefits over continuous corn. What is the economic and environmental impact upon available biomass for biofuels and the need for feed and food? Alfalfa leaves can contain as much as 30% or more protein as a fraction of the total dry matter. Typically during plant development, the stem becomes an increasing proportion of the total biomass; being lower in protein, the total plant protein decreases [20]. Harvesting schemes currently in place requires cutting the alfalfa at early-bud stage of development to keep the fiber content as low as possible and the protein content as high as possible. The down side to this harvesting practice is the need for frequent trips over the field to catch plant development at the early-bud stage. This may be reasonable for feed production for ruminant animals, but does not lend itself to practices that would be widely adopted in corn-alfalfa rotations. However, due to the high protein content of the leaves, separation of leaves from stems results in a rich source of protein for a potentially wide range of uses (Figure 2).

Earlier work using a dry fractionation system to separate leaves from stems resulted in an alfalfa leaf meal (pellets) with an estimated value of $200/ton [21]. However, there are few, if any, existing processing plants in North America today to determine if the value would be more or less than this predicted value [22]. A newly proposed system for harvesting alfalfa separates the leaves from the stems as they are harvested in the field, producing two components.

One fraction is rich in protein (leaves) and the other is rich in fiber (stems) [23]. The leaf fraction could be used in a wide range of applications including direct ensiling for high-protein feed, or dehydrated as alfalfa meal or other value-added products requiring high-protein materials [22]. The stems could be used as a source of biomass for biofuel production or for feed depending upon the needs of fiber in the ruminants diet. Because the alfalfa leaf does not change appreciably in protein content over the development of the plant, harvest can be delayed to allow greater amounts of total biomass accumulation [24]. According to Shinners, the advantages of field harvesting and fractionation include 1) production of a high-value protein fraction that avoids losses due to weather, 2) fractionation occurs at harvest so no further processing steps or equipment are needed, 3) capital costs of fractionation equipment are low, 4) fractionation occurs on the farm so only the desired fractions need leave the farm, and 5) ruminant feeds can be recombined to produce high-quality rations[22]. This system would provide an alternative to the harvesting/marketing system that is available today for

http://dx.doi.org/10.5772/59883

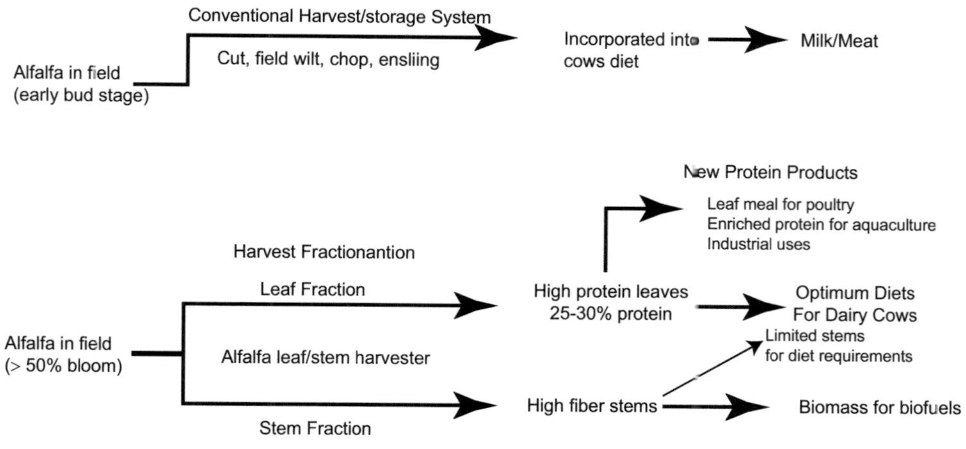

Figure 2. A comparison of the conventional harvest system for alfalfa compared to the proposed system of harvest and fractionation of leaves and stems into two component streams. This harvest system creates a high protein fraction and a high fiber fraction that allows better utilization of materials grown to fit specific needs whether it is animal feed or high fiber material for biofuels production. It is envisioned that the high protein leaf fraction could be utilized for a wide range of different animal production systems from dairy cows to poultry to enriched protein for aquaculture. The stems would be used for meeting fiber needs of ruminants (less than what would actually be produced per acre) to providing a feedstock for biofuels.

alfalfa and may provide the farmer with a cash crop incentive to produce more alfalfa in conjunction with corn (See Figure 3).

It is envisioned harvesting alfalfa using in field fractionation creates two product streams to enhance the total value of the alfalfa crop. Prototype machines have been built to effectively remove the leaves from stems creating two alfalfa components at harvest [23]. One of the real advantages of this type of harvest system is the ability to open the harvest window to avoid bad weather and to decrease the total number of harvests. A prototype leaf stripper was used to harvest alfalfa leaves and stems during the summer of 2013 to test the feasibility of creating high quality diets for dairy cows when harvesting late in plant development (full bloom stage). The idea is to decrease the number of harvests per season to limit production costs, but be able to recombine the two fractions in appropriate amounts of stems and leaves to meet the needs of a high producing dairy cow. Results of feeding trial indicated total milk production and quality of the milk remained the same and excess stems could be used for other applications such as biofuel production [25]. Although this was centered around a feeding trial it demonstrated the feasibility of having a viable harvest system that creates two value components from the alfalfa plant. Energy inputs into such a harvest system are less than what is required under the normal production scenarios [22]. Separation of leaves from the stems also allows additional in field processing to render the stems more digestible. Maceration breaks the stem material open allowing easier access of enzymes or microbes to enhance degradability/ digestibility [26]. Processing the stems separately from the leaves does not risk the loss of protein from the leaf due to juicing this material during the maceration process. Hence the high protein fraction is preserved and the high fiber fraction is processed in the field requiring less post harvest processing at the biofuel production sites.

(A) (B)

Figure 3. Prototype alfalfa leaf stripper. A. Process of stripping the leaf fraction from alfalfa plants. In this prototype machine, harvesting stems was a separate activity from harvesting of the leaf fraction. The stem fraction was left standing in the field until leaves had been removed and then stems were cut and chopped for ensiling. Next generation harvesters would combine these two operations into a single pass over the field. B. Alfalfa stems with 80-90% of the leaves removed.

The genetic make up of alfalfa has been studied over the past 20 years to maximize quality and digestibility. A key component of this research in the past has been genetic selection for alfalfa germplasm that can withstand frequent cuttings as opposed to the accumulation of large amounts of biomass. Now there is interest to exploit the genetic potential to increase more biomass then is currently available for alfalfa. Efforts to genetically select for a biomass-type alfalfa that produces larger stems and more branching with greater total yields has been successful[13, 24, 27]. According to Lamb et.al.,[24, 27] alfalfa genetically selected for increased biomass production and managed to maximize yields resulted in a 40% increase in tons per acre. Revised management techniques amounted to decreased stand density providing more space for individual plant growth and development coupled with a delayed harvest i.e., switching from early bud stage to plants at 50% bloom or later. This provides the biomass alfalfa plant to accumulate higher amounts of total plant material, both leaves and stems. With the larger more robust stems lodging is minimized compared to the typical hay type alfalfa [13]. Coupled with a new harvesting technique of in-field fractionation, this could improve the amount of biomass for biofuels while still producing a high-protein fraction for value-added products. The theoretical ethanol yield for alfalfa stems would be 137 gal/acre compared to 174 gal/acre for corn stover assuming only half of the stover is removed to maintain soil health and long term productivity[13]. Including the grain for ethanol production (473 gal/acre), corn far outpaces the amount of ethanol potential from alfalfa. However, the estimated protein yield per acre would be 0.49 tons/acre for alfalfa leaves, zero for the corn stover and 0.34 tons/acre for corn grain [13]. In the face of growing world populations protein production will be of increasing concern. In terms of outright biomass production, the system of crop rotations between corn and alfalfa lags behind year after year of corn production. From an economic perspective alfalfa-corn rotations provide several advantages in the corn production following alflalfa; 1) yield benefit of $30 to 60/acre, 2) lower fertilizer nitrogen inputs required (2 year time frame) $75 to 150/acre, and 3) no insecticide required the first of corn production $15/acre [13]. This results in an accumulative savings potential of $120 to 225/acre. The rotation system does provide for a more sustainable system, both from an environmental and economic standpoint, primarily from decreasing the application of commercial fertilizers by 75% over two years of production. These economic values do not take in to account the impact of carbon sequestration that would help offset aggressive removal of corn stover during that phase of the rotation cycle.

3. Alternatives for biofuel production

Current technologies rely primarily on the yeast-ethanol platform to create liquid fuels. The process has been well studied and continues to undergo development to utilize more of the cell wall sugars in addition to the cellulosic glucose. Much of the current biofuel industry is based on yeast fermentation of glucose that is derived from starch primarily from corn grain, although any cereal grain could be used. Brazil has adopted a slightly different approach and has based much of its ethanol production on sugarcane using yeast fermentation. These systems are not sustainable in the long run due to ever increasing populations with increasing demands for food. Capturing biomass for conversion to biofuels is a big part of the vision for decreasing dependence upon fossil fuels. Biomass to biofuels does not directly compete with production needs for food and feed and provides opportunities to maximize utilization of our landscape in ways that are sustainable and improves productivity. However, converting biomass to biofuels efficiently is a critical part of the story.

At this time ethanol production is the main form of biofuel product proposed for biomass[1-2]. This system utilizes yeast-based fermentation using primarily glucose as the substrate for ethanol production. The challenge in using corn stover or any other source of biomass in this process is the complexity of the plant cell wall. Cell walls are complex matrices composed of largely of cellulose microfibrils embedded in a matrix of structural polysaccharides. Once cell walls have reach their maximum size lignification occurs producing a hydrophobic polymer that drives the water from free spaces within the wall as it fills in these open areas (Figure 4) imparting additional strengthen to the wall. This process creates regions within the cell wall that are difficult to hydrolyze especially once the wall has been dried. A comparison of alfalfa stem cell wall composition with that of corn stover provides similar proportions of glucose on a kilogram of dry matter basis (Table 1). To render the glucose available for fermentation current technologies for ethanol production rely heavily on pretreatments to release sufficient amounts of the cellulosic portion of the wall for enzymatic conversion to glucose [28]. Pre-treatments are designed to disrupt the cell wall matrix allowing cellulytic enzymes access to the cellulosic components while minimizing the formation of degradation products. Typically dilute acids combined with high temperatures are the most common form of biomass pre-treatment [28]. In the case of grasses pretreatments effectively disrupt cross-linking of cell wall

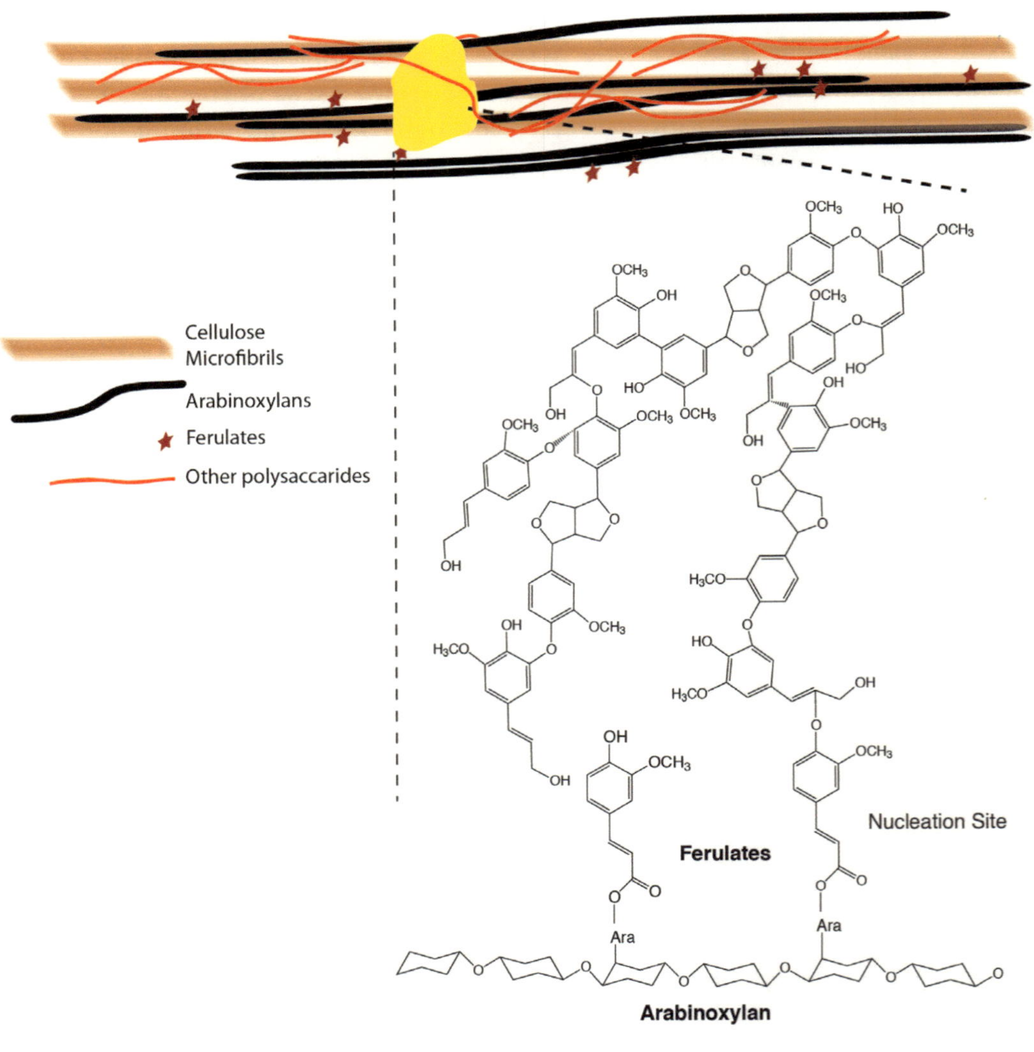

Figure 4. Cell wall model showing formation of lignin in grass wall matrix. Lignin in grasses is attached to ferulates that are shuttled out into the wall attached to arabinosyl side chains of arabinoxylans. This creates a tightly integrated wall matrix of lignin with wall structural polysaccharides. Similar cross-linking most likely occurs in dicot walls except the ferulates are not likely to be the most prominent anchor points to the wall carbohydrates. Treatment of walls with hot dilute acid solutions removes most of the non-cellulosic polysaccharides opening up the matrix to be more easily degraded by the addition of cellulosic enzyme cocktails.

arabinoxylans via ferulate dimers and to lignin via ferulate bridges (Figure 4) [29]. Acid treatments easily hydrolyzed arabinofuranose side chains of arabinoxylans, including those with attached ferulates allowing the wall to relax and expand for easier access by wall hydrolyzing enzymes. Treatment of alfalfa stems with low levels of acid during ensiling increased the amount of ethanol that could be produced [30]. However, best ethanol production was obtained after washing stem material after the acid treatment to remove degradation products that would interfere with yeast fermentation. A problem with acid hydrolysis of cell

Cell wall Component	Alfalfa Stem (N=153)	Corn	
		Stover (N=32)	Cob (N=56)
	% Dry Matter		
Glucose	18-37	23-34	20-33
Other Hexoses	21-41	26-36	23-34
Xylose	5-13	15-23	18-33
Other Pentoses	6-15	18-27	22-35
Lignin	7-22	6-12	3-15

Table 1. Cell wall composition of alfalfa stems compared to corn stover and corncobs. Other hexoses include the C6 sugars galactose and mannose and other pentoses refers primarily to the C5 sugar arabinose. Data from [13] and [55].

walls especially at high temperatures is the production of furfurals that inhibit yeast. The advantage of coupling dilute acid with ensiling is avoiding the need for high temperatures. Instead utilizing the longer-term storage of the biomass to allow limited degradation of the polysaccharides while minimizing the formation furfurals and other degradation products[30]. There may be highly effective means of solubilizing the cell wall (e.g., complete acid hydrolysis of all cell wall polysaccharides to monomeric sugars), but such methods are prohibitively expensive or make it difficult to remove byproducts. To prevent unwanted microbial fermentation of the released sugar, yeast-based fermentation must be maintained in a sterile environment. Providing and maintaining a sterile environment must be factored into the sequence of events from pretreatment to fermentation; it can be achieved, but at an additional cost to the overall process. From a utilization of the total biomass standpoint yeast fermentation leaves a 20 to 40% of potentially fermentable carbohydrates behind (Table 1) simply because yeast cannot deal effectively with them. This leaves a good deal of potential energy forming material off the table.

Ethanol is not the only biofuel under consideration as a product for biomass. Alternative systems for the conversion of biomass to biofuel are the syngas platform (details of this system can be found on the National Renewable Energy Laboratory website: www.nrel.gov/biomass/biorefinery.html) and the carboxylate platform. The syngas platform requires large inputs of energy to produce effective amounts of a useful biofuel. The carboxylate platform requires undefined mixed bacterial cultures under anaerobic conditions [31] (Figure 5). One of the big advantages of this system is the flexibility of the undefined mixed bacterial cultures to handle a wide range of substrates going into the system. More importantly they do not require a sterile environment in which to function. Popular sources of mixed anaerobic cultures are sewage sludge digesters and marine sediments[31-32]. The carboxylate platform works by the process of anaerobic degradation of carbohydrates to produce volatile fatty acids primarily acetic (C2), propionic (C3), and butyric (C4) acids although other VFAs can be produced.

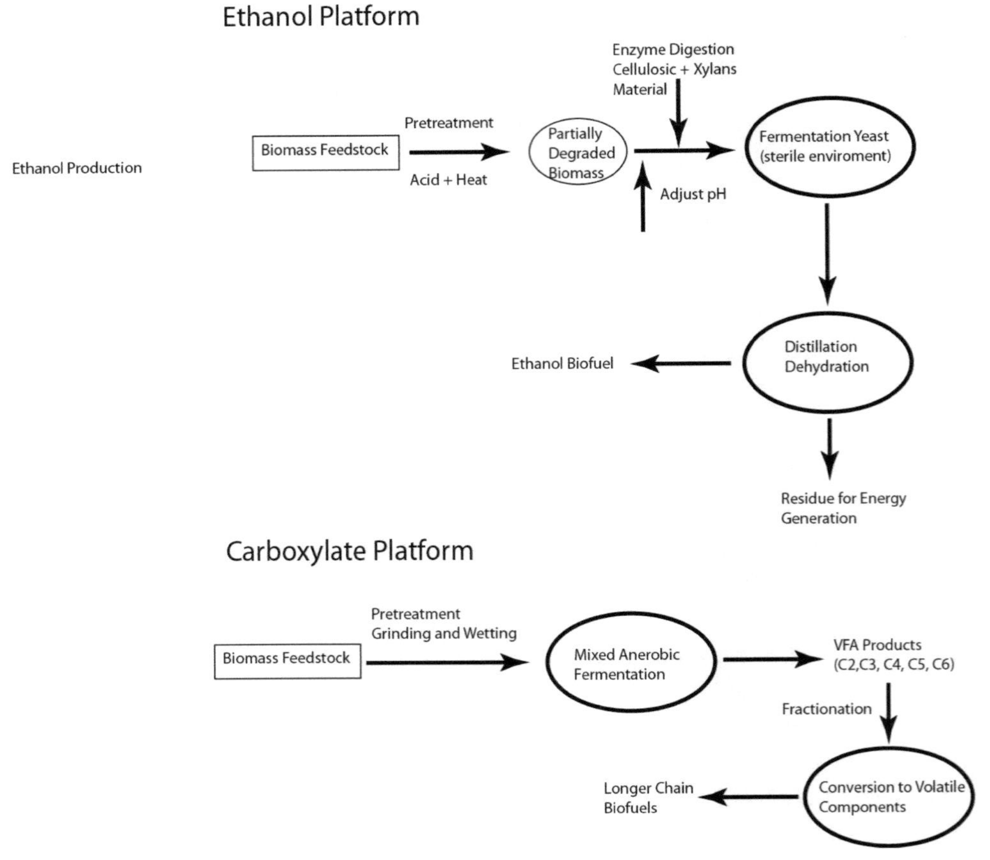

Figure 5. Schematic of biofuel production systems. Ethanol platform is the typical process proposed for conversion of plant biomass to liquid fuels. The carboxylate platform is an alternative method of producing liquid fuels using mixed anaerobic bacteria (ruminal microbes have advantages over typical systems utilizing sewage sludge or marine sediments) to produce VFAs for conversion to volatile fuel components. Cultures can be manipulated to produce higher ratios of C5, C6 and C7 VFAs for more energy dense molecules.

An advantage of the carboxylate platform is the general low inputs needed to obtain materials that can be modified to produce biofuels or bio-refinery products. Pre-treatments are minimized and may be confined to particle size reduction or mild chemical treatments providing the greatest advantages[31]. Most importantly the carboxylate platform does not require an antiseptic environment in which to operate, greatly simplifying handling of raw materials going into digesters. Significant work has been done on carboxylate platforms utilizing mixed cultures from sewage sludge treatments [31, 33]. Such systems have a great deal of flexibility when it comes to handling a wide range and complexity of crop residues or other carbon based materials from agricultural practices. These organic materials may be relatively abundant and of relative low value in their present form before fermentation to VFAs. A disadvantage of the sewage sludge inoculum is the general slow conversion rate and methanogens producing large amounts of methane[31]. In the case of manure or other organic waste digesters where time is

http://dx.doi.org/10.5772/59883

not a limiting factor this is quite acceptable and the methane can be easily captured and used as an energy source. With the right type of microbial mix, it is possible to produce longer-chain carboxylates caproate (C6) and caprylate (C8) from acetate in addition to the typical acetate, propionate, and butyrate through a process referred to as reverse β-oxidation[34]. The potential down side of this approach is the process tends to be slow and requires inhibition of methanogens to force the system to produce larger quantities of the longer-chain VFAs, e.g., n-caproate (C6) and n-caprylate (C8). Inhibition of methanogens can be efficiently achieved with compounds like bromoethane sulfonic acid, but this is relatively expensive and would be prohibitive on a large scale[31].

An alternative source of anaerobic microbes for the carboxylate platform for the conversion of plant biomass would be the cow's rumen. In comparison to waste stream anaerobic microbes, the rumen is a more specialized system having evolved to extract nutrient value out of a wide range of plant materials [35]. Although cell wall degradation and total feed utilization by dairy and beef cows can be improved, the microbial community in these ruminants has evolved to degrade fibrous plant material relatively quickly to supply needed nutrients to the animal [36]. The rumen is a mixed culture of anaerobic organisms effectively degrades carbohydrates, proteins, and fats present in feed mixtures to produce short-chain VFAs. The efficiency of this ruminal system appears to be much greater than what is in the typical waste stream systems[37]. The advantage of a ruminant-based carboxylate platform is the ability to degrade all the organic materials (polysaccharides, proteins, fats, and oils) with the exception of the lignin within short time periods of 24-72 hours. High producing ruminants like the dairy cow must be able to extract sufficient energy from feed materials within 48 hours to support her maintenance and milk production. Cow ruminant microbial communities have evolved over time to handle a diversity of substrates (i.e., easily degraded starch to more recalcitrant fiber materials). Ruminal microbial communities are quite complex with redundancy in the types of hydrolytic abilities that may come into play as substrates change coming into the cow [36]. Due to the relatively short incubation times slower growing acetogens (convert C3-C6 VFAs to acetate) and the methanogens (convert acetate to methane) do not have a chance to become well established. This in turn restricts methane production (8-15% of total energy) in this type of carboxylate platform avoiding the need to add specific methane inhibitors [36]. The small amount of methane that is produced could be captured and utilized as an energy input to maintain incubation temperatures.

Recently Weimer et.al., 2014 [38] demonstrated the ability of rumen microbial cultures to produce large amounts of valeric and caproic in short time periods of 48-72 hour incubations. It has been demonstrated that the addition of dilute amounts of ethanol to mixed culture fermentations in the carboxylate platform results in the extension of the short chain VFAs to medium length molecules thus capturing the fuel value of ethanol in a form that could be more easily recovered [34, 39]. What is unique and promising about the work of Weimer et.al., is the ability to speed up this process using ruminal mixed culture fermentations as opposed to the typical source of sewage digesters [38]. In addition they found that supplementing the mixture with ruminal derived *Clostridiumum kluyveri* an ethanol-utilizing bacteria resulted in production levels of 4.9-6.1 g/L of caproate in 48-72 hours using either switchgrass or alfalfa stems as the substrate. The level of caproate production seen by the Weimer group is similar to what others have achieved [34, 40], but in a 10 to 30 times less time frame for incubation. Being able to generate longer VFAs increases the energy density in each molecule increasing the value of the material for liquid fuels. In addition, the longer chain VFAs are easier to extract from the fermentation media decreasing recovery costs[38-39]. For any biomass to biofuel production process a key element is being able to produce sufficient amounts of fuel molecules in short periods of time and with limited inputs. The carboxylate platform based on ruminal microbes supplemented with additional strains of more specialized bacteria (e.g., *Clostridiumum kluyveri*) appears to hold a great deal of promise for biomass conversion. Little sample preparation was needed to treat the switchgrass and alfalfa stems for biofuel production using the ruminal microbial system. The fermentation process described here could be combined with other platforms that produce ethanol. For example concept of consolidated bioprocessing (CBP) [36, 41] is considered as a possible avenue for the production of ethanol from biomass to avoid the need for the addition of expensive hydrolytic enzymes. In most cases the CBP system does not produce sufficient ethanol to be cost effective [41]. However, coupled with a ruminal microbial based carboxylate platform the limited ethanol production could be effectively utilized to produce longer chain VFAs increasing energy density of each molecule [38].

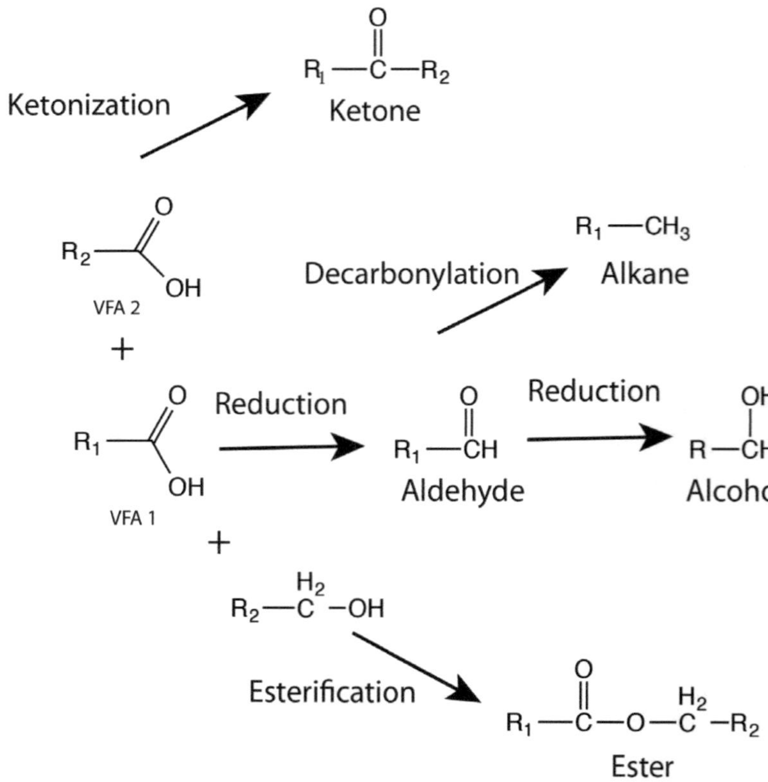

Figure 6. Multiple pathways for converting VFAs to volatile compounds that can serve as biofuels or as intermediates for the formation of additional organic compounds.

http://dx.doi.org/10.5772/59883

Volatile fatty acids must be converted to a form that increases their volatility to be good energy molecules. The medium length VFAs can be recovered by extraction [42] to allow additional modifications. Conversion of VFAs can be accomplished in different ways depending upon the tis desirable end product and its potential use. Possible conversion practices could utilize pure cultures of specific bacteria, electrochemical and thermochemical process. Useful end products that could be used for energy, solvents, or other biorefinery intermediates include ketones, aldehydes, alcohols, and alkanes (Figure 6). Due to the flexibility in the type of end product there are several avenues available to reach the desired outcome. Conversion process can be accomplished in a multitude of different ways using a single or multiple steps to reach desired products. Products such as ketones from VFAs using catalytic coupling [43] or ketones and secondary alcohols as produced in the MixAlco process [33]. The formation of volatile esters can be formed as demonstrated by Lange et.al., [44],Levy et.al., [45] or using microbial systems [46]. Production of alkanes can be achieved by decarboxylation of using pure cultures of microbes [47] or the use of electrochemical process using the Kolbe and/or the Hoefer-Moest processes [48]. The conversion of VFAs especially the medium length (C4-C6) increases volatility and at the same time decreases miscibility with water improving extraction process to isolate the biofuel molecules. The added advantage of VFA production (C2-C6 or longer) coupled with conversion technologies is the flexibility to produce a wide range of molecules that can be used for higher energy density fuel molecules or as starting molecules for other organic materials.

Typically biomass to biofuel systems are envisioned with a centrally located processing plant to handle large amounts of biomass. Unlike the grain ethanol production systems in which the grain is of relatively high density in terms of potential energy per volume, biomass tends to be much bulkier unless it is pelletized to increase bulk density [49]. When one is considering the utilization of corn stover and/or alfalfa stems these materials can be field processed into relatively high-density bales to improve the efficiency of shipping [50]. This is just one step in the complete process of collecting and moving biomass to centralized points for conversion to biofuels [51]. The challenge is keeping the collection, improving bulk density, and transportation costs to minimal levels to help final economic returns and the minimizing the carbon footprint associated with biomass to biofuels[50]. Perhaps it would be feasible to consider on farm conversion at least for the initial steps of the conversion process. In this scenario the harvested plant material (corn stover, alfalfa stems, switchgrass, etc) would be stored on the farm more with an ensiling process compared to dry storage. This provides an opportunity to add enzymes or dilute chemicals to enhance the subsequent digestion of the materials. Size reduction could also be incorporated into the process and storing materials wet eliminates the need for rehydration for fermentation. It could be envisioned that small on farm digesters could be used to process the biomass materials to produce VFAs (select additions of pure cultures and ethanol to create products for special uses) that would be recovered and transported to conversion sites. Processing on farm eliminates the need for consolidating biomass for shipment to centralized processing plants and open opportunities for other types of storage that could enhance conversion efficiency. Recovery of the VFAs or conversion on site to intermediates followed by extraction results in a improvements in energy density and allows materials to be shipped greater distances for further processing into molecules that provide the greatest benefit either as biofuels or as precursors for other organic based materials.

One of the challenges of any biomass conversion platform is dealing with the fermentation residual materials. Lignin is a primary component of the fermentation waste and in many schemes it is recovered and burned to supply energy for other steps in the complete process. With the carboxylate platform based upon mixed ruminal microbes, one of the by products could be the microbial protein as a value-added material. In the normal rumination process, formation of microbial protein is an important component to supply needed protein to the animal. In dairy production, microbial protein helps supply critical amino acids required for milk production, especially methionine and lysine that are often low or lacking in many forage-based diets [52]. Harvesting the microbial protein after biomass conversion to biofuels could provide an important protein supplement for dairy cow diets that is enriched in methionine and lysine. The microbial proteins would be insoluble along with the typical insoluble materials, i.e., lignin and other cell wall components. Recovery of these insoluble materials would be relatively straightforward. As an alternative the lignin-microbial-carbohydrate residue from the fermentation process could be used to replace phenolic-formaldehyde based adhesives[53]. Many of the ruminal microbes contain glycocalyx materials surrounding the individual cells that help them adhere to plant materials during digestion. The glycocalyx is a glycoprotein-polysaccharide complex that surrounds the cell membrane of some bacteria[54]. It has also been demonstrated that the lignin-microbial residues from ruminal fermentations, as proposed for the carboxylate platform, could be used to replace phenol-formaldehyde compounds as adhesives in the production of plywood composites[53]. Up to 70% of the typical phenol-formaldehyde formulation could be replaced by the more environmentally friendly residues that are byproducts of ruminal-based fermentations. Even if it would not be possible to replace all of the phenol-formaldehyde adhesive, decreasing significant amounts of this material would provide for healthier composites by decreasing the amount of formaldehyde outgassing that are a human health concern[53]. Key to the effectiveness of fermentation residues is creating the correct balance of lignin, the blend of rumen microbes and the types of glycocalyx material, and other minor phenolic materials in the plant materials.

4. Conclusion

This chapter is not meant to be a comprehensive assessment of biomass to biofuels, but rather a look at unconventional approaches that would enhance the sustainability of the entire process. To meet the goals of biofuel production by 2030 will require optimizing land use for food, feed, and bioenergy production. It should be approached from a standpoint of developing a viable biofuel production system that increases the amount of energy stored in the molecules making up the biofuels, i.e., longer-chain molecules, more energy per unit of fuel. To be sustainable into the future we must be willing to develop alternative systems that supply a range of biomaterials. Although the producing energy alternatives is of major concern at the present time we should be evaluating and developing bioenergy systems that allow flexibility not only in terms of feedstock going in, but the products coming out. Development of biomass to biofuels systems should look at how we can maximize the value of the total process, that is, optimize land use, embrace farming systems that decrease or eliminate soil/nutrient losses, improve economics of production, utilization of value-added products, and total energy production versus inputs. The entire process must also be sustainable from an environmental standpoint and provide economic advantages to the producer. Our vision into the future should be one of maximizing the productivity of each acre of farmland while meeting the needs for feed, food, and energy along with improving the soil for future generations. Decisions made today should not be overly influenced solely by short term economic gains.

Author details

Ronald Hatfield*

Address all correspondence to: ronald.hatfield@ars.usda.gov

USDA-Agricultural Research Service, U.S. Dairy Forage Research Center, Madison, WI, USA

References

[1] Lewis, N. S.; Nocera, D. G., Powering the planet: Chemical challenges in solar energy utilization. *Proceedings of the National Academy of Sciences of the United States of America* 2006, 103, (43), 15729-15735.

[2] Dhugga, K. S., Maize biomass yield and composition for biofuels. *Crop Science* 2007, 47, (6), 2211-2227.

[3] Young, H.; Sommerville, C., Developemnt of feedstocks for cellulosic biofuels. *F1000 Reports Biology* 2012, 4, 10.

[4] Somerville, C.; Youngs, H.; Taylor, C.; Davis, S. C.; Long, S. P., Feedstocks for Ligno-cellulosic Biofuels. *Science* 2010, 329, (5993), 790-792.

[5] Schmer, M. R.; Vogel, K. P.; Mitchell, R. B.; Perrin, R. K., Net energy of cellulosic ethanol from switchgrass. *Proceedings of the National Academy of Sciences of the United States of America* 2008, 105, (2), 464-469.

[6] Tilman, D.; Hill, J.; Lehman, C., Carbon-negative biofuels from low-input high-diversity grassland biomass. *Science* 2006, 314, (5805), 1598-1600.

[7] Casler, M. D.; Vogel, K. P., Selection for Biomass Yield in Upland, Lowland, and Hybrid Switchgrass. *Crop Science* 2014, 54, (2), 626-636; bPrice, D. L.; Casler, M. D., Divergent Selection for Secondary Traits in Upland Tetraploid Switchgrass and Effects on Sward Biomass Yield. *Bioenergy Research* 2014, 7, (1), 329-337; cPrice, D. L.; Casler,

M. D., Predictive Relationships between Plant Morphological Traits and Biomass Yield in Switchgrass. *Crop Science* 2014, 54, (2), 637-645.

[8] Wilhelm, W. W.; Johnson, J. M. F.; Hatfield, J. L.; Voorhees, W. B.; Linden, D. R., Crop and soil productivity response to corn residue removal: A literature review. *Agronomy Journal* 2004, 96, (1), 1-17.

[9] Wilhelm, W. W.; Johnson, J. M. E.; Karlen, D. L.; Lightle, D. T., Corn stover to sustain soil organic carbon further constrains Biomass supply. *Agronomy Journal* 2007, 99, (6), 1665-1667.

[10] Lane, J., A looming cellulosic feed stock shortage? In *Biofuels Digest*, Ascension Publishing Inc.: 2014; pp 1-4.

[11] Smil, V., *Enriching the Earth: Fritz Haber, Carl Bosch, and the Transformation of World Food Production*. MIT Press: Cambridge, MA, 2000.

[12] Smil, V., Detonator of the population explosion. *Nature* 1999, 400, (6743), 415-415.

[13] Jung, H. G. In *Alfalfa: A Comapnion Crop With Corn*, Alfalfa/Corn Rotations for Sustainable Cellulosic Biofuels Prodution 2010; available online: http://www.alfalfa-forage.org: 2010.

[14] Su, Y. Z., Soil carbon and nitrogen sequestration following the conversion of cropland to alfalfa forage land in northwest China. *Soil & Tillage Research* 2007, 92, (1-2), 181-189.

[15] Baker, J. In *Soil Carbon*, Alfalfa/Corn Rotations for Sustainable Cellulosic Biofuels Prodution Johnston Iowa, 2010; Johnston Iowa, 2010; p 13.

[16] Russelle, M. P.; Birr, A. S., Large-scale assessment of symbiotic dinitrogen fixation by crops: Soybean and alfalfa in the Mississippi river basin. *Agronomy Journal* 2004, 96, (6), 1754-1760.

[17] Vadas, P. A.; Barnett, K. H.; Undersander, D. J., Economics and Energy of Ethanol Production from Alfalfa, Corn, and Switchgrass in the Upper Midwest, USA. *Bioenergy Research* 2008, 1, (1), 44-55.

[18] Holin, F., Jump start Alfalfa by Interseeding it Into Corn. *Hay and Forage Grower* Feburary 9, 2014, 2014, p 1.

[19] Grabber, J. H., Interseeding Alfalfa With Corn. In Madison, 2014.

[20] Albrecht, K. A.; Wedin, W. F.; Buxton, D. R., Cell-wall composition and digestibility of alfalfa stems and leaves. *Crop Sci.* 1987, 27, (4), 735-41.

[21] Gray, A.; Kaan, D., Feasibility Study;Alfalfa Leaf Meal as a Value-added Crop and and Alfalfa Stems as Biomass Fuel. In 1996; Vol. National Technical Information Service Document No. PB97-105548.

[22] Shinners, K. In *Harvest, Storage,and Fractionation of Alfalfa*, Alfalfa/Corn Rotations for Sustainable Cellulosic Biofuels Prodution 2010; available online: http://www.alfalfa-forage.org: 2010.

[23] Shinners, K. J.; Herzmann, M. E.; Binversie, B. N.; Digman, M. F., Harvest fractionation of alfalfa. *Transactions of the Asabe* 2007, 50, (3), 713-718.

[24] Lamb, J. F. S.; Jung, H. J. G.; Sheaffer, C. C.; Samac, D. A., Alfalfa leaf protein and stem cell wall polysaccharide yields under hay and biomass management systems. *Crop Science* 2007, 47, (4), 1407-1415.

[25] Hatfield, R. D.; Hall, M. B.; Muck, R. E.; Radloff, W. J.; Shinners, K. J., Recombined, late harvested ensiled alfalfa leaves and stems give comparable performance to nomally harvested alfalfa silage. *Journal of Dairy Science* 2014, Book of Abstracts, 70.

[26] Hong, B. J.; Broderick, G. A.; Koegel, R. G.; Shinners, K. J.; Straub, R. J., Effect of shredding alfalfa on cellulolytic activity, digestibility, rate of passage, and milk production. *J. Dairy Sci* 1988, 71, 1546-1555; bKoegel, R. G.; Straub, R. J.; Shinners, K. J.; Broderick, G. A.; Mertens, D. R., An Overview of Physical Treatments of Lucerne Performed at Madison, Wisconsin, for Improving Properties. *Journal of Agricultural Engineering Res.* 1992, 52, (3), 183-191.

[27] Lamb, J. F. S.; Sheaffer, C. C.; Samac, D. A., Population density and harvest maturity effects on leaf and stem yield in alfalfa. *Agronomy Journal* 2003, 95, (3), 635-641.

[28] Alvira, P.; Tomas-Pejo, E.; Ballesteros, M.; Negro, M. J., Pretreatment technologies for an efficient bioethanol production process based on enzymatic hydrolysis: A review. *Bioresource Technology* 2010, 101, (13), 4851-4861.

[29] Ralph, J.; Grabber, J. H.; Hatfield, R. D., Lignin-ferulate crosslinks in grasses: active incorporation of ferulate polysaccharide esters into ryegrass lignins. *Carbohydrate Research* 1995, 275, (1), 167-178; bRalph, J.; Hatfield, R. D.; Grabber, J. H.; Jung, H. G.; Quideau, S.; Helm, R. F., Cell wall cross-linking in grasses by ferulates and diferulates. In *Lignin and Lignan Biosynthesis*, Lewis, N. G.; Sarkanen, S., Eds. American Chemical Society: Washington, DC, 1998; Vol. 697, Amer. Chem. Soc. Symp. Ser., pp 209-236.

[30] Zhou, S. F.; Weimer, P. J.; Hatfield, R. D.; Runge, T. M.; Digman, M., Improving ethanol production from alfalfa stems via ambient-temperature acid pretreatment and washing. *Bioresource Technology* 2014, 170, 286-292.

[31] Agler, M. T.; Wrenn, B. A.; Zinder, S. H.; Angenent, L. T., Waste to bioproduct conversion with undefined mixed cultures: the carboxylate platform. *Trends in Biotechnology* 2011, 29, (2), 70-78.

[32] Chang, H. N.; Kim, N. J.; Kang, J.; Jeong, C. M., Biomass-derived Volatile Fatty Acid Platform for Fuels and Chemicals. *Biotechnology and Bioprocess Engineering* 2010, 15, (1), 1-10.

[33] Holtzapple, M. T.; Granda, C. B., Carboxylate Platform: The MixAlco Process Part 1: Comparison of Three Biomass Conversion Platforms. *Applied Biochemistry and Biotechnology* 2009, 156, (1-3), 525-536.

[34] Steinbusch, K. J. J.; Hamelers, H. V. M.; Plugge, C. M.; Buisman, C. J. N., Biological formation of caproate and caprylate from acetate: fuel and chemical production from low grade biomass. *Energy & Environmental Science* 2011, 4, (1), 216-224.

[35] Mertens, D. R., Creating a system for meeting the fiber requirements of dairy cows. *Journal of Dairy Science* 1997, 80, (7), 1463-1481.

[36] Weimer, P. J.; Russell, J. B.; Muck, R. E., Lessons from the cow: What the ruminant animal can teach us about consolidated bioprocessing of cellulosic biomass. *Bioresource Technology* 2009, 100, (21), 5323-5331.

[37] Weimer, P. J. In *The relevance of ruminant animals to chemical conversion and biofuels technologies.*, Proceedings 2nd International Conference on Microbiology and Biotechnology, Minas Gerias, Brazil, 2013; al., H. C. M. e., Ed. Minas Gerias, Brazil, 2013.

[38] Weimer, P. J.; Nerdahl, M.; Brandl, D. J., Production of medium-chain volatile fatty acids by mixed ruminal microrganisms is enhanced by ethanol in co-culture with Clostridium kluyveri. *Bioresource Technology* 2014, in press.

[39] Agler, M. T.; Spirito, C. M.; Usack, J. G.; Werner, J. J.; Angenent, L. T., Chain elongation with reactor microbiomes: upgrading dilute ethanol to medium-chain carboxylates. *Energy & Environmental Science* 2012, 5, (8), 8189-8192; bVasudevan, D.; Richter, H.; Angenent, L. T., Upgrading dilute ethanol from syngas fermentation to n-caproate with reactor microbiomes. *Bioresource Technology* 2014, 151, 378-382.

[40] Grootscholten, T. I. M.; dal Borgo, F. K.; Hamelers, H. V. M.; Buisman, C. J. N., Promoting chain elongation in mixed culture acidification reactors by addition of ethanol. *Biomass & Bioenergy* 2013, 48, 10-16.

[41] Lynd, L. R.; Weimer, P. J.; van Zyl, W. H.; Pretorius, I. S., Microbial cellulose utilization: Fundamentals and biotechnology (vol 66, pg 506, 2002). *Microbiology and Molecular Biology Reviews* 2002, 66, (4), 739-739.

[42] Singhania, R. R.; Patel, A. K.; Christophe, G.; Fontanille, P.; Larroche, C., Biological upgrading of volatile fatty acids, key intermediates for the valorization of biowaste through dark anaerobic fermentation. *Bioresource Technology* 2013, 145, 166-174.

[43] Gaertner, C. A.; Serrano-Ruiz, J. C.; Braden, D. J.; Dumesic, J. A., Catalytic coupling of carboxylic acids by ketonization as a processing step in biomass conversion. *Journal of Catalysis* 2009, 266, (1), 71-78.

[44] Lange, J. P.; Price, R.; Ayoub, P. M.; Louis, J.; Petrus, L.; Clarke, L.; Gosselink, H., Valeric Biofuels: A Platform of Cellulosic Transportation Fuels. *Angewandte Chemie-International Edition* 2010, 49, (26), 4479-4483.

[45] Levy, P. F.; Sanderson, J. E.; Kispert, R. G.; Wise, D. L., Biorefining of Biomass to Liquid Fuels and Organic-Chemicals. *Enzyme and Microbial Technology* 1981, 3, (3), 207-215; bSanderson, J. E.; Barnard, G. W.; Levy, P. F., Conversion of Biomass-Derived Organic-Acids to Liquid Fuels by Electrochemical Oxidation in Aqueous-Solutions. *Journal of the Electrochemical Society* 1981, 128, (3), C123-C123.

[46] Park, Y. C.; Shaffer, C. E. H.; Bennett, G. N., Microbial formation of esters. *Applied Microbiology and Biotechnology* 2009, 85, (1), 13-25.

[47] Schirmer, A.; Rude, M. A.; Li, X. Z.; Popova, E.; del Cardayre, S. B., Microbial Biosynthesis of Alkanes. *Science* 2010, 329, (5991), 559-562.

[48] Levy, P. F.; Sanderson, J. E.; Wise, D. L., Development of a Process for Production of Liquid Fuels from Biomass. *Biotechnology and Bioengineering* 1981, 239-248; bKuhry, A. B.; Weimer, P. J. Biological/Electrolytic Conversion of Biomass to Hydrocarbons. US8518680 B2, 2013.

[49] Kaliyan, N.; Morey, R. V.; Schmidt, D. R., Roll press compaction of corn stover and perennial grasses to increase bulk density. *Biomass & Bioenergy* 2013, 55, 322-330.

[50] Morey, R. V.; Kaliyan, N.; Tiffany, D. G.; Schmidt, D. R., A Corn Stover Supply Logistics System. *Applied Engineering in Agriculture* 2010, 26, (3), 455-461.

[51] Sokhansanj, S.; Kumar, A.; Turhollow, A. F., Development and implementation of integrated biomass supply analysis and logistics model (IBSAL). *Biomass & Bioenergy* 2006, 30, (10), 838-847.

[52] Reynal, S. M.; Broderick, G. A., Optimal nutrient intake and digestion for ruminal microbial protein and milk yields in lactating dairy cows. *Journal of Animal Science* 2006, 84, 81-81; bBroderick, G. A.; Reynal, S. M.; Patton, R. A.; Heimbeck, W.; Lodi, P., Use of plasma concentrations to estimate bioavailability of methionine in rumen-protected products fed to dairy cows. *Journal of Dairy Science* 2010, 93, 236-236.

[53] Weimer, P. J.; Koegel, R. G.; Lorenz, L. F.; Frihart, C. R.; Kenealy, W. R., Wood adhesives prepared from lucerne fiber fermentation residues of Ruminococcus albus and Clostridium thermocellum. *Applied Microbiology and Biotechnology* 2005, 66, (6), 635-640; bWeimer, P. J.; Conner, A. H.; Lorenz, L. F., Solid residues from Ruminococcus cellulose fermentations as components of wood adhesive formulations. *Applied Microbiology and Biotechnology* 2003, 63, (1), 29-34.

[54] Weimer, P. J.; Price, N. P. J.; Kroukamp, O.; Joubert, L. M.; Wolfaardt, G. M.; Van Zyl, W. H., Studies of the extracellular glycocalyx of the anaerobic cellulolytic bacterium Ruminococcus albus 7. *Applied and Environmental Microbiology* 2006, 72, (12), 7559-7566.

[55] Hatfield, R. D., Carbohydrate composition of alfalfa cell walls isolated from stem sections differing maturity. Journal of Agricultural and Food Chemistry 1992, 40, (3), 424-430.